Form and Function in Birds

Form and Function in Birds

Volume 1

Edited by

A. S. KING

*Department of Veterinary Anatomy,
University of Liverpool,
Liverpool,
England*

J. McLELLAND

*Department of Anatomy,
Royal (Dick) School of
Veterinary Studies, University of
Edinburgh, Scotland*

1979

ACADEMIC PRESS

A Subsidiary of Harcourt Brace Jovanovich, Publishers

London · New York · Toronto · Sydney · San Francisco

ACADEMIC PRESS INC. (LONDON) LTD.
24/28 Oval Road,
London NW1

United States Edition published by
ACADEMIC PRESS INC.
111 Fifth Avenue
New York, New York 10003

British Library Cataloguing in Publication Data

Form and functions in birds.
Vol. 1
1. Birds—Physiology
I. King, Anthony, Stuart
II. McLelland, John
598.2′1 QL698 79-50523

ISBN 0-12-407501-0

Typeset by Eta Services (Typesetters) Ltd., Beccles, Suffolk
Printed in Great Britain

Contributors

AKESTER, A. R. *Sub-department of Veterinary Anatomy, University of Cambridge, Cambridge, England* (p. 381)

DUNCKER, H.-R. *Zentrum für Anatomie und Cytobiologie der Justus-Liebeg-Universität Giessen, Aulweg 123, D-6300 Giessen, West Germany* (p. 39)

GILBERT, A. B. *Reproductive Physiology Section, Agricultural Research Council, Poultry Research Centre, King's Buildings, West Mains Road, Edinburgh EH9 3JS, Scotland* (p. 237)

HODGES, R. D. *Wye College, University of London, Near Ashford, Kent TN25 5AH, England* (p. 361)

JOHNSON, O. W. *Department of Biology, Moorhead State University, Moorhead, Minnesota 56560, U.S.A* (p. 183)

KING, A. S. *Department of Veterinary Anatomy, University of Liverpool, Brownlow Hill and Crown Street, P.O. Box 147, Liverpool L69 3BX, England* (p. 1)

KING, D. Z. *Department of Histology, University of Liverpool, Liverpool, England* (p. 1)

MCLELLAND, J. *Department of Anatomy, Royal (Dick) School of Veterinary Studies, Summerhall, Edinburgh EH9 1QH, Scotland* (p. 69)

Preface

Although much is known about the functional anatomy of birds, most of this information is still widely dispersed in the original literature. *Form and Function in Birds* attempts to provide in three volumes a definitive and extensively illustrated account of avian morphology, surveying the principal features of avian structure and providing an insight into how these structures work in the living bird. Evolutionary factors influencing avian morphology have also been included where they especially help to clarify avian form. The overriding objective has been to show why birds are built as they are, and to assemble this information into a reference work for all biologists who research or teach with avian material.

Within the limits of these general objectives the authors were free to approach their subjects in their own way, but as editors we have attempted to achieve a moderate uniformity of style and format, to eliminate extensive overlapping, and to check many of the citations from the literature. Consequently our authors have been asked to endure fairly extensive editing, and we thank them for their cooperation and forbearance.

We would have liked to match as closely as possible the subjects within each volume, but this has proved extremely difficult. Some contributors found themselves under unexpectedly severe pressure from other commitments. For this reason it became impossible to assemble all the chapters simultaneously without imposing an intolerable delay on those authors who had completed their chapters on time. We therefore decided to bring out the three volumes as the material became available. In Volume 1 the material should be up-to-date as far as 1977, and in many instances the first part of 1978. Volume 2 is expected to follow a year after Volume 1, but Volume 3 is likely to be delayed until 1982 or 1983.

The anatomical terminology, macroscopic, mesoscopic and microscopic, is strictly based on the *Nomina Anatomica Avium* (Academic Press, London). The rapid expansion of research on birds in recent years has led to a proliferation of synonyms in the literature. The adoption of a standard set of unambiguous anatomical terms should enhance international scientific communication and improve data-retrieval systems. Therefore in *Form and Function in Birds* the official Latin term of the N.A.A. for each structure has been given where it is first described. Thereafter the English form of the term has generally been used.

The scientific taxonomic nomenclature is that of J. J. Morony, W. J. Bock and J. Farrand (published by the Department of Ornithology of the American Museum of Natural History, New York, in 1975). The English vernacular names of birds are those listed by E. S. Gruson (*Check List of the Birds of the World*, Collins, London, 1976). The names of the common laboratory and domestic birds refer to the following species: duck or *Anas*, domestic forms of *Anas platyrhynchos*; goose or *Anser*, domestic forms of *Anser anser*; pigeon or *Columba*, domestic forms of *Columba livia*; turkey or *Meleagris*, domestic forms of *Meleagris gallopavo*; chicken, domestic fowl or *Gallus*, domestic forms of *Gallus gallus*; quail, domestic forms of the genus *Coturnix*.

We wish to thank the staff of the Academic Press, London, for their expert assistance in producing this book.

September 1979

A. S. KING
and
J. McLELLAND

Contents

CONTRIBUTORS v

PREFACE vii

LIST OF ABBREVIATIONS x

1. AVIAN MORPHOLOGY: GENERAL PRINCIPLES
A. S. King and D. Z. King 1

2. COELOMIC CAVITIES
H.-R. Duncker 39

3. DIGESTIVE SYSTEM
J. McLelland 69

4. URINARY ORGANS
O. W. Johnson 183

5. FEMALE GENITAL ORGANS
A. B. Gilbert 237

6. THE BLOOD CELLS
R. D. Hodges. 361

7. THE AUTONOMIC NERVOUS SYSTEM
A. R. Akester. 381

SUBJECT INDEX 443

List of Abbreviations

The following abbreviations of Latin terms have been used, the plural form being in brackets, (as in the *Nomina Anatomica Avium*, Academic Press, London and New York, 1979).

A. (Aa.) = Arteria
Ant. = Anterior
Artc. (Artcc.) = Articulatio
Caud. = Caudalis
Cran. = Cranialis
G. (Gg.) = Ganglion
Gl. (Gll.) = Glandula
Lat. = Lateralis
Lig. (Ligg.) = Ligamentum
M. (Mm.) = Musculus
Maj. = Major
Med. = Medialis
Min. = Minor
N. (Nn.) = Nervus
Nuc. = Nucleus
Post. = Posterior
Proc. (Procc.) = Processus
R. (Rr.) = Ramus
Rdx. (Rdxx.) = Radix
Sut. (Sutt.) = Sutura
Synd. (Syndd.) = Syndesmosis
Synos. (Synoss.) = Synostosis
Tr. (Trr.) = Tractus
V. (Vv.) = Vena
Vas l. (Vasa l.) = Vas lymphaticum

To E. C. Amoroso, biologist and teacher, counsellor and friend.

"When I look back at the three or four choices in my life which have been decisive, I find that, at the time I made them, I had very little sense of the seriousness of what I was doing and only later did I discover what had seemed an unimportant brook was, in fact, a Rubicon."

W. H. Auden

1

Avian morphology: general principles

A. S. KING AND D. Z. KING*

Department of Veterinary Anatomy and Department of Histology, University of Liverpool, England*

CONTENTS

I. Introduction 2

II. Evolution of birds 2
   A. Synapsida 2
   B. Archosauria 4
   C. *Archaeopteryx* 5
   D. The ancestry of *Archaeopteryx* 6
   E. Aves 7
   F. Flightless birds 8
   G. Evolutionary relationships between Aves, Reptilia and Mammalia . . 11

III. The evolution of endothermy 11

IV. The influence of endothermy on morphology 14
   A. Locomotory apparatus 15
   B. Central nervous system 15
   C. Cardiovascular system 16
   D. Respiratory system 17
   E. Chewing adaptations 18
   F. Insulation 18
   G. The evolution of flight 20

V. Structural adaptations for flight 23
   A. Limitations upon body weight 23
   B. Modifications of body form 24
   C. Skeleton 26
   D. Flight muscles 29
   E. Respiratory system 30
   F. Cardiovascular system 32
   G. Special senses and brain 33

References 34

I. Introduction

So restrictive are the anatomical requirements of flight that, even when the flightless birds are included, the entire class of Aves presents greater uniformity of general structure than many single orders of fishes, amphibians and reptiles (Marshall, 1962, p. 555). It has been said that in the whole array of birds there is less variation in structure than in the 90 or so species of Primates and 290 species of Carnivora (Yapp, 1970, p. 40).

Paradoxically, however, the power of flight has also led to an immense diversity of anatomical detail. The ability to fly has enabled birds to penetrate a very wide variety of habitats. The result has been an extensive adaptive radiation, especially for locomotion and feeding. The multiplicity of these and other adaptations accounts for the relatively large number of species, about 8500, which occurs in the class Aves. In contrast the modern reptiles total about 6000 species (Bellairs and Attridge, 1975, p. 17), whilst the mammals (Yapp, 1970, p. 40) and amphibians (Bellairs and Attridge, 1975, p. 17) have even fewer species.

The word morphology is used advisedly in the title of this chapter, not simply as a synonym for structure but to signify the logical basis for structure (so reflecting the true spirit of its etymology). Hence the chapter examines the general factors, phylogenetic and physiological, which influence the form of the avian body. Although the adaptations for flight were based on an anatomical foundation derived from the reptilian ancestry of birds, it is possible that the anatomy of the immediate reptilian ancestors of birds was in a state of active change through the evolution of endothermy. This survey therefore begins with an outline of the evolution of birds, proceeds to a study of the origin and structural consequences of endothermy, and ends with a consideration of the anatomical adaptations to flight.

II. Evolution of birds

During the late Carboniferous period the labyrinthodont amphibians gave rise to the first primitive reptiles, the cotylosauria. From these stem reptiles radiated a great diversity of reptilian types, so that by the end of the Triassic almost all the major groups of reptile had made their appearance (Romer, 1966, p. 107). These may be classified (Fig. 1.1) into the five subclasses, Lepidosauria, Archosauria, Ichthyosauria, Euryapsida and Synapsida (Romer, 1966, p. 112). Of the remaining orders the Cotylosauria and Chelonia are sometimes included in the additional subclass Anapsida (Porter, 1972, pp. 199–232; Bellairs and Attridge, 1975, pp. 63, 71).

A. Synapsida

Among the earliest reptilian groups to appear in the upper Carboniferous (Romer, 1966, p. 173) were the mammal-like reptiles or Synapsida. These flourished during the Permian period and early Triassic, and were the dominant

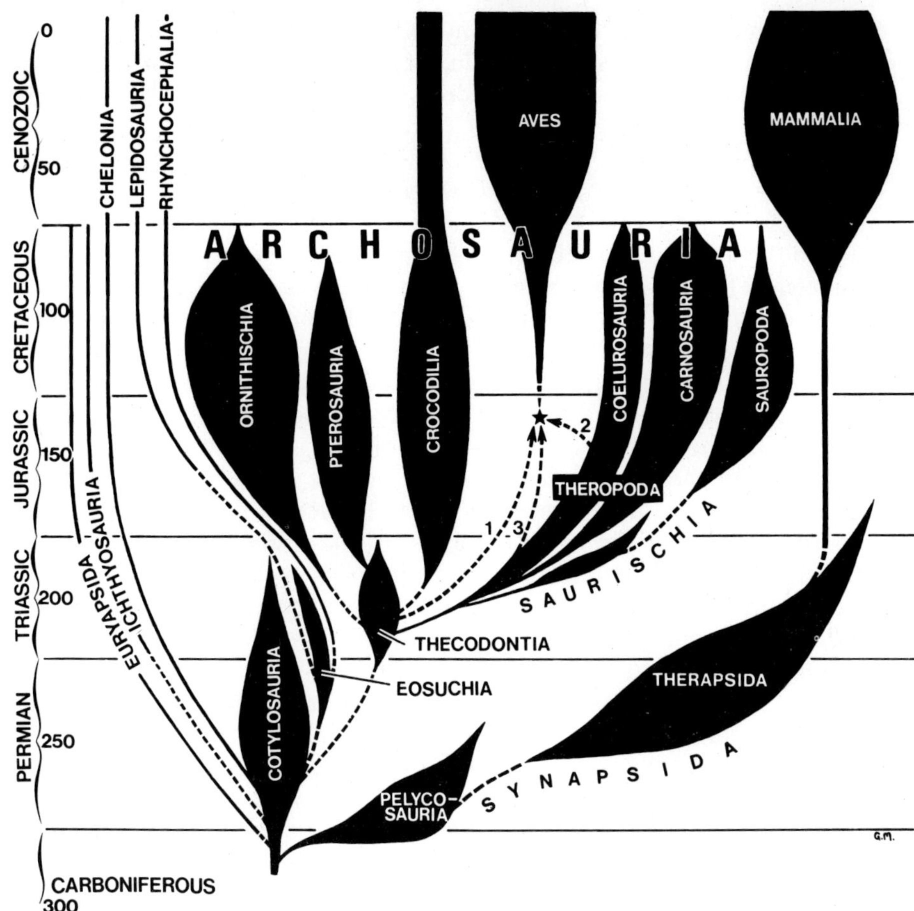

Fig. 1.1. The phylogeny of reptiles, mammals and birds. The five evolutionary lines on the left of the diagram are not drawn in proportion to the size of their populations. 1, 2 and 3 indicate three possible sources of *Archaeopteryx*. 1, is the view of Broom (1913), Heilmann (1926), Swinton (1960), Romer (1966) and de Beer (1975). 2, is the derivation proposed by Huxley (1868) and Ostrom (1973, 1975). 3, represents a possible source from late Triassic coelurosaurs. The dates are based on Romer (1966) and Bellairs and Attridge (1975). ★, *Archaeopteryx*.

reptiles for more than 70 million years. The early primitive forms (the Pelycosauria) gave rise to the more advanced Therapsida, and in the radiation and development of therapsids can be traced nearly the entire evolutionary change from reptiles into mammals (Romer, 1966, p. 184). By the late Triassic the typical therapsids had disappeared, leaving forms which were on the borderline between reptiles and mammals (e.g. *Diarthrognathus*); by the end of this period, mammals were probably in existence, for example *Sinoconodon*, a triconodont in the late Triassic of China (Romer, 1966, pp. 197, 199). For the next 100 million years, until the late Cretaceous, the mammalian organization was almost completely suppressed. Most of the forms that existed were small (about the size of rats and mice), mainly insectivorous, and perhaps nocturnal and more or less arboreal (Young, 1962, p. 546; Romer, 1966, p. 202). This long period saw the great radiation of the dinosaurs and their related groups, and it was only at the close of the Cretaceous, when the ruling reptiles failed, that the mammals gained an ascendant position.

B. Archosauria

While the therapsids were declining yet simultaneously providing the evolutionary material for the later advent of mammals, a reptilian line, the Archosauria, was beginning a radiation which was to dominate the vertebrate scene for 120 million years. In this radiation the first group to appear (in the early Triassic) was the Thecodontia. These reptiles possessed a two arched (diapsid) skull and many, though not all, showed a tendency towards bipedalism (Romer, 1966, p. 136). The origin of the thecodonts is obscure, but is generally believed to have been from the subclass Lepidosauria via the order Eosuchia (Carroll, 1969, p. 8). However, no intermediate forms between the Eosuchia and the Thecodontia are known. The small carnivorous reptile *Euparkeria*, of the suborder Pseudosuchia (Lower Triassic), can be regarded as a typical example of a primitive thecodont (Romer, 1966, p. 138). It was about two feet long, lightly built in both skull and post-cranial skeleton, and was probably a "facultative biped" rising onto its hind legs when a turn of speed was required (Bellairs and Attridge, 1975, p. 117). Romer (1966, p. 139) considered that this animal exhibited a structure from which nearly all later archosaurs could be derived. The thecodonts diversified during the Triassic, but became extinct by the end of that period. However, before their final disappearance they had given rise, directly or indirectly, to the four other orders of Archosauria, i.e. the Crocodilia, Ornithischia, Saurischia, and Peterosauria, and also to birds.

The two orders Ornithischia and Saurischia constitute the group that is popularly known as the "dinosaurs". The ornithischians were all herbivorous. They had a characteristic pelvis with the pubis lying parallel to the ilium, as in birds. Some were bipedal although the forelimbs were not particularly reduced, but many had returned to a quadrupedal gait. In spite of their name ("bird-hipped"), the Ornithischia are not generally considered to be ancestral to the Aves, though Galton (1970) suggested that the common ancestor of birds and ornithischians may have been a cursorial biped of the Middle Triassic.

The saurischians can be divided into two subsuborders, Sauropodomorpha and Theropoda. In both of these suborders the pubis is directed cranially and ventrally. The sauropods were all herbivorous and included the gigantic *Brontosaurus*, *Diplodocus* and *Brachiosaurus*; the theropods were entirely carnivorous and consisted of two infraorders, Coelurosauria and Carnosauria (Bellairs and Attridge, 1975, p. 129). The well-known *Tyrannosaurus* was a carnosaur.

The Coelurosauria were present in the Triassic, and are considered to be more primitive than the Carnosauria (Romer, 1966, p. 150). The coelurosaurs were relatively small, and very lightly built with thin-walled bones (Romer, 1966, p. 150). The skull was small, the orbits large, and the neck long and slender. *Coelophysis* was about eight feet long, and *Procompsognathus* about half that size (Romer, 1966, p. 151). *Compsognathus* from the Upper Jurassic was turkey-sized (Bellairs and Attridge, 1975, p. 133). All the animals in this infraorder were cursorial predators.

Before leaving the classification of reptiles the concept "sauropsid" should be mentioned. As proposed by Huxley (1867) the Sauropsida included the birds and all of the reptiles. Noting that the sauropsids therefore included the therapsids, Goodrich (1916) proposed that the reptiles should be divided into three groups, the "Protosauria" consisting of stem reptiles, the "Theropsida" comprising the synapsid reptiles and mammals, and the "Sauropsida" composed of the other reptiles and birds; unfortunately this attempt to divide the reptiles into two main evolutionary lines leaves some reptilian groups in a problematical position, notably the ichthyosaurs and plesiosaurs, and particularly the Chelonia which are not closely related to the other modern reptiles (Bellairs and Attridge, 1975, pp. 66–67). Nevertheless the term "sauropsid" remains in fairly general use, usually more in the sense proposed by Goodrich; but Huxley's usage has not been abandoned (e.g. Ostrom, 1975). The existence of two different meanings of sauropsid is unfortunate, since it is not always clear which interpretation authors intend and this uncertainty diminishes the value of a potentially useful concept.

C. Archaeopteryx

From the later Jurassic lithographic limestone of Germany have come five more or less complete fossil remains of the earliest known bird, *Archaeopteryx*. The conditions of fossilization clearly showed the feathers to be identical with those of modern flying birds, so *Archaeopteryx* can definitely be allocated to the class Aves. It is also the general opinion today that *Archaeopteryx* was capable of some sort of flight, as well as being arboreal. However, the remains of these small bipedal animals, which were about the size of the Common Magpie (*Pica pica*) (de Beer, 1975), show a mosaic of reptilian and avian features (de Beer, 1954, 1975). The reptilian characteristics include teeth; a long tail of 20 vertebrae; simple articulations between the vertebrae (amphicoelous); a short sacrum, involving no more than 6 vertebrae; free metacarpal bones in the hand, with claws on all three digits; free metatarsals in the foot; simple ribs without uncinate processes; and gastralia in the ventral body wall. The features listed by de Beer (1954) as "absolutely characteristic of birds" are the feathers, which were arranged as primaries on the hand and wrist, secondaries on the forearm, and contour feathers all over the body; the paired clavicles, which were joined in the midline to form the characteristic furcula or wish-bone; the pubis, which according to de Beer (1954) was directed caudally although Ostrom (1973) thought it possible that in the Berlin specimen the pubis had been dislocated and was originally directed ventrally or cranioventrally; and lastly, the foot, in which the first digit was opposed to the other three digits as in modern perching birds. The coracoid was non-avian in form, being short and square, not strut-like, and making only a cartilaginous or membranous contact with the sternum (Ostrom, 1976). The sternum is of particular interest because of its association with the flight musculature. De Beer (1954) believed that the sternum could be identified, but showed no sign of a keel: Ostrom (1976) maintained that no sternum had been preserved and that it was therefore almost certainly cartilaginous, but he agreed that it probably lacked a keel. Of the skeletal features mentioned above, the only one that is exclusively avian is the furcula (Ostrom, 1975). The possible significance of the anatomy of *Archaeopteryx* for flight is discussed in Section IV.G.

D. The ancestry of Archaeopteryx

It has long been accepted that birds are descended from reptiles, but the identity of the particular group which provided the true avian progenitors is still controversial. An extensive survey of the many contributions to this debate is beyond the scope of the present chapter, and for full details the reader is referred to the elegant review by Ostrom (1975). The following account is limited to the main lines of argument only.

In 1868 Huxley compared *Archaeopteryx* with the small contemporary dinosaur *Compsognathus*, and reasoned that, although the one was not descended from the other, the two animals were sufficiently alike to suggest a close relationship. The concept of a dinosaurian ancestry of birds persisted until the suspicion arose that the dinosaurs were a much more varied and diverse group than had hitherto been supposed. It was then considered that the dinosaurs might well be too specialized and that the ancestors of birds should be sought in a more primitive group.

Broom (1913) was apparently one of the first to suggest that the Pseudosuchia, a suborder of the Thecodontia, fulfilled the necessary requirements. Heilmann, in "The Origin of Birds" (1926), described *Archaeopteryx* in detail and compared all its known features with corresponding structures found in different groups of reptiles. In particular, he discussed the similarities between *Archaeopteryx* and the Coelurosaurs. He noted that, with the passage of time, the forelimb of the carnosaur theropods had shortened in relation to the hind limb, whereas in the coelurosaur theropods the forelimb had increased in length compared with the hindlimb thus coming more into line with *Archaeopteryx*. He also stressed the resemblance between the tridactyl manus of the coelurosaurs and the wing of *Archaeopteryx*. For these reasons he acknowledged that "it would seem a rather obvious conclusion that it is amongst the coelurosaurs that we are to look for the bird ancestor". Nevertheless, Heilmann (p. 182) concluded that because the clavicles were not present in coelurosaurs, as well as other details, the true ancestors of birds would prove to be a group of reptiles "closely akin" to coelurosaurs. The Pseudosuchia satisfied his criteria as progenitors of the avian line of evolution, and he suggested that the departure of both the coelurosaurs and the proavian from the Pseudosuchia possibly took place at the beginning of the Triassic. De Beer (1975) considered that the pseudosuchian *Aetosaurus* of the lower Triassic could qualify as an ancestor of *Archaeopteryx*, but Swinton (1960, p. 8) concluded that although the skull of *Aetosaurus* had strong affinities with *Archaeopteryx*, the postcranial skeleton had not. The pseudosuchian *Euparkeria* is most often proposed as a possible ancestor of *Archaeopteryx*, the dinosaurs, and the Crocodilia; Romer (1966, p. 138) considered *Euparkeria* to be "an almost ideal ancestor for later archosaurian types".

In 1973 and 1975 Ostrom surveyed the fossil evidence afresh. He compared the skeleton of *Archaeopteryx* with that of coelurosaurs and pseudosuchians, including some specimens not known to earlier workers such as Heilmann; although the dinosaurs under discussion were over two metres long (except *Compsognathus*), the similarities in structure were marked. Consequently he rejected the conclusion of Heilmann and subsequent workers and revived Huxley's tentative proposition of a century earlier that coelurosaurs could indeed be the direct ancestors of *Archaeopteryx*. He overcame the problem of the missing clavicles in coelurosaurs by pointing out that clavicles have in fact been reported in *Segisaurus*, *Oviraptor* and *Velociraptor*, all of them Coelurosaurs; furthermore he suggested that, as the clavicles are membrane bones, they might have been present in other coelurosaurs but in an unossified state and therefore not preserved in a fossil specimen. Ostrom's conclusion was that birds originated from a coelurosaur theropod ancestor in the middle or late Jurassic.

However, it should be noted that Heilmann (1926, p. 190), Simpson (1946), de Beer (1975) and Brodkorb (1971, p. 25) had rejected these proposed coelurosaur ancestors as "too specialized and too late". Indeed Bock (1965) postulated that the time span required for the appearance of a vertebrate class may be 50 million years or more. If this is so, then coelurosaurs suitable as ancestors of *Archaeopteryx* had to be in existence during the late Triassic or early Jurassic. This would, of course, disqualify late Jurassic coelurosaurs such as *Comsognathus* and *Ornitholestes*, which were contemporaries of *Archaeopteryx*, but there was no shortage of late Triassic coelurosaurs. Romer (1966, p. 150) stated that small coelurosaurs of slender build were already common in the late Triassic, and listed five examples from Europe and America; some of these were about one metre long, which more closely approaches the size expected for a near ancestor of *Archaeopteryx*.

E. Aves

Not only are the origins of *Archaeopteryx* itself controversial, but there is also much argument as to whether or not this primitive bird was the direct ancestor of modern birds. After *Archaeopteryx* the next known fossil bird is *Gallornis*, a possible representative of the flamingos. It appeared probably not later than 25 million years, and perhaps only 3 million years, after *Archaeopteryx*; Brodkorb (1971, pp. 34–35) considered this period to be too short to evolve a wader from a primitive tree-dwelling gliding bird. Furthermore Brodkorb (1971, p. 37) interpreted the two fossil feathers from the Koonwarra claystones of Australia as evidence for an essentially world-wide distribution of birds in the early Cretaceous, and concluded that *Archaeopteryx* could hardly have been the ancestor of any known Cretaceous bird. Bock (1969) also believed that indirect evidence suggested "that modern birds . . . had evolved by early Cretaceous times". As Swinton (1960, p. 12) pointed out, it would indeed be a truly remarkable coincidence if this unique and most ancient bird in our inadequate geological record turned out to be the progenitor of all subsequent birds.

Nevertheless, Ostrom (1974a) took the view that *Archaeopteryx*, as the only available Jurassic avian fossil, must be fitted into an evolutionary theory of the origin of birds in general, and in 1975 he expressed his considered opinion in the following words: "I personally believe that *Archaeopteryx* lies very close to bird origins and *probably is directly ancestral to all later birds*" (our italics).

By the end of the Cretaceous, various aquatic birds were well established, but ducks, gulls, petrels and terrestrial birds had not yet entered the record (Brodkorb, 1971, p. 40). On the other hand, there was a great radiation of both land and water birds during the early Tertiary. Thus by the end of the Eocene, or by the beginning of the Oligocene at the latest, 26 of the currently recognized 32 orders of birds were established (Brodkorb, 1971, p. 42). Finally during the Pleistocene glaciations of the Quaternary period the extinction of many species took place, and in due course the contemporary avifauna became established (Brodkorb, 1971, p. 45).

F. Flightless birds

Early in the nineteenth century, the large flightless birds were grouped together and given the name Ratitae. This convenient term, though no longer appearing in formal classifications of birds, is still used. The "ratites" can be considered to include the Struthioniformes (Ostrich, Africa; fossil forms, Eurasia), Rheiformes (rheas, S. America), Casuariiformes (emus and cassowaries, Australia and New Guinea), Apterygiformes (kiwis, New Zealand), Dinornithiformes (moas, New Zealand), and Aepyornithiformes (elephant birds, Madagascar). Moas and elephant birds are extinct.

The outstanding feature of these flightless birds is the possession of a sternum without a keel, hence the name Ratitae, or raft-like. All other birds with a keeled sternum have been called Carinatae.

The palate of the ratites has been considered to be of a special type. Huxley (1867) described it as one in which the vomer received the caudal extremities of the palatines and the rostral ends of the pterygoid bones. He gave it the name "dromaeognathus". Pycraft (1901) introduced the term palaeognathous for the type of skull in which, "in the adult, the pterygopalatine connection is by symphysis or ankylosis and not by articulation"; he considered the ratites and the tinamous (Tinamidae) to be palaeognathous, and the carinates neognathous. Bock (1963) found it impossible to define the palaeognathous palate morphologically in one sentence, and listed six characteristics.

The ratites have other features in common: the feathers are of the downy type; the rectrices are missing or irregularly arranged, and the pygostyle is small or underdeveloped; the coracoids and scapulae are fused, and the clavicles absent; the wing is reduced in size and may be vestigial or even non-existent; the sutures of the skull remain open and are not obliterated as in the carinates; the caeca are large; the males have a protrusible phallus (Marshall, 1962, p. 606; Yapp, 1970, p. 46; de Beer, 1975). The tinamous and ratites share a very large ilioischiatic fenestra (Cracraft, 1974), a unique rhamphothecal structure (Parkes and Clark, 1966), large caeca (Gadow, 1891, p. 691), and a protrusible phallus (Gerhardt, 1933, p. 307). With the exception of the kiwis which are the size of a fowl, all the ratites are big. The largest were the extinct forms, *Dinornis maximus* (a moa) and *Aepyornis* (an elephant bird) both being 3 m tall. The Ostrich (*Struthio camelus*) is the largest living bird being nearly 2·5 m high, followed by the Emu (*Dromaius novaehollandiae*) which reaches about 2·0 m. All the ratites are cursorial animals; even the nocturnal kiwis (*Apteryx*) can run fast (Marshall, 1962, p. 608), and the Ostrich has been credited with 80 km/hr (Hildebrand, 1974, p. 490).

These common features of the ratites can be interpreted in a number of ways. The similarities of the limbs may be attributed to convergence brought about by the loss of flight and the cursorial way of life. The palate was considered by the early workers to be primitive, but de Beer (1975) firmly believed the palaeognathous palate to be an example of neoteny. He also regarded the downy feathers and the open sutures of the skull as neotenous. Bock (1963) and Cracraft (1974) considered the ratites to be advanced in many morphological features including the palate.

The structure of the forelimb in general and the fused carpals and metacarpals in particular, the presence of an alula in rheas, the size of the brain, and the arrangement of the feathers, all indicate that these are birds that have descended from flying ancestors (Yapp, 1970, p. 47; Cracraft, 1974).

Unfortunately the relationships of the large flightless birds have had an impact on the classification of birds in general that seems out of all proportion to their real significance. According to McDowell (1948), Merrem in 1813 placed four orders into the Ratitae, the Ostrich, rheas, cassowaries and the Emu. Huxley (1867) recognized only three orders of birds altogether, Saururae (containing only *Archaeopteryx*), Ratitae and Carinatae. His Ratitae included five groups, *Struthio*, *Rhea*, *Casuarius* with *Dromaius*, *Dinornis* and *Apteryx*. Pycraft (1901) believed these birds to be primitive and polyphyletic, and yet placed them all in one taxon, the Palaeognathae, on the basis of their palaeognathous skull. In his opinion this arrangement was preferable to classifying them as Ratitae, since it avoided the problem that a few "carinates" have a "ratite" sternum, and enabled the tinamous to be classified as Palaeognathae despite their possession of a keeled sternum which prevented their inclusion among the Ratitae. Pycraft recognized seven orders in the Palaeognathae, namely Casuarii (emus and cassowaries), Struthiones, Rheae, Crypturi (tinamous), Dinornithes, Aepyornithes and Apteryges.

After a comprehensive study of the so-called palaeognathous birds, McDowell (1948) concluded that they did not represent a uniform primitive avian group, basing his opinion on the recognition of four different types of palaeognathous palate. He advocated abolishing the term Palaeognathae, and incorporating the palaeognathous birds into the Neognathae, under the Neornithes. In addition he suggested that classification would be simplified by the introduction of four new groups based on the four types of palate: Rheiformes and Tinamiformes; Apterygiformes, Dinornithiformes and Aepyornithiformes; Struthioniformes; and Casuariiformes.

De Beer (1975) pointed out that other lines of evidence, besides merely osteology, could provide information about the relationships of the "ratites". For example, the structure and chemical composition of the eggshell indicate that the Emu and cassowaries are more closely related to each other than to the rest. Electrophoresis of proteins suggests that the Ostrich is also related to cassowaries, the Emu, and possibly rheas. The host-specific bird louse, *Struthiolipeurus* is found only on the Ostrich and rheas. Such investigations suggest a monophyletic origin of ratites. Modern anatomical studies confirm this concept and firmly place the tinamous and ratites in one group (Bock, 1963; Cracraft, 1974).

Until recently, the fact that the tinamous and ratites form an outstanding example of discontinuous distribution was a great obstacle to the concept of a monophyletic origin for this group. However, the re-emergence and acceptance of the theory of continental drift now makes it easier to envisage a common ancestry for both ratites and tinamous. This origin must have occurred before the disruption of the southern continent of Gondwanaland which began during the late Jurassic period (see Cracraft, 1974, for review).

It is widely accepted that the ratites and tinamous arose from flying stock very early in the evolution of birds, and accumulating evidence seems to indicate that they are all related, although their precise relationships remain unclear. Because of this uncertainty about their relationships, there would seem to be advantages in classifying these birds under separate orders until more is known about their evolutionary history. Thus Wetmore (1960) and Storer (1971, p. 9) listed five separate orders of large flightless birds under the super-order Neognathae. On the other hand, Marshall (1962, p. 606) classified the tinamous and ratites under the super-order Palaeognathae, and Cracraft (1974) has suggested placing them together under the order Palaeognathiformes.

Throughout the world isolated regions can still be found which provide sheltered habitats, free from predators and in particular free from placental carnivores. Oceanic islands, such as Madagascar, Mauritius and the Galapagos, are obvious examples, and in such secluded places occur birds with virtually no indigenous enemies and therefore no further need to fly. However, once the power of flight has been relinquished it is no longer possible for a bird to return to an aerial mode of life, so should predators subsequently appear the flightless bird is highly vulnerable. Unfortunately predatory man has been responsible, either directly or indirectly, for the extinction of many flightless birds, including, the moas (*Dinornis*) and dodos (*Raphus*).

There is a small number of extant virtually flightless birds other than the ratites. Examples occur in unrelated orders, e.g. in the Anseriformes, Psittaciformes and Gruiformes (Thomson, 1964, p. 308). Amongst the Rallidae, which are essentially ground dwellers, there are a few species which live in secluded habitats and have dispensed with flight altogether. As in the ratites, the loss of flight in these scattered species has led to predictable structural changes, such as the reduction of the wing as in the Weka (*Gallirallus australis*) and the Galapagos Flightless Cormorant (*Nannopterum harrisi*), and degenerate hair-like plumage as in the Inaccessible Island Flightless Rail (*Atlantisia rogersi*) (Thomson, 1964, pp. 152, 677).

The penguins (Spheniscidae) constitute an exceptional category of flightless birds since their wings are used for propulsion through the water instead of flying. The wings are flattened and the bones much widened forming a rigid, powerful flipper (Hildebrand, 1974, p. 584). The body is large in comparison with wing size, and streamlined to reduce water resistance. The feathers are small and closely packed. Some authors (e.g. Marshall, 1962, p. 610) consider penguins sufficiently different from the remaining Neognathae to place them in a separate super-order, the Impennae.

G. Evolutionary relationships between Aves, Reptilia and Mammalia

Having probably arisen from thecodonts or coelurosaurs in the Triassic, birds have comparatively recent evolutionary associations with reptiles. Therefore it is to be expected that the anatomy of birds will include reptilian features. Since the Crocodilia arose from the thecodonts in the Triassic period, the evolutionary relationship between crocodiles and birds is relatively close. In contrast, the phylogenetic relationships between birds and the non-crocodilian living reptiles are much more remote, their lines of evolution being separate since the Carboniferous and Permian diversification of the stem reptiles. Consequently, if evolutionary relationships are to be used as a guide to anatomical comparisons, then the first place to look for similarities between living birds and living reptiles would be among the contemporary crocodilians.

The phylogenetic relationship of mammals to contemporary reptiles is much more remote. The Cotylosauria or stem reptiles, and the Synapsida or mammal-like reptiles, were the earliest known reptilian groups to appear in the Pennsylvanian. During the Permian they were still the dominant groups. From the cotylosaurs were derived all the subsequent reptilian forms, including the archosaurs; the synapsids were destined to give rise to the mammals (Romer, 1966, pp. 173, 323). Thus, the mammalian line departed from the reptilian-bird line at a very early stage in the evolution of higher vertebrates; it is therefore unlikely that the anatomy of the descendants of these two lines of evolution will show anything other than the most general similarities.

III. The evolution of endothermy

Since all contemporary reptiles are essentially ectothermic (i.e. they rely on external sources of heat to provide an adequate body temperature, Cowles 1947), the term "reptile" has become strongly linked to ectothermia. Being reptiles, the Archosauria (and hence the dinosaurs) have traditionally been regarded as ectothermic also. Nevertheless, recent researches have raised the possibility that some degree of endothermy might have been widespread in mesozoic and even late palaeozoic reptiles; not only the mammal-like therapsids, but most of the archosaurs may have been able to maintain an elevated body temperature by internal heat production. This suggestion has been reached through four main sources of evidence, namely ecology, bone structure, bioenergetics and biomechanics.

The ecological approach depends on an analysis of the relationships between the temperature gradients in the land masses in past epochs, and the distribution of animals in these land masses. This evidence has been reviewed by Bakker (1975). It is suggested that any large reptiles which existed in zones with cold winters were probably endothermic. Small ectothermic reptiles may be able to survive cool winter conditions by hibernating, but very large ectothermic reptiles would be unable to achieve an optimal body temperature during the short winter day and would have great difficulty in finding suitable places to hibernate. The distribution of the first of the synapsids, the pelicosaurs, was evidently confined to the region of the Permian equator. On the other hand a diverse population of late Permian therapsids survived the snowy winters of Gondwanaland. Furthermore a rich thecodont fauna existed in the southern regions of Gondwanaland in the Triassic, some forms extending into its southern limits. Some dinosaurs even reached the Cretaceous Arctic Circle.

The second source of evidence throwing light on the presence or absence of endothermy in fossil vertebrates is the histological structure of their bones. Essentially this depends on assessing the degree of development of secondary Haversian bone. In mammals and birds the presence of abundant secondary Haversian systems indicates rapid remodelling of bone, and therefore the capacity for continuous and speedy transfer of calcium from bone to plasma as required by the very active metabolism of these fully endothermic vertebrates. Another indicative feature of bony architecture is the presence or absence of so-called "growth rings". These are concentric laminae in the cortical bone, which are associated with seasonal variations in the rate of growth (Riqlès, 1969, 1974; Enlow, 1969, p. 65). Growth rings occur in contemporary reptiles and amphibians which are, of course, essentially ectothermic: they are generally absent in birds and mammals, except hibernating forms (Riqlès, 1969, 1974). Growth rings may therefore be indicators of ectothermy rather than endothermy.

Widely distributed, densely arranged Haversian systems do not occur throughout the cortex of the bones of contemporary reptiles (Enlow, 1969, p. 73). Thus the Squamata have compact bone which is virtually non-vascular, Haversian systems being absent; moreover there are relatively few cancellous trabeculae in the midshaft of the long bones, with little evidence of remodelling (Enlow, 1969, pp. 47, 61–62). In contemporary Chelonia and Crocodilia, Haversian systems are restricted to localized areas of the cortex, and the corticomedullary junction; cancellous bone is widely distributed in most medullary areas of the long bones, but secondary rebuilding is limited or totally lacking (Enlow, 1969, pp. 47, 53–54, 59).

Of the fossil reptiles, the pelicosaurs possessed growth rings (Riqlès, 1969, 1974) and in general had no structural features distinguishing their bones from those of present day reptiles (Enlow, 1969, p. 74). In the therapsids, however, dense Haversian bone, although less widespread than in the dinosaurs, is found in abundance in the cortex of certain long bones (Riqlès, 1969, 1974). The histological data suggest a very early (Permian) onset of endothermy, among both synapsids and archosaurs (Riqlès, 1974). Densely arranged Haversian systems certainly were widely distributed throughout the cortex of the compact bone of the saurischian and ornithischian dinosaurs (Enlow, 1969, p. 73) and the intensity of this Haversian reconstruction appears to be a truly major exception among the reptiles (Riqlès, 1969, 1974).

Through the presence of dense Haversian bone and the absence of growth rings, the bone of dinosaurs and therapsids appears closely comparable to that of many birds and mammals but quite distinct from that of the majority of other reptiles and amphibians (Riqlès, 1969, 1974). Since these characteristics of mammalian bone are associated with an active metabolism correlated with endothermic homeothermy, Riqlès (1969, 1974) suggested that the dinosaurs and therapsids were endothermic. However, Bouvier (1977) objected to such interpretations of the histological evidence, arguing that Haversian bone is not a reliable indicator of thermoregulatory mechanisms: secondary Haversian bone is found among both endotherms and ectotherms, and not all endotherms possess extensive Haversian reconstruction of their compacta. Indeed a great many mammals lack an extensive distribution of secondary osteones (Enlow, 1969, p. 58). Nevertheless, as Enlow (1969, p. 73) has stated, "the occurrence of widely distributed densely arranged Haversian systems throughout the cortex is a structural characteristic not found in any known amphibian or reptile either fossil or recent *except for dinosaurs*". Whilst this no doubt falls short of proving that dinosaurs were endothermic, Enlow (1969, p. 74) has pointed out that the possession of extensive Haversian rebuilding nevertheless indicates that mammals, birds and dinosaurs shared some important physiological factor. If not endothermy, what might this factor have been? McNab (1978) has suggested that it may have been thermal constancy (i.e. homeothermy), rather than endothermy. This would explain why Haversian substitution is not characteristic of small endotherms, and yet is found in large ectotherms to some extent.

The third line of enquiry into the possible presence of endothermy in fossil reptiles comes from a study of the energy budgets of carnivorous animals. The proposition is that the ratio of prey to predator population is a constant, related to the metabolic rate of the predator regardless of the body size of the animals in the prey–predator system (Bakker, 1972, 1975). Thus a stable population of ectothermic predators with a relatively low metabolic rate would be sustained by a relatively small stable population of prey; i.e. the ratio of prey to predator population would be relatively low. On the other hand, a stable population of endothermic predators with a high metabolic rate would have to be supported by a relatively large stable population of prey; i.e. the ratio of prey to predator population would be relatively high. For example Bakker cited a contemporary carnivorous ectothermic lizard as having a prey to predator ratio of about 2·5:1. Thus this ectothermic carnivorous population requires only about 2·5 times its population of prey animals to support it. In contrast an endothermic carnivorous mammalian population, such as lion or weasel, has a prey to predator ratio of from 33:1, to 100:1.

The prey to predator ratios of the fossil reptiles have been reviewed by Bakker (1975). In the pelicosaurs the ratio ranged from about 1·7:1, to 2·9:1. In the late Permian, the ratio for the earliest therapsids rose abruptly to about 6:1 or 11:1, and remained at about this level until the end of therapsid domination in the Triassic. For the thecodonts, the ratio can apparently be estimated only for isolated species, one of them giving a ratio of about 10:1, which is within the endothermic range. Among the dinosaurs the range was evidently as high as for Cenozoic mammalian communities, ranging from about 33:1, to 100:1. However, these bioenergetic estimates have not escaped criticism. For example Tracy (1976) pointed out that one cannot discount the possibility that a large carnivorous dinosaur such as *Tyrannosaurus* might have been a secondary carnivore and/or cannibal, whereas in calculations Bakker had implicitly assumed it to be exclusively a primary carnivore eating only large herbivorous dinosaurs. As an example, Tracy suggested that the application of Bakker's logic to the garter snake *Thamnophis sirtilis*, which is a contemporary secondary carnivore, would show it to possess a clearly endothermic prey to predator ratio although the animal is of course essentially ectothermic.

The fourth source of evidence is biomechanical. In an attempt to account for the development of thermogenic mechanisms in pre-avian and pre-mammalian reptiles, Heath (1968) had called attention to a possible relationship between the adoption of an erect stance and endothermy, suggesting that as a result of this shift in posture "mammal-like reptiles probably acquired *accidentally* a system to provide heat even at rest". The fully erect stance was evidently considered by Heath to require continual tonic activity of the antigravity muscles of the limbs and trunk. However, it seems unlikely that the development of endothermy can be explained simply by the adoption of a fully erect posture. Data assembled by Bennett and Dalzell (1973) show that the energetic cost of the erect stance is quite low; the metabolic rate of the sheep, cow and man when standing is only about 9% greater than when lying down, and in the horse the metabolic rate is the same in both postures. Although it is generally agreed that no direct causal relationship has been established between the erect posture and endothermy, it is still noteworthy that all modern terrestrial vertebrates with erect posture are endothermic and all living ectotherms are sprawlers (Ostrom, 1974b). Since the dinosaurs had a fully erect posture (Bakker, 1971b), the presence of this high correlation between the erect posture and endothermy in living vertebrates has been regarded as an indication that most dinosaurs operated on a relatively high metabolic level and very possibly were more endothermic than any living reptile (Ostrom, 1974b). Nevertheless this biomechanical evidence has been severely criticized by Feduccia (1973) and Bennett and Dalzell (1973) as highly speculative and failing to establish a physiological connection between endothermy, metabolism and posture.

Taken in their totality these various lines of enquiry have led Ostrom (1969, 1974b), Ricqlès (1969, 1974) and Bakker (1971b, 1975) to the hypothesis that the dinosaurs were probably endothermic. That many of the arguments are speculative is freely admitted (Ostrom, 1974b), as is the likelihood that the matter is beyond proof anyway; no part of the evidence is decisive and many individual items are open to alternative interpretations. On the other hand there seems to be no evidence which decisively disproves the hypothesis of dinosaur endothermy: evidently it is no easier to prove that dinosaurs were ectotherms than it is to prove that they were endotherms. Perhaps the dinosaurs (and therapsids) had achieved endothermy—but only at an intermediate level (Riqlès, 1974). Varying degrees of endothermy occur in mammals, where the Metatheria are intermediate between the Prototheria and Eutheria in body temperature and metabolism (Dawson and Hulbert, 1970). The debate continues.

IV. The influence of endothermy on morphology

The possession of an advanced level of endothermy as in living mammals and birds offers the opportunity for greatly increased physical activity. This section examines the possible impact of endothermy on the design requirements for locomotion, neural control, circulation, respiration, digestion and insulation, and ends with a consideration of the evolution of flight.

A. Locomotory apparatus

The expenditure of energy involves elaborate chemical reactions most of which are sensitive to temperature. According to van't Hoff's generalization, the rates of most chemical reactions double with each increase of 10°C. The significance of temperature in muscular activity is shown by studies of the flight muscles of insects (Heinrich and Bartholomew, 1972), which are said to be the most metabolically active tissue known. However, the power output needed by a moth for hovering cannot be achieved until its muscles are at a temperature of 35–38°C, and therefore a nocturnal moth must warm up its flight muscles before take-off by vibrating its wings.

A fully endothermic homeothermic animal can be relatively independent of the environmental temperature, having the capacity to obtain full muscular power instantly and virtually at any time including during resting periods between bouts of activity (Heinrich, 1977). In birds and mammals the metabolic cost of maintaining endothermic homeothermy is expensive, requiring at least 90% of the total energy metabolism to be devoted to the regulation of body temperature. However, the immediate availability of muscular power makes possible the evolution of modes of life which are based on active and sustained locomotion.

According to Bakker (1971a) most dinosaurs were similar to modern running mammals in their erect posture and in their locomotor anatomy and limb proportions. The coelurosaur Ornithomimidae of the late Cretaceous (e.g. *Dromiceiomimus*) had such lightly built but powerful hindlimbs that Russell (1972) regarded them as showing a level of cursorial adaptation seldom equalled and giving speeds probably superior even to *Struthio* which reaches 80 km/h. In contrast, the large contemporary reptiles, such as crocodiles, have relatively shorter limbs, less limb muscle, and lower top speeds (Bakker, 1975). Dodson (1974) agreed with Bakker that the structure of the limbs of dinosaurs showed them to be active cursorial animals, but Bennett and Dalzell (1973) maintained that the anatomy of the dinosaur scapula and glenoid cavity would preclude the degree of protraction of the forelimb required for fast locomotion. However, Bakker (1974) remained confident that his interpretation of the biomechanics of the dinosaur shoulder was right, permitting a parasagittal movement of the limb as in cursorial mammals with a total range about twice as great as that conceded by Bennett and Dalzell (1973).

B. Central nervous system

It was predicted by Heath (1968) that the adoption of a new posture would require changes in the neuraxis to handle increased sensory input and produce finer muscular control. The improved muscular performance made possible by the combination of endothermy and the erect posture (Bakker, 1971a, 1971b, 1974, 1975) may be expected to induce further neuroanatomical refinements. Also the erect posture presumably enabled better visual surveillance; when the erect posture was combined with rapid locomotion, the integration of visual with motor centres would become especially advantageous.

Comparisons by Jerison (1973, pp. 146–148, 163–172) of the weights of the brains and bodies of living reptiles, birds and mammals with those of fossil reptiles do not, however, reveal a clear overall enlargement of the brain in the latter. The size of the brain in relation to body weight in eight herbivorous species of dinosaur, two carnivorous dinosaurs and five pterosaurs was within the range of living reptiles and smaller than that of living mammals and birds, although the pterosaurs were at the upper edge of reptilian brain development. On the other hand, Russell (1969, 1972) found evidence of relatively large brains in the small bipedal carnivorous Cretaceous coelurosaurs, *Stenonychosaurus*, *Saurornithoides* and *Dromiceiomimus*. These reptiles had binocular vision with eyes similar in size to, or even larger than, those of *Struthio* (which has the largest eyes among living terrestrial vertebrates). The volumes of the brain endocasts (which unlike Jerison's endocasts were not corrected for conversion to brain volume) were held to indicate a ratio of brain to body volume about seven times that of an alligator, and an absolute brain size somewhat greater than that of living ratite birds of similar body weight (e.g. *Dromaius*). Noting the uncertainties about endocasts of therapsids (which are caused by the incomplete ossification of the skull) Jerison (1973, p. 152) would accept only the possibility of some lateral expansion of the cerebellum in this group. On the other hand Valen (1960) concluded that in advanced carnivorous therapsids such as the cynodonts the cerebral hemispheres were not much larger than in undoubted reptiles. However, the cerebellum was considerably enlarged and a pons was present (though it is absent in extant reptiles), thus indicating cerebellar integration of the motor pathways and therefore confirming Heath's prediction of improved muscular coordination.

To sum up, the evidence seems to indicate that the herbivorous dinosaurs resembled living reptiles in the ratio of brain size to body size, but the pterosaurs, some carnivorous dinosaurs, and perhaps some advanced carnivorous therapsids, did show signs of breaking away from the lower vertebrate level which characterizes fish, amphibians and reptiles in general.

C. Cardiovascular system

High internal heat production and increased and continuous muscular activity together demand the effective transport of nutrients and removal of waste products. Short bursts of intense muscular activity are possible in ectothermic animals through the anaerobic metabolism of glycogen by white muscle fibres, but this creates an oxygen debt which must be repaid later. Sustained muscular activity requires the aerobic metabolism of fatty acids by red muscle fibres. Rapid repayment of oxygen debt and high levels of continuous aerobic muscular metabolism are only possible if the circulatory system is efficient.

A substantial head of pressure within the systemic blood vascular system is essential for the rapid filtration of fluid and dissolved substances from the blood plasma, through the capillary walls, and into the tissue fluid. A high perfusion pressure is also required to enable the rapid redistribution of blood from one region of capillary beds to another, according to the varying demands of the tissues. But, as pointed out by Seymour (1976), an equally high perfusion pressure within the pulmonary circulation would lead to excessive filtration of fluid into the airways of the exchange tissue. It therefore appears likely that the development of endothermy would lead to the evolution of a four-chambered heart with the systemic and pulmonary circulations completely separated. Such separation seems especially probable in giant sauropods in which the head was as much as 650 cm above the heart, requiring an aortic pressure of about 500 mmHg to supply the brain with blood (Seymour, 1976).

Among the living homeothermic endotherms (mammals and birds) systolic pressures exceeding 100 mmHg are commonplace (Lindsay *et al.*, 1971, pp. 405–410) and the four-chambered heart is universal. In contrast, among the living ectotherms (fish, amphibians and reptiles) systolic pressures seldom exceed 50 mmHg (Lindsay *et al.*, 1971, pp. 410–411) and only Crocodilia have achieved complete separation of the ventricles.

D. Respiratory system

The high metabolic intensity of an endothermic animal results in a substantial increase in oxygen consumption at rest (White, 1978). For example at similar body mass and body temperature, under basal or standard conditions, the metabolic rate of lizards is only 0·28 that of eutherian mammals and 0·17 that of passerine birds (Dawson and Hulbert, 1970). Indeed both the resting and active metabolic rates of reptiles are only about one-tenth those of mammals (Gans, 1978). Thus, as a result of adopting endothermy, mammals and birds have higher metabolic rates and maintain such rates over more protracted periods; this presumably placed a premium on the development of more effective mechanisms for gaseous exchanges (Gans, 1978).

Respiratory adaptations which have been regarded as detectable in fossils are the presence of a diaphragm or air sacs. The possible presence of a diaphragm in cynodonts was postulated by Brink (1955) and Attridge (1956), mainly on the grounds that in these fossil forms a differentiation between thoracic ribs and progressively shorter lumbar ribs begins at the level where the diaphragm occurs in mammals, whereas in an average unspecialized reptile, ribs extend throughout the trunk (Brink, 1955). However, in no cynodont is there any direct anatomical evidence for a diaphragm, such as exostoses for the origin of the crura on the ventral aspect of the lumbar vertebrae (Jenkins, 1970).

The prospect of detecting the presence of air sacs in fossil bones would seem to be rather more promising. Pneumatic bones might well be recognizable in fossil skeletons, and if present in the postcranial skeleton would be virtually certain evidence for the existence of air sacs. Nevertheless considerable caution is needed in identifying pneumatic bones and it cannot be assumed that a dried bone was necessarily pneumatic simply because it was apparently hollow (Bellairs and Jenkin, 1960, p. 293; Feduccia, 1973). Some remarkable mistakes have been made in determining the presence of air sacs even within the bones of contemporary birds (King, 1966, p. 223). The literature on fossil reptiles (e.g. Huxley, 1868; Romer, 1966, pp. 45, 145, 155; Swinton, 1960, pp. 6, 11; Brodkorb, 1971, p. 26; Bakker, 1971b; Jerison, 1973, p. 162) contains many scattered allusions to pneumatic bones particularly in pterosaurs and the vertebrae of dinosaurs, but nearly always without giving any original references except that the monograph of Seely (1901) is sometimes cited for pterosaurs. In the vertebrae of several large dinosaurs described by Cope (1878), extensive excavations, hollow areas, and "pneumatic foramina" were present in the centra except those of the tail which were nearly solid, and this evidence seems reasonably convincing. However, the vertebrae of some living lizards have quite large foramina which could easily be mistaken for pneumatic foramina, but in fact transmit the venous drainage of the vertebral marrow (A. d'A. Bellairs, personal communication).

If air sacs really were present in pterosaurs and dinosaurs, they might have been capable of drawing air through the exchange tissue of the lung. Improved arteriolization of the blood might then have occurred as in birds (see Section V.E), but this would have required the development of an advanced avian parabronchial type of lung. Yet there is no justification for Bakker's (1974) assumption that the presence of air sacs necessarily guarantees the simultaneous evolution of an advanced avian type of lung capable of unidirectional air flow.

E. Chewing adaptations

The large energy budget of a continuously active endothermic vertebrate demands a high rate of fuel supply and hence the rapid digestion of food. The evolution of grinding mechanisms within the alimentary tract is therefore to be expected.

The masticatory apparatus of therapsids rivalled that of modern mammals in the complexity of the pattern of tooth crowns and jaw movement (Crompton *et al.*, 1978). Many herbivorous dinosaurs, including most of the ornithopods and all of the horned dinosaurs, developed powerful dental batteries capable of cutting and grinding like the teeth of living artiodactyls and perissodactyls; other herbivorous dinosaurs such as the brontosaurs only had beaks or strong incisiform teeth (Bakker, 1971b). Birds maintain a very high energy output without teeth, but do so with the aid of the gizzard (*ventriculus*) which, with the stones that it contains in most graminivores and herbivores, and in some omnivores (Ziswiler and Farner, 1972, pp. 377–378), constitutes the powerful gastric mill. Cineradiography has shown similar gastric mills in some contemporary Crocodilia (Bakker, 1971a). Whilst many doubtful gastric stones have been found scattered among dinosaur skeletons (Desmond, 1975, p. 133), in a few dinosaurs, including at least one sauropod and another non-sauropod herbivore, convincing clusters of small stones have been found among the ribs (Brown, 1941).

F. Insulation

Having achieved endothermy an animal must develop mechanisms for conserving and dissipating heat before it can be assured of being homeothermic in environmental conditions where the ambient temperature fluctuates. Hair or feathers can provide insulation which is capable of rapid adjustment.

The first appearance of hair is conjectural. The general opinion seems to be that the fossil skin impressions of dinosaurs lack any indications of either hair or feathers (Bakker, 1971b, 1972; Desmond, 1975, p. 121). Well preserved remains of the skin of hadrosaurs (Ornithischia) and sauropods show only a clear mosaic of small non-overlapping tubercles about the size of a pinhead (Brown, 1941). On the other hand, hair was almost certainly present in some pterosaurs (Bramwell, 1970; Desmond, 1975, p. 170).

The evolution of feathers is as little documented as the evolution of hair. Since both feathers and reptilian scales are epidermal, it is tempting to regard feathers as mobile elongated scales with frayed edges (Regal, 1975), but they might well be morphologically and embryologically distinct (de Beer, 1975). There is no direct fossil evidence for the transformation of scales into feathers, but Swinton (1960, p. 11) believed that the essential pattern for feathers was already imprinted in the hereditary scale equipment of coelurosaurs such as *Scleromochlus*. The overlapping keeled scales of *Longisquama*, which trap an insulating layer of air next to its body, are regarded by Bakker (1975) as a perfect ancestral stage for the insulation of birds.

The feathers of *Archaeopteryx* were identical in structure to those of modern birds (de Beer, 1975) and were therefore already highly specialized. Along the line of ancestors leading to *Archaeopteryx* there must have been forms bearing the progressively evolving stages of "proto-feathers" (Riqlès, 1974), even though this evolutionary sequence has not been found in the fossil record. The nature and function of the first feathers has been much debated (see Bock, 1969 and Ostrom, 1974a for discussion), but it seems likely that contour feathers evolved before down feathers, serving for insulation not flight. This appears to be consistent with the hypothesis stated above (p. 14) that endothermy may have been widespread among the dinosaurs, including the late Triassic and early to mid-Jurassic ancestors of *Archaeopteryx*. It would seem to be certain, however, that at least the immediate feathered ancestors of *Archaeopteryx must* have been endothermic: endothermy necessarily preceded feathers, because the possession of an insulating covering of feathers by an ectotherm would deprive it of heat gain from solar radiation (Ostrom, 1974a). Admirably designed though they were for insulation, the first feathers—if they were contour feathers—were also ideally preadapted for flight. All that was necessary to produce an aerofoil was to enlarge the feathers of the wing and tail and attach them firmly to the skeleton (Ostrom, 1974a).

Archaeopteryx probably had an arboreal habitat. The ambient temperature in trees would be lower and more variable than on the ground, and, once gliding between the branches had been achieved (see below), the animal would have been constantly moving in and out of the cool shade of the evergreen foliage. The possession of insulation in the form of feathers secondarily adapted for flight would have enabled *Archaeopteryx* to penetrate fully a new arboreal Jurassic niche which was out of the reach of the carnivorous dinosaurs and rich in new reserves of food supply especially insects. Its endothermic homeothermy would have allowed the animal to extend its activity into early morning and evening, even into the dark, and during the cloudy, rainy, or windy weather of temperate regions; furthermore its eggs could be removed from the ground and incubated in an arboreal nest (Bock, 1969).

G. The evolution of flight

For those vertebrate lines that began to explore aerial habitats, the possession of endothermy and the greater muscular power that goes with it must have been crucial to meeting the high energy requirements encountered during the evolution of steady horizontal flight. Aerial experiments have been made by many vertebrates, including some living amphibians (*Rhacophorus*) and reptiles (*Draco volans*) (Pennycuick, 1972, p. 5), but such ectothermic forms would be restricted to gliding or, at the best, short bursts of muscular activity.

Two theories have been advanced to explain the evolution of flight, one "cursorial" (Ostrom, 1974a), and the other "arboreal" (Bock, 1965, 1969; Brodkorb, 1971, p. 34; de Beer, 1975). According to the original cursorial theory, the bipedal "Proavis" ran along the ground extending and beating its feathered arms until lift occurred. As de Beer (1975) pointed out, when expressed in this form the cursorial theory disregards "the adaptive criterion, which is of paramount importance in all cases of evolution". In a search for an adaptive advantage, Ostrom (1974a) suggested that the running *Archaeopteryx* could have used its extended feathered arms to catch flying insects. This elongated, predatory, grasping forelimb capable of strong adduction was thus preadapted as a "proto-wing". Vigorous flapping during the catching of prey would have produced some degree of lift. Improvements of the wing, such as enlargement of the wing feathers and elongation of the wing with greater pectoral muscles, would have enabled leaping attacks, with some degree of lift, and possibly flying after escaping insects. Subsequent advances in structure could culminate in take-off and the beginnings of powered flight.

One objection to this concept is the drag caused by extending the wings while running (Galton, 1970) and the likelihood of damaging the feathers when slapping down insects (Harrison, 1976). It also fails to account satisfactorily for the long feathered tail, which in *Archaeopteryx* had an area of about 140 cm^2, approaching the dimension of one of the wings (Bramwell, 1971). Such an appendage would be nothing but a disadvantage to a cursorial animal trying to run at high speed, but would be well adapted for control and stability in flight (Pennycuick, 1972, p. 26). Brodkorb (1971, p. 34) argued that the long flat tail of *Archaeopteryx* is not that of a ground-dweller, since the soft wide rectrices would soon have become frayed on the ground. Ostrom's (1974a) reply to this was that several cursorial predaceous birds have long tail plumage but are not particularly good flyers, including the Secretary Bird (*Sagittarius serpentarius*), Roadrunner (*Geococcyx californiana*) and the Savannah Hawk (*Heterospizias meridionalis*). On the other hand the loss of flight in flightless birds has generally been followed by reduction of the tail (Romer, 1966, p. 169). Moreover it might be asked, how many modern ground-living birds catch insects by running along the ground with their wings extended, in the supposed manner of "Proavis" and *Archaeopteryx*?

The cursorial theory was regarded by de Beer (1975) as "utterly untenable" because the caudally directed first digit of the pes would make fast running impossible, and secondly because in cursorial animals it is a rule that the hind limbs are longer than the forelimbs. Galton (1970), and particularly Ostrom (1974a), countered the first of these objections by pointing out that a partly or fully reversed (i.e. apposable) hallux apparently occurred in nearly all theropod dinosaurs, including many very speedy cursorial bipedal coelurosaurs. The fact that it was evidently present in all the carnivorous theropods suggests that its initial function was for grasping prey (Galton, 1970; Ostrom, 1974a). Doubtless *Archaeopteryx* adapted it for grasping branches (Galton, 1970), though not with great efficiency since its hallux appeared to be slightly elevated above the plane of the cranially directed digits and was shorter than them, whereas in contemporary passerines the hallux is nearly the length of the third (the longest) digit and the digits are roughly in the same plane (Bock and Miller, 1959). The second criticism was answered by Galton (1970) by showing that in *Archaeopteryx* the general proportions of the hind limb as a whole were, by mammalian standards, markedly cursorial.

The arboreal theory of the evolution of flight proposes as an initial stage a small bipedal lightly feathered "Proavis", with the forelimbs adapted for predaceous grasping. Urged on by the search for new food supplies and possibly by the need to escape from larger dinosaurs, "Proavis" made tentative forays into the trees. It may have jumped into low branches much as the domestic fowl leaps onto its perch, or the predatory digits of the hand may have helped it to climb the trunks of trees (de Beer, 1975), the terminal phalanx of the third digit of *Archaeopteryx* being extremely curved with a needle-like point.

Leaping between branches and parachuting to the ground followed (Bock, 1969). These actions required longer stiffer feathers on the outstretched forelimbs and tail to increase the aerofoil surface and control the orientation of the body. Angled parachuting would lead gradually to gliding, enabling greater distances to be covered. Flapping during gliding would extend the distance still further and make changing of direction possible. After the elements of flapping had been mastered, the perfected flight of modern birds could be gradually evolved. This process of conversion from reptiles to birds probably took place slowly, the many morphological changes occurring not abruptly or simultaneously but at different times and rates (de Beer, 1954). Galton (1970) suggested that the earliest stage may well have been a fast moving ground-living bipedal archosaur which he named "Pre-proavis", the second stage being the arboreal "Proavis".

The general concensus seems to be that *Archaeopteryx* itself had not completed the evolution of perfected flight, but may have managed weak, ungainly, flapping flight, or may even have been only a barely adequate glider (Ostrom, 1974a). The following anatomical features (Ostrom, 1974a; de Beer, 1975) have been put forward as evidence for weakness in flight. Quill knobs (*papillae remigiales*) for the attachment of the secondary flight feathers were lacking on the bones of the forearm (Ostrom, 1974a). The lack of fusion of the carpal and metacarpal bones prevented firm attachment of the primary flight feathers. The pectoral crest of the humerus was small. The coracoid bones were arranged as in theropods and appear not to have braced the shoulder joint against the sternum. The absence of any sternal carina may have meant that the volume of the pectoral muscles was small; because of the shortness of the coracoid, the pectoral muscles must also have been short. For these reasons the pectoral muscles may have had insufficient power on the downstroke for steep take off or hovering flight. The "biceps tubercle" of the coracoid bone appears to be the precursor of the avian acrocoracoid process (Ostrom, 1976), but it was still well distal to the glenoid cavity and there was no triosseal canal, so the supracoracoid muscle could not have powered the upstroke in flapping flight (Walker, 1972). Whereas the glenoid cavity of birds, bats and pterosaurs faces laterally, thus permitting good abduction of the humerus for powered flight, Bakker and Galton (1974) believed that in *Archaeopteryx* the glenoid cavity was still directed ventrally as in dinosaurs; this led them to doubt whether the shoulder joint would have allowed even a gliding form of flight. The humerus, ulna and carpometacarpus appear to lack the specialized features that are involved in the perfected folding and powerful extension of the manus which characterize modern birds (Ostrom, 1976). It was concluded by Ostrom (1976) that flight-related skeletal features were in general absent from the fore limb and pectoral girdle of *Archaeopteryx*.

However, it has been pointed out by Yalden (1971) that the skeletal evidence for flight weakness in *Archaeopteryx* may not be entirely conclusive. The absence of a large sternal carina is not significant, since some species of bat fly well without one (see Section V.D). Although the coracoid bones and hence the pectoral muscles were undeniably short, this does not establish a lack of power since a short broad muscle may produce more power than a longer one of the same mass. The pectoral crest may not have been small after all; on the contrary, by new measurements Yalden showed it to be both very wide and very long compared with modern birds, and regarded this as compelling evidence that the animal was capable of flapping flight. Moreover, Ostrom (1976) considered that the glenoid cavity *is* directed laterally, more or less as in modern carinates.

It is believed that none of the bones of *Archaeopteryx* was pneumatic (Swinton, 1958, p. 23; de Beer, 1975). But this, of course, is not proof of the absence of air sacs, since several species of modern birds, all of which have air sacs, are said to have no pneumatic bones (King, 1966, p. 221); however, it does raise the possibility that *Archaeopteryx* had not developed air sacs and might therefore have had inferior mechanisms for gaseous exchanges.

It was originally believed that the cerebellum of *Archaeopteryx* was too small for the three dimensional neuromuscular control required for flight (de Beer, 1954). However, Jerison (1973, pp. 183–188) has shown that in the endocast on which this was based, there was a misinterpretation of the midline leading to a serious underestimate of the size of the brain. The brain of *Archaeopteryx* is now believed to be intermediate between living reptiles and birds in size and shape, and was therefore showing an evolutionary response to the new niche which had been occupied (Jerison, 1973, pp. 197–199).

Some calculations of the flying characteristics of *Archaeopteryx* have shown marked inferiority to modern birds and indicated an uncomfortably high landing speed (Heptonstall, 1970). Other calculations, which included a reduction in the estimated body weight from 500 g down to 200–250 g (Yalden, 1971; Bramwell, 1971) were less pessimistic, and it was emphasized that as *Archaeopteryx* was arboreal it only had to glide until it stalled on the tree canopy with very little forward speed.

V. Structural adaptations for flight

All classes of vertebrates have experimented with parachuting and gliding, but only bats, birds and the pterosaurs have mastered steady horizontal and climbing flight.

A. Limitations upon body weight

Flight imposes on birds a maximum body size. Considerations of the effects of changes of scale reveal the general relationships between size and flight performance. These have been analysed by Pennycuick (1972, pp. 35–43), and the following summary is based on that source except where stated otherwise.

The wing loading is the weight of the bird divided by the area of its wings. Small birds tend to have a relatively low wing loading of about 0·11–0·23 g/cm^2, but many larger birds have much higher values up to 2·5 g/cm^2 (Hildebrand, 1974, pp. 613, 614). Because their wing loading is higher, larger birds in general have to fly faster than smaller ones. However, it can be shown that if bird A weighs twice as much as bird B it will require not twice as much power to fly at its minimum power speed, but 2·25 times as much. If it is assumed that the proportion of the body weight allocated to the flight musculature is much the same in all birds (see Section V.D), then in the larger birds each gram of muscle must produce more power than in the smaller birds. Furthermore, the maximum power available is determined not only by the total weight of the flight muscles but also by the frequency of flapping, but for mechanical and aerodynamic reasons the frequency of flapping *decreases* as the bird increases in size. In fact if bird A weighs twice as much as bird B it will have only 1·59 times as much power from its flight muscles, even though it needs 2·25 times as much power to fly. Therefore, as birds get heavier, their reserves of power diminish until eventually a point is reached when their flight muscles are unable to provide sufficient power for flapping flight.

Whilst acknowledging that the exact maximum weight compatible with horizontal flapping flight cannot be calculated, Pennycuick empirically estimated the limit to be about 12 kg; as was pointed out by Marshall (1962, p. 557), an 80 kg angel would evidently have considerable difficulty in getting off the ground. Twelve kilograms is about the weight of the Mute Swan (*Cygnus olor*), but the Whooper Swan (*Cygnus cygnus*) is stated to weigh about 17 kg (Thomson, 1964, p. 751). Furthermore to be effective a bird should be able to do a good deal more than simply fly horizontally; it will have to carry food for itself and sometimes for its young, and must be capable of rapid take-off and steep climbing to escape predators.

Some pterosaurs and birds succeeded in partly overcoming these restrictions on body size. The Cretaceous pterosaur *Pteranodon* is believed to have had a wing span of about 8 m and a weight of about 18 kg (Bramwell, 1971). The giant Cretaceous pterosaur recently discovered in western Texas had an estimated wing span of about 15 m (Lawson, 1975), and was presumably even heavier than *Pteranodon*. The pleistocene condor, *Teratornis incredibilis*, which is believed to have been the largest flying bird of all time (Wetmore, 1955, p. 51), had an estimated mass of about 20 kg (Pennycuick, 1972, p. 39). However, these really gigantic flyers, and other very large pterosaurs and extinct flying birds, were almost certainly soarers, drawing energy from winds or thermals. Dependance on soaring calls for a reliable climate, with constant wind conditions and predictable temperatures for thermals. Throughout almost the whole of the Mesozoic a broad belt of equable climate extended widely in both directions from the equator (Newell, 1972). The onset of rougher weather at the end of the Cretaceous could have brought about the extinction of such highly specialized soarers as *Pteranodon* (Bramwell, 1971).

B. Modifications of body form

1. Trunk

General streamlining of the body is essential for fast flight. Also the body must be compact and firm to enable the thrust of the wings to be transmitted to a point near its centre of mass, thus enabling the wings to propel the body as a unit without deforming it or making it flap (Hildebrand, 1974, p. 261). The general firmness of the trunk is accentuated by varying degrees of fusion between the vertebrae.

2. Wing

As a result of the almost total commitment of the forelimb to flight, nearly all birds possess two completely independent locomotory systems, wings for flight (or swimming) and hindlimbs for walking, running, or swimming. This is a point of superiority over bats, in which the wing is attached to the hindlimb thus greatly restricting mobility on the ground.

There are four general types of wing, with many intermediate forms (Savile, 1957). The elliptical wing typifies many Passeriformes and some Galliformes and Columbidae which have to manoeuvre through restricted openings in vegetation. The shape is short and broad with a low aspect ratio of between 3 and 6 (aspect ratio = wing span divided by average width, or span^2 divided by area), the outline closely resembling that of the mark II Spitfire of 1940–45. Wing loading is only moderate or low. The wing beat is farily fast and the amplitude moderately great. The alula is large and the primaries may separate to form additional wing slots to prevent stalling at low speeds (Hildebrand, 1974, p. 618). Manoeuvrability is good (Savile, 1957).

The broad soaring wing, which occurs in New World vultures (Cathartidae), eagles (*Aquila*, etc.) and pelicans (Pelecanidae), is moderately long and broad with medium aspect ratios of 5 to 7 and only moderate wing loading. The alula and wing slots are prominent. These wings are adapted for soaring at low speed.

The high speed wing, which characterizes swifts (Apodidae), falcons (Falconidae), hummingbirds (Trochilidae), and to a lesser extent ducks (*Anas*, etc.) and terns (*Sterna*), is relatively small with a moderately high aspect ratio of between 5 and 9. Drag is low and wing loading high. The wing beat is rapid and the amplitude small. The wing tip is tapered and may be swept back. There are no tip slots, except in falcons which close them in fast flight (Savile, 1957).

The long soaring wing is confined among birds to oceanic soaring species such as albatrosses (Diomedeidae), the Gannet (*Morus bassanus*) and gulls (*Larus*) (Savile, 1957), but probably occurred also in some large pterosaurs (Hildebrand, 1974, p. 620). It is long, slender and pointed with a high aspect ratio of between 8 and 18 and high wing loading. The high ratio of lift to drag allows soaring at high speed. There are no tip slots, but the alula may be of substantial size. Such wings have an inherently high efficiency and satisfactorily meet the contradictory requirements of gliding and flapping; on the other hand their relative fragility and clumsiness demand a habitat free from obstacles (Brown, 1961, p. 300).

In all of these wings, lengthening is an obvious adaptation to flight. Lengthening increases the area swept out during flapping, thus bringing the induced power down to manageable levels for low speeds (Pennycuick, 1972, p. 15). The wing surface consists of the primary and secondary feathers, the shafts of which radiate from the manus and ulna respectively. The curvature of these feathers gives the wing a distinct camber. The forces and moments developed over the expanded vanes are transmitted to these bones and ultimately concentrated at the head of the humerus. Consisting of barbs and interlocking barbules, the vane of each flight feather is almost airtight; it is light yet stiff enough to be supported at one end only (in contrast to the bat's membrane which stretches between the tips of the five digits of the manus and continues to the skeleton of the hindlimb); it can be readily repaired after minor damage, the barbules being rehooked by simply passing the vane through the bird's beak (de Beer, 1975), again a marked improvement on the more easily torn wing of the pterosaur and bat; and if severely damaged a whole feather may eventually be replaced by moulting.

It is important that the mass of the wing should be kept to a minimum, since the lower the mass the less the inertia during the wing stroke. As pointed out by Norberg (1970) it is particularly necessary to reduce the mass of the wing tip, since the amplitude and therefore the speed of movement of the various parts of the wing during the wing stroke increase with the distance from the shoulder joint; bats overcome this difficulty by tapering the bones towards the wing tip, but in birds as just stated the lateral extremity of the wing consists of feathers only.

The leading edge of the wing between the shoulder and carpal joint is formed by a membraneous fold of skin, the prepatagium, the cranial edge of which is tensed by the strong prepatagial elastic ligament. The prepatagium is also kept taut by the tensor prepatagialis muscle which inserts on the medial end of the elastic ligament (Baumel, 1979). A smaller fold, the metapatagium, completes the caudal edge of the wing between the trunk and the region of the elbow joint, and is kept under tension by the pars metapatagialis of the latissimus dorsi muscle (vanden Berge, 1979).

The mechanics of the wing vary with different types of flapping flight (Brown, 1961, pp. 294–299; Hildebrand, 1974, pp. 611–615). In fast level flight propulsion comes solely from the downstroke, virtually all of it being provided by the lateral half of the wing. The upstroke is passive, the wing being lifted by air pressure rather than muscular effort, the only function of the upstroke being to support the bird. Much lift is generated by the slower-moving medial part of the wing on both the upstroke and downstroke.

The slow ascending and descending flight of small woodland birds also appears to be relatively uncomplicated. Marked lift and propulsion come from the downstroke, the upstroke again being merely a recovery stroke (Hildebrand, 1974, p. 613). The steep ascending and descending flight of strong flyers of medium size such as pigeons (Columbidae), ducks (Anatidae), hawks (Accipitridae) and pheasant (Phasianidae) is much more complex. The downstroke provides much lift but little propulsion. The upstroke is distinguished by a very rapid rotation at the shoulder and extension of the arm, producing a backward flick of the primaries which provides almost all of the propulsion (Brown, 1961, p. 296). In hovering flight as in Trochilidae much lift is obtained on both the forward stroke and the back stroke (Hildebrand, 1974, p. 611).

3. *Tail*

In a flying animal a tail would appear to be a useful adaptation for stability in pitch (i.e. stability about a horizontal axis normal to the flight path), much as pitch can be regulated in a conventional tailed aircraft by adjusting the tailplane. *Archaeopteryx* and the earlier pterosaurs had long tails which seem well adapted for this purpose; later forms of both groups more or less dispensed with the tail, suggesting that tailless control is more efficient aerodynamically even if more difficult to operate (Pennycuick, 1972, p. 26). Reduction of the tail renders the bird unstable in flight, but instability is advantageous; it increases manoeuvrability, reduces the stalling speed of a large bird thus making it easier to land, and in extreme cases lowers the stalling speed to make hovering flight possible. Birds which normally have a tail can fly without it (Smith, 1952).

In modern carinates the bones of the tail have been reduced to about five to eight free caudal vertebrae, followed by the pygostyle (pygostylus) consisting of three to six fused caudal vertebrae (Baumel, 1978) and providing a firmer and more readily controllable base for the attachment of the tail feathers (Ostrom, 1976). The main tail feathers, the rectrices, are attached to the tissue covering the pygostyle, radiating from it like the blades of a fan which can be spread transversely and cocked up or down (Bellairs and Jenkin, 1960, p. 251).

In flight (Pennycuick, 1972, pp. 27–28), the tail is spread and depressed at very low speeds, particularly during landing and take-off. This provides additional wing area and sucks air downward over the central region of the wing thus increasing lift, thereby enabling controlled flight at speeds lower than otherwise possible.

C. *Skeleton*

1. *General reduction of weight*

It is often stated that the bones of birds are very light as an adaptation for flight. Frequently this characteristic is said to be linked with pneumatization of the skeleton, although pneumatization is supposed to be completely absent in some species (King, 1966, p. 221). In general the avian skeleton probably is lighter than that of mammals, but facts are hard to find.

Headley (1895, pp. 110–115) suggested that pneumatization of an avian long bone results in a relative increase in its girth, and a reduction in the thickness of its wall as well as disappearance of bone marrow. These changes could increase the strength and reduce the weight, such modifications being particularly significant in large birds where the ratio of the cavity volume to the surface area of the bone would be relatively great (Bellairs and Jenkin, 1960, p. 292). However, Headley's measurements are not entirely convincing.

It was claimed by Strong (1919) that in the albatrosses (Diomedeidae), where the degree of pneumatization is "extraordinary", "there is no solid bone much thicker than writing paper except in the leg bones", the skull bones consisting of thin layers of compact bone separated by a diploe reduced to slender spicules and devoid of marrow. Similar remarks have been made about pterosaurs such as *Pteranodon*, in which the "skeleton was reduced to paper-like thickness" (Bramwell, 1971). D'Arcy Thompson's (1942) well known drawing of the metacarpal bone of a vulture is often produced as evidence for the general lightness, tubular form, and internal strutting of avian bones. The skeleton of a pigeon was found by Welty (1962, p. 3) to constitute 4·4% of the total body weight, whereas that of a white rat measured 5·6%. The same author cited a report that the skeleton of a frigatebird (Fregatidae) with a seven-foot wingspan weighed only 114 g; this was less than the plumage, which together with the breast muscles comprised 47% of the total body weight. Hildebrand (1974, p. 621) noted that the skeleton of an eagle accounted for less than 7% of the total body weight, and considered this to be about half the value for man. From observations on more than 20 species of bird and several species of mammal Chappel (1978) found the femur, tibia, ulna and radius to be about 10–15% less dense in birds than in mammals. The density of the avian humerus varied greatly. Thus among strong flyers it was 0·67 g/cm^3 in *Falco tinnunculus* and 0·69 in *Columba livia*, but 1·43 in *Hirunda rustica*; among cursorial species the values were 0·58 in *Carpococcyx renauldi* but 1·67 in *Gallirallus australis*. The density of the humerus of the rat was 1·47.

2. *Skull and teeth*

It seems reasonable to assume that birds and pterosaurs were subject to selection pressures to decrease the weight of the skull for flight. Welty (1962, p. 3) reported that the skull of a pigeon accounted for only 0·21% of the total body weight, whereas the skull of a rat accounted for 1·25%. Certainly the honeycomb of air spaces in the skull bones and the delicate spicules which support them are well known features of the avian skull. Four different designs of air spaces were described by Bühler (1972). In his material extreme weight reduction occurred in the Red-necked Nightjar (*Caprimulgus ruficollis*) in which the outermost lamella of the parietal bone was only 20 μm thick; the entire skull and lower jaw weighed only 0·45 g, although its length was 5 cm and its breadth 3 cm. Another device for saving weight was suggested by Bock (1974,

p. 157); the kinetic maxilla of birds redistributes stress from a potentially weak area (skull roof) to a strong area (skull floor), enabling the bone of the roof and lateral wall of the skull to be thinned.

It is tempting to assume that the loss of teeth, with attendant reduction of the jaws and chewing muscles, are direct adaptations to flight by reducing the weight of the head, as suggested by Welty (1962, p. 6) and others. This interpretation is consistent with the presence of teeth in early representatives of both birds and pterosaurs and their subsequent disappearance in both groups (Brodkorb, 1971, p. 26). The replacement of some or all of the teeth by a horny beak is, however, far from being an exclusively avian feature, since it occurred in many dinosaurs including the coelurosaur ornithomimids, in the great majority of the Ornithischia, and in the therapsid dicynodonts (Romer, 1966, pp. 151, 156, 184).

Dilger (1957) considered that weight saving might be a partial explanation, but related the loss of teeth mainly to the need for speed in both ingestion and digestion as a result of the high metabolic rate of birds. The more finely the food is divided the faster it is digested. The development of the muscular gizzard into the gastric mill enables birds to omit the preliminary chewing and to ingest food rapidly, storing it in the oesophagus or crop for subsequent processing by the gizzard. The widespread occurrence of gizzard stones in birds (see Section IV.E of this chapter) and the massive muscular development of the wall of the gizzard in many species argue against the concept that the loss of teeth and reduction of chewing musculature simply saves weight. On the other hand the development of the gastric mill does have the advantage of bringing these heavy structures closer to the centre of gravity.

3. Bones of pectoral girdle and wing

The pectoral girdle is strongly built, and is virtually immovable in both birds and pterosaurs though not in bats (Hildebrand, 1974, p. 623). The massive coracoid acts as a strut bracing the shoulder joint from the sternum. Ligaments firmly attach the blade-like scapula to the ribs (Bellairs and Jenkin, 1960, p. 252), which are strengthened by their uncinate processes and their ossified sternal components; together, the scapula and ribs further help to brace the shoulder joint from the sternum. In most flying birds the apophysis furculae is close to the apex carinae and strongly attached to it by ligaments. In several Procellariiformes and most Pelecaniformes the apophysis forms a strong synovial articulation (and in extreme cases a synostosis) with the apex carinae (Baumel, 1979); at least in these latter species the furcula clearly forms an additional rigid brace holding the shoulder joint away from the sternum, but in general it may function like a spring to maintain the proper transverse spacing of the shoulder joints (Ostrom, 1976). The coracoid, the scapula and ribs, and the furcula prevent the shoulder joint from being pulled towards the sternum when the powerful pectoral muscle contracts during flight. During gliding, the same bones simply suspend the sternum, which acts as a platform carrying the weight of the body (d'Arcy Thompson, 1942, p. 1013). In "carinates" the prominent carina of the sternum greatly increases the surface area available for the origin of the pectoral muscle.

The glenoid cavity becomes directly laterally in birds, bats and pterosaurs, thus making possible the adduction and abduction of the forelimb which are essential for flight (Bakker and Galton, 1974). In most carinates the tendon of insertion of the supracoracoid muscle passes through the triosseal canal and raises the wing. The pectoral crest of the humerus is prominently developed for the insertion of the pectoral muscle.

The ulna is larger than the radius. The ulna and radius together form a unit which is mediolaterally slightly convex (the concavity facing ventrally), the two bones being separated by a relatively large distance; these relationships greatly increase the resistance to bending forces in the plane in which the two bones lie (Norberg, 1970). In most birds the ulna carries two lines of quill knobs to which the follicle of each secondary flight feather is attached by two ligaments (Edington and Miller, 1941). Some of the distal carpal bones fuse with the metacarpal bones, forming a single rigid bone the carpometacarpus, which acts as a solid platform for the attachment of the primary flight feathers (Ostrom, 1976). A few flight feathers are also attached to the phalanges of the alula, helping to prevent stalling at slow speed. There is a general simplification of the manus by reduction and fusion of bones.

The shoulder joint enables the humerus to move in nearly all directions, but the movements of the joints of the wing distal to the shoulder joint are restricted to flexion and extension in the plane of the wing surface, with minimal pronation and supination. The surface of the wing is therefore stiffened against the resistance of the air during the downstroke and upstroke. Furthermore, extension of the elbow automatically extends the manus; this mechanism works like a pair of "drawing parallels", the radius acting as a rod which pushes the manus into extension when the elbow is extended and pulls the manus into flexion when the elbow joint is flexed (Fisher, 1957).

D. Flight muscles

The flight muscles have been adapted to transmit large amounts of power to the wings. In fast powered flight the work is done mainly by the pectoralis muscle during the downstroke (see Section V.B, ii). The mass of this muscle is believed to constitute about 15% of the total body weight in birds generally (Pennycuick, 1972, p. 22), ranging from 10 to 21% (Hartman, 1961). The higher values are found in birds with high wing loadings for their size (Pennycuick, 1972, p. 22).

Kuroda (1961) confirmed early reports of the existence of subdivisions of the pectoralis muscle which may have specialized functions. For instance, in all soaring birds, regardless of their taxonomic position, the pectoralis muscle seems to be divided into superficial and deep parts. The deep part, which accounts for about 8–12% of the total mass of the muscle (Pennycuick, 1972, p. 54), is absent in non-soaring birds and is believed to form a specialized fast acting muscle which adjusts the position of the wing against the varying force of the wind in order to keep the wing motionless in the horizontal position while gliding. It is particularly well developed in pelagic soarers (e.g. Procellariiformes) to take advantage of changes in air flow produced by the sea waves.

In fast level flight the upstroke is essentially passive, but in the steep jump take-offs of strong flyers the upstroke provides not only lift but most of the propulsion; in hovering flight not only the forward stroke but the back stroke supplies much lift (see Section V.B, ii). In birds which are specialized for steep take-off or hovering the supracoracoid muscle, which raises the wing by re-directing its tendon through the triosseal canal, is relatively well-developed. The data collected by Hartman (1961) indicate that the ratio of the mass of the pectoral to that of the supracoracoid muscle is lowest in hummingbirds, where it is often only about 2:1; in birds which fly explosively for short distances such as tinamous (*Tinamus major*) or quails (*Odontophorus guttatus*) the ratio is about 3:1; birds such as pigeons (e.g. *Columba livia*) that take off steeply have a ratio which is typically about 6:1; in fast flyers such as terns (e.g. *Sterna hirundo*) the ratios are about 10:1; in soaring birds it sometimes goes as high as 20:1, as in the Osprey (*Pandion haliaetus*). Great variations occur, however, among birds which are closely related and have similar modes of flight.

Both the pectoral and the supracoracoid muscle arise in part from the carina. The carina may support the pectoral muscle so that it avoids exerting pressure on the supracoracoid muscle while the latter is contracting (Pennycuick, 1972, p. 22); thus the carina may not be simply an adaptation for increasing the size of the pectoral muscle (Pennycuick, 1972, p. 23). Moreover some bats (e.g. *Macrotus californicus*) have quite a large ventral median ridge on the sternum; but in other species (e.g. *Plecotus auritus*) the keel is very low, and the large right and left pectoral muscles arise from a median ligamentous raphe and therefore pull against one another (Norberg, 1970). The keel in pterosaurs was lower than in birds, but the flight muscles may have been just as well developed as in birds by arising from a median raphe like that of bats; indeed the ligamentous origin of the pectoralis muscle in bats has the advantage of being lighter than the bony keel of the avian sternum (Norberg, 1970).

E. Respiratory system

Bird flight requires a metabolic rate well above the maximum attainable during exercise by small terrestrial mammals (Thomas and Suthers, 1972). For example the Budgerigar (*Melopsittacus undulatus*) when flying horizontally at its most economical speed, uses oxygen at about 1.5 times the rate of a mouse of the same weight running hard in an exercise wheel (Tucker, 1968a); when ascending, or when flying rapidly, slowly, or in turbulent air, the Budgerigar increases its oxygen consumption to over 3 times that of the rodent (Tucker, 1968c).

At its most economical speed during horizontal steady state flight at sea level, the Budgerigar increases its metabolic rate (oxygen consumption) by about 13 times above its standard metabolic rate; for short periods of maximum effort, as in turbulent air, it can increase its metabolic rate to about 20–30 times the standard metabolic rate (Tucker, 1968c). However, a well-trained human athlete can expend power for a few minutes at 15–20 times his standard metabolic rate (Tucker, 1969). Moreover during a flight lasting a few minutes at sea level a bat can increase its metabolic rate to more than 30 times its metabolic rate when resting quietly in a chamber at an ambient temperature of 30°C (Thomas and Suthers, 1972). These comparisons suggest that the avian respiratory system has no structural adaptations which are an exclusive prerequisite for flight.

Nevertheless, there is good evidence that birds are far more tolerant of high altitude than mammals. Although migrating birds generally fly below 1500 m, flights at altitudes of 6100 m (20 000 ft) or more have been repeatedly observed by radar (Lack, 1960; Nisbet, 1963). An extreme example is the Bar-headed Goose (*Anser indicus*). This species migrates directly across the Himalayan summits with no more than brief rests on the way, and is probably the only known animal which regularly goes from near sea-level to altitudes of 9200 m in such a short period of time (Black *et al.*, 1978). As already stated, economical bird flight at sea level evidently requires an increase in oxygen consumption of about 13 times the standard value, an increase which would represent heavy work in a man. Yet, far from being able to do heavy work, an unacclimatized man is in a state of incipient hypoxic collapse after 10 min at 6100 m (see Tucker, 1968b, for review). In a hypobaric chamber, mice were comatose at 6100 m, but House Sparrows (*Passer domesticus*) could fly and gain altitude; at 7620 m they could still fly, but lost altitude (Tucker, 1968b).

Comparisons of tidal volume, respiratory frequency and minute volume, in mammals and non-passerine birds as calculated by Lasiewski and Calder (1971) from allometric equations, have shown that the ventilation rate (minute volume expressed as ml/min) is about 25% lower in birds than in mammals. On the other hand the resting or maintenance metabolic rates of non-passerine birds are similar to those of mammals (Lasiewski and Dawson, 1967). Lasiewski and Calder (1971) and Schmidt-Nielsen (1975, p. 39) concluded that, if the resting oxygen consumption is similar in non-passerine birds and mammals and yet the ventilation rate is lower in the birds, this can only mean that birds remove a greater proportion (about 6%) of the oxygen in the respired air than do mammals (only about 4%). Similarly the arterial CO_2 tension in birds is about 30% lower than in mammals (Lasiewski and Calder, 1971), and since the ventilation rate is lower in birds than mammals this indicates that CO_2 wash out is more extensive in birds. Furthermore the arterial PCO_2 (about 28 mmHg) of birds (Calder and Schmidt-Nielsen, 1968) is lower than the PCO_2 of avian end expired air (about 40 mmHg) (Scheid and Piiper, 1972), whereas in mammals these two values are in equilibrium. As Schmidt-Nielsen (1975, p. 40) noted, these functional differences between birds and mammals point inescapably to structural differences in either air flow or blood flow in the lungs.

The outstanding anatomical differences between the respiratory tracts of birds and mammals lie firstly in the total elimination of blind-ending tubes within the airways of the avian lung and their replacement by anastomosing parabronchi and air capillaries, and secondly in the development of air sacs. The bellows-like function of the air sacs makes possible the continuous suffusion of air *through* the anastomosing airways of the lung, as opposed to the tidal flow *in and out* of the blind-ending bronchial tree which characterizes mammals. Several independent lines of investigation (reviewed by Piiper and Scheid, 1973, pp. 167–168; Schmidt-Nielsen, 1975, pp. 41–42; King, 1975, pp. 1915–1916; Dawson, 1975; Fedde, 1976, p. 129; McLelland, 1978) have now shown the flow through the great majority of the avian parabronchi to be unidirectional in the caudal–cranial direction, probably during both inspiration and expiration. It is also now agreed that the relationship between the air in the parabronchi and the blood in the pulmonary arterial vasculature is essentially cross-current; thus the blood approaches the parabronchus essentially at right angles to the air flowing along the parabronchial lumen. This relationship was established experimentally by Scheid and Piiper (1972) and confirmed anatomically by Abdalla and King (1975). The cross-current arrangement permits a greater degree of removal of oxygen from the respired air than in mammals, and the greater elimination of carbon dioxide from the blood. Thus the avian lung has a higher efficiency of gas exchange, since less ventilation is needed to achieve a certain arterialization and with equal ventilation a higher degree of arterialization is obtained (Scheid and Piiper, 1970). These structural modifications of the avian respiratory system therefore appear to constitute major anatomical adaptations enabling birds to perform heavy work at high altitudes.

Various possible evolutionary advantages arise from this facility (see Tucker, 1968b, for review). Firstly it enables a few highly specialized species to occupy mountain habitats. Secondly, high altitude flight could allow migrating birds to fly above the cloud cover in order to navigate from the sun and stars (although radar observations show that migrating birds normally remain below clouds at medium and high levels, Nisbet, 1963). Thirdly the energetic cost of migration can be reduced by flying over rather than round mountains, and by taking advantage of jet-flow air streams and updrafts associated with mountain ranges.

F. Cardiovascular system

Meeting the high oxygen demands of flapping flight, especially at high altitude, depends not only on respiratory but also on circulatory efficiency. The oxygen transport characteristics of the blood would be one possible point of adaptation; thus bats have a high haematocrit (about 60%) and a correspondingly high oxygen capacity (27·5 vol. %), the respective values in small terrestrial mammals apparently being about 45% and 18 vol. % (Thomas and Suthers, 1972). However, the high haematocrit in bats can presumably be achieved only at the cost of increased blood viscosity and cardiac work. Humans residing at about 5300 m have values of about 59% and 30 vol. % (Prosser and Brown, 1961, p. 223), similar to that of the bat. In the House Sparrow (*Passer domesticus*) at sea level the haematocrit is about 48 and the oxygen capacity about 19 vol. % (Tucker, 1968b). In seven species of bird the oxygen dissociation curve was similar in shape and position to that of the mammal, indicating a similarity in blood-oxygen affinity between birds and mammals (Lutz *et al.*, 1974); there have been various other reports that bird blood has a lower affinity for oxygen than mammalian blood (e.g. Wells, 1976), but Schmidt-Nielsen (1975, p. 45) concluded that the oxygen transport characteristics of bird blood are similar to those of mammals in general. Nevertheless, it was pointed out by Holle *et al.* (1977) that the possibility cannot be ruled out that among birds there are variations due to species, age and seasonal or hormonal factors. This prediction has evidently turned out to be correct for the Bar-headed Goose, a bird which regularly ascends to exceptionally high altitudes (see p. 31). The haemoglobin of this species appears to have a relatively greater affinity for oxygen than that of *Anas platyrhynchos*. Thus in hypoxic conditions the values for the oxygen content of arterial and venous blood are considerably higher in the Bar-headed Goose than in the duck (Black *et al.*, 1978). At 9150 m, the difference between arterial and venous oxygen content in this goose is almost twice that in the duck; this gives the goose a substantial advantage, since the delivery of O_2 to the tissues will be much greater for a given cardiac output and arterial PO_2.

To find evidence of more efficient oxygen transport in birds it is necessary to consider the heart itself. The relative weight (% of body weight) of the heart of *Passer domesticus* is 2·7 times greater than that of the laboratory mouse (Tucker, 1968b). The heart rate in birds in general is considered to be high (Jones and Johansen, 1972, p. 202), being 190 min^{-1} in *Anas*, 244 in *Columba*, 257 in *Cygnus olor*, 347 in *Falco peregrinus*, 401 in *Larus canus*, 700 in *Fringilla coelebes* (Kruta *et al.*, 1971, pp. 342–343) and 1200 in hummingbirds (Lasiewski, 1964). Calculations by Tucker (1968b) suggested that the flying Budgerigar (*Melopsittacus undulatus*) has a cardiac output greater than 3·75L kg min^{-1}, more than seven times the maximum for man and the dog. The relative enlargement of the heart coupled with its fast rate of beating appears to be a major structural adaptation for the improved transport of oxygen; doubtless it contributes substantially to the ability of birds, when flying at high altitudes, to satisfy the oxygen demands of vigorously exercising muscles by means of blood which has a lower oxygen content.

Tucker (1968b) calculated that a hypothetical sparrow flying at 6100 m would require a cardiac output of at least 2·78L kg min^{-1} to meet an increase in oxygen consumption of eight times the resting level, the arterial blood then being 63% saturated. As just stated, measurements on *Melopsittacus undulatus* show that cardiac outputs greater than this can in fact be achieved. In contrast, the estimated stroke volume in the bat is only intermediate between that of small terrestrial mammals and birds (Thomas and Suthers, 1972). Therefore, although the heart rate in the flying bat is high, comparable to that of a flying bird of similar size, the cardiac output of the bat is relatively smaller than that of the bird. The bat compensates for this by its higher haematocrit, which thus enables its blood to transport a larger amount of oxygen per unit stroke volume than in a bird (Thomas and Suthers, 1972).

G. Special senses and brain

The importance of vision in the diurnal aerial habitat occupied by birds and pterosaurs is shown by the great enlargement of the orbit in these animals (Jerison, 1973, pp. 163, 197). The importance of the olfactory sense, on the other hand, should decrease in flying vertebrates, since odours would be quickly dispersed high among the trees (Bock, 1969).

These changes in the special senses had repercussions in the brain. In both pterosaurs and birds there was a corresponding enlargement of the visual region of the midbrain, leading to the formation of the massive optic lobes, the homologues of the rostral colliculi. The olfactory lobes were proportionately reduced (Jerison, 1973, p. 171).

The evolution of the locomotor system in birds must have had profound effects on the development of the brain. If the arboreal theory of the evolution of flight be accepted (see Section IV.G), the small bipedal "proavis", having entered the trees by leaping or climbing, must have encountered problems of balance more critical than those of its quadrupedal ancestors. The importance of balance increased as locomotion penetrated into three dimensions. Throughout these early evolutionary steps, one would expect the development of vestibular and visual integration with the cerebellum, and of cerebellar regulation of somatic motor pathways, with resultant enlargement of the cerebellum (Bock, 1969), as in *Archaeopteryx* (see Section IV.G).

Once the aerial mode of life had become really crucial it presented selection pressures so drastic as to require further major modifications of the avian brain (Jerison, 1973, pp. 156–157). The development of stable flapping flight required the integration by the cerebellum of the continuous proprioceptive input from the muscles with the output from the motor centres, in order to control accurately the repeated contraction and relaxation of opposing muscle groups. With reduction of the tail, stable flight progressed into unstable flight with enhanced manoeuvrability (see Section V.B, iii). To be maintained in flight and manoeuvred, the aerodynamically unstable bird now required a control system capable of continuous supervision and instantaneous minor corrections of the wings and tail (Brown, 1961, pp. 301–302). These flight perfections of the brain would again have been focussed on the cerebellum and its integration with the vestibular, visual and somatic motor systems. Thus the decerebrate pigeon can fly when thrown into the air, but removal of the cerebellum disorganizes its locomotion completely (Jerison, 1973, p. 158).

The net result of these developments was a clear overall enlargement of the avian brain in relation to body weight, of the same general order as that which has occurred in mammals (Jerison, 1973, p. 169). Although so similar to birds, the pterosaurs apparently failed to achieve a really definite enlargement of the brain as a whole, the increase in their optic lobes and cerebellum presumably being compensated approximately by the reduction in the olfactory lobes. In the relationship between brain size and body weight, the pterosaurs were inferior to mammals and birds and were still clearly reptilian, though they were at the upper edge of reptilian brain development (Jerison, 1973, p. 172).

References

Abdalla, M. A. and King, A. S. (1975). The functional anatomy of the pulmonary circulation of the domestic fowl, *Resp. Physiol.* **23**, 267–290.

Attridge, J. (1956). The morphology and relationships of a complete therocephalian skeleton from the Cistephalus Zone of South Africa, *Proc. R. Soc. Edin.* B **66**, 59–93.

Bakker, R. T. (1971a). Ecology of the brontosaurs, *Nature, Lond.* **229**, 172–174.

Bakker, R. T. (1971b). Dinosaur physiology and the origin of mammals, *Evolution, Lancaster, Pa.* **25**, 636–658.

Bakker, R. T. (1972). Anatomical and ecological evidence of endothermy in dinosaurs, *Nature, Lond.* **238**, 81–85.

Bakker, R. T. (1974). Dinosaur bioenergetics—a reply to Bennett and Dalzell and Feduccia, *Evolution, Lancaster, Pa.* **28**, 497–503.

Bakker, R. T. (1975). Dinosaur renaissance, *Scient. Am.* **232**, 58–78.

Bakker, R. T. and Galton, P. M. (1974). Dinosaur monophyly and a new class of vertebrates, *Nature, Lond.* **248**, 168–172.

Baumel, J. J. (1980). Osteologia. *In* "Nomina Anatomica Avium" (J. J. Baumel, A. S. King, A. M. Lucas, J. Breazile and H. E. Evans, Eds). Academic Press, London and New York.

Bellairs, A. d'A. and Attridge, J. (1975). "Reptiles." Hutchinson University Library, London.

Bellairs, A. d'A. and Jenkin, C. R. (1960). *In* "Biology and Comparative Physiology of Birds" (A. J. Marshall, Ed.), Vol. 1. Academic Press, New York and London.

Bennett, A. F. and Dalzell, B. (1973). Dinosaur physiology: a critique, *Evolution, Lancaster, Pa.* **27**, 170–174.

Berge, J. vanden (1980). Myologia. *In* "Nomina Anatomica Avium" (J. J. Baumel, A. S. King, A. M. Lucas, J. Breazile and H. E. Evans, Eds). Academic Press, London and New York.

Black, C. P., Tenney, S. M. and van Kroonenburg, M. (1978). Oxygen transport during progressive hypoxia in Bar-headed Geese (*Anser indicus*) acclimatized to sea level and 5600 meters. *In* "Respiratory Function in Birds, Adult and Embryonic" (J. Piiper, Ed.). Springer Verlag, Berlin.

Bock, W. J. (1963). The cranial evidence for ratite affinities, *Proc. Int. Ornithol. Congr.* 13th, 1962, **1**, pp. 39–54.

Bock, W. J. (1965). The role of adaptive mechanisms in the origin of higher levels of organization, *Syst. Zool.* **14**, 272–287.

Bock, W. J. (1969). The origin and radiation of birds, *Ann. N.Y. Acad. Sci.* **167**, 147–155.

Bock, W. J. (1974). The avian skeletomuscular system. *In* "Avian Biology" (D. S. Farner and J. R. King, Eds), Vol. IV. Academic Press, New York and London.

Bock, W. J. and Miller, W. de W. (1959). The scansorial foot of the Woodpeckers, with comments on the evolution of perching and climbing birds, *Am. Mus. Novit. No.* **1931**, 23–45.

Bouvier, Marianne (1977). Dinosaur Haversian bone and endothermy, *Evolution, Lancaster, Pa.* **31**, 449–450.

Bramwell, Cherrie D. (1970). The first hot-blooded flappers, *Spectrum: an Oxford J. Sci.* **69**, 12–14.

Bramwell, Cherrie D. (1971). Aerodynamics of *Pteranodon*, *J. Linn. Soc. (Biol)* **3**, 313–328.

Brodkorb, P. (1971). *In* "Avian Biology" (D. S. Farner and J. R. King, Eds), Vol. 1. Academic Press, New York and London.

Brink, A. S. (1955). Speculations on some advanced mammalian characteristics in the higher mammal-like reptiles, *Palaeont. Afr.* **4**, 77–96.

Broom, R. (1913). On the South-African pseudosuchian *Euparkeria* and allied genera, *Proc. Zool. Soc. Lond.* 619–633.

Brown, B. (1941). The last dinosaurs, *Nat. History* **48**, 290–295.

Brown, R. H. J. (1961). *In* "Biology and Comparative Physiology of Birds" (A. J. Marshall, Ed.), Vol. 11. Academic Press, New York and London.

Bühler, P. (1972). Sandwich structures in the skull capsules of various birds, *Colloquium an Biologie und Bauen*, 1971, *Univ. Stuttgart*, **114**, 39–50.

Calder, W. A. and Schmidt-Nielsen, K. (1968). Panting and blood carbon dioxide in birds, *Am. J. Physiol.* **215**, 477–482.

Carroll, R. L. (1969). *In* "Biology of the Reptilia" (C. Gans, A. d'A. Bellairs and T. S. Parsons, Eds), Vol. 1. Academic Press, London and New York.

Chappel, B. M. (1978). On the relative density of avian and mammalian bones, *J. Anat., Lond.* **127**, 216.

Cope, E. D. (1878). On the vertebrata of the Dakota epoch of Colorado, *Proc. Am. phil. Soc.* **17**, 233–247.

Cowles, R. B. (1947). Comments by readers, *Science, N.Y.* **105**, 282.

Cracraft, J. (1974). Phylogeny and evolution of the ratite birds, *Ibis* **116**, 494–521.

Crompton, A. W., Taylor, C. R. and Jagger, A. (1978). Evolution of homeothermy in mammals, *Nature, Lond.* **272**, 333–336.

Dawson, T. J. and Hulbert, A. J. (1970). Standard metabolism, body temperature, and surface areas of Australian marsupials, *Am. J. Physiol.* **218**, 1233–1238.

Dawson, W. R. (1975). Avian physiology, *A. Rev. Physiol.* **37**, 441–465.

de Beer, G. (1954). *Archaeopteryx* and evolution, *Advmt. Sci., Lond.* **42**, 160–170.

de Beer, G. (1975). "The Evolution of Flying and Flightless Birds." Oxford University Press, Oxford.

Desmond, A. J. (1975). "The Hot Blooded Dinosaurs—A Revolution in Palaeontology." Blond and Briggs, London.

Dilger, W. C. (1957). The loss of teeth in birds, *Auk* **74**, 103–104.

Dodson, P. (1974). Dinosaurs as dinosaurs, *Evolution, Lancaster, Pa.* **28**, 491–504.

Edington, G. H. and Miller, Agnes E. (1941). The avian ulna: its quill-knobs, *Proc. R. Soc. Edin.* (B) **61**, 138–148.

Enlow, D. H. (1969). *In* "Biology of the Reptilia" (C Gans, A. d'A. Bellairs and T. S. Parsons, Eds), Vol. 1. Academic Press, London and New York.

Fedde, M. R. (1976). Respiration. *In* "Avian Physiology" (P. D. Sturkie, Ed.), 3rd edn. Springer, New York.

Feduccia, A. (1973). Dinosaurs as reptiles, *Evolution, Lancaster, Pa.* **27**, 166–169.

Fisher, H. I. (1957). Bony mechanism of automatic flexion and extension in the pigeon's wing, *Science, N.Y.* **126**, 446.

Gadow, H. (1891). *In* "Vögel. Dr. Bronn's Klassen und Ordnungen des Thier-Reichs" (H. Gadow and E. Selenka, Eds), Vol. 6, division 4, Anatomischer Teil. Winter, Leipzig.

Galton, P. M. (1970). Ornithischian dinosaurs and the origin of birds, *Evolution, Lancaster, Pa.* **24**, 448–462.

Gans, C. (1978). Ventilation mechanisms: problems in evaluating the transition to birds. *In* "Respiratory Function in Birds, Adult and Embryonic" (J. Piiper, Ed.). Springer Verlag, Berlin.

Gerhardt, U. (1933). Kloake und Begattungsorgane. *In* "Handbuch der vergleichenden Anatomie der Wirbeltiere" (L. Bolk, E. Goppert, E. Kallius and W. Lubosch, Eds), Vol. 6. Urban and Schwarzenberg, Berlin.

Goodrich, E. S. (1916). On the classification of the Reptilia, *Proc. R. Soc.* (B) **69**, 261–276.

Harrison, C. J. O. (1976). Feathering and flight evolution in *Archaeopteryx*, *Nature* **263**, 762–763.

Hartman, F. A. (1961). Locomotor mechanisms of birds, *Smithson. misc. Collns.* **143**, 1–49.

Headley, F. W. (1895). "The Structure and Life of Birds." Macmillan, London.

Heath, J. E. (1968). The origins of thermoregulation. *In* "Evolution and Environment" (Ellen T. Drake, Ed.), pp. 259–278. Yale University Press, New Haven and London.

Heilmann, G. (1926). "The Origin of Birds." Witherby, London.

Heinrich, B. (1977). Why have some animals evolved to regulate a high body temperature? *Am. Naturalist* **111**, 623–640.

Heinrich, B. and Bartholomew, G. A. (1972). Temperature control in flying moths, *Scient. Am.* **226**, 71–77.

Heptonstall, W. B. (1970). Quantitative assessment of the flight of *Archaeopteryx*, *Nature, Lond.* **228**, 185–186.

Hildebrand, M. (1974). "Analysis of Vertebrate Structure." J. Wiley and Son, New York.

Holle, J. P., Meyer, M. and Scheid, P. (1977). Oxygen affinity of duck blood determined by *in vivo* and *in vitro* technique, *Resp. Physiol.* **29**, 355–361.

Huxley, T. H. (1867). On the classification of birds and on the taxonomic value of the modification of certain of the cranial bones observable in that class, *Proc. zool. Soc. Lond.* 415–472.

Huxley, T. H. (1868). On the animals which are most nearly intermediate between birds and reptiles, *Ann. Mag. nat. Hist. series* **4**, 66–75.

Jenkins, F. A. (1970). Cynodont postcranial anatomy and the "prototherian" level of mammalian organization, *Evolution, Lancaster, Pa.* **24**, 230–252.

Jerison, H. J. (1973). "Evolution of the Brain and Intelligence." Academic Press, New York and London.

Jones, D. R. and Johansen, K. (1972). The blood vascular system of birds. *In* "Avian Biology" (D. S. Farner and J. R. King, Eds), Vol. 2. Academic Press, New York and London.

King, A. S. (1966). Structural and functional aspects of the avian lungs and air sacs, *Int. Rev. gen. exp. Zool.* **2**, 171–267.

King, A. S. (1975). Aves respiratory system. *In* "Sisson and Grossman's The Anatomy of the Domestic Animals" (R. Getty, Ed.), Vol. 2. Saunders, Philadelphia.

Kruta, V., Seliger, V. and Woodbury, R. A. (1971). *In* "Respiration and Circulation" (P. L. Altman and D. S. Dittmer, Eds), *Fedn. Am. Socs. exp. Biol.*, Bethesda.

Kuroda, N. (1961). A note on the pectoral muscles of birds, *Auk* **78**, 261–263.

Lack, D. (1960). The height of bird migration, *Br. Birds* **53**, 5–10.

Lasiewski, R. C. (1964). Body temperatures, heart and breathing rate and evaporative water loss in hummingbirds, *Physiol. Zool.* **37**, 212–223.

Lasiewski, R. C. and Calder, W. A. (1971). A preliminary allometric analysis of respiratory variables in resting birds, *Resp. Physiol.* **11**, 152–166.

Lasiewski, R. C. and Dawson, W. R. (1967). A re-examination of the relationship between standard metabolic rate and body weight in birds, *Condor* **69**, 13–23.

Lawson, D. A. (1975). Pterosaur from the latest Cretaceous of West Texas: discovery of the largest flying creature, *Science, N.Y.* **187**, 947–948.

Lindsay, H. A., Freix, E. D., Hoversland, A. S., Link, R. P. and Heisler, C. R. (1971). *In* "Respiration and Circulation" (P. L. Altman and D. S. Dittmer, Eds). *Fedn. Am. Socs. exp. Biol.*, Bethesda.

Lutz, P. L., Longmuir, I. S. and Schmidt-Nielsen, K. (1974). Oxygen affinity of bird blood, *Resp. Physiol.* **20**, 325–330.

McDowell, S. (1948). The bony palate of birds. Part I. The palaeognathae, *Auk* **65**, 520–549.

McLelland, J. (1978). Respiration in birds, *Pavo* **16**, 1–11.

McNab, B. K. (1978). The evolution of endothermy in the phylogeny of mammals, *Am. Nat.* **112**, 1–21.

Marshall, A. J. (1962). *In* "A Textbook of Zoology," (Parker and Haswell, Eds), Vol. 2. Macmillan, London.

Newell, N. D. (1972). The evolution of reefs, *Scient. Am.* **226**, 54–65.

Nisbet, I. C. T. (1963). Measurements with radar of the height of nocturnal migration over Cape Cod, Massachusetts, *Bird-Banding* **34**, 57–67.

Norberg, Ulla M. (1970). Functional osteology and myology of the wing of *Plecotus auritis* Linnaeus (Chiroptera), *Ark. Zool.* **22**, 483–543.

Ostrom, J. H. (1969). Terrestrial vertebrates as indicators of Mesozoic climates, *Proc. North Am. Paleontol. Conv.* pp. 347–376.

Ostrom, J. H. (1973). The ancestry of birds, *Nature, Lond.* **242**, 136.

Ostrom, J. H. (1974a). Archaeopteryx and the origin of flight, *Q. Rev. Biol.* **49**, 27–47.

Ostrom, J. H. (1974b). Reply to "Dinosaurs as Reptiles", *Evolution, Lancaster, Pa.* **28**, 491–493.

Ostrom, J. H. (1975). The origin of birds, *A. Rev. Earth Planet. Sci.* **3**, 55–77.

Ostrom, J. H. (1976). Some hypothetical anatomical stages in the evolution of avian flight. *In* "Collected Papers in Avian Paleontology Honoring the 90th Birthday of Alexander Wetmore" (Storrs L. Olson, Ed.).

Parkes, K. C. and Clark, G. A. (1966). An additional character linking ratites and tinamous, and an interpretation of their monophyly. *Condor* **68**, 459–71.

Pennycuick, C. J. (1972). "Animal Flight." Arnold, London.

Piiper, J. and Scheid, P. (1973). Gas exchange in avian lungs: models and experimental evidence. *In* "Comparative Physiology" (L. Bolis, K. Schmidt-Nielsen and S. H. P. Maddrell, Eds). North Holland Publishing Co., Amsterdam.

Porter, K. R. (1972). "Herpetology." Saunders, Philadelphia.

Prosser, C. L. and Brown, F. A. (1961). "Comparative Animal Physiology", 2nd edn. Saunders, Philadelphia.

Pycraft, W. P. (1901). On the morphology and phylogeny of the Palaeognathae (Ratitae and Crypturi) and Neognathae (Carinatae), *Trans. zool. Soc. Lond.* **15**, 149–267.

Regal, P. J. (1975). The evolutionary origin of feathers, *Q. Rev. Biol.* **50**, 35–66.

Riqlès, A. J. de (1969). L'histologie osseuse envisagee comme indicateur de la physiologie thermique chez les tetrapodes fossiles, *C.R. Acad. Sci. Paris* **268**, 782–785.

Riqlès, A. J. de (1974). Evolution of endothermy: histological evidence, *Evol. Theory* **1**, 51–80.

Romer, A. S. (1966). "Vertebrate Paleontology", 3rd edn. The University of Chicago Press, Chicago.

Russell, D. A. (1969). A new specimen of *Stenonychosaurus* from the Oldman formation (Cretaceous) of Alberta, *Can. J. Earth Sci.* **6**, 595–612.

Russell, D. A. (1972). Ostrich dinosaurs from the late Cretaceous of Western Canada, *Can. J. Earth Sci.* **9**, 375–402.

Savile, D. B. O. (1957). Adaptive evolution in the avian wing, *Evolution, Lancaster, Pa.* **11**, 212–224.

Scheid, P. and Piiper, J. (1970). Analysis of gas exchange in the avian lung: theory and experiments in the domestic fowl, *Resp. Physiol.* **9**, 246–262.

Scheid, P. and Piiper, J. (1972). Cross current gas exchange in avian lungs: effects of reversed parabronchial air flow in ducks, *Resp. Physiol.* **16**, 304–312.

Schmidt-Nielsen, K. (1975). Recent advances in avian respiration. *In* "Avian Physiology" (M. Peaker, Ed.). Academic Press, London and New York.

Seely, H. G. (1901). "Dragons of the Air." Methuen, London.

Seymour, R. S. (1976). Dinosaurs, endothermy and blood pressure, *Nature, Lond.* **262**, 207–208.

Simpson, G. G. (1946). Fossil penguins. A note on *Archaeopteryx* and *Archaeornis*, *Bull. Am. Mus. Nat. Hist.* **87**, 92–95.

Storer, R. W. (1971). Classification of birds. *In* "Avian Biology" (D. S. Farner and J. R. King, Eds), Vol. 1. Academic Press, New York and London.

Strong, R. M. (1919). Adaptation in bone architecture, *Sci. month.* **8**, 71–80.

Swinton, W. E. (1958). "Fossil Birds." Trustees of the British Museum, London.

Swinton, W. E. (1960). The origin of birds. In "Biology and Comparative Physiology of Birds" (A. J. Marshall, Ed.), Vol. 1. Academic Press, New York and London.

Thomas, S. P. and Suthers, R. A. (1972). The physiology and energetics of bat flight, *J. exp. Biol.* **57**, 317–335.

Thompson, D'A. W. (1942). "Growth and Form." Cambridge University Press, Cambridge.

Thomson, A. L. (1964). *In* "A New Dictionary of Birds" (A. L. Thomson, Ed.). Nelson, London.

Tracy, C. R. (1976). Tyrannosaurs: evidence for endothermy? *Am. Nat.* **110**, 1105–1106.

Tucker, V. A. (1968a). Upon the wings of the wind, *New Sci.* **38**, 694–696.

Tucker, V. A. (1968b). Respiratory physiology of house sparrows in relation to high-altitude flight, *J. exp. Biol.* **48**, 55–66.

Tucker, V. A. (1968c). Respiratory exchange and evaporative water loss in the flying budgerigar, *J. exp. Biol.* **48**, 67–87.

Tucker, V. A. (1969). Energetics of bird flight, *Scient. Am.* **220**, 70–78.

Valen, L. V. (1960). Therapsids as mammals, *Evolution, Lancaster, Pa.* **14**, 304–313.

Walker, A. D. (1972). New light on the origin of birds and crocodiles, *Nature, Lond.* **237**, 257–263.

Wells, R. M. G. (1976). The oxygen affinity of chicken haemoglobin in whole blood and erythrocyte suspensions, *Resp. Physiol.* **27**, 21–31.

Welty, J. C. (1962). "The Life of Birds." Saunders, Philadelphia.

Wetmore, A. (1955). Paleontology. *In* "Recent Studies in Avian Biology" (A. Wolfson, Ed.). University Illinois Press, Urbana.

Wetmore, A. (1960). A classification for the birds of the world, *Smithson. misc. Collns.* **139**, 1–37.

White, F. N. (1978). Circulation: a comparison of reptiles, mammals, and birds. *In* "Respiratory Function in Birds, Adult and Embryonic" (J. Piiper, Ed.). Springer Verlag, Berlin.

Yalden, D. W. (1971). The flying ability of *Archaeopteryx*, *Ibis* **113**, 349–356.

Yapp, W. B. (1970). "The Life and Organisation of Birds." Edward Arnold, London.

Young, J. Z. (1962). "The Life of Vertebrates." Clarendon, Oxford.

Ziswiler, V. and Farner, D. S. (1972). Digestion and the digestive system. *In* "Avian Biology" (D. S. Farner and J. R. King, Eds), Vol. 2. Academic Press, New York and London.

2
Coelomic cavities

HANS-RAINER DUNCKER

Zentrum für Anatomie und Cytobiologie der Justus-Liebig-Universität Giessen, Aulweg 123, D-6300 Giessen, Germany

---

CONTENTS

I. Introduction 39

II. Phylogenetic and functional relationships 40

III. Subdivisions of the avian coelomic cavity 49
A. Pericardial cavity 52
B. Pleural cavities 55
C. Subpulmonary cavities 57
D. Dorsal and ventral hepatic peritoneal cavities 59
E. Intestinal peritoneal cavity 61

References 67

I. Introduction

Vertebrate anatomists have given relatively little attention in the past to the coelom and its subdivisions. Probably the main reason for this is because within most classes of vertebrates, the structure of the coelom is very constant. Nevertheless, reptiles, which demonstrate a great diversity in the form of their coelom, have been neglected. Probably the least investigated aspects of the coelom are the evolutionary significance of the subdivisions and their functional importance. The subdivisions of the coelom are most complex in some reptilian families and in birds so much so that an understanding of their anatomy depends to a great extent on a sound knowledge of their development.

Of the available accounts of the coelomic cavities in vertebrates generally, those by Goodrich (1930, pp. 613–656), Broman (1937) and Nelsen (1953) are essential reading. Of the older literature restricted to birds, a number of papers

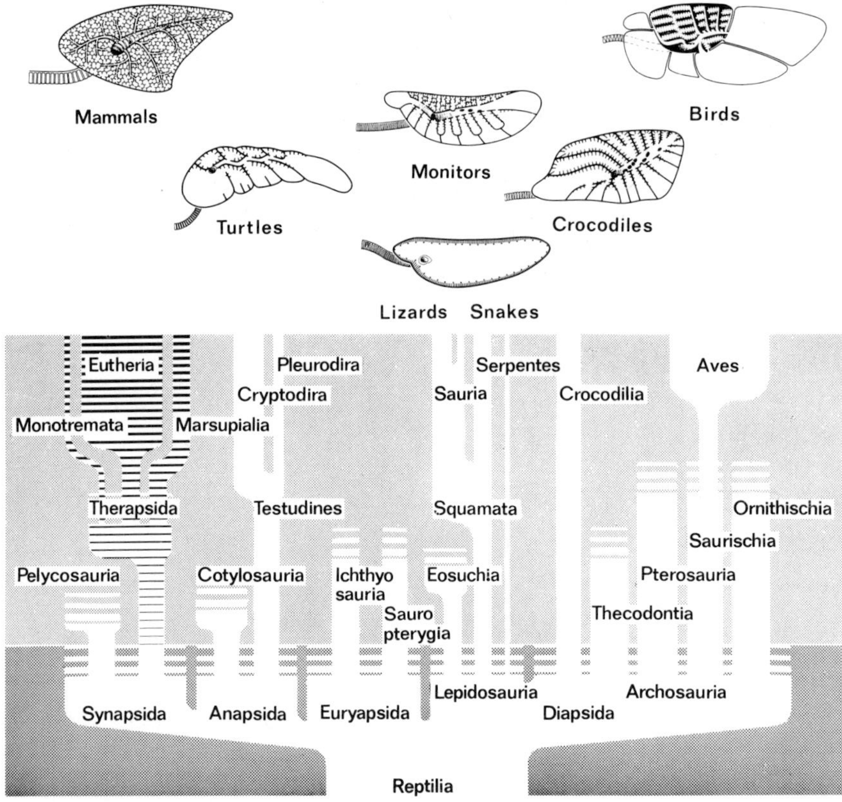

Fig. 2.1. Phylogenetic tree demonstrating the relationship between the different reptilian orders. Also shown are the types of lungs, evolved by reptiles, birds and mammals: heterogeneous in reptiles and birds, homogeneous in mammals. From Duncker (1978b).

are important for the development of our knowledge and for establishing the nomenclature, e.g. the monographs by Sappey (1847) and Campana (1875) on the avian respiratory apparatus, and the contributions of Huxley (1882), Butler (1889) and Beddard (1896). Particularly valuable has been the developmental study by Poole (1909). More recently the coelomic cavities in domestic birds have been described in some detail by Bittner (1925), Petit (1933), Kern (1963), Goodchild (1970) and McLelland and King (1970, 1975). The following account is based on the above literature and on the observations of the author in birds and reptiles* (Duncker, 1971, 1978a,b).

II. Phylogenetic and functional relationships

The highly complex arrangement of the coelom in birds may be best understood by examining first the wide range of forms the coelomic cavity takes in reptiles and the way these variations are correlated with differences in the functional anatomy of the respiratory system (Fig. 2.1).

* Supported by grants from the Deutsche Forschungsgemeinschaft

In its simplest form as in small living lizards (Sauria) and their relatives the coelomic cavity is divided into two parts, a pericardial cavity lying ventrally at the level of the pectoral girdle and a pleuroperitoneal cavity (Fig. 2.2). Embryologically, a primary septum transversum develops where the pericardium meets the yolk stalk (Fig. 2.3). The developing liver invades this septum transversum, thereby acquiring its topographical relationship to the pericardial cavity (Fig. 2.4). In these small lizards, further development of the septum transversum into a partition separating the pleural cavity from the peritoneal cavity does not occur. Thus, the liver can elongate far caudally and is connected only at its cranial tip to the pericardium (Fig. 2.6). This connection between liver and pericardium is in many cases only maintained by the caudal vena cava (Fig. 2.2).

In accordance with these structural relationships, the lungs, arising from the very short primary bronchi dorsal to the pericardium, are freely movable in the pleuroperitoneal cavity. They are attached to the vertebral column lateral to the dorsal mesogastrium via the dorsal mesopneumonium. In addition, each lung often has a ligamentous connection to the ventral mesentery, the connection sometimes extending onto the liver. The unicameral (single-chambered) lungs, which consist of only one large compartment, generally occupy the cranial part of the pleuroperitoneal cavity, i.e. the part supported by ribs. In some forms, however, such as skinks (Scincidae) and snakes (Serpentes), the lungs can extend much further caudally, although the caudal dilatation of the lung is largely non-respiratory. In these reptiles the body wall is supported by ribs as far caudally as the lungs extend.

The primary septum transversum in some reptiles such as chameleons

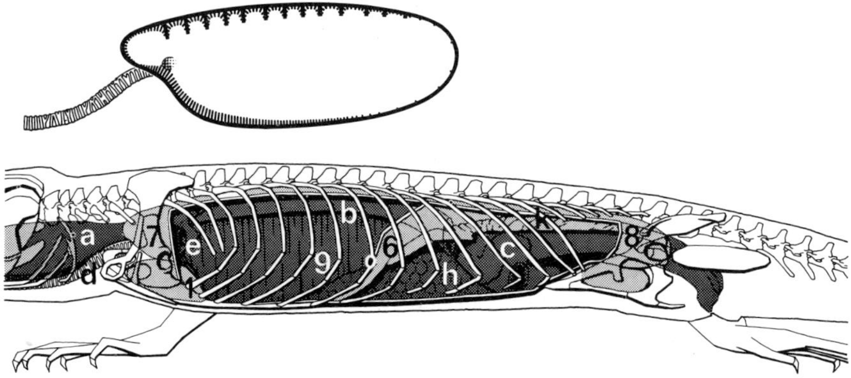

Fig. 2.2. Green Lizard (*Lacerta viridis*). Schematic drawing of the unicameral (single-chambered) lung and coelomic cavities (left lung, left liver lobe and stomach removed). Two coelomic cavities are present, the pericardial cavity, at the level of the pectoral girdle, and the pleuroperitoneal cavity. The pleuroperitoneal cavity is divided into right and left parts cranially by the mediastinal mesentery and caudally by the liver and its dorsal and ventral mesenteries. Caudal to the liver the ventral part of the pleuroperitoneal cavity is undivided. The large mesosalpinx (mesoviduct) with its attachment to the lateral body wall is shown. 1, pericardial cavity; 6, cranial mesoviduct; 7, pleuroperitoneal cavity, cranial extension; 8, pleuroperitoneal cavity, caudal extension; 9, mediastinal mesentery; a, pharynx; b, eosophagus in the mediastinum; c, intestinal tract; d, trachea; e, primary bronchus, cut at the lung hilus; h, liver, cut medially; k, oviduct.

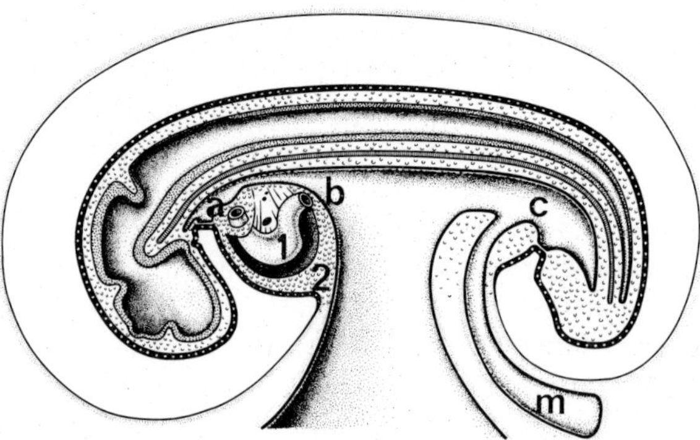

Fig. 2.3. Schematic drawing of a generalized early amniote embryo within its amnion, with large yolk stalk (vitelline duct). As a result of the enormous growth of the brain, the unpaired cranial part of the coelomic cavity comes to lie under the cranial foregut as the pericardial cavity, the pericardium forming with the yolk stalk the septum transversum. 1, primitive tubular heart in the developing pericardial cavity; 2, septum transversum; a, pharynx; b, foregut; c, hindgut; m, allantois. From Duncker (1978a).

(Chamaeleonidae) develops further to form a postpulmonary septum (pulmonary fold) (Fig. 2.7). The postpulmonary septum separates the cranial part of the lungs from the cranial parts of the liver, and is connected to the pericardium. A characteristic feature of this small form of the postpulmonary septum is the presence of a connection between it and the liver at the *area nuda*, through which the caudal vena cava enters into the pericardial cavity. In chameleons the cranial nephric folds in the adult female are developed bilaterally as the cranial part of the mesosalpinx (*mesoviductus dorsalis*). They are attached to the caudal part of the dorsal abdominal wall and are cranially continued on the lateral thoracic wall into the lateral portions of the small postpulmonary septum (Fig. 2.7).

A complete postpulmonary septum, separating the lung or pleural cavity from the peritoneal cavity, occurs in turtles (Testudines) (Fig. 2.8), monitor lizards (Varanidae) (Fig. 2.9) and crocodiles (Crocodilia) (Fig. 2.10). It arises medioventrally out of the septum transversum and dorsolaterally out of the cranial nephric folds (Figs 2.4, 5). Dorsomedially it is closed by portions of the dorsal mesentery. The lateral parts of the septum possibly receive contributions from the body wall. Since the lungs in these reptiles fuse with the body wall, and also ventrally and caudally with the postpulmonary septum, much of the pleural cavity is eliminated. In monitor lizards (Fig. 2.9) and crocodiles (Fig. 2.10) the lungs occupy the entire cranial part of the thoracoabdominal cavity. The postpulmonary septum is consequently pushed caudally, so that its most cranial part lies caudal to the caudal extremity of the sternum. From here it ascends dorsally and caudally along a line of attachment to the lateral and dorsal parts of the body wall. Through development, the septum is attached cranially to the pericardium and caudally to the liver at the *area nuda* the extent of which varies in different species. In monitor lizards and crocodiles the caudal displacement or "descent" of the heart, by which it comes to lie in the middle of the thoracoabdominal cavity and not at the level of the pectoral girdle, is a result of the expansion of the lungs (Figs 2.9, 10). In turtles, however, the lung is restricted to the more dorsal part of the thorax (Fig. 2.8). The heart in these reptiles therefore has a more cranial position, although the basic topographical relationships between the heart, the postpulmonary septum and the liver are not disturbed.

The attachment of the lungs in turtles, crocodiles and monitor lizards to the dorsal and lateral parts of the thoracic wall and to the postpulmonary septum can be correlated with their internal structure. In these reptiles the lung is generally subdivided by numerous partitions into a large number of heterogeneous compartments which are connected to a single long intrapulmonary bronchus (Figs 2.8, 9, 10). Usually the partitions are most numerous in the mediodorsal part of the lung around the primary bronchus, so that this part of the lung consists of numerous very small compartments and therefore has a large surface area for gas exchange. In contrast, the most cranial part of the lung, as well as the ventral and caudal parts, possess relatively few partitions and therefore consist of large compartments and have a small surface area for gas exchange. These large ventral and caudal compartments have primarily non-respiratory functions, with the result that the multi-chambered lung has a high compliance.

Since the lung is fused with the thoracic wall and the postpulmonary septum, its small dorsomedial compartments are inflated to some extent during all phases of the respiratory cycle. The large dilatations, especially the ventral and caudal compartments, may be totally collapsed at rest. During normal respiration the cranial and mediodorsal parts of the lung undergo a limited ventilation produced by movements of the cranial ribs. Although the large ventral and caudal compartments may be inflated to a greater extent, this only occurs in special situations such as during defensive behaviour, vocalization or swimming. If the lungs were not fused to the thoracic wall and postpulmonary septum, then

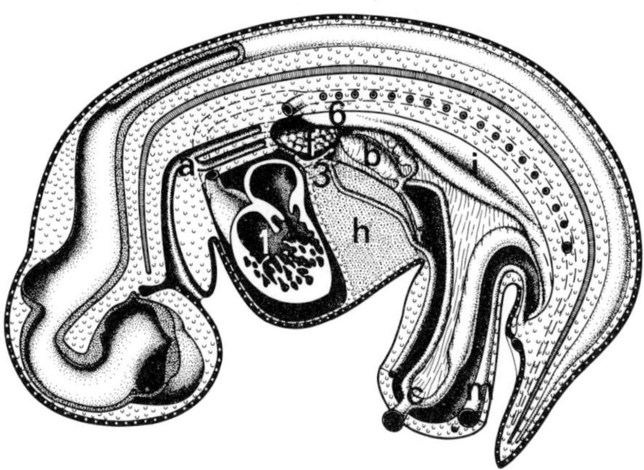

Fig. 2.4. Schematic drawing of a generalized later amniote embryo. The lung buds lie in the pericardioperitoneal canal, which still connects pericardial and peritoneal cavities. The liver has totally invaded the septum transversum. At the dorsal margin of the septum transversum the postpulmonary septum has begun to sprout opposite the cranial nephric fold. 1, heart in pericardial cavity; 3, dorsal part of septum transversum with sprouting postpulmonary septum; 6, cranial nephric fold; a, pharynx; b, stomach with dorsal and ventral mesenteries; c, intestinal tract; f, lung bud; h, liver and gallbladder cut medially in septum transversum; i, kidney; m, allantois. From Duncker (1978a).

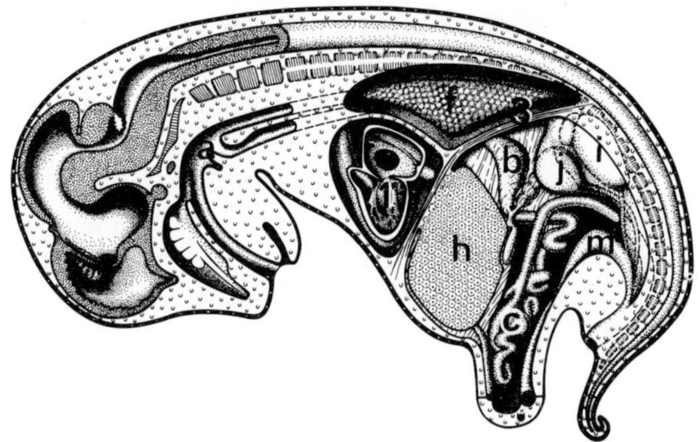

Fig. 2.5. Schematic drawing of a generalized late amniote embryo with closed postpulmonary septum. The pleural cavity is also closed off from the pericardial cavity. The liver is secondarily separated from the former septum transversum and ventral body wall by outgrowing recesses, the ventral mesentery remaining as the only connection. With further lung growth the pleural cavity may occupy the entire thoracic cavity. 1, heart in pericardial cavity; 3, postpulmonary septum; a, pharynx, from which the trachea originates; b, stomach with dorsal and ventral mesenteries; c, intestinal tract; f, lung; h, liver, cut medially and showing its connection to the postpulmonary septum, the *area nuda*; i, kidney; j, gonad; m, urinary bladder. From Duncker (1978a).

taking into account surface tension, the large compartments would be overexpanded, whilst the small compartments would collapse, even during inspiration.

Another serosal septum which occurs in some reptilian families is the posthepatic septum. In large teiids (Macroteiidae), which have freely movable lungs and no postpulmonary septum, this posthepatic septum divides the pleuroperitoneal cavity into a cranial part, occupied by the lungs and liver, and a caudal part in which lie the intestines and gallbladder (Fig. 2.6). The posthepatic septum appears to be a caudal division of the primary septum transversum (Fig. 2.4), being split off from its caudal side by recesses which extend from above between the liver and the septum. On both sides the septum attaches dorsally to the cranial extension of the mesosalpinx which develops out of the cranial nephric fold. This in turn is responsible for the dorsal closure of the posthepatic septum. However, the closure can be incomplete, and is frequently total on one side only.

The coelomic septa in reptiles consist mainly of strong collagenous connective tissue lined by mesothelium. The lungs of some turtle species are covered dorsocranially and ventrally by striated muscle. In the postpulmonary septum, as in the mesenteries, there are a few smooth muscle cells. Larger numbers of smooth muscle cells are present in the posthepatic septum of teiids, where they tend to be arranged in scattered bundles running in different directions, especially in the lateral and ventral parts of the septum.

In crocodiles in which both postpulmonary and posthepatic septa occur (Fig. 2.10), the dorsal and ventral mesenteries in the liver region subdivide the hepatic coelomic cavity into right and left chambers. The entire margin of the posthepatic septum where it attaches to the body wall is connected to a thick sheet of striated muscle which originates from the musculature of the internal abdominal wall attaching to the most caudal ribs of the abdomen and, via an aponeurosis, to the pelvis (Fig. 2.10). This so-called *musculus diaphragmaticus* extends ventral to the posthepatic septum and attaches to the pericardium. It is innervated by lumbar spinal nerves and functions to retract the posthepatic septum. Due to the solid nature of the liver, the muscle also acts as a retractor of the postpulmonary septum. Retraction of all these structures can be likened to the movement of a piston in a cylinder and has the effect of inflating the lung. This mechanism therefore forcefully supports the respiratory movements produced by the thoracic ribs and their musculature. The crocodilian lung which is partly adherent to the walls of the pleural cavity has a structure which is highly adapted to this mode of ventilation. Thus, the dorsoventral gradient in the amount of partitioning of the lung tissue is not so extensive as in monitor lizards. However, the most cranial and caudal parts of the crocodilian lungs are made up of larger compartments with less gas exchange surface (Fig. 2.10).

The foregoing demonstrates the diversity of the subdivisions of the coelomic cavity in reptiles and the way the different forms of coeloms are associated with the internal structure of the lungs and with the very special modes of ventilation. Birds have adopted the basic structural principles of the lung and coelomic subdivisions of reptiles (Fig. 2.11). This is particularly exemplified by the development of the postpulmonary and posthepatic septa and the adhesion of the lung to the dorsal wall of the thorax. However, the subdivisions of the coelom in the ancestors of crocodiles and birds by the postpulmonary and the

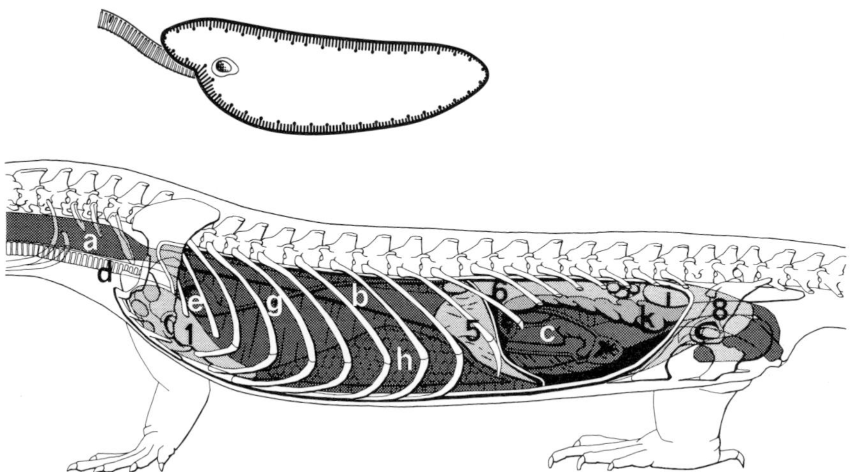

Fig. 2.6. Teju (*Tupinambis nigropunctatus*). Schematic drawing of the unicameral lung and coelomic cavities (left lung, left liver lobe and stomach removed). The pericardial cavity lies at the level of the pectoral girdle. The posthepatic septum divides the pleuroperitoneal cavity into a cranial pleuroperitoneal and a caudal peritoneal cavity. The posthepatic septum is dorsally closed by the cranial mesosalpinx (mesoviduct). The mediastinal mesentery and the liver with its dorsal and ventral mesenteries divide the pleuroperitoneal cavity into right and left parts. 1, pericardial cavity; 5, posthepatic septum; 6, cranial mesoviduct; 8, peritoneal cavity, caudal extension; a, pharynx; b, eosophagus in mediastinum; c, intestinal tract; d, trachea; e, primary bronchus, cut at the lung hilus; g, mesopneumonium; h, liver, cut medially; i, kidney; k, oviduct. From Duncker (1978a).

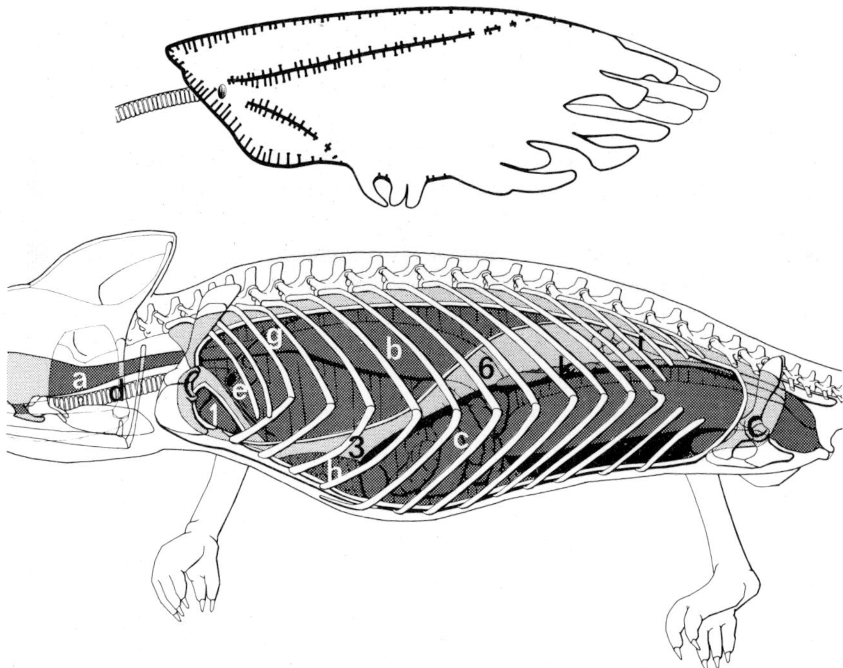

Fig. 2.7. Chameleon (*Chamaeleo chamaeleon*). Schematic drawing of the plaucicameral lung and coelomic cavities (left lung and left liver lobe removed). The pericardial cavity is situated cranially at the level of the pectoral girdle and is connected to a small post-pulmonary septum which separates the small liver ventrally from the large lung. The postpulmonary septum is connected laterally to the cranial mesosalpinx (mesoviduct) and is caudally open. The cranial part of the pleuroperitoneal cavity is divided by the mediastinal mesentery into right and left parts. The large lung extends from the cranial part of the pleuroperitoneal cavity into the caudal part which is ventrally undivided. 1, pericardial cavity; 3, postpulmonary septum; 6, cranial mesoviduct; a, pharynx; b, oesophagus and stomach in mediastinum; c, intestinal tract and rectum, suspended by the dorsal mesentery; d, trachea; e, primary bronchus, cut at the lung hilus; g, mesopneumonium; h, liver, cut medially; i, gonad and kidney; k, oviduct. From Duncker (1978a).

posthepatic septa have led to quite different phylogenetic developments in both groups. Thus, crocodiles possess only a thin, little differentiated postpulmonary septum. Furthermore, the diaphragmatic muscle, which is connected to the posthepatic septum (Fig. 2.10), is the main inspiratory muscle together with the intercostal muscles. Birds, in contrast, have a respiratory system which includes constant-volume lungs for gas exchange, and air sacs which act as bellows for ventilation (Fig. 2.11); it is upon the air sacs that the respiratory movements of the thoracoabdominal musculature act. This lung-air sac system is characterized by a very high compliance. Its high degree of differentiation results in the splitting of the postpulmonary septum into the horizontal and oblique septa by the ingrowth of air sacs. The dorsal derivative of the postpulmonary septum, the horizontal septum, encloses the pleural cavity (Fig. 2.11). The highly heterogeneous partitioned lung is extensively fused to the horizontal septum and to the outer wall of the pleural cavity. The volume-constancy of the pleural cavity depends on its very dorsal situation in the thorax, on the coarse structure of the horizontal septum, and on the action of the costoseptal muscles which lie in the lateral margin of the horizontal septum. The thoracic air sacs are situated between the horizontal and oblique septa in the extracoelomic subpulmonary cavity. Through secondary expansion of the cranial group of air sacs, the subpulmonary cavity is extended far cranially up to and in front of the shoulder girdle. The abdominal air sac invades the intestinal peritoneal cavity secondarily. The posthepatic septum of birds, in contrast to that of crocodiles, has no special functional significance.

A consideration of the development of the mammalian respiratory system and the way its structure is related to the subdivisions of the coelomic cavity is complicated by the fact that no surviving representatives of the reptilian ancestors are available. Thus, it can only be assumed from the lung structure of living reptiles that the existence of the pleural cavity in mammals is connected with the development of an equally subdivided lung (Fig. 2.12). Only a freely movable lung which is homogeneously partitioned can be inflated uniformly during inspiration. In contrast to the single primary bronchus of reptiles, a bronchial tree is present which allows all the alveoli of the lung to be supplied equally. For the forceful inflation of such a lung with a very low compliance a

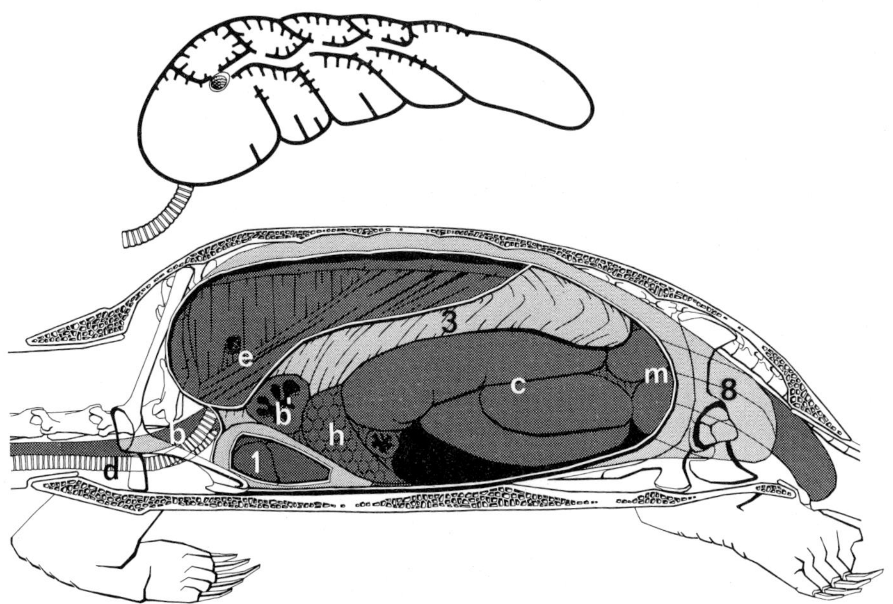

Fig. 2.8. Red-eared Turtle (*Pseudemys scripta elegans*). Schematic drawing of the multicameral lung and coelomic cavities (left lung, left liver lobe and stomach removed). The pericardial cavity lies cranially near the pectoral girdle and is dorsally connected to the postpulmonary septum, which divides totally the pleural cavities from the peritoneal cavity. Both pleural cavities are medially separated by a mediastinum, containing the retractor muscles of the neck. Caudal to the liver the peritoneal cavity is undivided. 1, pericardial cavity; 3, postpulmonary septum; 8, peritoneal cavity, caudal extension; b, oesophagus; b′, entrance into the stomach; c, intestinal tract; d, trachea; e, primary bronchus, cut at the lung hilus; h, liver, cut medially; m, urinary bladder and accessory bladder. From Duncker (1978a).

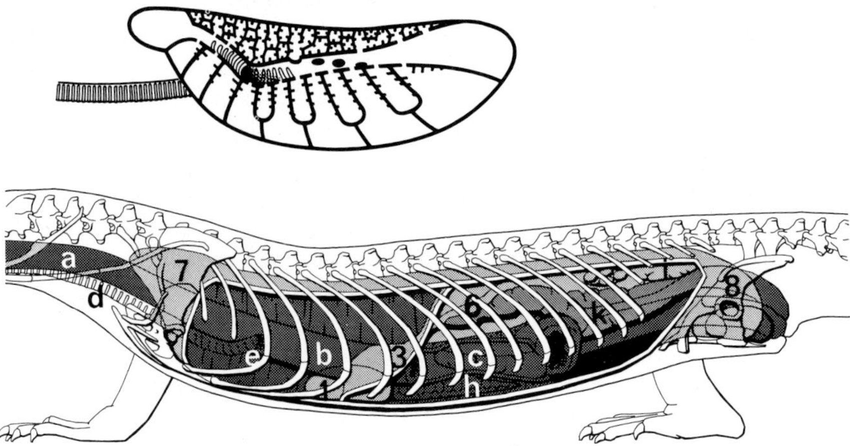

Fig. 2.9. Monitor Lizard (*Varanus exanthematicus*). Schematic drawing of the multicameral lung and coelomic cavities (left lung, left liver lobe and stomach removed). The pleural cavities are totally separated from the peritoneal cavity by the postpulmonary septum, which is dorsally closed by the cranial part of the mesosalpinx (mesoviduct), the former cranial nephric fold. The heterogeneous lungs occupy the entire thoracic region, displacing the postpulmonary septum caudally, and thus forcing the caudal displacement of the pericardial cavity, which is ontogenetically connected to the postpulmonary septum. 1, pericardial cavity; 3, postpulmonary septum; 6, cranial mesoviduct; 7, pleural cavity, cranial extension; 8, peritoneal cavity, caudal extension; a, pharynx; b, oesophagus in mediastinum; c, intestinal tract; d, trachea; e, primary bronchus, cut at the lung hilus; h, liver, cut medially; i, kidney; k, oviduct. From Duncker (1978a).

highly muscular postpulmonary septum, the diaphragm, is developed (Fig. 2.12). Like the postpulmonary septum of other vertebrates, it originates from the primary septum transversum, supported by the cranial nephric folds and by the dorsal mesentery (Fig. 2.5). This represents the basic structure of the diaphragm, to which are added larger parts of the body wall, the so-called pleuroperitoneal folds, incorporating the striated musculature and its innervation, the phrenic nerve. In this way the reptilian postpulmonary septum has undergone a special phylogenetic development adapted functionally to the structural needs of the mammalian lung.

The postpulmonary septum in reptiles, birds and mammals is not a completely homologous structure. However, in all these vertebrates the postpulmonary septum is derived basically from the septum transversum, the cranial nephric folds and the dorsal mesentery (Figs 2.3, 4). The septum in birds and mammals, but not in reptiles, also incorporates striated muscle from the body wall, although the distribution of the muscle and its innervation is markedly different in the two classes. In birds the horizontal septum contains the costoseptal muscles which are innervated by thoracic spinal nerves, and are almost certainly derived from the body wall. The oblique septum has a well-developed tensor muscle which, however, consists of smooth muscle cells and is most likely to be an internal development of the septum. In mammals the diaphragm contains striated muscle which comes from the body wall via the pleuroperitoneal folds. Because of the very early embryological development of this special differentiation of the mammalian postpulmonary septum, the innervation is derived from a specialized spinal nerve, the phrenic nerve.

Only the basic structure, therefore, of the postpulmonary septum in the different groups of vertebrates is homologous. The further differentiation of the septum in the various groups, especially the incorporation of musculature from the body wall in birds and mammals and the splitting of the septum into two parts in birds, are specific developments related to the functional needs of each group. These specific differentiations are not comparable, and the structures derived from them, especially the muscle tissue, are not homologous.

III. Subdivisions of the avian coelomic cavity

In all birds the coelom is subdivided in a manner rather similar to that in many reptiles and especially to crocodiles, since both postpulmonary and posthepatic septa are present. However, the resemblance in the subdivision of the coelomic cavity between crocodiles and birds is restricted to these general features only,

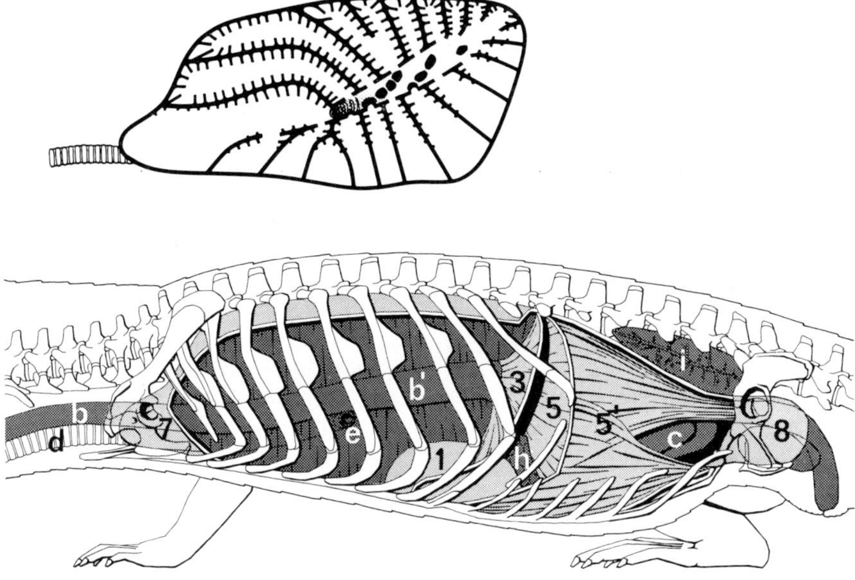

Fig. 2.10. Caiman (*Caiman crocodilus*). Schematic drawing of the multicameral lung and the coelomic cavities (left lung and left liver lobe removed). The heterogeneous lungs totally fill the thoracic region, thereby displacing the postpulmonary septum caudally. The pericardial cavity, because of its connection to the postpulmonary septum, has also moved caudally. The peritoneal cavities for the lobes of the liver are separated from the remaining peritoneal cavity by the posthepatic septum. The diaphragmatic muscle is attached to the margins of the posthepatic septum and ventrally to the pericardial wall. 1, pericardial cavity; 3, postpulmonary septum; 5, posthepatic septum; 5′, diaphragmatic muscle; 7, pleural cavity, cranial extension; 8, peritoneal cavity, caudal extension; b, oesophagus; b′, oesophagus in mediastinum; c, intestinal tract; d, trachea; e, primary bronchus, cut at the lung hilus; h, liver, cut medially; i, kidney. From Duncker (1978a).

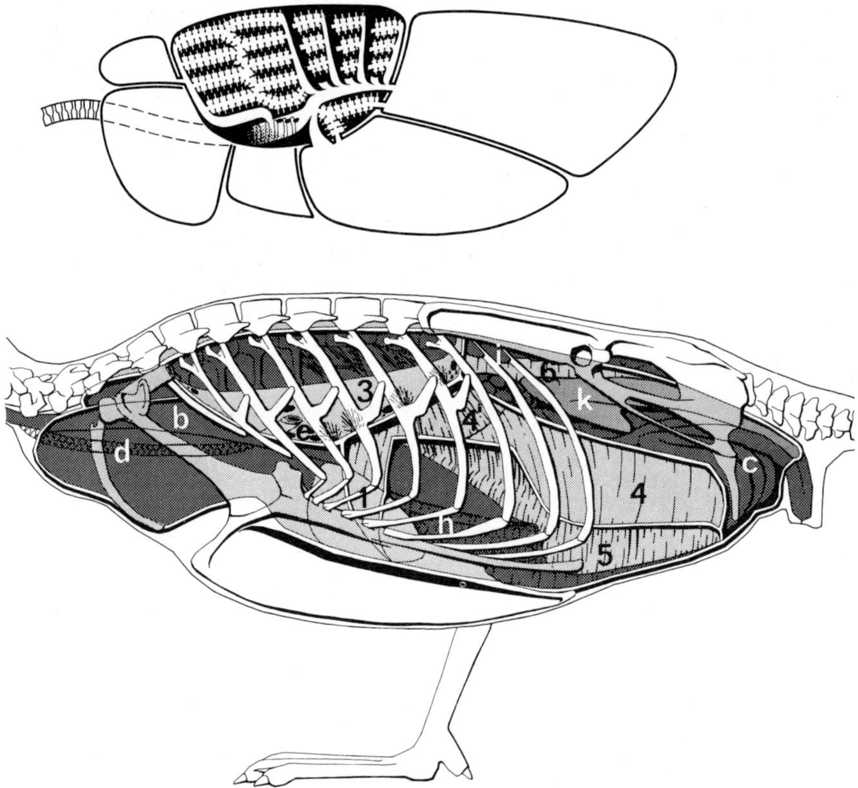

Fig. 2.11. Mallard (*Anas platyrhynchos*). Schematic drawing of the lung air sac system and coelomic cavities (left lung removed, left oblique septum cut at the attachment of the post-hepatic septum, left liver lobes removed). The pericardial cavity lies cranially on the sternum. The pleural cavity is ventrally bordered by the horizontal septum, beneath which lies the subpulmonary cavity with the air sacs which are separated medially from the intestines by the oblique septum. Horizontal and oblique septa constitute a uniform septum (the original postpulmonary septum) at the caudal lung margin, which is dorsally closed by the cranial part of the mesoviduct. The hepatic peritoneal cavities for the lobes of the liver are separated from the intestinal peritoneal intestinal cavity by the posthepatic septum. 1, pericardial cavity; 3, horizontal septum; 4, oblique septum; 4′, m. septi obliqui; 5, posthepatic septum; 6, cranial mesoviduct; b, oesophagus; c, intestinal tract; d, trachea; e, primary bronchus, cut at the lung hilus; h, liver, cut medially; i, kidney; k, oviduct. From Duncker (1978a).

the respiratory systems and the related septal subdivisions of the coelom in the ancestors of the two groups having evolved along different lines.

The first subdivision of the avian coelom is the pericardial cavity (*cavitas pericardialis*) which lies in the cranial part of the thorax attached to the sternum (Figs 2.11, 14, 15, 16). The pericardium (*pericardium*) is connected dorsally to the hilus of the lung and to the horizontal septum (*septum horizontale*) and laterally to the oblique septum (*septum obliquum*) (Figs 2.14, 15, 16). Cranially and laterally the pericardial cavity is surrounded by the cranial group of air sacs. Its caudal part bulges between the liver lobes.

The right and left pleural cavities (*cavitates pleurales*) enclose the relatively small lungs and occupy the most dorsal part of the thorax (Figs 2.11, 13). Each pleural cavity is bounded by the parietal pleura (*pleura parietalis*) which lines medially the vertebral column, dorsally the transverse processes of the vertebrae, and dorsolaterally the ribs and intercostal muscles. Ventrally the pleural cavity is bounded by the horizontal septum, which originates from the ventral margin of the vertebral bodies or their ventral processes and is fixed laterally to the ventral extremities of the vertebral ribs. Originating at the lateral attachment of the horizontal septum are the costoseptal muscles (*mm. costoseptales*) which constitute the lateral margin of the septum (Fig. 2.13). The volume-constancy of the pleural cavity in all respiratory phases depends on both the structure of the coarse horizontal septum and the action of the costoseptal muscles (Soum, 1896; Fedde *et al.*, 1964). In late ontogeny the pleural cavity is greatly reduced by the fusion of the parietal pleura to the visceral pleura (*pleura visceralis*).

The subpulmonary cavity is not a coelomic cavity, but rather a major subdivision of the thoracoabdominal cavity. As such it is an important constituent of the respiratory system. It lies on each side ventral to the horizontal septum and pleural cavity, extending caudally into the cranial part of the abdomen

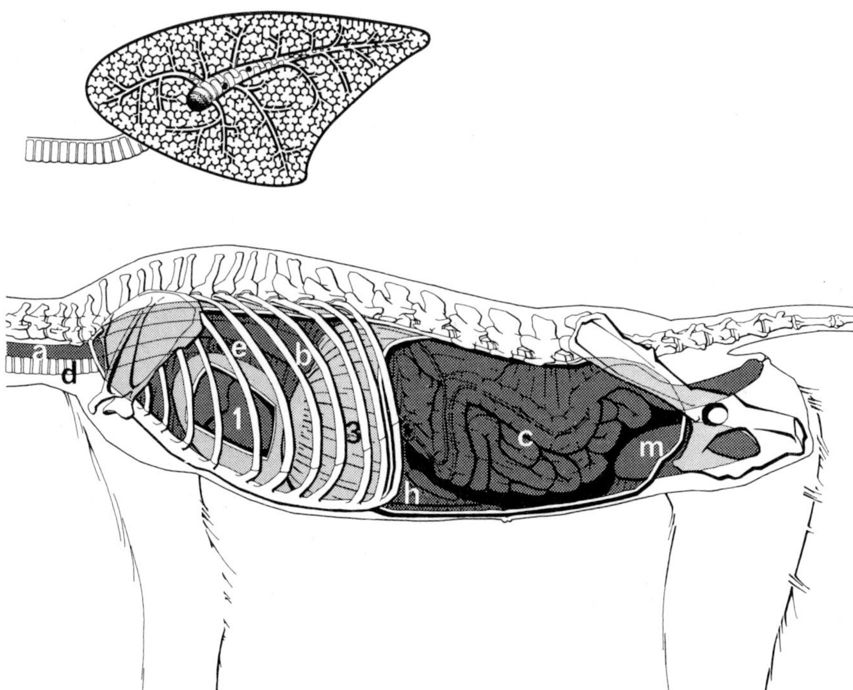

Fig. 2.12. Rhesus Monkey (*Macaca mulatta*). Schematic drawing of the lung and coelomic cavities (left lung, left liver lobe and stomach removed). The pleural cavities are separated from the peritoneal cavity by the postpulmonary septum which has developed into the diaphragm. The thoracic region is filled by the lungs, which are responsible for the descent of the diaphragm and the pericardial cavity. Both pleural cavities are separated medially by the mediastinum. The peritoneal cavity caudal to the liver possesses a broad ventral communication between the right and left sides. 1, heart in pericardial cavity; 3, diaphragm; a, oesophagus; b, oesophagus in mediastinum; c, intestinal tract, suspended by the dorsal mesentery; d, trachea; e, primary bronchus, cut at the lung hilus; h, liver, cut medially; m, urinary bladder. From Duncker (1978a).

(Figs 2.13, 14, 15, 16). This cavity is specific to birds and is formed by the splitting of the postpulmonary septum by air sacs which totally fill it. Although the subpulmonary cavity is extracoelomic, its development has an important influence on the further subdivision of the coelom and therefore is included in this account. The subpulmonary cavity is lined dorsally by the horizontal septum, laterally by the thoracic wall consisting of the vertebral and sternal ribs and the lateral margin of the sternum, and ventromedially by the oblique septum. Due to the development of the cranial group of air sacs, this subpulmonary cavity is also cranially elongated, lateral and cranial to the pericardial cavity. The cranial group of air sacs surrounds the trachea, the oesophagus and the cranial vessels of the heart (Figs 2.13, 14, 15, 16). Caudal to the pericardial cavity the subpulmonary cavity is separated from the oesophagus, proventriculus and liver by the oblique septum (Figs 2.14, 15, 16).

The right and left hepatic peritoneal cavities (*cavitates peritoneales hepaticae*) represent the cranioventral part of the peritoneal cavity (*cavitas peritonealis*), and are separated from the rest of the peritoneal cavity by the posthepatic septum (*septum posthepaticum*), which descends in a ventrocaudal direction from about the level of the pulmonary hilus to the caudal margin of the sternum (Figs 2.14, 15). The lateral walls of the hepatic cavities are formed by the oblique septa, and the ventral wall by the parietal peritoneum (*peritoneum parietale*) lining the sternum. The right and left cavities are separated by the dorsal and ventral mesenteries (*mesenterium, mesenterium ventrale*). This part of the mesentery includes cranially the proventriculus (Figs 2.17, 18). The right and left hepatic ligaments (*ligg. hepatica*) connect the visceral peritoneum (*peritoneum viscerale*) lining the liver lobes with the oblique septa on both sides and separate the small cranially situated dorsal hepatic cavities (*cavitates peritoneales hepaticae dorsales*) from the large ventral hepatic cavities (*cavitates peritoneales hepaticae ventrales*).

The intestinal peritoneal cavity (*cavitas peritonealis intestinalis*) is lined dorsally by the parietal peritoneum of the dorsal body wall including the pelvic girdle, laterally by the oblique septa and the parietal peritoneum of the lateral body wall, and ventrally by the posthepatic septum and the parietal peritoneum of the ventral body wall. This single peritoneal cavity contains the gastro-intestinal tract suspended by the dorsal mesentery, the gallbladder attached to the posthepatic septum, the gonads suspended by the mesorchia (*mesorchia*) in the male and the mesovarium (*mesovarium*) in the female, the left oviduct suspended by the mesosalpinx (*mesoviductus dorsalis*), and, except in kiwis (Apterygidae), the abdominal air sacs suspended by parietal peritoneum (Figs 2.14, 15, 16).

A. Pericardial cavity

In all birds the pericardial cavity is situated ventrally in the midline in the cranial part of the thorax. In birds with short and rounded bodies such as pigeons (Columbidae) (Fig. 2.15) and galliform and passerine species (Fig. 2.16) it extends cranially to the level of the joints between the sternum and coracoid bones. When the body is more elongated as for example in ducks (Anatidae),

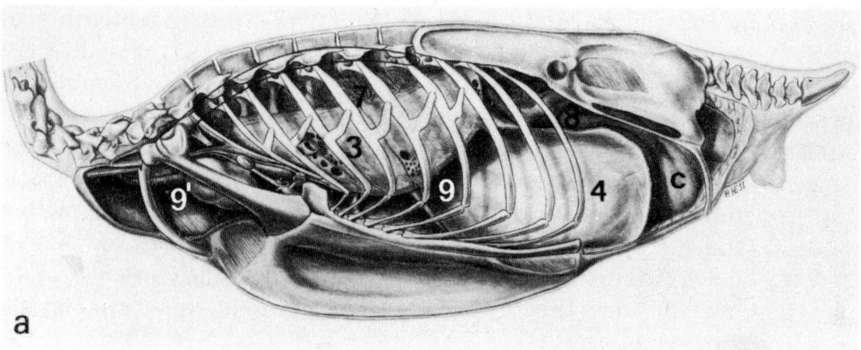

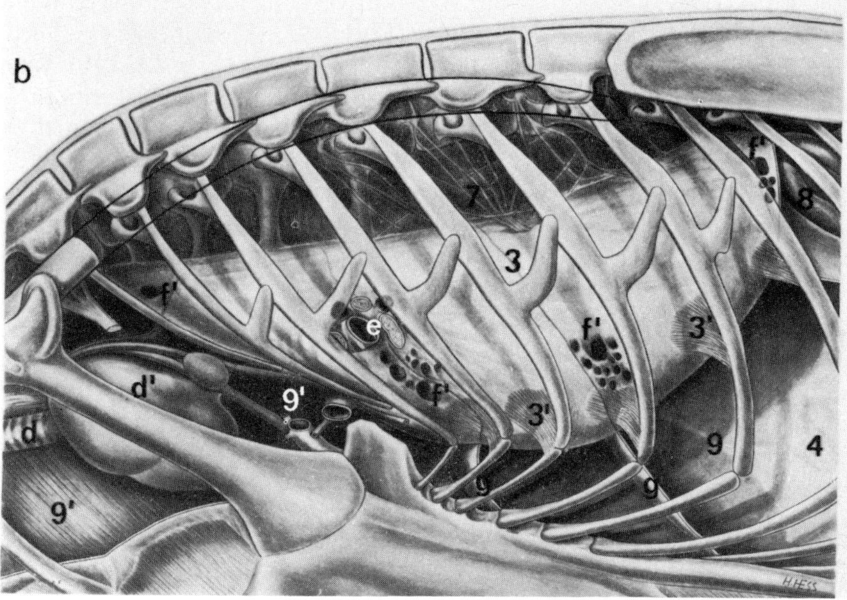

Fig. 2.13. Mallard (*Anas platyrhynchos*). (a) Left lateral view of the thoracoabdominal cavity showing the horizontal and oblique septa and the pleural cavity (thoracic and abdominal walls excised except for ribs, left lung removed from the pleural cavity). (b) Enlargement of thoracic region in (a). The pleural cavity is limited by the parietal pleura lining medially the vertebral column and dorsolaterally the vertebral transverse processes, the vertebral ribs and the muscles between them; ventrally, the pleural cavity is limited by the horizontal septum which has openings for the structures passing through the pulmonary hilus and for the air sac ostia. The costoseptal muscles lie at the lateral margin of the horizontal septum. Caudally the unsplit horizontal and oblique septa are attached dorsally, the oblique septum extending far caudally, to attach to the lateral body wall. The oblique septum separates the air sacs in the subpulmonary cavity, ventral and caudal to the lung, from the peritoneal cavity medially. Only the abdominal air sac fills the dorsal part of the peritoneal cavity. 3, horizontal septum; 3′, costoseptal muscles; 4, oblique septum; 7, pleural cavity; 8, peritoneal cavity; 9, subpulmonal cavity; 9′, subpulmonal cavity, secondary cranial extension; c, intestinal tract; d, trachea; d′, bulla of syrinx; e, primary bronchus, cut at the pulmonary hilus; f′, air sac ostia; g, septa between air sacs. From Duncker (1971).

divers (Gaviidae) and gannets (Sulidae), the cavity does not reach so far cranially. Caudally the pericardial cavity generally ends about the middle of the sternum, although in small birds with large hearts, e.g. hummingbirds (Trochilidae) and some passerine species, it may extend further caudally.

Dorsally the pericardium is connected to the hilus of the lung at the point of entrance of the pulmonary vein (Figs 2.15, 18). It has also a small connection to the horizontal septum. The pericardium is in contact dorsally with the oesophagus and the primary bronchi, and the cranial and caudal venae cavae (Figs 2.15, 17, 18). Cranially and laterally it is related to the clavicular and the cranial thoracic air sacs (Figs 2.14, 15, 16). Cranially the pericardium is penetrated by the great arteries.

On the lateral side of the heart the oblique septum arises from the pericardium (Figs 2.14, 15). Dorsal to the heart, the right and left oblique septa unite and attach to the ventral border of the vertebral column between the pulmonary hilus and the caudal margin of the lung (Figs 2.17, 18). Through this median septum the oesophagus passes into the peritoneal cavity (Figs 2.17, 18). Caudal to the origins of the oblique septa from the pericardium, the pericardial cavity bulges into the hepatic peritoneal cavity and is surrounded by the cranial parts of the liver (Figs 2.17, 18). Dorsally in this region is the *area nuda* where the liver is in direct connection with the right oblique septum; via this area the caudal vena cava enters the pericardial cavity (Figs 2.14, 15). Caudally the pericardium continues as the hepatopericardial ligament (*lig. hepatopericardiacum*) into the ventral mesentery between the lobes of the liver.

The pericardium ventrally is attached to the sternum by connective tissue. Because of its origin from the pericardium the oblique septum is thereby also fixed to the cranial part of the sternum. In a number of large species, diverticula from the clavicular air sacs extend caudally between the pericardium and sternum (Fig. 2.17), thus raising the pericardium. However, the connection of the pericardium to the sternum is maintained by connective tissue bridges which also provide the cranioventral fixation of the oblique septum. Further caudally the oblique septum generally inserts directly onto the lateral margin of the sternum.

In contrast to other birds, in all passerine species the oblique septum unites in the midline caudal to the pericardial sac with the oblique septum of the opposite side. Between the sternum on the one side and the pericardium and the united oblique septa on the other extends a large diverticulum of the clavicular air sac up to the caudal margin of the sternum, where the oblique septa are attached (Figs 2.16, 18). In these birds the pericardium is strongly attached to the sternum cranially and laterally by broad connective tissue bridges (Figs 2.16, 18), and in this way the oblique septa are cranially fixed in position.

The pericardial wall is lined internally by the parietal serous pericardium (*pericardium serosum parietale*). At the point where the great vessels penetrate the pericardial sac the parietal pericardium is reflected as visceral pericardium or epicardium (*pericardium serosum viscerale*). In the space between the great arteries and veins the pericardial cavity forms the small transverse pericardial sinus (*sinus transversus pericardialis*).

B. Pleural cavities

The pleural cavities are symmetrically developed on both sides of the body (Figs 2.17, 18) and occupy the most dorsal part of the thorax (Fig. 2.13). Each cavity extends between the levels of the first movable vertebral rib cranially and the last rib caudally, the most caudal part of the pleural cavity therefore lying below the ilium. The outer wall of the pleural cavity is formed by parietal pleura, the inner wall by visceral pleura. However, the full development of the pleural cavity with a complete and continuous cleft between the parietal and visceral pleural membranes is only found in early ontogeny.

The cavity is bordered medially by the parietal pleura lining the vertebral column, the dorsomedial margin lying at the base of the neural arches ventral to the transverse processes (Figs 2.13, 17, 18). The medial wall of the cavity is extended ventrally in some groups of birds by ventral processes of the vertebrae and by a septum between them. When the medioventral border of the lung reaches further ventrally than the vertebral column, and ventral processes of the vertebrae are missing, the right and left pleural cavities are separated ventrally by an interpleural septum which forms the common origin of the horizontal and oblique septa (Fig. 2.18). The cavity is bounded dorsolaterally by the parietal pleura lining the vertebral transverse processes, the vertebral ribs and the related musculature. Compared to the pleural cavity of other vertebrates, that of birds extends much further dorsally reaching directly beneath the vertebral transverse processes (Fig. 2.13). Consequently the craniodorsally flattened dorsal parts of the vertebral ribs cut ventrally into the cavity and subdivide its dorsal region.

The ventral wall of the pleural cavity is supported entirely by the horizontal septum which stretches between the vertebral column and the thoracic wall. The attachment to the thoracic wall follows the line of an arc, starting high up on the first movable vertebral rib, reaching its most ventral point at the level of the first two or three articulations between the vertebral and sternal ribs, and then ascending to its attachment dorsally on the last or second to last rib below the ilium. When viewed laterally, therefore, the pleural cavity has a triangular shape, except in galliform and many passerine species in which it is more quadrangular. Caudal to the lung, the horizontal and oblique septa are not separated from each other and form a vertical septum which represents the original postpulmonary septum (Figs 2.13, 17).

The lateral part of the horizontal septum caudal to the pulmonary hilus is supported by the costoseptal muscles which arise from the ribs and project in the shape of a fan into the thin aponeurotic septum (Figs 2.13, 14, 15, 16, 17, 18). These muscles contract in expiration and relax in inspiration (Soum, 1896; Fedde *et al.*, 1964). and thus maintain the constant volume of the pleural cavity in spite of the respiratory movements of the cranial vertebral ribs which tend to alter slightly the position of the lateral attachment of the horizontal septum. Movements of the more caudal vertebral ribs do not affect the septum because of the ascending line of attachment of the septum to the thoracic wall.

The horizontal septum is penetrated by a number of structures including those passing through the hilus of the lung, and the ostia of the cervical, clavicular and cranial and caudal thoracic air sacs (Figs 2.13, 14, 15, 16, 17, 18).

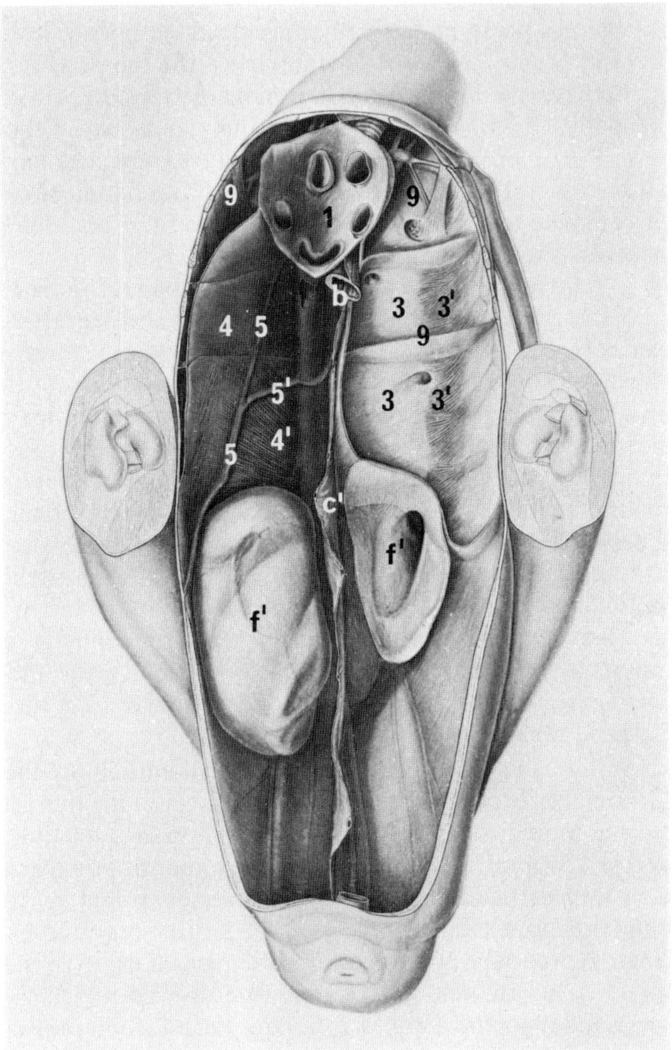

Fig. 2.14. Greater Rhea (*Rhea americana*). Ventral view of the thoraco-abdominal cavity (abdominal wall and intestinal tract removed). The pericardial cavity is surrounded by the clavicular air sac cranially and laterally. On the right side of the dissection the oblique septum is visible, the ventral margin of which has been cut off. The oblique septum passes into the pericardium cranially and contains on its surface the lines of attachment of the posthepatic septum and the right hepatic ligament. Attached to the dorsomedial part of the septum is the caudal vena cava whilst in the caudal part is the muscle of the oblique septum which is partly covered by the abdominal air sac bulging into the intestinal peritoneal cavity. On the left side of the dissection the subpulmonary cavity is opened and the oblique septum and the wall of the abdominal air sac are largely removed. This has exposed the clavicular and thoracic air sacs in the subpulmonary cavity, and the horizontal septum with the air sac ostia and the broad costoseptal muscles. 1, pericardial cavity; 3, horizontal septum; 3′, costoseptal muscles; 4, oblique septum; 4′, muscle of oblique septum; 5, posthepatic septum; 5′, right hepatic ligament; 9, subpulmonary cavity; b, oesophagus; c′, dorsal mesentery; f′, abdominal air sac. From Duncker (1971).

Caudally the perpendicular septum formed by the unsplit horizontal and oblique septa is penetrated by the ostium of the abdominal air sac (Figs 2.13, 14, 17). For the position of the horizontal septum in relation to neighbouring structures see Figs 2.13, 17 and 18.

A pleural cavity lined by the parietal and visceral pleural membranes is present in the early embryo. During the second half of incubation intrapulmonary bronchi penetrate the surface of the lung and invade the primary postpulmonary septum, thus forming the air sacs and splitting the primary septum into the horizontal and oblique septa. During this period, fusion of the parietal and visceral pleural membranes occurs starting with the fusion of the lung to the horizontal septum. At the end of incubation the pleural cavity is reduced to a number of very small spaces some of which may persist into adult life, especially in the medial and dorsolateral parts of the cavity. The fusion of the visceral and parietal pleural membranes, together with the partitioning of the dorsal part of the pleural cavity by the ribs, is sufficient to prevent movement of the lung against the parietal pleura so that even small volume changes in limited parts of the lung do not occur.

C. *Subpulmonary cavities*

The right and left extracoelomic subpulmonary cavities (the subpulmonary chambers of Huxley, 1882) are formed in the embryo by the growth of the thoracic air sacs into the postpulmonary septum on each side, splitting the septum into the horizontal and oblique septa (Figs 2.13, 14, 15, 16, 17, 18). The cavities are secondarily extended cranially by the development of the clavicular and cervical air sacs.

The medial wall of each subpulmonary cavity is formed by the oblique septum which arises from the vertebral column and cranially by the pericardium (Figs 2.14, 15, 16, 17, 18). Laterally the oblique septum in most taxonomic groups is connected to the sternum. In the passerine species, however, the septum instead unites with that of the opposite side to form a sac which surrounds the liver and proventriculus (Figs 2.16, 18). The fusion of the two oblique septa continues to the caudal margin of the sternum, to which the septa are attached. The space between the fused right and left septa and the sternum is occupied by a large unpaired caudal diverticulum of the clavicular air sac. In these passerine species, the functionally important cranial fixation of the oblique septa is achieved via their attachment to the pericardium (Figs 2.16, 18). The lateral wall of the subpulmonary cavity is formed by the body wall including the ventral parts of the vertebral ribs, the sternal ribs and the related musculature, and extends caudal to the lung and the horizontal septum for a distance of one to several intercostal spaces (Figs 2.13, 14, 15, 16, 17, 18). The precise form of the subpulmonary cavity varies in relation to the conformation of the body and rib cage.

The portion of the subpulmonary cavity between the horizontal and oblique septa contains only the cranial and caudal thoracic air sacs except in kiwis (Apterygidae) in which the abdominal air sac is also present. Cranially, the subpulmonary cavity extends lateral to the pericardial cavity to reach the caudal

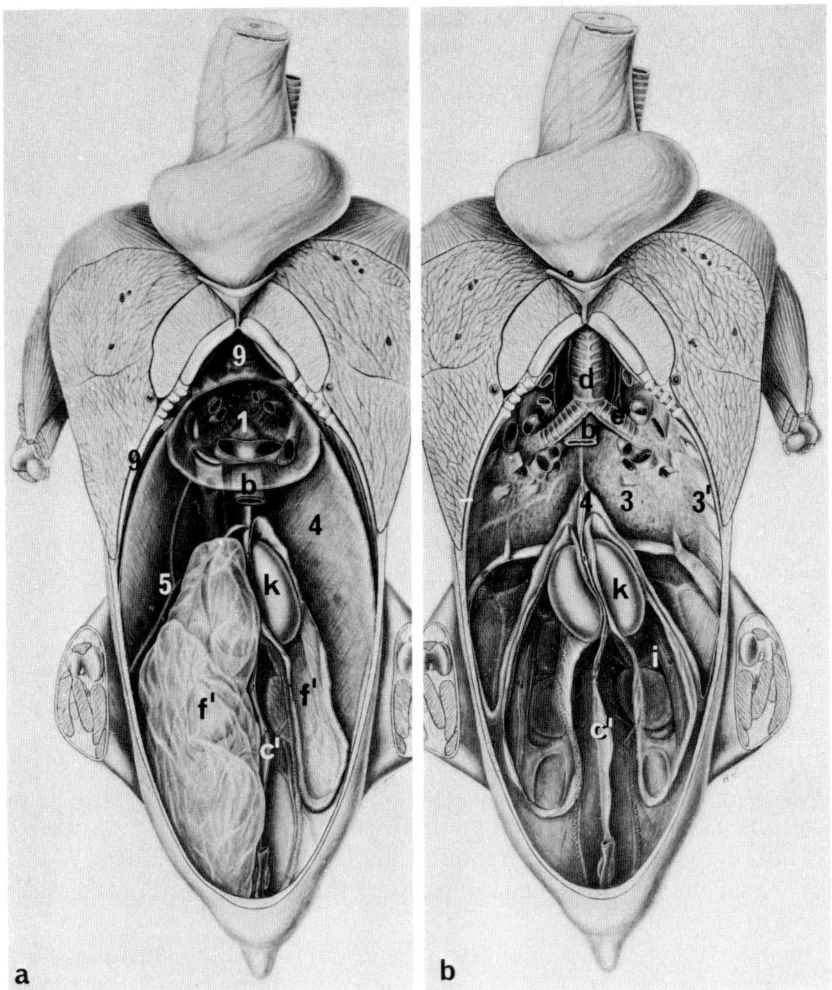

Fig. 2.15. Domestic Pigeon (*Columba*). Ventral view of the thoracoabdominal cavity. (a) Sternum and ventral abdominal wall, heart and intestinal tract removed. Cranial and lateral to the pericardial cavity lie the subpulmonary cavities with the clavicular air sac. The oblique septum continues ventrally on both sides into the pericardial wall, its ventral attachment to the sternal margin having been removed with the sternum. On the right side of the dissection the attachments of the posthepatic septum and the caudal vena cava to the oblique septum are visible. In the intestinal peritoneal cavity the two large, thin walled abdominal air sacs are visible, the left one having been opened to show the left testis bulging into its dorsal wall. Between the two abdominal air sacs is the cut margin of the dorsal mesentery. (b) As in (a), except that the dorsal pericardial wall and the oblique septum on both sides have been removed; the cut margins of the oblique septum can be identified cranial to the opened abdominal air sacs. The subpulmonary cavities are united ventral to the trachea. Cranial to the pericardial cavity they contain the trachea, oesophagus, blood vessels and nerves. On both sides, the horizontal septum, covering the ventral surface of the lung, contains the costoseptal muscles at its lateral margin. The cut edges of the oblique septa cranially fuse with the dorsal mesentery. Through the opened abdominal air sacs the attachment of the oblique septa to the parietal peritoneum of the dorsal body wall and to the ventral surface of the kidneys is visible. 1, pericardial cavity; 3, horizontal septum; 3′, costoseptal muscles; 4, oblique septum; 5, posthepatic septum; 9, subpulmonary cavity; b, oesophagus; c′, dorsal mesentery; d, trachea; e, primary bronchus; f′, abdominal air sac; i, kidney; k, testis. From Duncker (1971).

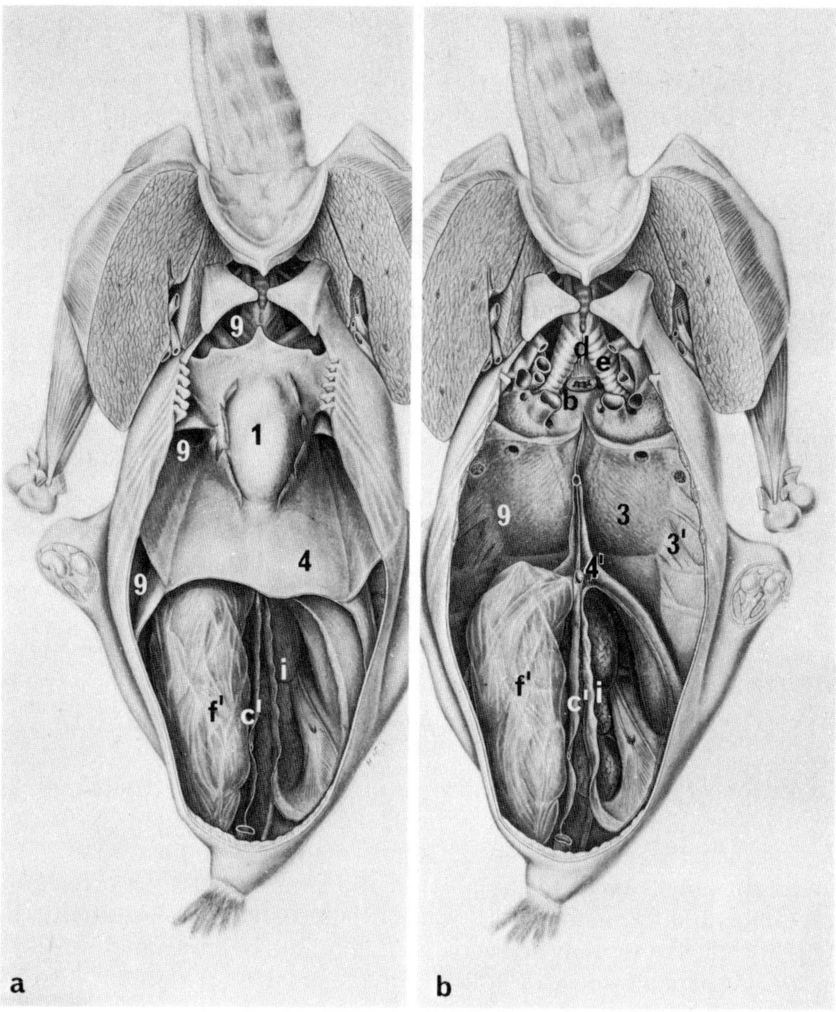

Fig. 2.16. Carrion Crow (*Corvus corone*). Ventral views of the thoracoabdominal cavity. (a) Sternum and ventral abdominal wall removed. The subpulmonary cavities which are ventrally united, surround the pericardial cavity cranially, laterally and ventrally. The pericardial cavity is attached to the sternum cranially and laterally by broad bands of connective tissue. It continues caudally and laterally into the oblique septa which are united caudal to the pericardium and dorsal to the sternum. The subpulmonary cavities contain the cranial group of air sacs the walls of which can be seen. In the peritoneal cavity the viscera have been removed so that the cut margin of the dorsal mesentery is visible. On both sides of the mesentery lie the abdominal air sacs, the left of which has been cut off along its dorsal line of attachment. (b) As in (a) except that the pericardial cavity and the united oblique septa have been cut off close to their dorsal origin. Thus, the dorsal part of the subpulmonary cavities are opened and the horizontal septa and the hilus regions of the lungs are visible. The trachea, oesophagus, blood vessels and nerves pass through the parts of the subpulmonary cavities lying cranial to the pulmonary hilus. The horizontal septa contain the air sac ostia and the costoseptal muscles. In the remaining dorsal parts of the oblique septa the muscle of the septa can be seen. Between the two oblique septa lies the dorsal mesentery. In the peritoneal cavity are the abdominal air sacs, the left one being cut off at its attachment to the dorsal abdominal wall and the surface of the kidney. 1, pericardial cavity; 3, horizontal septum; 3′, costoseptal muscles; 4, oblique septum; 4′, muscle of the oblique septum; 9, subpulmonary cavity; b, oesophagus; c′, dorsal mesentery; d, trachea; e, primary bronchus; f′, abdominal air sac; i, kidney. From Duncker (1971).

part of the neck, its precise point of termination, however, varying with the species. The cranial part of the subpulmonary cavity is ventrally continuous with that of the opposite side, the part dorsal to the trachea being separated from the other cavity by a median septum containing the oesophagus. The cranial part of the subpulmonary cavity contains, in addition to the cervical and clavicular air sacs, the trachea and primary bronchi, the oesophagus, the major blood vessels and nerves, and the sternotracheal muscles. Except for the sternotracheal muscles, all structures are held to the walls of the cavity by thin septa formed by the air sac walls. The air sac walls are always attached to the walls of the subpulmonary cavity, except for those sites where they surround passing structures or are adjoined to one another, forming delicate membranes separating the individual air sacs.

The oblique septum consists in all birds of coarse connective tissue and in most species is supported at its caudal attachment to the vertebral column by smooth muscle (*m. septi obliqui*) which extends up to one third of the width of the septum (Figs 2.14, 17). The muscle is richly innervated by adrenergic fibres, and by varying the tension in the oblique septum may influence the volume changes of the thoracic air sacs.

D. Dorsal and ventral hepatic peritoneal cavities

The peritoneal cavity is divided into an unpaired intestinal cavity and paired hepatic cavities, the intestinal cavity being separated from the hepatic cavities by the posthepatic septum (Figs 2.14, 15, 17, 18). The posthepatic septum originates dorsally, caudal to the hilus of the lung between the origins of the oblique septa, and forms the dorsal wall of the hepatic peritoneal cavity. From its dorsal origin the septum descends gradually in a ventrocaudal direction to reach the ventral body wall a variable distance caudal to the sternum. The lateral walls of the hepatic peritoneal cavity are formed by the oblique septa and the parietal peritoneum of the lateral body wall. Ventrally the cavity is bounded by the parietal peritoneum lining the sternum and the ventral abdominal wall. In passerine species, however, in which the oblique septa have united dorsal to the sternum, the floor of the cavity is formed by the oblique septa (Fig. 2.18).

The hepatic peritoneal cavity is occupied mainly by the liver and in its craniodorsal part also by the oesophagus. The gallbladder, if present, lies caudal to the posthepatic septum in the intestinal peritoneal cavity although it is always attached to the posthepatic septum and via the ventral mesentery to the liver. In the hepatic peritoneal cavity the liver is supported by the dorsal and ventral parts of the ventral mesentery which divide the cavity into right and left portions (Figs 2.17, 18). The oesophagus lies in the dorsocranial part of the cavity, incorporated into the dorsal mesentery. Further caudally it passes through the posthepatic septum, to which it is attached either directly or via the ventral mesentery. The ventral mesentery of the liver is continued cranially as the hepatopericardial ligament. In the craniodorsal part of the hepatic cavity the visceral peritoneum of each liver lobe is connected to the oblique septum by the hepatic ligament (Figs 2.14, 15). The hepatic ligament totally separates off a small dorsal recess, the dorsal hepatic peritoneal cavity, from the large ventral hepatic peritoneal cavity. The dorsal cavity is restricted to the cranial part of the hepatic peritoneal cavity. The right dorsal cavity is caudally completely separated from the intestinal peritoneal cavity by the posthepatic septum. The left dorsal hepatic cavity in *Gallus* and *Anas* communicates caudally and dorsally with the intestinal cavity. This connection between the two cavities is only covered by the wall of the abdominal air sac. The caudal extension of the ventral hepatic peritoneal cavity reaches far more caudally in most birds than the caudal margin of the liver, which regularly lies at the caudal end of the sternum.

The liver in its hepatic coelomic cavity is covered by visceral peritoneum. The lateral walls of the hepatic cavities are covered by parietal peritoneum. The parietal and visceral peritoneal membranes are continuous at the dorsal and ventral attachments of the ventral mesentery and at the lateral attachments of the hepatic ligaments. Embryologically the hepatic peritoneal cavities develop in the form of recesses in the primary septum transversum between the liver and the oblique septum, between the liver and the sternum, and between the liver and the posthepatic septum. These recesses separate the lobes of the liver from the walls of the peritoneal cavity. However, in all birds these recesses do not separate a small cranial connection of the liver to the right oblique septum, the *area nuda* of the liver, which is a derivative of the ventral part of the primary septum transversum. In some birds, such as rheas (Rheidae) and penguins (Spheniscidae), the connection is broad and contains the caudal vena cava. In other birds like *Gallus* and *Anas* it is very narrow and the caudal vena cava is almost totally incorporated into the liver and separated from the oblique septum by a recess. In the caudal part of the *area nuda* the hepatic veins enter the caudal vena cava, whilst in the cranial part, which continues onto the dorsolateral part of the pericardium, the caudal vena cava penetrates the pericardial wall.

E. *Intestinal peritoneal cavity*

The intestinal peritoneal cavity contains the gastrointestinal tract from the proventriculus to the rectum, the gonads, the spleen and in all birds except the kiwis (Apterygidae) the abdominal air sacs (Figs 2.17, 18). The intestinal peritoneal cavity is dorsally lined by the parietal peritoneum of the dorsal body wall. Here, retroperitoneally lie the kidneys which are attached to the pelvic girdle ventrally (Figs 2.15, 16, 17, 18). The dorsal wall of the cavity begins cranially at the caudal margin of the lung which is covered by the transverse and perpendicularly oriented part of the unsplit oblique and horizontal septa (Fig. 2.17). Attaching to the unsplit septum are the kidney and adrenal gland. Laterally this septum is penetrated by the primary bronchus to form the ostium into the abdominal air sac (Figs 2.17, 18). The lateral wall of the intestinal peritoneal cavity is formed cranially by the oblique septum and caudally by the parietal peritoneum of the lateral body wall (Figs 2.13, 14, 15, 16, 17, 18). The ventral wall is cranially formed by the caudoventrally directed posthepatic septum and caudally by the parietal peritoneum of the ventral body wall.

The entire gastrointestinal tract from the proventriculus to the rectum is connected in the midline to the dorsal abdominal wall by the dorsal mesentery

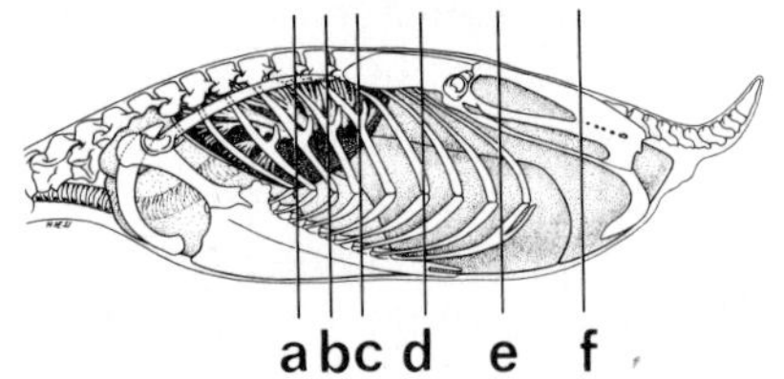
a b c d e f

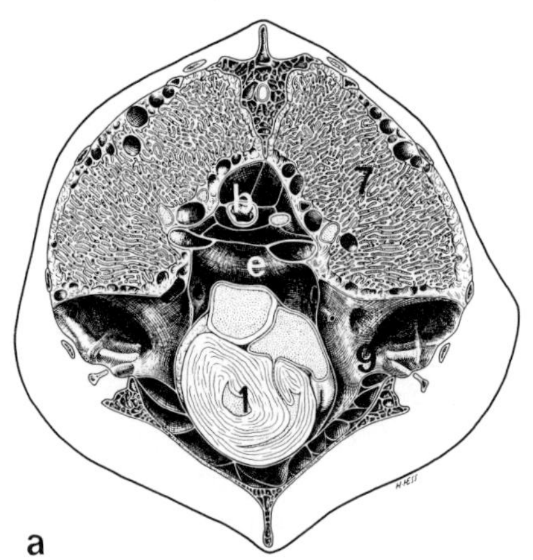
b
7
e
9
1
a

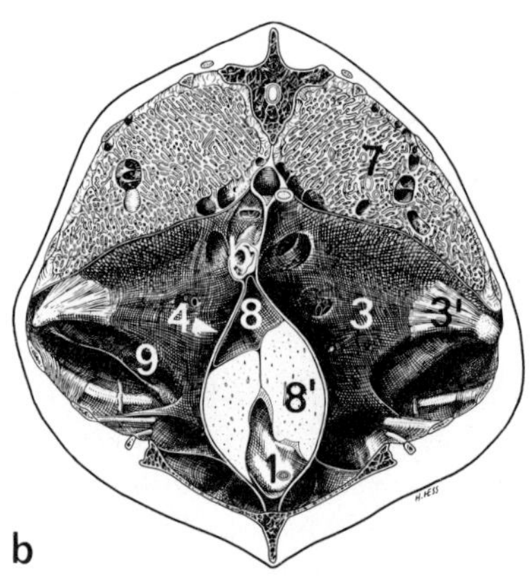
7
b
4
8
3
3'
9
8'
1
b

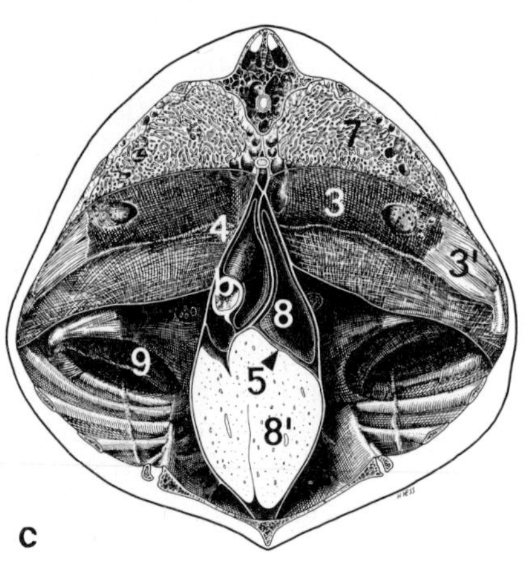
7
3
4
3'
8
9
5
8'
c

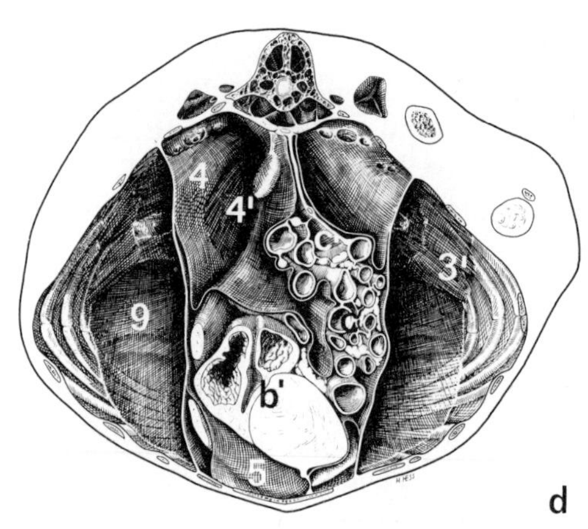
4
4'
c
3'
9
b'
5
d

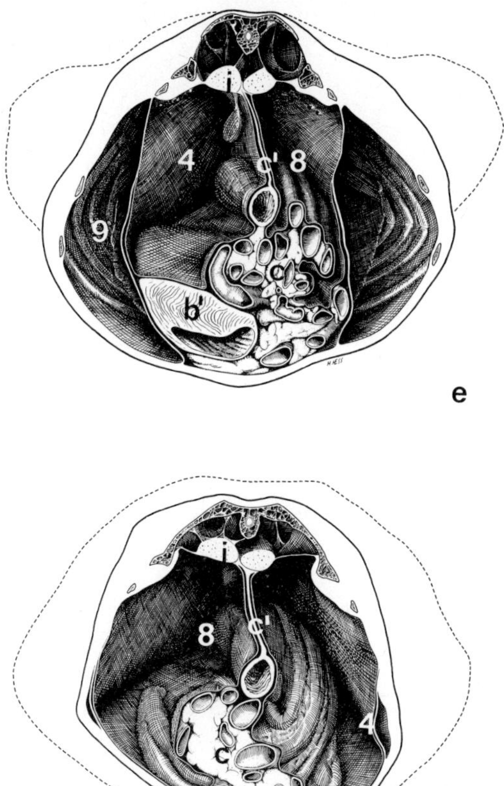

f

Fig. 2.17. Mute Swan (*Cygnus olor*). Caudal views of transverse sections through the trunk. The positions of the sections relative to the skeleton, lungs and air sacs are indicated on the lateral view. (a) Section through the region of the pulmonary hilus, cutting the lungs and pleural cavities, the heart and pericardial cavity, and the united subpulmonary cavities. (b) Section just caudal to the pericardial cavity through the caudal part of the lung, and through the cranial parts of the hepatic peritoneal cavities lying between the oblique septa. The horizontal septa, air sac ostia and costoseptal muscles can be seen. (c) Section through the caudal parts of the lungs, the subpulmonary cavities with the cranial and caudal thoracic air sacs, and the cranial parts of the intestinal peritoneal cavity. This region of the intestinal peritoneal cavity contains the proventriculus and the cranial parts of the abdominal air sacs and is separated from the hepatic peritoneal cavities by the posthepatic septum. (d) Section caudal to the lung, through the subpulmonary and intestinal peritoneal cavities and the caudal tip of the left hepatic peritoneal cavity. Dorsally, the unsplit horizontal and oblique septa are penetrated by the ostia for the abdominal air sacs. The air sacs fill the dorsal region of the intestinal cavity on either side of the intestines, which are suspended by the dorsal mesentery. Within and ventral to the synsacrum, pneumatic spaces are visible. (e) Section caudal to the hip joints through the caudal parts of the subpulmonary cavities. The fusion of the walls of the abdominal air sacs with the parietal peritoneum of the dorsal body wall and with the oblique septa is shown. Above the kidneys, large pneumatic spaces are developed. (f) Section caudal to the subpulmonary cavities. The abdominal air sacs fill most of the intestinal peritoneal cavity. 1, heart and pericardial cavity; 3, horizontal septum; 3′, costoseptal muscles; 4, oblique septum; 4′, muscle of oblique septum; 5, posthepatic septum; 7, pleural cavity with lung; 8, intestinal peritoneal cavity with abdominal air sac; 8′, hepatic peritoneal cavity with liver; 9, subpulmonary cavity; b, oesophagus or proventriculus; b′, gizzard; c, intestinal tract; c′, dorsal mesentery; e, primary bronchus; i, kidney. From Duncker (1971).

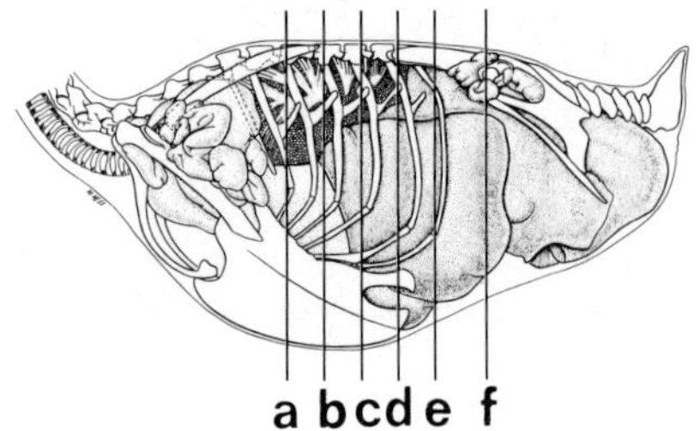
a b c d e f

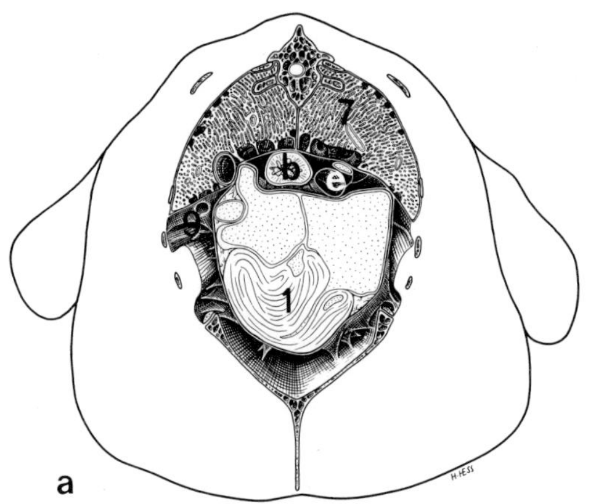
7
b
e
9
1
H.HESS
a

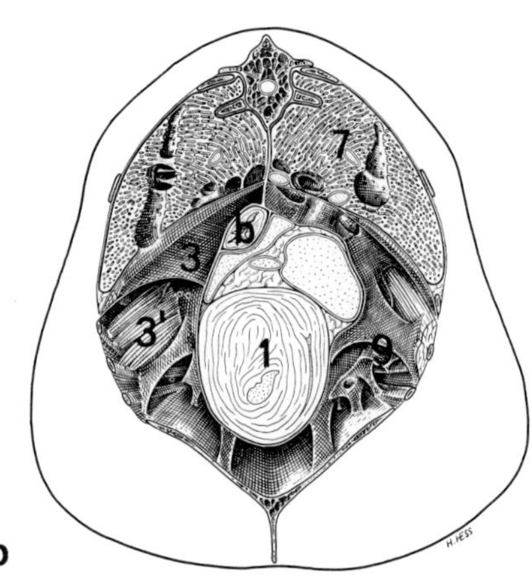
7
b
3
3'
1
9
H.HESS
b

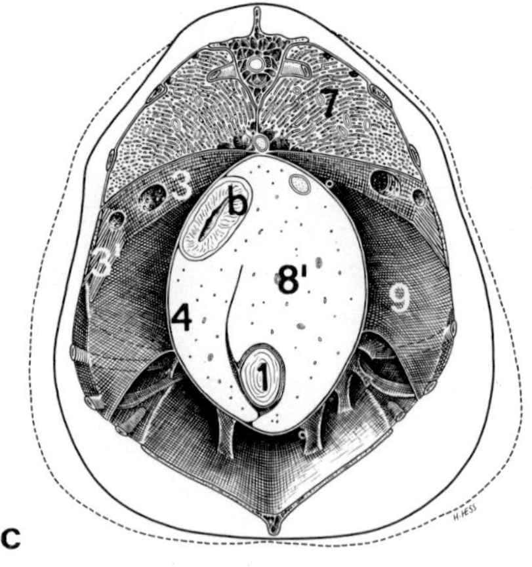
7
3
b
3'
8'
4
9
1
H.HESS
c

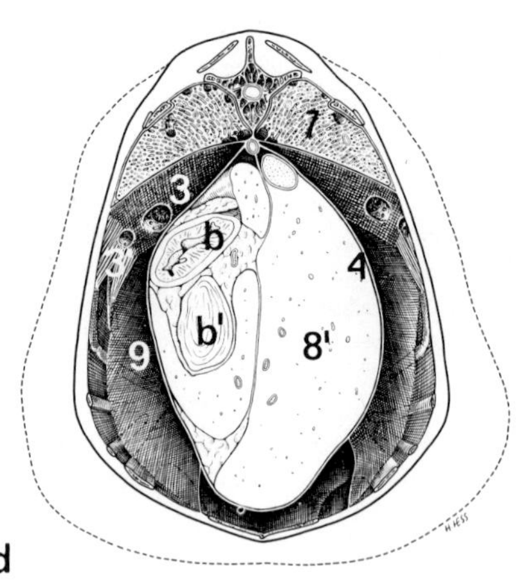
7
3
b
3'
4
b'
8'
9
H.HESS
d

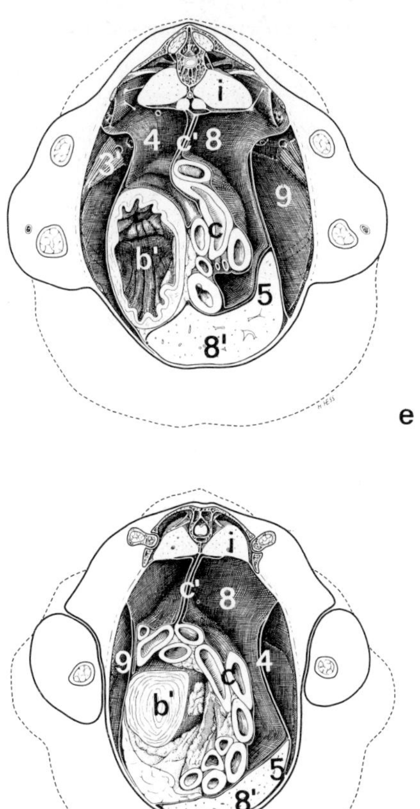

Fig. 2.18. Carrion Crow (*Corvus corone*). Caudal views of transverse sections through the trunk. The positions of the sections relative to the skeleton, lungs and air sacs are indicated on the lateral view. (a) Section through the region of the pulmonary hilus, cutting the lungs and pleural cavities, the heart and pericardial cavity and the ventrally united subpulmonary cavities. (b) Section through the middle of the pleural cavities, the pericardial cavity, and the cranial tips of the hepatic peritoneal cavities which are situated between the origins of the oblique septa. (c) Section through the caudal parts of the pleural cavities, the hepatic peritoneal cavities and the caudal tip of the pericardial cavity. The ventrally united oblique septa enclose the liver and proventriculus. The oblique septa are attached to the sternum by fibrous tissue. The horizontal septa contain air sac ostia and the costoseptal muscles. (d) Section through the most caudal part of the pleural cavities and through the hepatic and intestinal peritoneal cavities which are surrounded by the oblique septa. (e) Section caudal to the pleural cavity and to the caudal margin of the sternum, through the subpulmonary, intestinal peritoneal and hepatic peritoneal cavities. The oblique septa are attached caudal to the sternum to the ventrolateral body wall. Between the septa are the gizzard and intestines whilst dorsally lie the abdominal air sacs. The liver ventrally is separated by the posthepatic septum from the intestinal peritoneal cavity. Dorsal to the peritoneal cavity and the kidneys the pneumatic spaces within the synsacrum are visible. (f) Section through the caudal parts of the subpulmonary cavities and the hepatic peritoneal cavity. Dorsal to the liver lie the gizzard and intestines. The dorsal part of the intestinal peritoneal cavity is filled by the abdominal air sacs. 1, pericardial cavity with heart; 3, horizontal septum; 3′, costoseptal muscles; 4, oblique septum; 5, posthepatic septum; 7, pleural cavity with lung; 8, peritoneal cavity with abdominal air sac; 8′, hepatic peritoneal cavity with liver; 9, subpulmonary cavity; b, oesophagus or proventriculus; b′, gizzard; c, intestinal tract; c′, dorsal mesentery; e, primary bronchus; i, kidney. From Duncker (1971).

(Figs 2.14, 15, 16, 17, 18). The ventral mesentery is only present between the levels of the proventriculus and the proximal part of the duodenum and is connected near the midline to the dorsal side of the posthepatic septum. On the ventral side of the posthepatic septum the ventral mesentery continues into the dorsal part of the liver. It connects the liver and gallbladder to the duodenum and forms at its caudal end the hepatoduodenal ligament (*lig. hepatoduodenale*). Caudal to the hepatoduodenal ligament and the proximal part of the duodenum, a ventral mesentery is absent. Consequently the intestinal peritoneal cavity is single, its dorsal region only being divided into right and left parts by the dorsal mesentery (*mesoduodenum*, *mesojejunum*, *mesoileum* and *mesorectum*). The cranial region of the cavity is formed by a right and left recess, both of which are totally separated from one another by the dorsal and ventral mesenteries supporting the proventriculus, gizzard and proximal duodenum.

The lateral wall of the intestinal peritoneal cavity is formed to varying extents in different species by the oblique septum, by the parietal peritoneum of the body wall and, in birds with a large caudolateral extension of the posthepatic septum as in *Gallus*, by the lateral part of the posthepatic septum. The stomach, or in a number of species, its caudal part, the gizzard, is suspended by an attachment to the left lateral peritoneal wall. In both male and female birds the gonads protrude ventrally into the intestinal peritoneal cavity on either side of the dorsal mesentery, the left ovary suspended by the mesovarium and the right and left testes by the mesorchia. The left oviduct is suspended by the mesosalpinx (*mesoviductus dorsalis*) which extends cranially into the dorsal part of the unsplit horizontal and oblique septa, demonstrating even in adult birds the dorsal closure of the postpulmonary septum by the cranial nephric fold. For details of the mesenterial ligaments of the reproductive organs in *Gallus* see King (1975).

Except in kiwis (Apterygidae), on each side of the body the abdominal air sac penetrates the dorsal perpendicular unsplit portion of the horizontal and oblique septa and is suspended in the intestinal peritoneal cavity by parietal peritoneum. The wall of each abdominal air sac therefore is formed by air sac epithelium, a thin connective tissue layer and serosal mesothelium. However, the dorsal part of the air sac and its peritoneal covering fuses with the parietal peritoneum lining the dorsal and dorsolateral walls of the intestinal peritoneal cavity to an extent which is highly variable between different taxonomic groups. In penguins (Spheniscidae), rheas (Rheidae) (Fig. 2.14) and the ostrich (*Struthio camelus*) there is only a very small region of attachment near the ostium of the air sac. However, in other birds the area of attachment extends more caudally covering the ventral surface of the kidneys laterally and reaching medially often to the level of the dorsal mesentery (Fig. 2.16). Thus the area of attachment also involves the testis in a large number of species, so that it bulges considerably into the abdominal air sac (Fig. 2.15). The left ovary is never enclosed by the air sac wall, leaving a free passage for the ovulated eggs to the abdominal opening of the oviduct.

Laterally the abdominal air sac fuses to a varying extent with the dorsal parts of the oblique septum (Figs 2.17, 18), often the area of fusion continuing on to the gizzard which is attached to the lateral wall of the intestinal peritoneal cavity. The lateral fusion of the abdominal air sac ends cranial to the point where the oblique septum attaches to the lateral body wall. The abdominal air sacs fill the dorsal part of the intestinal peritoneal cavity dorsal and lateral to the intestines to an extent which depends on the size of the body and on the state of respiration (Figs 2.15, 16, 17, 18).

References

Beddard, F. E. (1896). On the oblique septa ("diaphragm" of Owen) in the passerines and in some other birds, *Proc. zool. Soc. Lond.* **1896**, 225–231.

Bittner, H. (1925). Beitrag zur topographischen Anatomie der Eingeweide des Huhnes, *Z. Morph. Ökol. Tiere* **2**, 785–793.

Broman, I. (1937). Cölom. *In* "Handbuch der vergleichenden Anatomie der Wirbeltiere" (L. Bolk *et. al.*, Eds), Vol. 3. Urban and Schwarzenberg, Berlin-Wien.

Butler, G. W. (1889). On the subdivision of the body-cavity in lizards, crocodiles and birds, *Proc. zool. Soc. Lond.* **1889**, 452–474.

Campana, A. (1875). "Recherches d'anatomie . . ., Premier memoire: Physiologie de la respiration chez les oiseaux, Anatomie de l'appareil pneumatique-pulmonaire etc. chez le poulet." Masson et Cie, Paris.

Duncker, H.-R. (1971). The lung air sac system of birds, *Ergeb. Anat. Entwickl.-Gesch.* **45**, 1–171.

Duncker, H.-R. (1978a). Coelom-Gliederung der Wirbeltiere – Funktionelle Aspekte, *Verh. Anat. Ges.* **72**, 91–112.

Duncker, H.-R. (1978b). General morphological principles of amniotic lungs. *In* "Proceedings in Life Sciences. Respiratory Functions in Birds, Adult and Embryonic" (J. Piiper, Ed.). Springer Verlag, Heidelberg.

Fedde, M. R., Burger, R. E. and Kitchell, R. L. (1964). Anatomic and electromyographic studies of the costo-pulmonary muscles in the cock, *Poult. Sci.* **43**, 1177–1184.

Goodchild, W. M. (1970). Differentiation of the body cavities and air sacs of *Gallus domesticus* post mortem and their location *in vivo*, *Br. Poult. Sci.* **11**, 209–215.

Goodrich, E. S. (1930). "Studies on the Structure and Development of Vertebrates." Reprinted 1958. Dover Publications, New York.

Huxley, T. H. (1882). On the respiratory organs of *Apteryx*, *Proc. zool. Soc. Lond.* **1882**, 560–569.

Kern, D. (1963). "Die Topographie der Eingeweide der Körperhöhle des Haushuhnes (*Gallus domesticus*) unter besonderer Berücksichtigung der Serosa- und Gekröseverhältnisse". Vet. Med. Diss. Gießen.

King, A. S. (1975). Aves. Urogenital system. *In* "Sisson and Grossman's The Anatomy of the Domestic Animals" (R. Getty, Ed.), Vol. 2. Saunders, Philadelphia.

McLelland, J. and King, A. S. (1970). The gross anatomy of the peritoneal coelomic cavities of *Gallus domesticus*, *Anat. Anz.* **127**, 480–490.

McLelland, J. and King, A. S. (1975). Aves. Coelomic cavities and mesenteries. *In* "Sisson and Grossman's The Anatomy of the Domestic Animals" (R. Getty, Ed.), Vol. 2, Saunders, Philadelphia.

Nelsen, O. E. (1953). The development of the coelomic cavities. *In* "Comparative Embryology of the Vertebrates". The Blakiston Company, New York.

Petit, M. (1933). Péritoine et cavité péritonéale chez les oiseaux, *Rev. vét. J. Méd. vét.* **85**, 376–382.

Poole, M. (1909). The development of the subdivisions of the pleuroperitoneal cavity in birds, *Proc. zool. Soc. Lond.* **1909**, 210–235.

Sappey, P. C. (1847). "Recherches sur l'Appareil Respiratoire des Oiseaux." Germer-Bailliere, Paris.

Soum, J. H. (1896). Recherches physiologiques sur l'appareil respiratoire des oiseaux, *Ann. Univ. Lyon* **28**, 1–126.

3
Digestive system

JOHN McLELLAND

Department of Veterinary Anatomy, University of Edinburgh, Scotland

CONTENTS

I. Oral cavity and pharynx 70
A. General morphology of oral cavity and pharynx 70
B. Tongue 74
C. Oral sacs 80
D. Salivary glands 81
E. Taste buds 84
F. Bill 84
G. Deglutition 89

II. Oesophagus 90
A. Gross morphology of oesophagus 90
B. Crop 92
C. Structure of oesophagus 96
D. Blood supply of oesophagus 97
E. Nerve supply of oesophagus 97
F. The oesophagus in digestion 97
G. Oesophageal sacs 100

III. Stomach 101
A. Definitions. General form of stomach 101
B. The stomach *in situ* 103
C. Proventriculus 103
D. Intermediate zone of stomach 112
E. Gizzard 113
F. Pyloric part of stomach 125
G. Blood supply of stomach 127
H. Nerve supply of stomach 128
I. Stomach motility 129
J. Gastric digestion 131

IV. Intestines 132
A. Composition of intestines 132
B. Intestinal dimensions 132
C. Gross form of intestines 134
D. The intestines *in situ* 141
E. Intestinal folds and villi 143
F. Structure of intestines 146
G. Blood supply of intestines 148
H. Nerve supply of intestines 149
I. Intestinal function 149

V. Pancreas 151
A. Pancreas size 151
B. External appearance of pancreas 151
C. Exocrine pancreas and pancreatic ducts 152
D. Blood supply of pancreas 154
E. Nerve supply of pancreas 155
F. Exocrine pancreatic secretion 155

VI. Liver 156
A. Physical characteristics of liver 156
B. Liver form 156
C. Hepatic blood supply 158
D. Internal structure of liver 159
E. Biliary system 161
F. Bile 164

References 164

I. Oral cavity and pharynx

The anatomy of the oral cavity (*cavitas oralis*) and *pharynx* is so diverse that a complete description in the space available is not possible. Much of the following account therefore, is concerned with birds generally and covers in detail only selected features. An excellent description of this region of the digestive tract is that by Göppert (1903) which forms the basis of the following account. Typical examples of the cavities are shown in Fig. 3.1. Since a soft palate and pharyngeal isthmus are absent, the two cavities form a common chamber. On the basis of the embryology of the visceral arches, the boundary between the cavities has been placed dorsally at the junction of the choana and the pharyngeal cleft, and ventrally at the base of the tongue between the entoglossal and rostral basibranchial bones (McLelland, 1978).

A. General morphology of oral cavity and pharynx

The palate (*palatum*) contains the *choana*, a longitudinal fissure which is somewhat enlarged caudally and through which the oral and nasal cavities communicate (Fig. 3.1a). In a small number of species, e.g. herons (Ardeinae), ducks (Anatidae) and the ratites, the choana is unusually short and restricted to the caudal part of the palate (Fig. 3.1e). The margins of the choana may be thickened as in the Gray Heron (*Ardea cinerea*). The portion of the palate rostral to the choana is relatively longer in long-billed birds than in birds with short bills. On each side of the choana, close to the rod-like palatine bones, the mucosa of the palate frequently forms a longitudinal ridge, the lateral palatine ridge (*ruga palatina lateralis*) (Fig. 3.1a). The part of the palate between the right and left ridges is unsupported by skeleton. In some species a longitudinal median ridge (*ruga palatina mediana*) is present rostral to the choana (Fig. 3.1a). The ridge formation on the palate of seed-eating passerines, e.g. the Fringillidae, Emberizidae, Ploceidae and Estrildidae, is highly complex dividing the palate into a series of arch-like grooves (Ziswiler, 1965). Typically in these birds, there is a longitudinal median ridge which is flanked by longitudinal intermediate ridges (*rugae palatinae intermediales*) and longitudinal lateral ridges, a variable number of secondary and tertiary ridges arising from the sides of the intermediate and lateral ridges. In some groups, e.g. the Emberizidae, the median ridge is expanded caudally into a buttress-like prominence (*torus palatinus*). The region of the palate between the right and left lateral ridges is the median palatine groove (*sulcus palatinus medianus*), that between the lateral ridge and the edge of the bill being the lateral palatine groove (*sulcus palatinus lateralis*). The median groove fails to extend caudally to the level of the choana. The lateral groove contains the lower mandible when the bill is closed, and during the husking process holds the seed. In the Hawfinch (*Coccothraustes coccothraustes*) the palate caudal to the longitudinal ridges has a pair of finely ridged knobs which meet in midline and overlie a similar pair of knobs on the lower mandible (Newton, 1967). The functional significance of the diverse forms of palate in granivorous passerines is discussed below in connection with bill adaptations.

The roof the pharynx contains the pharyngeal cleft (*fissura infundibularis*) (Fig. 3.1a), a median longitudinal fissure which is the common opening of the right and left pharyngotympanic tubes (*tubae pharyngotympanicae*). According to Heidrich (1908), the two pharyngotympanic tubes in *Gallus* open by a single duct (*tuba pharyngotympanica communis*) into the dorsal part of the pharyngotympanic infundibulum (*infundibulum pharyngotympanicum*) between paired infundibular folds (*plicae infundibulares*). The pharyngotympanic infundibulum opens into the pharyngeal cavity (*cavitas pharyngealis*) at the pharyngeal cleft. Caudal to the pharyngeal cleft the roof of the pharynx protrudes somewhat on each side of the midline and makes a sharp boundary with the oesophagus (Fig. 3.1e). Caudally-directed papillae either arranged regularly in rows or scattered apparently at random are characteristic features of the palate (*papillae palatinae*) and the roof of the pharynx (*papillae pharyngeales*) in the majority of species (Fig. 3.1c).

The tongue (*lingua*) (Fig. 3.1b) is attached to the floor of the oral cavity by the lingual frenulum (*frenulum lingualis*) and shows a great diversity of form which is reviewed below. When the bill is shut, the dorsal surface of the tongue (*dorsum linguae*) is applied to the palate and consequently its form generally reciprocates that of the palate. In species such as *Gallus* in which the palate is ridged laterally, the tongue lies within the space enclosed by the ridges. When the bill is shut the tongue commonly closes off the rostral part (*pars rostralis*) of the choana. In a small number of birds the floor of the mouth ventral to the tongue contains one or more sac-like diverticula, the oral sacs (*sacci orales*) which act either to carry food or as display chambers during the breeding season

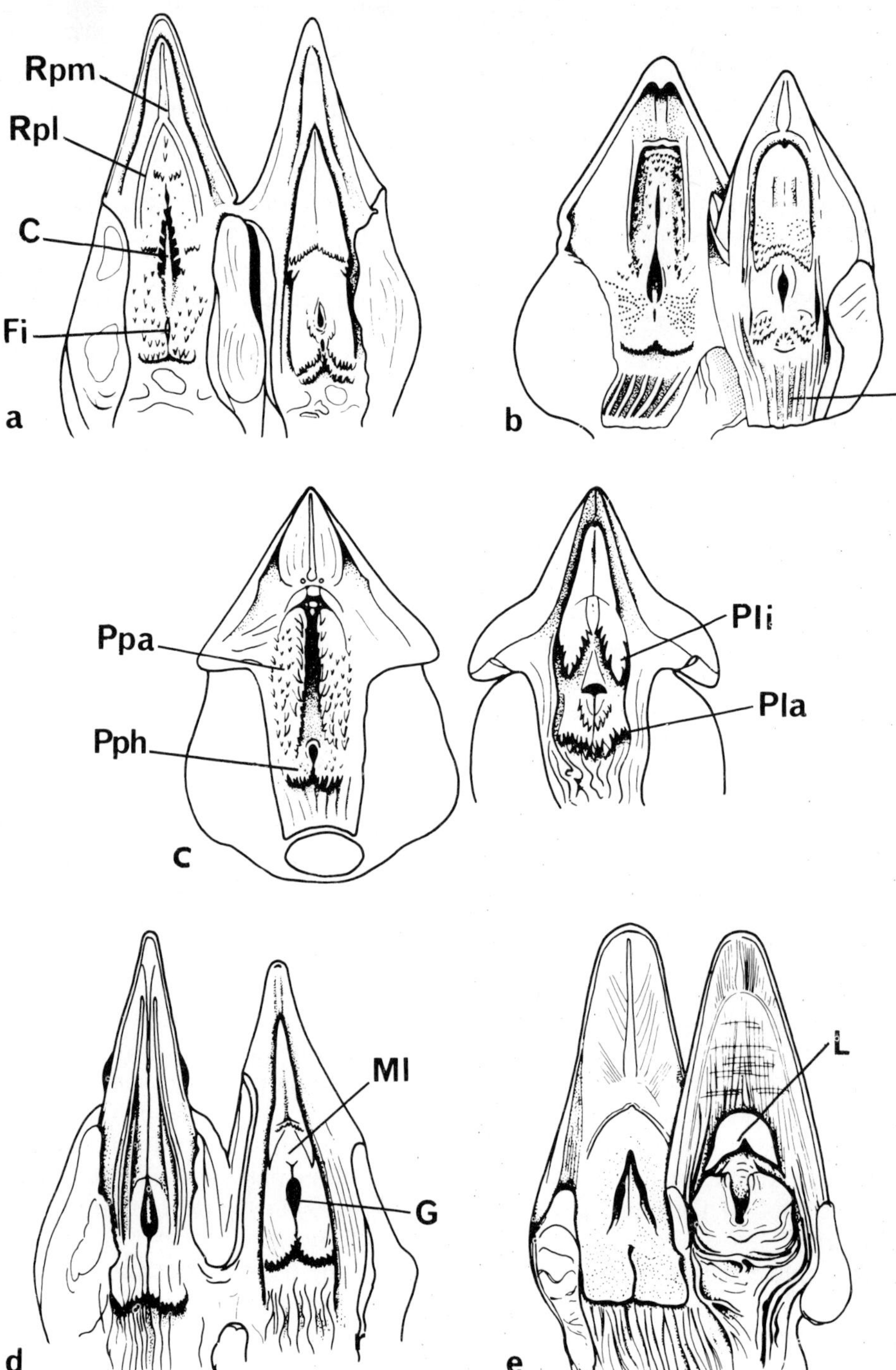

Fig. 3.1. Oral cavity and pharynx. (a) Black Grouse (*Tetrao tetrix*). (b) Common Kestrel (*Falco tinnunculus*). (c) House Sparrow (*Passer domesticus*). (d) *Columba*. (e) Ostrich (*Struthio camelus*). C, choana; Fi, infundibular fissure; G, glottis; L, tongue; Ml, laryngeal mound; O, oesophagus; Pla, laryngeal papillae; Pli, lingual papillae; Ppa, palatine papillae; Pph, pharyngeal papillae; Rpl, lateral palatine ridge; Rpm, median palatine ridge. From Göppert (1903).

(see Oral sacs below). Close behind the tongue lies the laryngeal mound (*mons laryngealis*), the *glottis* being situated on its rostral surface (Fig. 3.1d). In only a few species, e.g. *Ardea*, is there a large interval between the tongue and the larynx, the wide gap in *Ardea* probably being an adaptation for swallowing large portions of food. The glottis usually lies directly ventral to the caudal part of the choana as in the House Sparrow (*Passer domesticus*) and *Columba* (Fig. 3.1c,d) or somewhat caudal to it as in the Common Kestrel (*Falco tinnunculus*) (Fig. 3.1b). When the bill is shut, the larynx may be closed off from the oral cavity rostrally by the close application of the floor of the mouth and the tongue to the palate, and from the oesophagus caudally by the close application of the floor of the caudal part of the pharynx to the roof. Caudally-directed papillae on the tongue and on the floor of the pharynx are a characteristic feature of most species (Fig. 3.1c).

The oral cavity and pharynx are lined by a stratified squamous epithelium which in areas subject to abrasion is keratinized, the extent of the keratinization therefore varying considerably in different species. In *Gallus*, for example, the keratinization appears to be restricted to the caudally-directed papillae and to the apex and ventral surface of the tongue (*apex linguae*, *ventrum linguae*) (Hodges, 1974, pp. 36–38, 44). Olson *et al.* (1974) noted that the stratified squamous epithelium of the roof of the pharynx in *Meleagris* continued into the pharyngotympanic infundibulum for a short distance before changing into a simple columnar epithelium with a brush border. In newly-hatched birds, especially passerines, which are actively fed by the parents, the colour of the mouth lining and the pattern of markings of the lining is quite distinctive. In some species in addition, conspicuous flanges develop on the bill. When the nestling opens its mouth, the appearance of the lining combines with the exaggerated gaping to form a begging display which stimulates the parent to feed. Amongst the species with highly elaborate begging displays are grass finches (*Emberizoides* spp.) and the Bearded Reedling (*Panurus biarmicus*). The distinctive appearance of the oral cavity is transient and disappears when the chick is no longer dependant on the parents for food. In the connective tissue beneath the epithelium in most species is an almost continuous layer of salivary glands (*gll. oris*, *gll. pharyngis*) (see Salivary glands below), which have numerous openings into the oral and pharyngeal cavities. In the roof of the pharynx, in the region of the pharyngeal cleft and the pharyngotympanic infundibulum, are a large number of lymphatic nodules (*lymphonoduli pharyngeales*). Several parts of the oral cavity and pharynx appear to have a rich sensory innervation, especially in some species the bill, palate and tongue. For a detailed account of the distribution of the endings in songbirds, see Krulis (1978). Taste buds (*gemmae gustatoriae*) are relatively few in number and are restricted to certain well-defined regions (see Taste buds below). The upper and lower margins of the oral cavity are formed by the hard keratinized beak or bill (*rostrum*) (see Bill below) which functionally replaces the lips and teeth. The arteries of the oral cavity and pharynx in *Gallus* include the pterygopharyngeal and palatine branches of the maxillary artery and the descending oesophageal, laryngeal, lingual and sublingual branches of the mandibular artery. The veins are tributaries of the rostral cephalic vein and the interjugular anastomosis (McLelland, 1975).

B. Tongue

The tongue (*lingua*) (Fig. 3.2) is a highly diverse organ showing considerable variability in its size, form and structure much of which can be closely related to the feeding habit. The following brief account is based mainly on the profusely illustrated descriptions by Gardner (1926, 1927) and is concerned mainly with the general appearance of the tongue and the way it relates to function, rather than with specific detail. Harrison (1964) divided the adaptations of the tongue into three main categories; adaptations for collecting food, adaptations for manipulating food and adaptations for swallowing. In any species it is usually possible to identify one of these adaptations as being primarily responsible for the form of the tongue, although in most instances several adaptations exist side by side in the one organ.

1. Adaptations of tongue for collecting food

Tongues in this category are all protrusible organs and function primarily as probes, spears, brushes and capillary tubes.

(a) The tongue of woodpeckers (Picidae) is typically a long, narrow, rounded mobile organ usually capable of being thrust out of the mouth some considerable distance to procure insects and the sap of trees (Fig. 3.2a). The protrusibility is due to an exceptionally well-developed hyobranchial apparatus (*apparatus hyobranchialis*). The longest tongues occur in the insectivorous species, e.g. the Green Woodpecker (*Picus viridis*) and the Common Flicker (*Colaptes auratus*) in which the hardened apical portion is usually armed laterally with barbs, and the dorsal surface roughened by numerous minute, caudally-directed, spinous papillae. During protrusion, the tongue is coated with mucous which is secreted by the enlarged mandibular salivary gland, and firmly held to the surface of the tongue by the dorsal papillae. In all insectivorous woodpeckers the tongue functions either as a "lime twig" for catching ants or as a spear for impaling larger insects (Steinbacher, 1964). A much shorter form of tongue is characteristic of the sapsucker woodpeckers, e.g. Williamson's Sapsucker (*Sphyrapicus thyroideus*), in which the apical portion is bordered by fine bristles and which is used as a brush to extract sap by capillary traction. For details in individual species consult Lucas (1896, 1897), Leiber (1907), Gardner, (1926, 1927), Scharnke (1931a), Steinbacher (1934, 1935, 1941, 1955, 1957, 1964) and Beeker (1953).

A functionally similar form of tongue to that of woodpeckers is the probe-like organ of some passerine species. In titmice (Paridae) the double rostral extremity of the hyobranchial apparatus combines with two lateral projections to give the apical portion of the tongue the appearance of a four-tined pitchfork. Similarly in nuthatches (Sittidae) the apex of the tongue is formed by 6 or 7 tangled tips which are used to extract insects and their eggs from the crevices of tree bark (Fig. 3.2h).

(b) The tongues of most flower frequenting species, i.e. flowerpeckers (Dicaeidae), sunbirds (Nectariniidae), honeyeaters (Meliphagidae), honey-creepers (Drepanididae and Emberizidae), white-eyes (Zosteropidae), hummingbirds (Trochilidae), nectar-feeding warblers (Parulidae) and some orioles (Icteridae) are usually highly protrusible organs capable of being thrust rapidly in and out of the corolla of a flower to obtain pollen, nectar and small insects (Fig. 3.2c,d,e,f,m). In relation to the diverse nature of the diet, the rostral portion of the tongue exhibits many specializations. These include curling, fraying and splitting in various combinations to produce tongues of widely different patterns ranging from the split tongue with little curling and fraying characteristic of the Dicaeidae, to the highly adapted, curled, frayed and tubular tongues of the Meliphagidae. A basically similar adaptation of the tongue also occurs in flower frequenting lories (Trichoglossidae) in which the apex of the tongue is curled and carries a stiff brush of several hundred bristles. All tongues of this type function mainly as a brush for licking or as a capillary tube for sucking up fluid. For details in individual species consult Lucas (1896, 1897), Gadow (1883, 1891a, pp. 666–667), Beddard (1891), Gardner (1926, 1927), Scharnke (1930, 1931a,b, 1932, 1933), Stadtmüller (1938), Steinbacher (1951) and Weymouth *et al.* (1964).

2. Adaptations of tongue for manipulating food

Tongues in this category are non-protrusible structures which generally closely conform to the shape of the mandible and function basically to hold and manipulate food.

(a) The tongues of many piscivorous species are adapted to hold the slippery prey by means of numerous stiff, sharp, caudally-directed papillae. These papillae may be scattered over the entire surface of the tongue as in penguins (Spheniscidae), restricted to the lateral margins as in shearwaters and fulmars (Procellariidae) (Fig. 3.2k,q), concentrated in a caudal patch as in loons (Gaviidae) (Fig. 3.2i), or distributed in a double row on the dorsal surface as in Mergansers (Anatidae) (Fig. 3.2b).

(b) The relatively thick, soft tongue of birds of prey including owls (Strigidae), is a rasp-like instrument in which the rostral portion is often exceptionally hard and rough (Fig. 3.2o) and sometimes is curled or trough-like as in eagles and vultures (Accipitridae) (Fig. 3.2r).

(c) The tongues of most ducks, geese and swans (Anatidae) show a common pattern which is an adaptation, in conjunction with the lamellae of the bill, either for straining minute food particles from water or for tearing grass and weeds (Fig. 3.2l). Basically the tongue is a thick, fleshy rectangular organ, the rostral portion forming a scoop-like process and the lateral margins carrying a double row of overlapping bristles which are interspersed caudally with about 3–11 tooth-like processes. These bristles and processes interdigitate with the lamellae of the bill. On the dorsal surface of the tongue is a shallow median groove (*sulcus lingualis*) the margins of which caudally bear rough tooth-like projections. At the root of the tongue is a fleshy eminence or cushion (*torus linguae*).

In straining ducks the tongue is primarily a filtering organ which is first depressed to allow water to fill the dorsal groove and then pressed against the palate causing the water to be ejected laterally between the bristles which retain

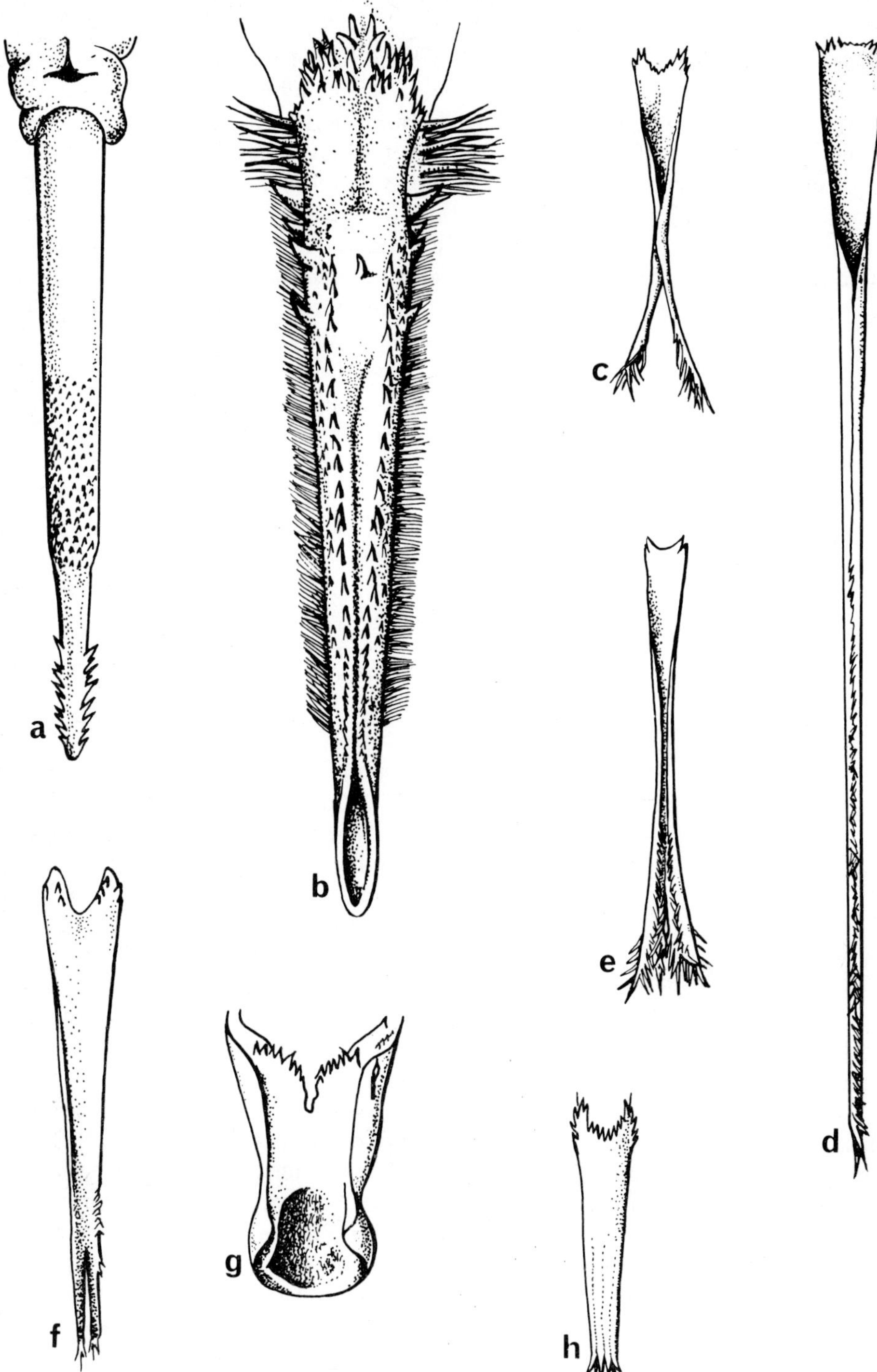

Fig. 3.2. Tongues. (a) White-headed Woodpecker (*Picoides albolarvatus*). (b) Red-breasted Merganser (*Mergus serrator*). (c) Bananaquit (*Coereba flaveola*). (d) *Hemignathus* sp. (e) *Myzomela* sp. (f) Troupial (*Icterus icterus*). (g) Golden-capped Parakeet (*Aratinga auricapilla*). (h) White-breasted Nuthatch (*Sitta carolinensis*). (i) Common Loon (Gavia

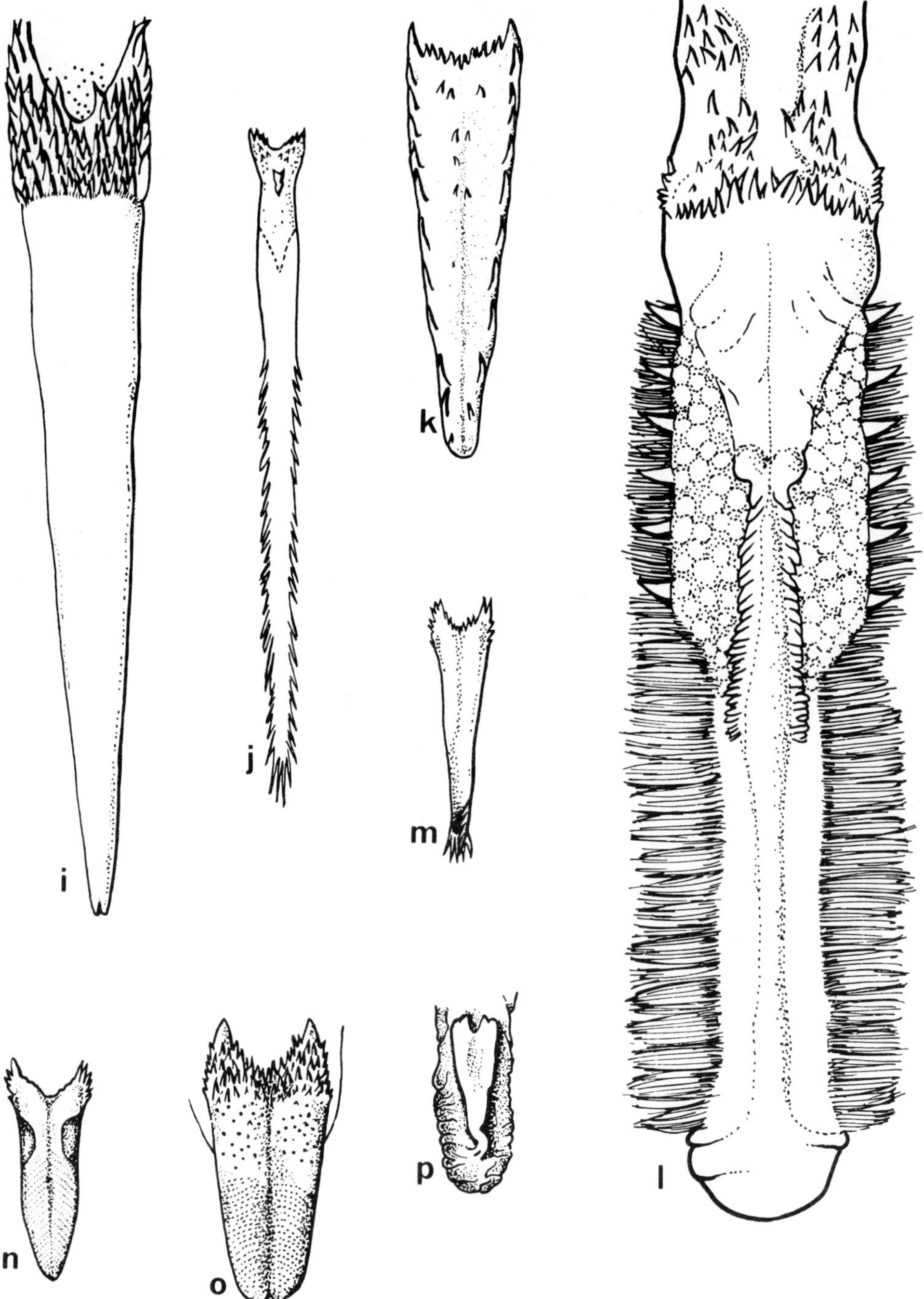

immer). (j) Collared Aracari (*Pteroglossus torquatus*). (k) Sooty Shearwater (*Puffinus griseus*). (l) Cinnamon Teal (*Anas cyanoptera*). (m) Cape May Warbler (*Dendroica tigrina*). (n) Cassin's Finch (*Carpodacus cassinii*). (o) American Kestrel (*Falco sparverius*). (p) Anhinga (*Anhinga anhinga*). (q) Northern Fulmar (*Fulmarus glacialis*). (r) Bearded Vulture (*Gypaetus barbatus*). (s) American Robin (*Turdus migratorius*). (t) Western Gull (*Larus occidentalis*). Scale approximately ×2·5, except for (i), (q) and (t) which are approximately ×1·5. From Gardner (1926).

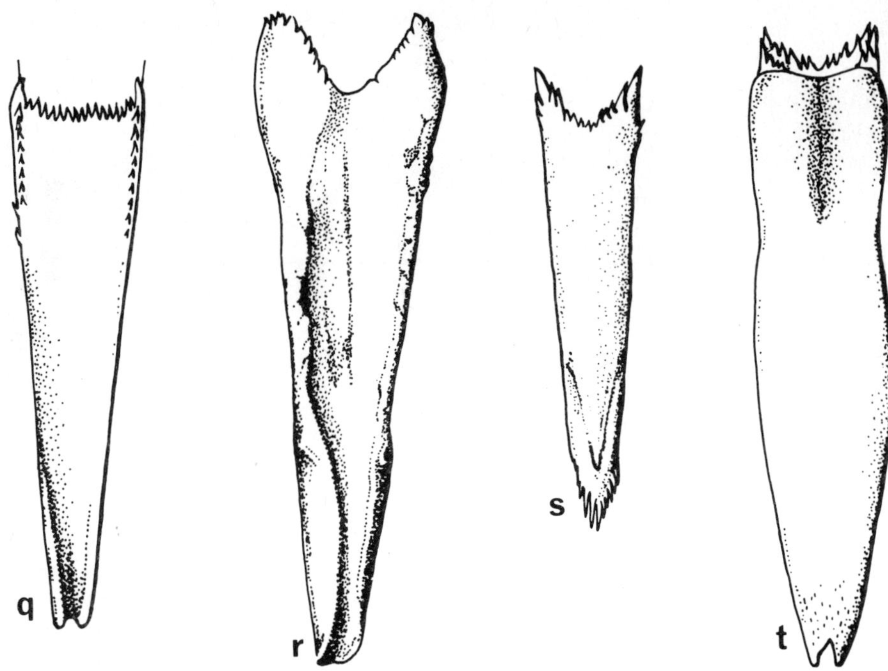

Fig. 3.2. (*continued*)

the solid food particles. The mechanics of feeding in the Mallard (*Anas platyrhynchos*) have been intensively investigated by Zweers *et al.* (1977). These workers compared the straining action of the feeding apparatus in this species to the action of a suction pump. Within the oral cavity are two pistons, a rostral piston formed by the so-called right and left lingual bulges, i.e. the parts of the tongue immediately caudal to the scoop-like apex, and a caudal piston formed by the lingual cushion. The rostral and caudal pistons retract and protract simultaneously. During the straining cycle the lingual bulges may be elevated or depressed and therefore act as a piston with a closed or open valve. The lingual cushion, however, during straining always lies against the roof of the mouth and therefore acts as a piston with a closed valve. In the first stages of straining, elevation of the rostral piston against the palate divides the oral cavity into a rostral chamber and right and left caudal chambers, the caudal chambers being closed caudally by the lingual cushion. Simultaneously the bill opens and the tongue retracts. This causes water to be sucked into the rostral chamber, whilst water already present in the caudal chamber is expelled via the corners of the mouth where filtering takes place. The straining cycle ends with depression of the lingual bulges, protraction of the tongue and closure of the bill.

As a cropping organ in geese and swans, part of the tongue is converted into a coarse cutting margin which functions to hold and tear the vegetation. In the Whooper Swan (*Cygnus cygnus*) this is achieved by modification of the rounded processes at the edge of the dorsal groove to form a row of sharp spines, whereas in the Canada Goose (*Branta canadensis*) it is the lateral bristles that are altered and joined together to form a row of tooth-like projections.

(d) Seed and nut-eaters tend to have short, thick, fleshy tongues which are manipulated rather like a finger. In parrots (Psittacidae) the tongue is broadest at the apex and may be flat, grooved or tubular (Fig. 3.2g). In finches (Fringillidae) the tongue is cylindrical and sloping, the latter being an adaptation which allows the seeds to be rolled in the husking process (Fig. 3.2n).

3. *Adaptations of tongue for swallowing*

(a) Caudally directed papillae (*papillae linguales*) are a feature of the tongue of most species including those with major adaptations for procuring food or eating, and serve to assist swallowing. Usually these papillae are distributed in a transverse row at the root of the tongue but they may also be scattered along the lateral margins and on the dorsal surface. The caudally-situated papillae according to Gardner (1926, 1927) are the dominant adaptive feature of the slender, lanceolate tongue characteristic of a large number of omnivorous passerines, e.g. warblers (Muscicapidae) (Fig. 3.2s), vireos (Vireonidae), crows (Corvidae), wrens (Troglodytidae) and drongos (Dicruridae) which show little adaptation to a particular type of food apart from some splitting, fraying and curling at the apex. An essentially similar form of tongue occurs in galliform species. The basally-situated papillae help both to propel the food caudally and to prevent its regurgitation.

(b) In many birds the tongue is a rudimentary organ, often as an adaptation allowing bulky food to be swallowed whole and quickly. Amongst the species with a reduced form of tongue are *Pelecanus*, *Sula*, *Phalacrocorax* and *Anhinga* (Fig. 3.2p), the ratites and tinamiforms, some sphenisciforms and tubinarine species, *Numenius*, and the Ciconiidae, Threskiornithidae, Bucerotidae, Upupidae, Alcedinidae and Caprimulgidae.

Not included in this brief survey are the tongues of species such as gulls (Laridae) (Fig. 3.2t), plovers (Charadriidae) and rails (Rallidae) which do not appear to be adapted in any special way to the feeding habit. Possibly in many birds the primary role of the tongue is not directly concerned with picking up food or its subsequent manipulation but is instead as an organ of touch which is used to seek out food. Certainly nerve endings (Herbst corpuscles) have frequently been reported in the tongues of many species. Also omitted from the account are the tongues of all those birds such as the fruit-eating trogons (Trogonidae), motmots (Momotidae) and toucans (Ramphastidae) (Fig. 3.2j) which have a highly complex form but for which no explanation can be provided in terms of diet.

For a description of the bones (*apparatus hyobranchialis*) and musculature (*musculi linguales*) of the tongue in *Gallus* refer to McLelland (1975). Amongst the other common constituents are the lingual group of salivary glands, and in some species with relatively thick tongues, e.g. parrots (Psittacidae) and ducks (Anatidae), there are also quantities of fat and cavernous vascular tissue (Stadtmüller, 1938). A detailed account of the structure of the tongue in domestic birds is provided by Preuss *et al.* (1969).

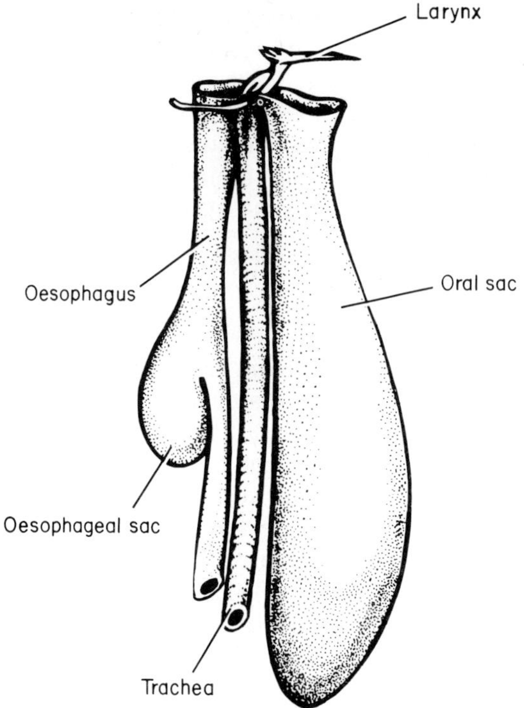

Fig. 3.3. Lateral view of the oral and oesophageal sacs in the adult male Great Bustard (*Otis tarda*). From Garrod (1874a).

C. *Oral sacs*

In a small number of species the oral cavity contains one or more sac-like diverticula, the oral sacs (*sacci orales*), which are usually concerned with carrying food or acting as display chambers during the breeding season.

Amongst the most organized forms of food-carrying sacs are those which have been described ventral to the tongue in the Rosy Finch (*Leucosticte arctoa*) by Miller (1941), in the Pine Grosbeak (*Pinicola enucleator*) by French (1954), and in the nutcrackers, *Nucifraga columbiana* and *Nucifraga caryocatactes* by Bock *et al.* (1973). In both *Leucosticte* and *Pinicola* the sacs are paired structures which are restricted to mature birds during the breeding season. In *Nucifraga* species, however, the sac is unpaired and is fully developed throughout the year in both mature and immature birds. Bock *et al.* (1973) have provided a very complete account of the sac in Clark's Nutcracker (*Nucifraga columbiana*). In this species the sac fills most of the space ventral to the tongue between the rami of the mandible. The very large V-shaped opening of the sac into the oral cavity extends from a point rostral to the root of the tongue, along both sides of the root to the caudal extremity of the glottis. Ventrally the sac is suspended by the mylohyoid and cucullus muscles. Attaching to the ventral surface are the genioglossal muscles which in this species fail to reach the tongue. When the sac is empty its inner wall is strongly folded allowing considerable expansion on filling with nuts. Since the sac lies ventral to most of the jaw and tongue apparatus, filling of it does not interfere with the function of these structures. Whilst a constrictor muscle is not associated with the opening of the sac, the size of the opening is controlled by the tongue which acts as a trap door. Emptying of the sac appears to be a passive action involving opening of the bill and up and down swinging movements of the head. The genioglossal muscles are preadapted and now function only to collapse the sac when it is empty. Bock *et al.* proposed that the sac in *Nucifraga* species has been evolved from the primitive forms of seed-carrying pouches that occur in several Corvidae, e.g. *Corvus* and *Picus*, by relaxation of the mylohyoid muscle rostral to the root of the tongue. In pelicans (Pelecanidae) the floor of the mouth is enormously enlarged to form a capacious bag-like structure which is used to catch fish. A similar but much less developed gular sac also occurs in cormorants (Phalacrocoracidae) and the Dovekie (*Alle alle*) (Storer, 1971).

A sac-like diverticulum of the oral cavity which is used as a display chamber during the breeding season has been described in several adult male bustards (Otididae) including *Otis tarda* and *Choriotis kori* by Murie (1868, 1869) and Garrod (1874a,b). The diverticulum (Fig. 3.3), unlike the food carrying sacs, develops in the lingual frenulum and lies ventral to the trachea in close contact with the skin of the neck. During the "show off", which occurs in the breeding season, the sac becomes inflated with air. Outside the breeding season the sac regresses (Gadow, 1891a, p. 667). Garrod (1874b) has suggested that in young male birds development of the sac occurs by inflation of the air passages during successive breeding seasons causing the thin median membrane between the two vertical lateral folds of the frenulum to be progressively stretched. A basically similar form of sac to that of bustards has also been described by Gadow (1891a, p. 622) in the male Musk Duck (*Biziura lobata*) although in this species the sac is never sufficiently large to come into contact with the skin.

D. *Salivary glands*

The salivary glands (*gll. oris, gll. pharyngis*) are compound tubular structures distributed in the walls of the oral cavity and pharynx. How well-developed the glands are appears to be related to diet, the glands in granivorous species with dry food, for example, being larger than those of birds of prey (Fahrenholz, 1937). In many fish-eaters in which the diet is naturally well-lubricated, the glands are very poorly developed, sometimes groups of glands being absent, e.g. in the Gray Heron (*Ardea cinerea*), the Black-necked Stork (*Ephippiorhynchus asiaticus*) and the Greater Flamingo (*Phoenicopterus ruber*), whilst there are reported to be no salivary glands in the Anhinga (*Anhinga anhinga*) and the Great Cormorant (*Phalacrocorax carbo*) (Antony, 1920).

For details of the organization of the glands in a wide range of species the account of Antony (1920) is extremely useful, and the extensive description by Fahrenholz (1937) is an excellent summary of the early literature. In their most characteristic arrangement (Fig. 3.4) the glands are distributed in the oral and pharyngeal cavities as follows: maxillary, palatine and sphenopterygoid glands (*gl. maxillaris, gll. palatinae, gll. sphenopterygoideae*) in the roof; the glands of

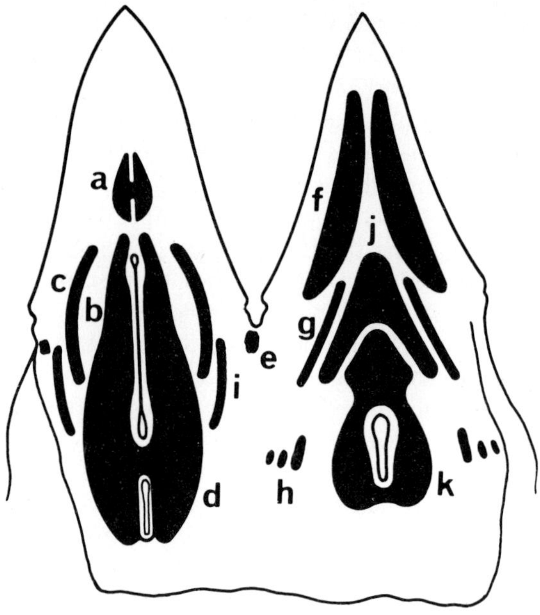

Fig. 3.4. Salivary glands of *Gallus*. (a) Maxillary gland; (b) medial palatine glands; (c) lateral palatine glands; (d) sphenopterygoid glands; (e) gland of the corner of the mouth; (f) rostral mandibular glands; (g, h, i) medial, intermediate and lateral groups of caudal mandibular glands; (j) caudal group of lingual glands; (k) cricoarytenoid glands. The rostral group of lingual glands is not shown. From Saito (1966).

the corner of the mouth (*gl. anguli oris*) in the cheeks (*buccae*); and the mandibular, lingual and cricoarytenoid glands (*gll. mandibularis*, *gll. linguales*, *gll. cricoarytenoideae*) in the floor. The palatine glands are divided in some species, e.g. *Gallus* into medial and lateral groups. The arrangement of the mandibular glands is highly variable. Thus in many granivorous and insectivorous passerines, e.g. the House Sparrow (*Passer domesticus*), the Canary (*Serenus canaria*), the Blackbird (*Turdus merula*) and the Common Starling (*Sturnus vulgaris*), there are external, intermediate and internal mandibular glands (*gll. mandibulares externae*, *intermediales*, *internae*). Another common arrangement occurs in *Columba*, *Anas* and *Gallus* in which the mandibular glands are divided into rostral and caudal groups (*gll. mandibulares rostrales*, *caudales*). The caudal group of glands is often further divided into lateral, intermediate and medial groups. In some species, e.g. the Common Kestrel (*Falco tinnunculus*) and the Eurasian Nightjar (*Caprimulgus europaeus*), the mandibular glands are undivided. The lingual glands are usually divided into rostral and caudal groups. In the Gray Heron (*Ardea cinerea*) the lingual glands are the sole salivary glands present. The maxillary glands and the glands at the corner of the mouth are generally monostomatic, whereas the remaining glands are usually polystomatic.

Each salivary gland consists of a variable number of lobules, a lobule being composed of many secretory tubules opening into a common cavity (McCallion and Aitken, 1953). The common cavity drains by a single duct which opens either directly into the lumen of the digestive tract or after anastomosing with the other ducts (*ductuli glandularium oralium, pharyngealium*) to form a common duct. In *Gallus*, the distal parts of the ducts are lined by a stratified squamous epithelium which changes to low columnar in the proximal parts of the ducts and in the common cavities (McCallion and Aitken, 1953). According to Chodnik (1948), in *Gallus* the most common cell type in the secretory tubules after feeding or a 24 h fast is a tall columnar cell which has a basal nucleus and is packed with secretory material staining positive for mucus. Following a meal, a relatively small number of these cells discharge their contents into the glandular lumen. The cells then become very narrow and small before again acquiring more secretory material. Fujii and Tamura (1966) demonstrated that the mucosubstance in *Gallus* is an acid mucopolysaccharide which is sulphated in some glands and non-sulphated in others. The presence of amylase in salivary glands has recently been investigated in several species by Jerrett and Goodge (1973) using a variety of techniques. They concluded that whilst the glands of *Gallus* and *Meleagris* did not display significant amylolytic activity, amylase is secreted by the salivary glands of the House Sparrow (*Passer domesticus*). They suggested that amylase secretion by the glands of *Gallus* and *Meleagris* may not be necessary since in these birds the large form of the crop allows food to be stored for a sufficient length of time to be acted on by plant amylase. In *Passer*, however, the fusiform crop has much less storage potential so that amylase secretion by the salivary glands may be a compensatory mechanism. Despite this, the functional significance of amylase is still unknown. Leasure and Link (1940) observed that the total volume of saliva secreted by *Gallus* over a period of 24 h ranged from 7 to 25 ml. Secretion of saliva is increased by parasympathetic stimulation (Chodnik, 1948).

In three taxonomically widely separate groups of birds, the mandibular salivary glands are greatly enlarged in adaptation to special requirements of the birds. Group (a): in woodpeckers (Picidae), the enlarged portion of the mandibular glands fills the space between the rami of the mandible and in the Green Woodpecker (*Picus viridis*) is about 7 cm long (Greschik, 1913; Antony, 1920). The sticky mucous secretion coats the apex of the tongue enabling the tongue to function as a probing "lime-stick" for capturing insects. Group (b): in the majority of swifts (Apodidae) the mandibular glands secrete a mucilaginous substance which is used as cement in the construction of nests. In some *Collocalia* species of S.E. Asia, the nest is built entirely of saliva and forms the edible nest of "bird's nest soup". The nest-cement is a glycoprotein. In connection with the formation of the nest, seasonal enlargement of the mandibular glands has been observed in both sexes of a number of species (Marshall and Folley, 1956; Johnston, 1958; Medway, 1962). In the Chimney Swift (*Chaetura pelagica*) for example, the glands reach a maximum size at the height of the breeding season when they each weigh about 70 mg and fill the space between the floor of the mouth and the skin. In contrast, the inactive gland weighs only about 3 mg (Johnston, 1958). Medway could find no correlation between the activity of the glands in Thunberg's Swiftlet (*Collocalia fuciphaga*) and the reproductive state of individual birds. He suggested that the glands may be responsive to testicular hormone and that the dominant environmental factor controlling the activity of the glands is probably the presence or absence of the nest. Group (c): in the Gray Jay (*Perisoreus canadensis*), parts of the mandibular glands (probably the medial mandibular glands), are enlarged semilunar structures about 24 mm in length and curved dorsally around the caudal region of the mandible (Bock, 1961). According to Dow (1965), the mucous secretion of the glands is used in the production of boli which are permeated and covered with saliva. The jays store the boli by sticking them to conifer needles and parts of trees, the storage of food above ground enabling the jays to occupy the Boreal region during winter when the ground is covered by deep snow.

E. Taste buds

By far the most extensive accounts of taste buds (*gemmae gustatoriae*) in birds covering many species are the early descriptions of Botezat (1904, 1906, 1910) and Bath (1906). The number of taste buds in birds is relatively small, the available data suggesting that less than 500 are present (compare the 17000 reported in the rabbit by Moncrieff, 1951, p. 172) and that considerable interspecific differences exist (see for example Bath, 1906). The distribution of the buds is also highly variable between species, although recent examinations of this aspect have been very few. Amongst the areas where the buds are located are the palate, the floor of the mouth, the base of the tongue and the floor of the pharynx. The buds appear to be commonly associated with the ducts of salivary glands. An excellent description of the distribution in the Mallard (*Anas platyrynchos*) has recently been provided by Berkhoudt (1977) who was able to correlate the distribution with the hypothetical pathway taken by food described by Zweers *et al.* (1977). The structure of taste buds is dealt with in the Special Sense Organs. For details of the sense of taste in birds consult the reviews of Wenzel (1973) and Kare and Rogers (1976). The latter concluded that whilst birds have a sense of taste "no pattern, whether chemical, physical, nutritional, or physiological, can be correlated consistently with the bird's taste behaviour".

F. Bill

The bill or beak (*rostrum*) consists of the bones of the upper and lower jaws and their thick horny covering or rhamphotheca (*rostrum maxillare*, *rostrum mandibulare*). The following short account is restricted almost entirely to the keratinized parts of the bill, most of the data being provided by the substantial surveys of Newton and Gadow (1896, pp. 32–36) and Mountford (1964).

The upper horny bill or rhinotheca covers the premaxillary bones and extends caudolaterally on the maxillary bones. The dorsal part on the midline is the culmen, the lateral cutting edges the upper tomia. Most of the upper bill is usually covered by hard keratin. In waders (shore-birds), i.e. Charadrii, however, the whole bill is relatively soft, whilst in the Anatidae only the tip or dertrum of the bill is hardened as the nail or neb. The soft base of the bill is the cere which in parrots (Psittacidae), pigeons (Columbidae) and birds of prey (Falconiformes and Strigiformes) is a swollen and highly sensitive structure. The cere is usually bare but in parrots it is feathered. The upper bill carries the nostrils or external nares. The base of the culmen is occasionally enlarged to produce various forms of ornamental excrescences including the so-called frontal shield of coots (*Fulica*) and the casque of hornbills (Bucerotidae) (Fig. 3.5g). In some species of pelican (Pelecanidae) the upper bill of the male during the breeding season develops several upward projecting ornamental appendages which are later lost.

The lower horny bill or gnathotheca covers the dentary bones. The prominent ventral ridge formed by the junction of the two halves of the jaw is the gonys (especially well-developed in gulls, Laridae), whilst the lateral cutting edges are the lower tomia.

In most birds the horny covering of the bill is in one piece. Some species, however, have a compound form of rhamphotheca. In albatrosses (Diomedeidae) and fulmars (*Fulmarus*), for example, both upper and lower bills consist of four sections, whilst in *Apteryx*, *Sula*, *Phalacrocorax*, *Pelecanus*, *Threskiornis*, *Platalea* and *Ardea* the upper bill is in three sections. The rostral part of the culmen in the embryo carries a small pointed protuberance, the egg tooth, which is used at hatching to break the shell (Wetherbee, 1959; Clark, 1961). Lucas and Stettenheim (1972, p. 17) noted that this protuberance is similar to the horny process in crocodiles and turtles but is unlike the egg-tooth of lizards and snakes. In a limited number of species, e.g. the Lapwing (*Vanellus vanellus*), the Oystercatcher (*Haematopus ostralegus*) and the Stone Curlew (*Burhinus oedicnemus*) a tooth-like process also occurs on the lower bill. The time of disappearance of the egg-tooth varies considerably between species. In galliform and passeriform birds, for example, the tooth is lost during the first week after hatching, but in many other birds, e.g. falconiforms, bustards (Otididae) and some penguins (Spheniscidae), it persists for a longer period. In the megapodes (Megapodiidae), which are reported to hatch by kicking their way out of the shell, the egg-tooth at hatching is vestigial or absent. A pair of sharp hooks occur on both the upper and lower bills of the nestlings of parasitic honeyguides (Indicatoridae) and are used to kill the nestlings of the host species. Normal wear and tear of the rhamphotheca is compensated for by new growth. Moulting of the bill occurs in puffins (*Fratercula*) at the end of the breeding season, when the nine brightly-coloured plates which form the outer covering of the bill are shed. Intraspecific differences in the colour of the bill between the male, female and young are relatively common. Differences in the shape of the bill between the male and the female of the same species, however, are uncommon.

The size and shape of the bill in newly-hatched birds is surprisingly uniform even in species like crossbills (*Loxia*) and skimmers (*Rynchops*) in which the bill of the adult has a characteristic appearance (Fig. 3.5d,e). The wide range of variations in the adult form are related to the feeding habit. The following short account covers some of the more distinct bill adaptations and is based mainly on Pycraft (1910), Mountford (1964), and Storer (1971) who should be consulted for further details.

(1) *Dagger-like bills*. Characteristic of species like herons and bitterns (Ardeidae) which require to grasp fast moving aquatic prey. This form of bill in the Anhinga (*Anhinga anhinga*) is used to impale fish.

(2) *Serrated bills*. The tomia of mergansers (Anatidae) carry numerous tooth-like serrations which are used for seizing and holding prey (Fig. 3.5a).

(3) *Spatula-like bills*. As in spoonbills (Threskiornithidae) (Fig. 3.5l). During feeding the open bill is immersed in water or mud in an almost vertical position and is raked vigorously from side to side to catch small fish and invertebrates.

(4) *Boat-shaped bills*. An enlarged form of bill which in the Whale-headed Stork (*Balaeniceps rex*) is adapted to hold slippery frogs (Fig. 3.5p).

(5) *Lamellated bills*. In the majority of ducks, geese and swans (Anatidae) one row of lamellae is present on the upper bill and two rows (dorsal and lateral) on the lower bill (Goodman and Fisher, 1962, pp. 37–44). In grazing species, e.g. the Canada Goose (*Branta canadensis*) and the Muscovy Duck (*Cairina moschata*) the lamellae are thick and flat, those in the dorsal row of the lower mandible having a sharp edge. When the lower bill is forcefully adducted therefore, a scissor-like cutting action results. Characteristically in these species, the bill has a narrow tip with a broad nail in adaptation for grasping. In contrast, the lamellae in straining species, e.g. the Mallard (*Anas platyrhynchos*) and the Ruddy Duck (*Oxyura jamaicensis*), are blade-like and act with the bristles and processes of the tongue to filter out solid food particles from water. By means of rapid opening and closing of the jaws, water and solid particles enter the mouth at the apex of the bill, the water only leaving the mouth at the sides. This straining mechanism reaches its highest development in the Northern Shoveler (*Anas clypeata*) (Fig. 3.5c). Characteristically in straining species the apex of the bill is broad and the nail narrow, a grasping action not being required.

The structure of bill nail was recently studied in several species of duck by Berkhoudt (1976) using scanning electron microscopy. The nail of the upper bill has a series of ventrally-directed pores through which protrude dermal papillae which are covered distally by caps of keratin. The papillae contain the sensory corpuscles of Herbst and Grandry. Sensitive dermal papillae also project through dorsorostrally-directed holes in the lower nail. For a highly detailed account of the distribution and organization of the Grandry and Herbst corpuscles in the beak of geese consult Gottschaldt and Lausmann (1974).

The large lamellated decurved bill (Fig. 3.5b) and the highly complex feeding mechanism of flamingos (Phoenicopteridae) has been described in considerable detail by Jenkin (1951). The bill of these birds consists of a large trough-like lower mandible into which the small upper mandible fits like a lid. In *Phoenicopterus* the upper mandible is shallow-keeled and the gape is situated at the side of the bill. In *Phoeniconaias* and *Phoenicoparrus* the upper mandible is deep-keeled and lies flush between the rami of the lower mandible, the gape being situated on the dorsal surface of the bill. In all species, lamellae extend from the edges of the gape and from the inner surface of the lower bill and the surface of the keel. These lamellae vary considerably in size, shape and number from one part of the bill to another and between bills of different species. During feeding the bill is inverted. Movement of water in and out of the mouth is achieved mainly by a piston-like action of the tongue, the solid food particles being filtered out by the lamellae. For details of the considerable differences in the filtering mechanism between the bottom-feeder *Phoenicopterus* and the surface feeders *Pnoeniconaias* and *Pnoenicoparrus* consult Jenkin (1951).

Amongst other species with lamellated bills are prions (*Pachyptila*) and storks

(*Anastomus* spp.). In *Anastomus*, a gap exists in the closed bill between the distal parts of the upper and lower mandibles except where the tips are opposed (Fig. 3.5f). The gap, according to Thomson (1964, p. 784), is an adaptation for securing slippery water snails.

(6) *Trowel-like bills.* A laterally compressed form of bill unique to skimmers (Rynchopidae), in which the lower mandible overshoots the upper mandible

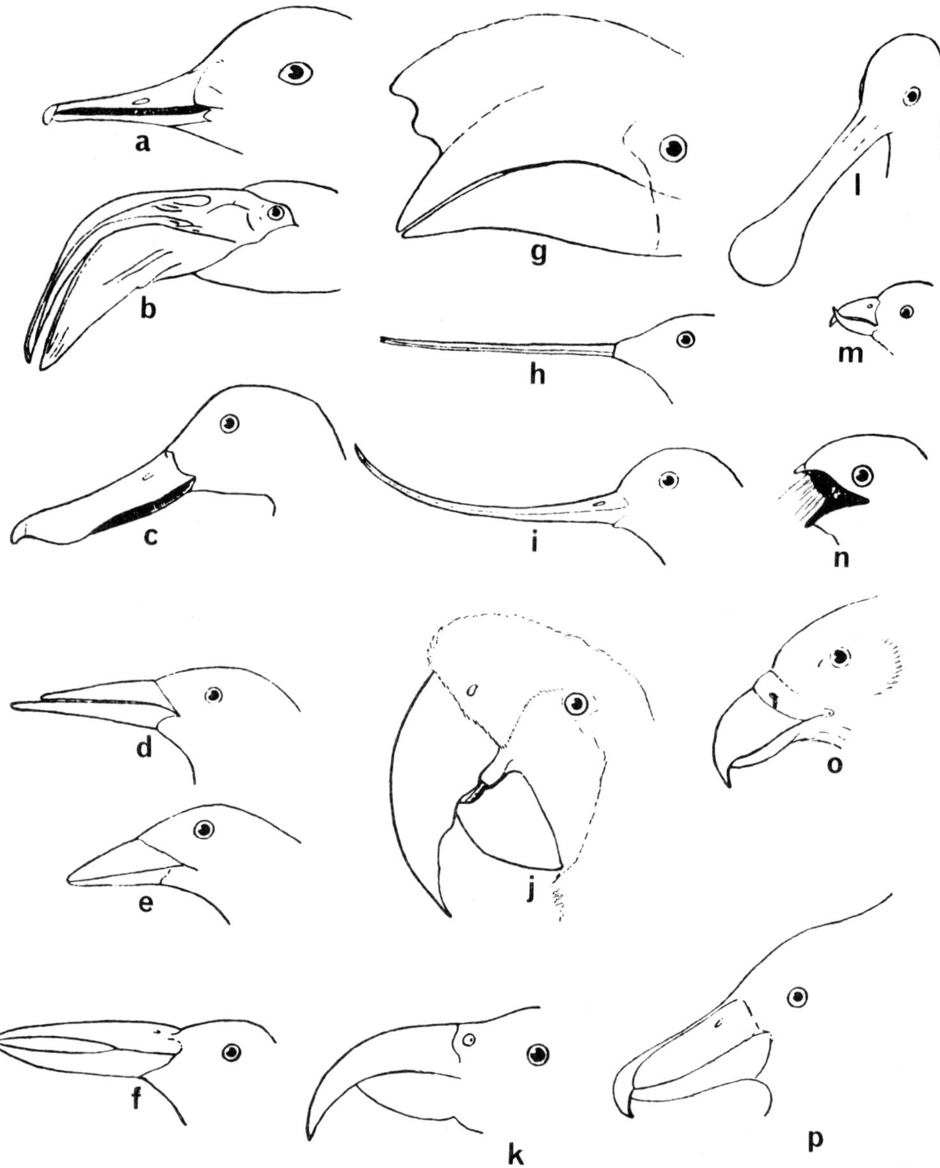

Fig. 3.5. Bills. (a) Merganser. (b) Flamingo. (c) Shoveler. (d) Skimmer (adult). (e) Skimmer (young). (f) Open-bill Stork (*Anastomus*). (g) Hornbill. (h) Hummingbird. (i) Avocet. (j) Parrot. (k) Parrot. (l) Spoonbill. (m) Crossbill. (n) Nightjar. (o) Eagle. (p) Whale-headed Stork (*Balaeniceps*). From Pycraft (1910).

(Fig. 3.5d). Feeding in these species takes place on the wing, the lower mandible being ploughed beneath the surface of the water. When the mandible comes into contact with a fish the bill is snapped shut.

(7) *Elongated, slender bills.* A highly flexible straight or decurved bill characteristic of waders (Charadrii) and adapted for probing in sand or mud. The dertrum of the upper bill typically contains numerous sensory nerve endings.

(8) *Strongly curved bills.* An elongated, slender recurved bill (Fig. 3.5i) is utilized by avocets (Recurvirostridae) in side to side sweeping movements to obtain small crustacea and insect larvae from the surface of shallow water.

(9) *Raptorial bills.* The hooked bill of falconiform and strigiform species which is adapted for shredding food (Fig. 3.5o). A rather similar form of bill occurs in the carnivorous shrikes (Lanniidae), the upper tomia of these birds carrying a pair of tooth-like projections.

(10) *Stout, curved bills.* An extremely powerful form of bill (Fig. 3.5j,k) characteristic of parrots (Psittacidae). The lower bill has a blunt, rostral border which presses against a horny prominence on the palate. Rostral to the prominence are often numerous rasp-like ridges. Both the prominence and ridges are used for reducing the stones of fruit as well as for sharpening the tip of the lower bill.

(11) *Chisel-shaped bills.* A strong elongated form of bill utilized by woodpeckers (Picidae) for uncovering wood-boring insects.

(12) *Conical bills.* The characteristic bill of seed-eating passerine species. Two basic methods are employed by these birds for breaking open and peeling the shell depending on the form of the seed, the detailed structure of the bill and palate being closely adapted to the method of opening (Ziswiler, 1964, 1965; Newton, 1967). With dicotyledonous seeds the shell is removed by cutting, with monocotyledonous seeds by crushing. In the Fringillidae, all of which feed on dicotyledonous seeds, the seed is wedged firmly in the groove of the palate between the lateral palatine ridge and the edge of the upper bill. Elongated seeds lie longitudinal to the bill, round seeds with the seam longitudinally orientated. Cutting is achieved by rapid rostro-caudal movements of the sharp-edged lower bill, the seed being supported by the point of the tongue. Large seeds require relatively large cutting forces and are therefore positioned in the caudal part of the groove where the groove is widest and the forces exerted by the lower bill greatest. With most seeds a single cutting procedure is insufficient and the seed is then transferred to the opposite side of the mouth for further treatment. Ziswiler noted that with particularly hard seeds as many as twenty such transfers could be involved. To remove the cut shell the lower bill is introduced between the shell and the kernel and the seed rotated by the tongue. Lateral to and fro movements of the lower mandible were observed during this procedure by Ziswiler. In the Emberizidae, the Ploceidae and most of the Estrildidae, all of which feed on monocotyledonous seeds, the shape of the seed determines where it is positioned in the mouth. Elongated seeds are orientated transversely to the bill, close to the buttress-like prominence of the palate which is a characteristic feature of these birds. Upward and downward movements of the lower mandible force the seed against the prominence, crushing the shell. Round seeds in contrast, may be held, as in the Passerinae, the Ploceinae and the Viduinae, in the lateral groove of the palate where they are crushed by up and down

movements of the blunt-edged lower mandible. With both shapes of seed, the tongue is used to remove the shell.

In the Hawfinch (*Coccothraustes coccothraustes*) the conical bill is especially well-developed in connection with its ability to deal with very large seeds such as those of cherries and olives. Cracking of these larger seeds takes place between a pair of knob-like projections on the palate and corresponding projections on the lower bill. This mechanism, as observed by Newton (1967), ensures that the strain of cracking the seeds is shared equally by the muscles on both sides of the jaw.

(13) *Crossed bills.* In adult crossbills (*Loxia*) the tips of the upper and lower mandibles are crossed (Fig. 3.5m) forming a unique device which allows crossbills, unlike other species, to open unripe conifer cones and extract the seed. The direction of crossing of the mandibles is variable and in an equal number of birds the lower mandible bends to the right and left. When the bill is open the tips of the mandibles lie directly opposite each other. The complex extraction procedure has been described by Robbins (1932), Tordoff (1954) and Newton (1967). In the present account the lower mandible is directed to an inner scale of the cone, the upper mandible to an outer scale. Three stages are involved in the procedure at any one of which the seed can be extracted. First the open bill is inserted from the side between the scales. The lower mandible is abducted sideways towards the body of the cone, causing the tip of the upper bill to raise the outer scale and make possible extraction of the seed. Second, if the seed cannot be removed, the scales may be prized further apart by closing the bill and rotating it sideways so that the tips of the crossed mandibles have a wedge-like action on the scales. Third, maximum separation of the scales is achieved if the rotated bill is next raised. The scales are now held apart by the upper bill alone, its tip pressing against the outer scale and the culmen against the inner scale. This arrangement allows opening of the lower mandible so that the prehensile tongue can be fully extended to extract the seed.

(14) *Short, pointed bills.* Characteristic of small insectivorous species such as warblers (Sylviinae and Parulidae).

(15) *Flower-probing bills.* An elongated straight, decurved or recurved form of bill (Fig. 3.5h) characteristic of hummingbirds (Trochilidae), sunbirds (Nectariniidae) and many other flower-frequenting species and used for probing into flower heads to obtain insects or nectar.

(16) *Wrybill.* The distal portion of the bill in the Wrybill (*Anarhynchus frontalis*) is bent to the right in adaptation for seeking out insects under stones.

(17) *Broad-gaped bills.* A form of bill characteristic of swifts (Apodidae) and swallows (Hirundinidae) as an adaptation for catching insects whilst feeding on the wing. A basically similar bill occurs in certain other insectivorous species including nightjars (Caprimulgidae) (Fig. 3.5n) and flycatchers (Muscicapinae and Tyrannidae), the gape of these birds being beset with numerous rostrally directed bristles or vibrissae for which a tactile function has been suggested.

G. Deglutition

Experimental studies on the process of deglutition in birds are restricted to a limited number of investigations in domestic species by Henry *et al.* (1933), Halnan (1949), Vonk and Postma (1949), Pastea *et al.* (1968a,b), White (1968, 1970, pp. 90–96) and Suzuki and Nomura (1975). The following data were obtained from the analysis in *Gallus* by White who employed cineradiography and endoscopic cine film. In *Gallus* the mechanisms involved in both eating and drinking appear to be similar (Fig. 3.6). During prehension the head is lowered. A pellet of food is grasped by the bill and carried by the tongue to the palate where it is held by the sticky saliva. This is accompanied by reflex muscular closure of the choana. Rapid and synchronous rostro-caudal movements of the tongue and larynx, aided by the caudally-directed papillae then propel the pellet caudally into the pharynx where it accumulates caudal to the laryngeal mound. After several pellets reach the pharynx, the head is raised and the food propelled into the oesophagus by rostro-caudal movements of the larynx.

In drinking, the bill is immersed as far as the corners of the mouth. Fluid enters the oral cavity by rapid rostro-caudal movements of the tongue and accumulates in the floor of the pharynx between the tongue and larynx. This is accompanied by a rostral movement of the larynx so that the airway between the choana and the trachea is maintained. Finally the head is raised and under the influence of gravity the fluid flows lateral to the larynx and into the oesophagus. This flow is accompanied by several rostro-caudal movements of the larynx. Before the bird drinks again, the larynx is raised caudally occluding the oesophagus. The activity of the jaw, hyobronchial and laryngeal muscles during deglutition was investigated by Suzuki and Nomura (1975).

According to Cade and Greenwald (1966), fluid may be taken up by birds in one of four ways as follows. (a) By scooping or sipping as in galliform and passeriform species and described above. (b) By lapping with the tongue as in parrots (Psittacidae). (c) By means of a highly protrusible grooved or tubular tongue as in hummingbirds (Trochilidae). (d) By a little understood sucking method as in pigeons (Columbidae) and the mouse birds *Colius indicus* and *Colius striatus* in which the bill is immersed throughout the entire deglutition process. Cade and Greenwald observed that, in contrast to popular opinion, sandgrouse (Pteroclidae) do not drink like pigeons, since after sucking fluid they raise their heads in order to swallow.

II. Oesophagus

A. Gross morphology of oesophagus

The macroscopic features of the oesophagus have been dealt with in a wide range of species by Gadow (1891a, pp. 671–672), Swenander (1899, 1902), Groebbels (1932, pp. 454–456), Niethammer (1933), Pernkopf and Lehner (1937), Portmann (1950) and Ziswiler and Farner (1972). In the cranial part of the neck the oesophagus (*pars cervicalis*) lies in the midline closely attached ventrally by connective tissue to the larynx and trachea (Fig. 3.7). The remainder of the cervical oesophagus is situated to the right of the midline between the jugular vein and thymus dorsally and the trachea ventrally. In some species part

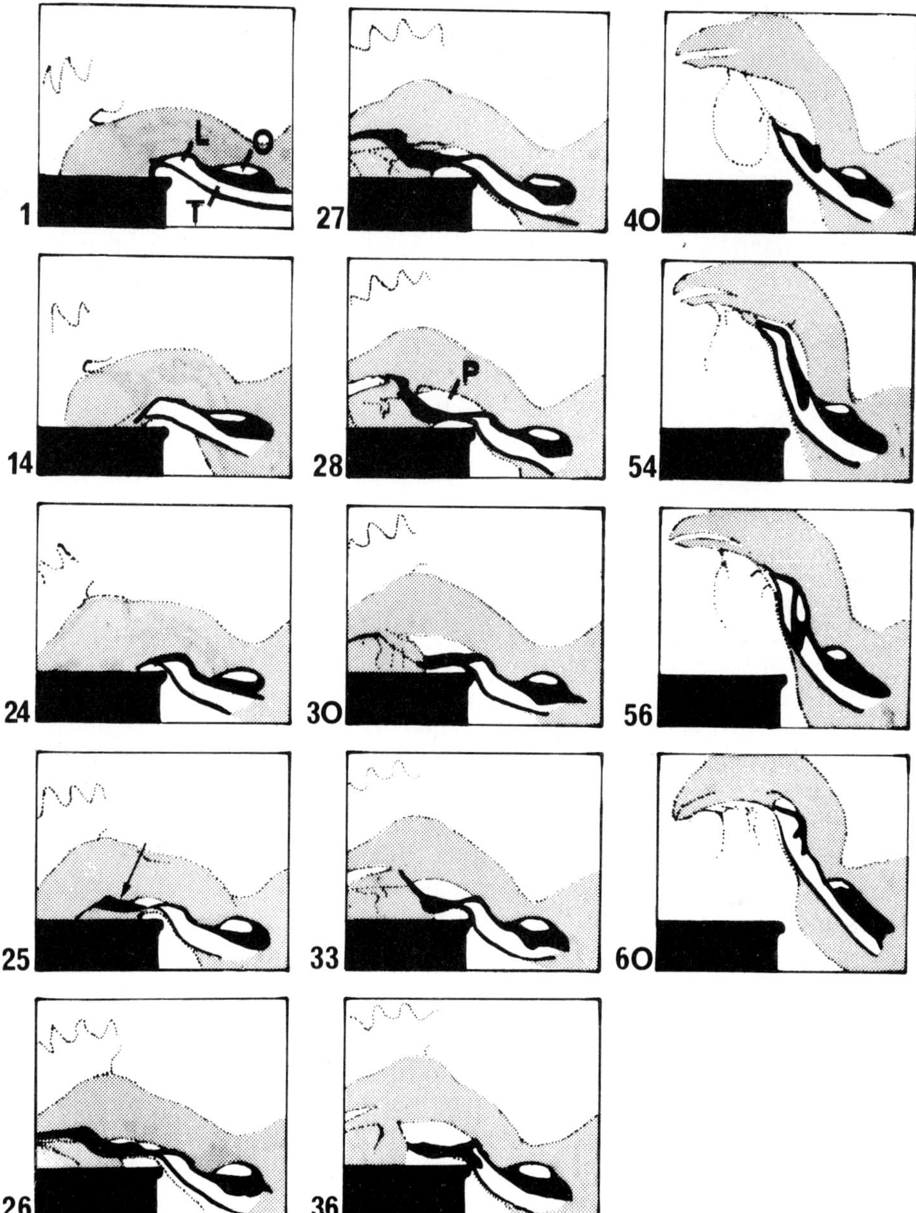

Fig. 3.6. Drawings made from a 16 mm radiographic cinefilm exposed at 16 frames/s showing an adult domestic fowl swallowing radio-opaque barium sulphate suspension. The numbers indicate the frame number in the film sequence. The beak is placed in the container of fluid and the tongue moved rapidly rostrally and caudally until the floor of the pharynx is filled with fluid (1–24). The head is then raised (25–40) and the fluid (indicated by arrow in 25) flows caudally into the oesophagus under the influence of gravity (36–60). L, larynx; O, oesophagus; P, pharynx; T, trachea. From White (1970).

of the cervical oesophagus is expanded to form the crop (*ingluvies*) (Fig. 3.7) in which food is stored, whilst in others there is a diverticulum, the oesophageal sac (*saccus oesophagealis*) (Fig. 3.3), which functions during the breeding season in display or for the production of mating calls. Details of both these structures are provided below. The position of the cervical oesophagus relative to the vertebral column was investigated in the domestic birds by Bego and Rapić (1964, 1965). When the neck is unextended, the cervical oesophagus is shorter than the S-shaped cervical region of the vertebral column, the cranial third of the oesophagus lying ventral to the column, and the caudal two-thirds lying to the right or dorsal to it. Since most of the cervical oesophagus is only loosely attached to the surrounding structures by connective tissue, its position relative to the vertebral column changes when the neck is extended. Then the entire length of the stretched cervical oesophagus remains a straight tube and is not affected by movements of the neck. The thoracic oesophagus (*pars thoracica*) extends caudally, dorsal to the trachea and syrinx and between the extra-pulmonary primary bronchi, and is closely related to the cervical, clavicular and cranial thoracic air sacs (Fig. 3.7). It opens into the proventriculus in the left part of the body cavity. However, it is not usually possible to determine externally the precise point where the oesophagus joins the stomach.

The diameter of the avian oesophagus is in general greater than that of other vertebrates. There is usually a close relationship between its calibre and the size of the items of food which are swallowed or the extent to which it is used to store food (Swenander, 1902). Thus the oesophagus is widest in birds such as *Alca* and *Larus*, procellariiform species, *Phalacrocorax*, *Ciconia*, *Ardea*, *Botaurus*, *Fulica* and *Gallinula*, the Accipitres and Strigiformes, and *Cuculus* and *Alcedo* which swallow very large pieces of food or store food throughout the whole length of the oesophagus. It is narrowest in birds like the parrots (Psittacidae) and insectivorous species, e.g. the Wryneck (*Jynx torquilla*) and the Common Swift (*Apus apus*), which swallow very small food items. The internal surface of the oesophagus is longitudinally folded (*plicae esophageales*) which helps to increase the distensibility of the tube. The size of the folds usually depends on the same factors which determine the calibre of the oesophagus and they are therefore best developed in birds like the *Alcae* and *Lari* and the Procellarii-formes and very small in the piciform species and the insectivorous passerines (Barthels, 1895; Swenander, 1902).

B. Crop

The cervical part of the oesophagus in a number of species has a distinct expansion, the crop (*ingluvies*), which functions as a highly distensible storage chamber for food (Fig. 3.7). The great diversity in the appearance of the crop has been described by Gadow (1891a, pp. 671–672), Swenander (1899, 1902), Magnan (1911a, pp. 53–57), Niethammer (1933) and Ziswiler (1967a,b). The simplest form is basically a spindle-shaped enlargement of the cranial, middle or caudal regions of the cervical oesophagus (Fig. 3.8a) as in *Casuarius*, *Uria*, *Fratercula*, *Phalacrocorax*, *Anas*, *Ciconia*, *Leptoptilos*, *Philomachus*, *Tringa*, *Numenius*, *Scolopax*, *Haematopus* and *Otis*, and the Trochilidae, Paradisaeidae,

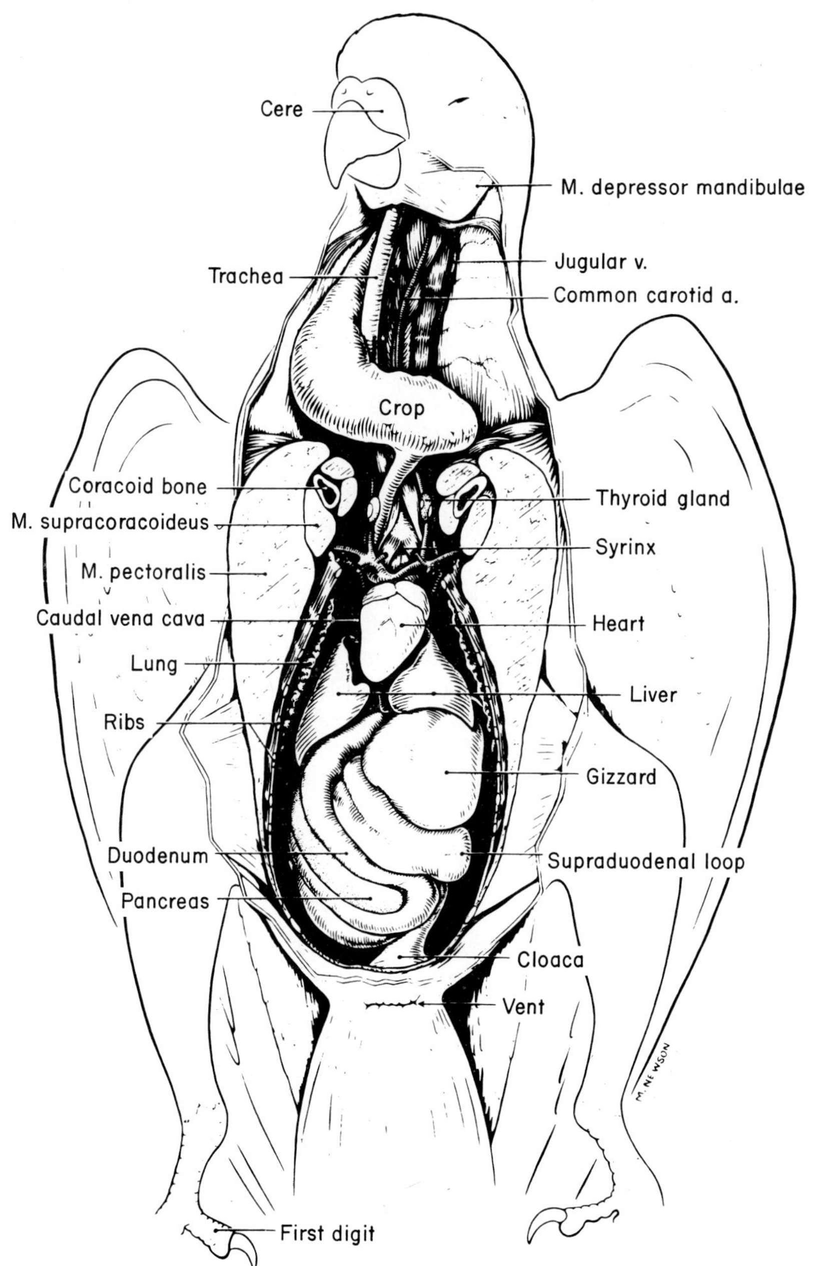

Fig. 3.7. Alimentary canal *in situ* of the Budgerigar (*Melopsittacus undulatus*), sternum and abdominal wall removed. From Evans (1969).

Fringillidae, Estrildidae and some Emberizidae. Filling of this enlargement with food in many passerines causes it to expand laterally or dorsally (Niethammer, 1933). A relatively unusual form of expansion of the filled crop occurs in the Redpoll (*Acanthis flammea*) (Fisher and Dater, 1961). In this species the full crop extends dorsally on the right side of the neck, and across the dorsal surface of the vertebral column to the left side of the neck. The right part of the expanded oesophagus is by far the largest and stretches the entire length of the neck. The connection between the right and left parts of the crop is slightly constricted in the dorsal midline by the dorsal spinal tract of feathers and by a pair of dermal muscles. Consequently, when the bird is viewed from the dorsal side, two crops appear to be present, one on either side of the midline. Fisher and Dater suggested that the dorsal position of the full crop is brought about by its large size, since a more ventrally positioned crop might impede flexion of the neck when the bird feeds. Expansion of the crop dorsal to the vertebral column has also been reported in several other passerine species (Eber, 1956; Farner, 1970). In two species of hummingbird (Trochilidae) the crop arises from the medial side of the oesophagus, and as in some of the passerines, when filled expands dorsal to the vertebral column (Niethammer, 1933).

The best-developed forms of crop are highly differentiated sac-like structures which arise as ventral or lateral diverticula of the caudal part of the cervical oesophagus in the Thinocoridae, some Emberizidae and Ploceidae, and in the Tinamiformes, Falconiformes, Psittaciformes, Galliformes and Columbiformes. In the Thinocoridae, Falconiformes and Galliformes (Fig. 3.8b,c) the crop lies on the right side of the neck and when full rests on the furcula (Swenander, 1902; Niethammer, 1933; Hanke and Niethammer, 1955). In psittaciform species (Figs 3.7, 8e) the crop is orientated transversely across both sides of the caudal part of the neck (Evans, 1969; Feder, 1969). The crop in pigeons (Columbidae) (Fig. 3.8d) is especially well-developed and consists of very large right and left lateral diverticula (*diverticulum dextrum ingluviale, diverticulum sinistrum ingluviale*) which are connected in the midline by a small median diverticulum (*diverticulum medianum ingluviale*). The inner surfaces of the lateral diverticula are strongly folded (*plicae ingluviales*) (Weber, 1962). Associated with some of the more highly-developed forms of crops like those of galliform and columbiform species are dorsal and ventral sheets of striated muscle fibres belonging to the clavicular part of the cucullaris muscles (Berge, 1979). These muscular sheets attach to the adventitia of the crop and probably act by forming a sling-like support for the organ

A unique form of crop occurs in the Hoatzin (*Opisthocomus hoazin*) and is remarkable for both its size and complexity (Gadow, 1891b; Böker, 1929). Unlike in other birds the crop consists of cervical and thoracic components (Fig. 3.9). The main compartment is an exceptionally large ventral diverticulum of the caudal part of the cervical oesophagus. This extensive sac stretches caudally on the ventral surface of the sternum so that the cranial part of the sternal keel is suppressed. Other alterations of the skeleton ascribed to the presence of such a large crop include fusion of the clavicles to the coracoid bones and sternum, and narrowing of the cranial part of the thorax. The wall of the cervical crop is extremely thick as a result mainly of the great development of the circular muscle layer. The inner surface is lined by a hard epithelium and is raised into approximately twenty parallel ridges (*rugae ingluviales*), each ridge measuring about 3 mm in height and 6 mm in width. The exceptionally long thoracic oesophagus is constricted in places and runs a tortuous course before joining the proventriculus. Böker (1929) identified in this part of the oesophagus four sections including a relatively short straight cranial section, a much longer S-shaped crop-like section, a shorter crop-like section, and an extremely short and narrow connecting section which opens into the proventriculus. The internal surface of the thoracic oesophagus is lined by longitudinal ridges like those in the cervical crop. The very large cervical crop of the Kakapo (*Strigops habroptilus*) is superficially very similar to that of the Hoatzin but its circular muscle layer is much thinner and the internal ridges are poorly developed. Moreover, the thoracic oesophagus in *Strigops* is short and straight with no crop-like expansions (Böker, 1929).

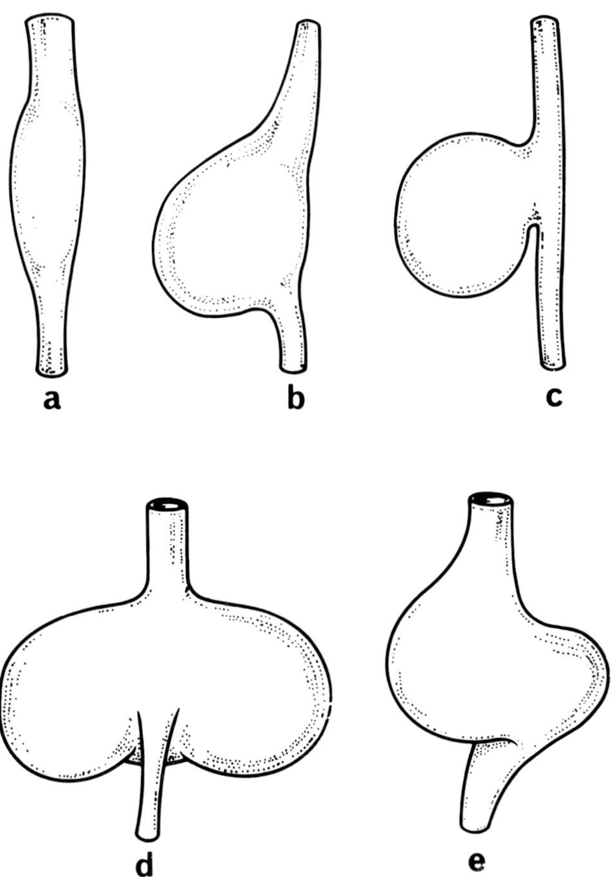

Fig. 3.8. Crops. (a) Great Cormorant (*Phalacrocorax carbo*). (b) Griffon Vulture (*Gyps fulvus*). (c) Peafowl (*Pavo cristatus*). (d) *Columba*. (3) Budgerigar (*Melopsittacus undulatus*). Ventral views except (d) dorsal view. (a), (b) and (c) from Pernkopf and Lehner (1937).

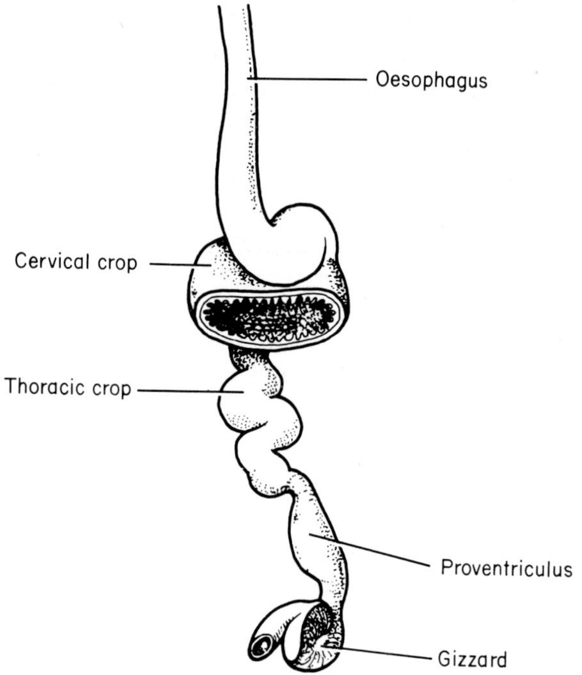

Fig. 3.9. Crop of Hoatzin (*Opisthocomus hoazin*), ventral view. Approximately ×0·66 natural size. From Pernkopf and Lehner (1937).

C. *Structure of oesophagus*

Amongst the accounts which provide comparative data on the microanatomy of the oesophagus those by Barthels (1895), Schreiner (1900), Swenander (1902), Kaden (1936), Niethammer (1933), Ziswiler (1967a,b) and Feder (1972a) are especially useful. The oesophageal wall consists basically of four layers: mucous membrane (*tunica mucosa esophagi*), submucosa (*tela submucosa esophagi*), muscle tunic (*tunica muscularis esophagi*), and an external fibrous layer (*tunica adventitia*). The mucous membrane consists of a longitudinal muscularis mucosae (*lamina muscularis mucosae*), a lamina propria and a stratified squamous epithelium. The epithelium varies considerably in thickness between species. Although it is reported to be cornified, the ultrastructural study by Kudo (1971b) in *Gallus* and *Columba* indicated that, in these species at least, the cornification is incomplete. Since chickens and pigeons feed on hard or coarse food, Kudo concluded that the hypothesis correlating dietary habits with the degree of cornification of the oesophageal epithelium is not applicable to birds. A characteristic feature of the oesophageal mucosa is the large number of mucous glands (*gll. esophageales*) which were investigated in considerable detail by Kaden (1936). The glands vary from tubular to alveolar, Ziswiler (1967a) noting that their structure may change from one part of the oesophagus to another in the same species. The glands are usually more numerous in the thoracic oesophagus and in some birds, e.g. *Melopsittacus*, are even completely absent from the cervical oesophagus (Feder, 1969). Frequently associated with the glands are nodules of lymphatic tissue (*lymphonoduli esophageales*) (Hodges, 1974, p. 46). The submucosa is very poorly developed. The muscle tunic contains only smooth muscle fibres which in the majority of species are circularly orientated (*stratum circulare*). However, an external longitudinal layer (*stratum longitudinale*) is also present in the Dromaiidae, Phasianidae, Rheidae, Phalacrocoracidae, Sulidae, Ardeidae and Tinamidae (Barthels, 1895; Schreiner, 1900; Swenander, 1902; Hanke, 1957). When the oesophagus has both circular and longitudinal muscle layers, the circular layer is always thickest, especially in the thoracic oesophagus (Swenander, 1902).

The microanatomy of the crop appears to be basically similar to that of the rest of the oesophagus, although in some of the more developed forms the glands (*gll. ingluviales*) are reported to be much less numerous. In *Gallus*, for example, glands are totally absent from the crop fundus (*fundus ingluvialis*) and only occur in a small area at the opening into the oesophagus (*ostium ingluviale*) (Hodges, 1974, p. 47). Histological changes in the oesophagus and crop associated with diet have been observed in *Anser* and in the Common Tern (*Sterna hirundo*) (Rybicki, 1959; Rybicki and Lubańska, 1959; Cymborowski, 1968).

D. *Blood supply of oesophagus*

The blood supply of the oesophagus and crop in *Gallus* has been described in considerable detail by Baumel (1975). The arteries include the oesophagotracheobronchial and ascending oesophageal branches of the common carotid artery, the ingluvial branches of the artery accompanying the vagus nerve, the descending oesophageal branch of the mandibular artery, and the oesophageal branch of the coeliac artery. The veins are tributaries of the mandibular and jugular veins.

E. *Nerve supply of oesophagus*

The gross anatomy of the nerve supply of the oesophagus and crop in *Gallus* has been reviewed by McLelland (1975). The cervical oesophagus is innervated by an oesophageal plexus which receives contributions from the glossopharyngeal and hypoglossal nerves and the recurrent branch of the vagus. The thoracic oesophagus is innervated by the vagus and by nerves from the coeliac plexus. The intramural nerves are distributed as myenteric, submucosal, muscle and mucosal plexuses (Iwanow, 1930; Kolossow *et al.*, 1932; Ábrahám, 1936). For a summary of the complex and often contradictory physiological and pharmacological literature on the innervation the reader is referred to Bolton (1971), Ziswiler and Farner (1972), Bennett (1974) and Sturkie (1976a).

F. *The oesophagus in digestion*

The main function of the oesophagus is to carry food between the pharynx and the stomach and to ensure that the stomach has an almost continuous supply of food. Most studies on this movement have been carried out on the domestic species, especially the fowl, and are reviewed by Hill (1971b), Ziswiler and Farner (1972), Hill and Strachan (1975) and Sturkie (1976a). Food is transported down the oesophagus by peristaltic waves, the contractions cranial to the crop being much more rapid than those in the oesophagus caudal to the crop (Pintea *et al.*, 1957). This motility appears to be controlled by cholinergic excitatory fibres, the oesophagus caudal to the crop also receiving an adrenergic inhibitory innervation which would account for its relatively slow rate of contraction (Ohashi, 1971). When the gizzard is empty, food in the oesophagus travels directly to the stomach, the entrance to the crop being closed by the contraction of the longitudinal muscle layer of the oesophagus. With relaxation of the muscle further food is diverted into the crop where it is stored (Ihnen, 1928; Ashcraft, 1930; Halnan, 1949; Vonk and Postma, 1949). In species which have no crop, food may be stored throughout the whole length of the oesophagus. Within a well-developed type of crop like that of the domestic fowl, the stored food undergoes softening and swelling. Entry of the food to the crop inhibits its contractions for a while. Using radiography, Pastea *et al.* (1968a) demonstrated that movements of the crop and gizzard are integrated and that the gizzard may have a pace-maker role. Hill and Strachan (1975), however, found that gizzard activity, in turn, is affected by the physiological state of the crop. They postulated that the inhibition of gizzard contractions which follows distension of the crop can be explained by the increase in proventricular acid-secretion demonstrated to occur with crop distension by Rouff and Sewing (1971), the resulting increase in acidity in the duodenum possibly triggering off an inhibitory mechanism. The movement of food from the crop back to the oesophagus was investigated by Macowan and Magee (1932). They found that a bolus is first formed in the crop by a strong wave of contraction which starts in the ventral wall. As the wave moves caudally, the indentation so formed gets deeper and when it reaches the caudoventral border of the crop it is joined by another constriction in the opposing wall of the oesophagus. Separation of the bolus from the rest of the food occurs between the constrictions, and the bolus then passes into the thoracic oesophagus. Return of the bolus, in part at least, to the crop, can occur by retroperistalsis (Vonk and Postma, 1949). Retroperistalsis is also involved in the regurgitation of food which has been stored in the adult crop or oesophagus for feeding to the young of some species including the Accipitridae, Ciconiidae, Spheniscidae, Columbidae, Psittacidae and some passerines (Fisher and Dater, 1961; Ziswiler, 1967a,b; Ziswiler and Farner, 1972).

The very large and muscular crop of the vegetarian Hoatzin (*Opisthocomus hoazin*), unlike in other species is the primary organ of physical digestion, the gizzard, which is usually the site for mechanical breakdown of food in birds, being very reduced (the weight of the cervical crop and thoracic oesophagus in the Hoatzin is fifty times greater than that of the gizzard) (Gadow, 1891b; Böker, 1929; Sick, 1964). Within the crop, the leaves on which the bird mainly feeds, are squeezed and ground down with the aid of the mucosal ridges and the hard epithelial lining in preparation for chemical digestion in the lower parts of the tract.

The oesophagus does not appear to be very important in the chemical phase of digestion (Ziswiler and Farner, 1972; Sturkie, 1976b). In the adults of two species the oesophagus proper is the source of nutritive fluid for the chicks. The oesophagus of the Greater Flamingo (*Phoenicopterus ruber*) produces seasonally in both male and female birds, a red-coloured juice which is regurgitated and fed to the young birds (Lang *et al.*, 1962; Lang, 1963; Wackernagel, 1964; Studer-Thiersch, 1967). The juice is a merocrine secretion of the acinar glands which are distributed throughout the whole length of the oesophagus and is composed of blood (0·5%), protein (8·7%), glucose (0·2%), fat (approximately 18%), and canthoxanthine, xanthophyll and traces of β-carotene. In the male Emperor Penguin (*Aptenodytes forsteri*) which incubates the single egg and carries out the initial brooding of the chick, the oesophagus produces a nutritive fluid which is fed to the newly hatched chick (Prévost, 1961; Prévost and Vilter, 1963). During this period the oesophageal wall is extremely thick and vascular and there is proliferation of the cells of both the epithelium and the glands. The nutritive fluid is produced by desquamation of the fat-laden epithelial cells and by glandular secretion rich in mucopolysaccharides and glycoproteins. The whitish fluid accumulates in the stomach prior to being regurgitated. Its dry content includes lipids (28%), proteins (59%), carbohydrates (7·8%) and ash (4·6%).

The crop in both male and female pigeons and doves (Columbidae), in addition to its other functions, produces during the breeding season crop milk, a holocrine secretion of the epithelium which is fed to the young. The production of this milk is almost totally restricted to the lateral diverticula, the walls of which become markedly thickened and vascular (Litwer, 1926; Beams and Meyer, 1931; Niethammer, 1931, 1933; Weber, 1962; Dumont, 1965). The first histological signs in the lateral diverticula of milk production occur about the 6th day of incubation and include an increase in the number of blood vessels in the lamina propria and proliferation of the overlying epithelium. At about the 10th day of incubation numerous fat droplets appear in the epithelial cells distal to the stratum germinativum. These fat-laden cells by about the 16th day desquamate to form a yellowish-white milk. Basically similar histological changes to these, although at a lower intensity, are present in the parts of the median diverticulum adjacent to the lateral diverticula, but no histological signs of milk production occur in the middle part of the median diverticulum. The secretion of crop milk continues until about two weeks after hatching. The milk is composed of proteins (12·4%), lipids (8·6%), ash (1·37%) and water (74%) (Vandeputte-Poma, 1968). Thus the composition resembles that of mammalian milk since it is very rich in fat and protein. Unlike mammalian milk, however, it lacks carbohydrates and calcium. Production of milk is, as in mammals, controlled by the hypophysial hormone, prolactin. The response of the crop epithelium to intramuscular or intradermal injections of prolactin is the method which is most frequently used to bioassay this hormone. For the first few days after hatching, the chicks receive only milk but after about the fourth day they are fed progressively greater amounts of other types of food. In the Wood Pigeon (*Columba palumbus*) there is a rhythm of feeding activity during the incubation and brooding periods in the adults which is probably typical of all pigeons in the wild (Murton, 1964). This rhythm is associated with the fact that the crop milk which is fed to the chicks during the first few days must be uncontaminated by other material, and it is essential that the crop of the parent bird is empty during the formation of milk. For this reason a pattern of alternate brooding by the parents has developed, one parent on its long on-nest period producing milk but not ingesting any food, whilst the other parent in the long off-nest period feeds when no milk is being produced.

G. Oesophageal sacs

During the breeding season, the cervical oesophagus of a number of species is inflated with air and is used in "showing-off" or as a resonating chamber for the production of mating calls. The anatomical basis for this phenomenon has been investigated in the following species: the Sage Grouse (*Centrocercus urophasianus*) (Clarke *et al.*, 1942; Honess and Allred, 1942), the American Bittern (*Botaurus lentiginosus*) (Chapin, 1922), the Australian Bustard (*Choriotis australis*) and the Great Bustard (*Otis tarda*) (Garrod, 1874a,b; Niethammer, 1937), the Painted Snipe (*Rostratula benghalensis*) (Niethammer, 1966), the Prairie Chicken (*Tympanuchus cupido*) (Lehmann, 1941), the Little Button-quail (*Turnix sylvatica*) (Niethammer, 1961) and the Amazon Umbrella-bird (*Cephalopterus ornatus*) (Sick, 1954). Generally the filling of the oesophagus with air is restricted to the male birds but in *Rostratula* and *Turnix* it occurs only in females. Enlargement of the oesophagus (*saccus esophagealis*) can be either in the form of a bilaterally symmetrical spherical widening of the tube as in *Centrocercus* (Fig. 3.10), or by distension of a crop-like diverticulum in the mid-cervical region as in *Otis* (Fig. 3.3). Few details are available of the structure of the enlargements or diverticula. In *Rostratula* the unfilled enlargement in the female oesophagus is twice the size of the equivalent region of the male oesophagus and the longitudinal mucosal folds and mucosal glands are more numerous (Niethammer, 1966). In contrast, the well-differentiated sac-like diverticulum of *Otis* has neither the mucosal folds nor the glands which occur in the other regions of the oesophagus. Glands are also absent in the oesophageal expansion of *Turnix* (Niethammer, 1961). The circular muscle of the oesophagus in both *Centrocercus* and *Otis* is especially well-developed immediately caudal to the expansion.

The process of inflation of the oesophagus is described in detail in the male Sage Grouse (*Centrocercus urophasianus*) by Honess and Allred (1942) and Clarke *et al.* (1942). In preparation for strutting and mating a number of changes occur in the neck region including alterations to the plumage, extension of the skin of the neck and breast, the development of yellow pigmentation on the bare areas of skin, and hypertrophy of the cervical musculature, connective tissue and the cervical portion of the oesophagus. During the strut, the bird produces a mating call which was described by Honess and Allred (1942) as a soft, hollow "plopping" sound. Air reaches the oesophagus by travelling from the air sacs and lungs via the trachea and glottis. It is prevented from escaping rostrally by the tongue which is raised dorsally, and by the close apposition of the floor and roof of the pharynx. The inflated part of the oesophagus is relatively elastic and in the breeding male has a capacity twenty-five times greater than that of the oesophagus in the non-breeding adult male. The inflated oesophagus appears as a bilobed structure since it is constricted in the ventral midline by the presence

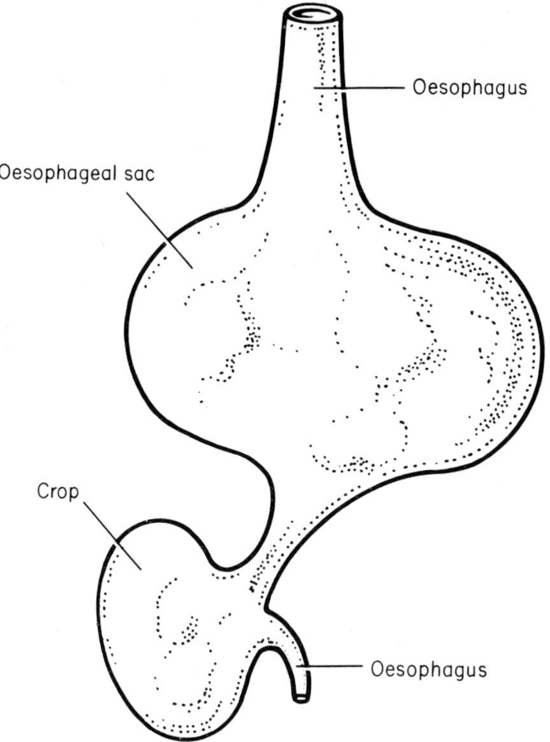

Fig. 3.10. Oesophageal sac and crop of the Sage Grouse (*Centrocercus urophasianus*), ventral view. Based on Honess and Allred (1942).

of the trachea, the sternohyoid muscles and the relatively inelastic skin of the ventral median part of the neck. Simultaneous contraction of cranial and caudal groups of striated neck muscles tighten the skin over the oesophagus. This compresses the air in the expanded oesophagus and distends the relatively elastic bare areas of skin to produce two hemispherical swellings, one on either side of the neck. Vibration of the membranes of the bare areas results in the "plopping" sound. Air is prevented from passing down the oesophagus from the expanded portion by contraction of the well-developed circular muscle. In deflation of the oesophagus, the air escapes via the mouth.

III. Stomach

A. Definitions. General form of stomach

The avian stomach (*gaster*) consists basically of two chambers which may or may not be externally distinguishable from each other (Fig. 3.11). The cranial chamber is the proventriculus (*proventriculus*) or glandular part (*pars glandularis*) which is continuously cranially with the oesophagus and secretes gastric juice. The caudal chamber is the gizzard (*ventriculus*) or muscular part (*pars muscularis*) which functions as the site of gastric proteolysis and in many species as the organ of mechanical digestion. Either chamber of the stomach may have an important storage function. Lying between the proventriculus and gizzard is the intermediate zone (*zona intermedia gastris*). The pyloric part (*pars pylorica*) connects the gizzard with the duodenum. The general appearance of the stomach varies considerably between different groups of birds and seems to be determined mainly by diet (Swenander, 1902). For extensive data on the weight of the stomach in a large number of species consult Magnan (1911a, pp. 58–72). Much of the total weight of the organ is due to the gizzard except in the carnivorous piscivores in which the weights of the proventriculus and gizzard are relatively similar. Comparisons of the weights of the proventriculus and gizzard between groups of birds with different feeding habits have shown the existence of an inverse relationship between the development of the two chambers (Magnan, 1911b).

Two basic types of stomach can usually be identified depending mainly on the extent to which the gizzard is adapted for the physical preparation of food. The first type is relatively undifferentiated and is characteristic of carnivorous and piscivorous species in which the primary requirement is for an expansible type of organ for holding food, the relatively soft nature of the diet necessitating little mechanical treatment. In these birds, both chambers are specialized for storage, the stomach being a relatively large sac-like structure in which the junction between the two chambers is often difficult to identify externally, especially when the chambers are dilated with food (Fig. 3.12). Typical examples of this type of stomach are shown in Fig. 3.11a,b,c,d,e. Usually one of the chambers is better developed than the other, as for example in the Great Cormorant (*Phalacrocorax carbo*) in which the proventriculus is the larger (Fig. 3.11a) and in the Little Owl (*Athene noctua*) in which the gizzard is the better developed (Fig. 3.11d). The gizzard in this undifferentiated type of stomach is always thin-walled and relatively poorly muscled.

The second type of stomach is highly differentiated and is characteristic of omnivorous, insectivorous, herbivorous and granivorous species in which the diet consists predominantly of tough food requiring mechanical treatment before being acted on by the gastric juice. In these species, therefore, the morphology of the stomach is the direct opposite of that in fish and meat eaters, since the primary requirement now is for a powerful triturating machine, the need for storage being either relatively unimportant or fulfilled by the oesophagus. This second type of stomach typically consists of a small, spindle-shaped proventriculus and a massively developed gizzard in which the muscle tunic is thick and unevenly developed, the junction between the proventriculus and gizzard being marked by a distinct constriction or isthmus (Figs 3.11j, 15).

In many groups of birds including fruit-eaters, grain-eating insectivores and species feeding mainly on shell-fish, the form of the stomach is somewhere between that of the two basic types described above and depends upon whether the role of the gizzard is mainly that of a storage organ or an organ concerned with the physical digestion of food. In these species the extent to which each chamber is used for storage depends on the development in the proventriculus of the glands which secrete gastric juice, and in the gizzard of the muscle tunic, since if either is especially well-developed more than usual expansion of the chamber is prevented. Examples of intermediate forms of stomach are shown in Fig. 3.11f,h.

B. The stomach in situ

The following account of the position and relations of the stomach is mainly applicable to species in which the proportions of the proventriculus and gizzard are those of the highly differentiated second type of stomach. When the proportions differ from this, the position and relations of the organ in many cases are also affected and therefore do not necessarily conform to the description given below (see for example the position of the stomach in procellariiform species, p. 106). The present description is taken mainly from the account in *Gallus* by Kern (1963), Schummer (1973, pp. 48–51) and McLelland (1975) and can be related to Fig. 3.7 of the abdominal contents in *Melopsittacus undulatus*.

The proventriculus extends craniocaudally somewhat ventrally and to the left on the left side of the body adjacent to the last few thoracic vertebrae and the most cranial lumbosacral vertebrae. Most of the proventriculus *in situ* is concealed by the liver, on the left lobe of which it makes an impression. Medially, the proventriculus is related to the spleen. Dorsally it is covered by the cranial thoracic and abdominal air sacs which separate it from the lung, the left testis, and in the female in lay, the left ovary and oviduct. In the sexually inactive female, the ileum and caeca may also lie dorsal to the proventriculus. The craniocaudal axis of the gizzard extends somewhat ventrally and to the right in the left ventral part of the body cavity, so that the ventral part of the organ may lie to the right of midline. In *Gallus*, the gizzard *in situ* is a much more prominent organ than the proventriculus because of its larger size and because most of its ventral and left surfaces lie directly on the body wall. Cranially the gizzard is related to the right and left lobes of the liver on which it makes impressions. The dorsal surface and the dorsal part of the right surface are covered by the left abdominal air sac which separates the gizzard from the left testis, ovary and oviduct, and the rectum and left caecum. In the male, and in the immature and sexually inactive female, part of the jejunum also lies dorsal to the gizzard. The ventral part of the right surface of the gizzard lies close to the duodenum and pancreas. Caudally, the gizzard is related to the duodenum, jejunum and the distal parts of the caeca.

C. Proventriculus

1. Size and shape of proventriculus

The data on the stomach weight provided by Magnan (1911a, pp. 58–72,b) show that the proventriculus is best developed in piscivores, carnivorous piscivores, frugivores and granivorous insectivores which feed on relatively large items or portions of food. Examples of stomachs with an obviously well-developed form of proventriculus are shown in Fig. 3.11a,b,c,g,i. Typically in

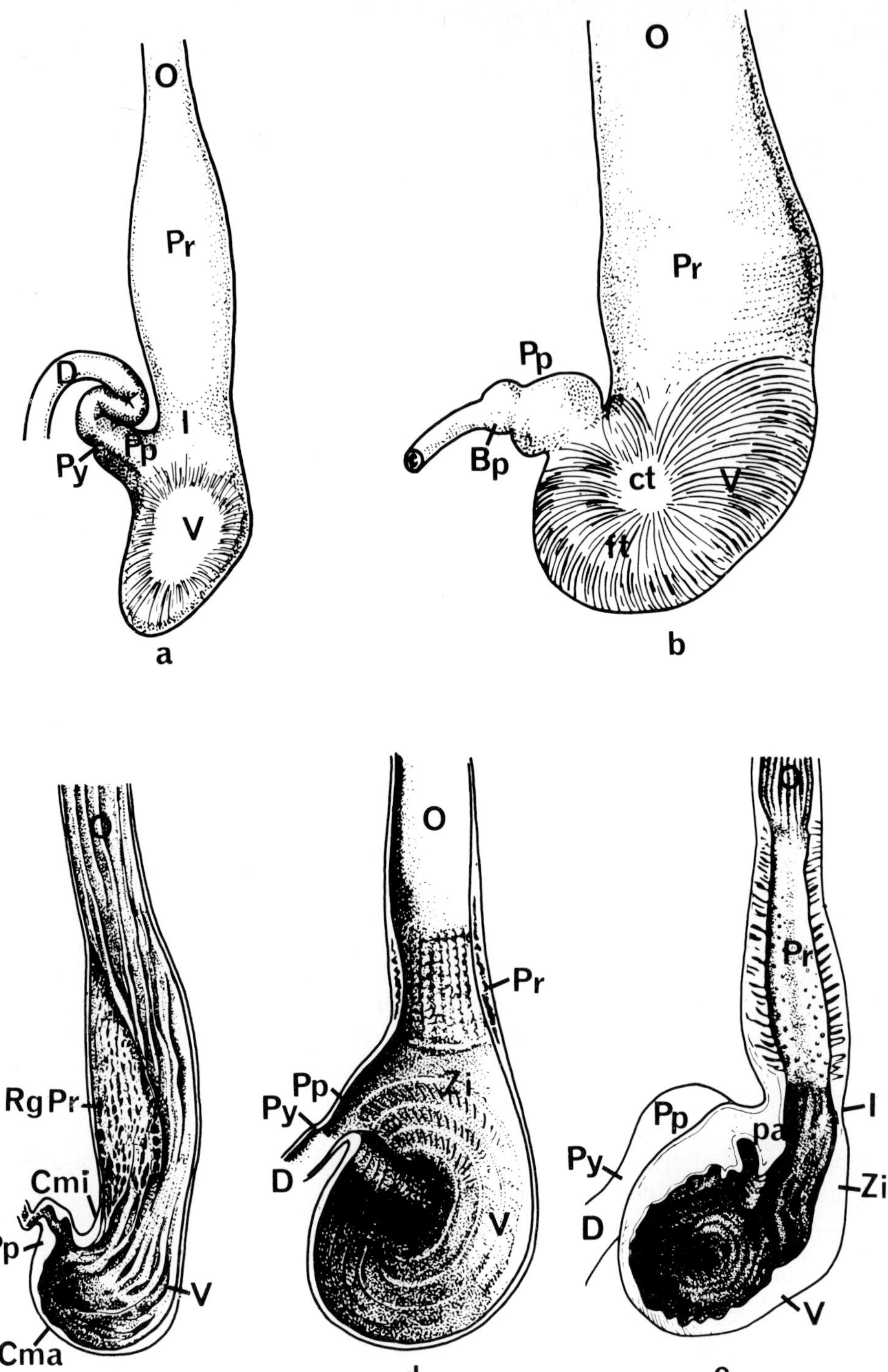

Fig. 3.11. Stomachs, ventral views. (a) Great Cormorant (*Phalacrocorax carbo*), ×0·5. (b) Gray Heron (*Ardea cinerea*), ×0·73. (c) Little Blue Penguin (*Eudyptula minor*), ×0·45. (d) Little Owl (*Athene noctua*), ×0·83. (e) Great-crested Grebe (*Podiceps cristatus*), ×0·68. (f) A gull (*Larus*), ×0·82. (g) Ostrich (*Struthio camelus*), ×0·21. (h) Orange-winged Parrot (*Amazona amazonica*), ×0·88. (i) Parkinson's Petrel (*Procellaria parkinsoni*), ×0·71. (j)

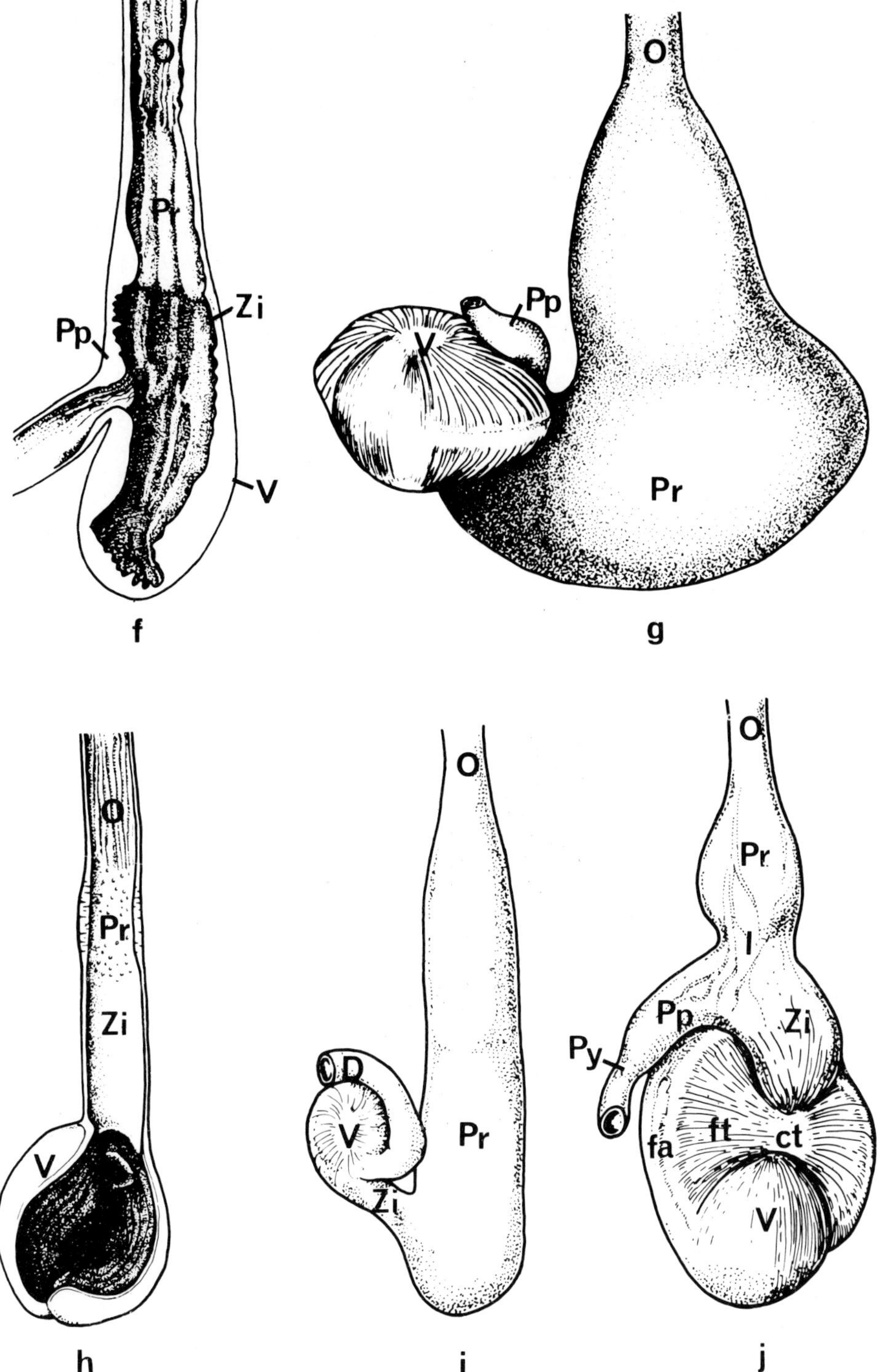

Peafowl (*Pavo cristatus*), ×0·37. Bp, pyloric bulb; Cma, greater curvature; Cmi, lesser curvature; ct, tendinous centre; D, duodenum; fa, annular surface; ft, tendinous surface; I, isthmus; O, oesophagus; pa, angular fold; Pp, pyloric part; Pr, proventriculus; Py, pylorus; RgPr, glandular region of proventriculus; V, gizzard; Zi, intermediate zone. From Pernkopf and Lehner (1937).

these species the proventriculus is an extensible sac-like structure which as in the Gray Heron (*Ardea cinerea*) (Fig. 3.11b) may not be easily distinguished externally from the gizzard (Magnan, 1912a). The lumen of this enlarged form of proventriculus is characteristically wide and in the very elongated organ which occurs in most piscivores may have a greater capacity than that of the gizzard. The proventriculus of procellariiform species is especially well-developed (Forbes, 1882a; Cazin, 1887b; Pernkopf and Lehner, 1937; Matthews, 1949). In these birds the oesophagus passes without marked narrowing into an enormously long, thin-walled proventriculus which extends to the caudal end of the abdominal cavity before bending cranially and to the right to join the relatively small, twisted gizzard (Fig. 3.11i). Since the opening to the duodenum is displaced to the left side of the gizzard the proximal part of the duodenum first ascends and then crosses to the right before forming the typical duodenal loop enclosing the pancreas. A very large proventriculus displacing a relatively small gizzard to its right side is also present in the Ostrich (*Struthio camelus*)

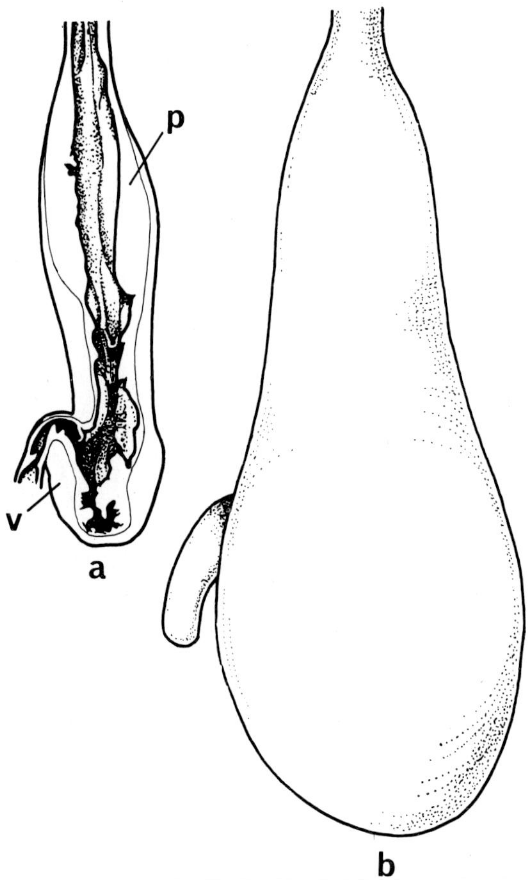

Fig. 3.12. Stomach of the Golden Eagle (*Aquila chrysaetos*), (a) contracted, and (b) dilated. P, proventriculus; V, gizzard. Approximately two-thirds natural size. From Swenander (1902).

(Pernkopf and Lehner, 1937). Unlike in the gizzard of tubinarine species, however, the opening to the duodenum in *Struthio* is dorsal (Fig. 3.11g).

Most other birds, i.e. carnivores, carnivorous insectivores, testacivores, insectivores, omnivores, granivores and herbivores, tend to have a relatively small proventriculus (Magnan, 1912). Sometimes as in the Wood Duck (*Aix sponsa*), the proventriculus is so reduced in size that it is difficult to distinguish externally from the oesophagus. Although the proventriculus of carnivorous species is relatively small it is nevertheless a highly distensible organ (Fig. 3.12), the external form being extremely variable and determined to a great extent by the nature of the food. Examples of stomachs with a strikingly small form of proventriculus are shown in Fig. 3.11d,h,j.

2. *Interior of proventriculus*

The transition from the surface relief of the oesophagus to that of the proventriculus is usually abrupt and generally the inner wall of the proventriculus is without the huge folds or ridges which characterize the oesophagus (Fig. 3.11e). In a few species, however, most of which are fish and meat eaters, carnivorous insectivores and omnivores, such mucosal ridges (*plicae proventriculares*) do occur (Fig. 3.11f), and serve to increase the storage capacity of the organ (Cazin, 1887b; Swenander, 1902; Magnan, 1912). The well-developed folding of the proventricular wall in procellariiform species according to Matthews (1949), is a device which enables a greater number of compound glands to be present. In the Greater Rhea (*Rhea americana*), the junction of the oesophagus and proventriculus is marked by a 1 cm high circular fold, within which the muscular tunic of the proventriculus is thickened (Feder, 1972b). Usually visible with the naked eye on the inner surface of the proventriculus are the openings of the compound glands which secrete gastric juice (Fig. 3.11e). Sometimes as in *Gallus* for example, the openings are situated on papilla-like elevations (*papillae proventriculares*) of the mucous membrane (Fig. 3.16). Between the openings, the mucosa is usually thrown into a variably developed system of anastomosing rugae and sulci (*rugae proventriculares, sulci proventriculares*); occasionally the rugae which arise on the papillae as in *Gallus*, *Meleagris* and *Anser* are arranged concentrically (Pernkopf and Lehner, 1937). For further details of the arrangement of the rugae in different species see Swenander (1902).

3. *Structure of proventriculus*

The microanatomy of the proventriculus has been investigated very extensively in *Gallus* by Chodnik (1947), Calhoun (1954, pp. 49–53), Aitken (1958), Toner (1963), Michel (1971), Hodges (1974, pp. 47–54) and Horváth (1974). Many useful comparative data are also available in the accounts of Cattaneo (1884), Cazin (1887a,b), French (1898), Schreiner (1900), Swenander (1902), Groebbels (1924), Matthews (1949), Luppa (1962), Ziswiler (1967a,b), Feder (1972b,c) and Ziswiler and Farner (1972). The wall of the proventriculus consists of four layers: the mucous membrane, submucosa, muscular tunic and serosa.

The mucous membrane (*tunica mucosa gastris*) is lined by a single layer of columnar cells which secrete mucus. In *Gallus*, Chodnik (1947) found the number of mucous granules in an epithelial cell varied with its functional state. Immediately following a meal, the number of granules markedly decreased, whilst two hours after feeding, the apical portion of the cell contained as many granules as when food with withheld. This mucous secretion acts as a protective lining for the surface of the epithelium. In the majority of accounts, both superficial and deep glands (*gll. proventriculares superficiales, gll. proventriculares profundae*) are described in the lamina propria. The superficial glands are reported to be of the simple mucous secreting tubular type and open into the lumen at the bases of the sulci. In *Gallus*, however, some doubt has recently been expressed by several authorities including Hodges (1974, p. 49) and Horváth (1974) about the correctness of identifying such structures as glands. Most of the lamina propria is taken up by the deep glands (Fig. 3.16) which are large compound structures secreting gastric juice. The distribution and structure of these glands are dealt with below. The muscularis mucosae (*lamina muscularis mucosae*) is organized into two layers, a variably developed inner layer lying between the compound glands and the surface epithelium and a relatively thin outer longitudinal layer lying between the compound glands and the submucosa (Swenander, 1902; Feder, 1972c). The existence of an inner layer in the muscularis mucosae probably explains why in many descriptions, e.g. Cazin (1887b), the compound glands are allotted to the submucosa. The two layers of the muscularis mucosae are interconnected by muscle bundles between the compound glands. As suggested already by Pernkopf and Lehner (1937), the splitting of the proprial muscle, is likely to be associated in some way with the mechanism of emptying of the glands.

The submucosa (*tela submucosa gastris*) is extremely thin and contains the submucosal nerve plexus.

The smooth muscle of the muscular tunic (*tunica muscularis gastris*) (Fig. 3.20) is usually arranged into a well-developed inner circular layer (*stratum circulare*) and a much thinner outer longitudinal layer (*stratum longitudinale*). In *Larus* the circular layer is reported to be thickened at the junction with the oesophagus (Schreiner, 1900). The longitudinal layer may be incomplete as in some seed-eating passerines (Ziswiler, 1967a) or totally absent as in the Northern Fulmar (*Fulmarus glacialis*), the King Eider (*Somateria spectabilis*) and the Gray Parrot (*Psittacus erithacus*) (Swenander, 1902). Between the circular and longitudinal muscle layers is the myenteric nerve plexus.

The microstructure of the proventriculus appears to be influenced by the type of food consumed (Rybicki, 1959; Rybicki and Lubánska, 1959; Cymborowski, 1968).

4. *Deep glands of proventriculus*

The deep proprial glands of the proventriculus (*gll. proventriculares profundae*) show considerable interspecific variation in their distribution (Fig. 3.13), Cazin (1887b) recognizing a connection between the distribution of the glands and the relative size of the organ. Thus when the proventriculus is small and capable of little enlargement, the glands are usually arranged close to one another in an almost uniformly thick belt extending over most of the organ (Fig. 3.13b). Occasionally as in the Wryneck (*Jynx torquilla*) the belt of glands is extremely narrow and then fails to reach the cranial part of the proventriculus (Swenander, 1902). When the proventriculus is enlarged and distensible, the glands tend to be regularly scattered with wide interstices or else are concentrated into discrete areas (*regiones glandulares*) of various shapes and sizes leaving part of the proventricular wall free. One relatively common form of aggregation of the glands is into longitudinal tracts (the Juga of Swenander, 1902) which appear on the inner surface of the proventriculus as broad elevations (Fig. 3.13a). According to Swenander (1902), two tracts of glands as in owls (Strigiformes) and the Red-breasted Merganser (*Mergus serrator*) is the most usual arrangement, but sometimes as in *Aquila chrysaetos*, *Pandion haliaetus*, *Buteo buteo*, *Falco peregrinus*, *Pernis apivoris*, *Mergus merganser*, *Rallus aquaticus*, *Somateria mollissima* and *Grus grus* the tracts are more numerous. In some species, e.g. the Golden Eagle (*Aquila chrysaetos*), the glandular tracts are unevenly developed (Swenander, 1902; Steinbacher, G., 1934). Instead of tracts, the glands are sometimes arranged into patch-like aggregations which usually have a well-defined circular, oval or triangular outline (Figs 3.11c, 13d,e). Two such patches have been reported in darters (Anhingidae) (Garrod, 1878b; Forbes, 1882b; Cazin, 1884, 1887b), the Greater Adjutant Stork (*Leptoptilos dubius*) (Newton and Gadow, 1896, p. 916), an ibis (Mitchell, 1895a) and the Great Cormorant (*Phalacrocorax carbo*) (Swenander, 1902), whereas a single patch of glands occurs in the Ostrich (*Struthio camelus*) (Cazin, 1887b; Duerdon, 1912; Pernkopf and Lehner, 1937), the Greater Rhea (*Rhea americana*) (Remouchamps, 1880; Cazin, 1887b; Pernkopf and Lehner, 1937; Feder, 1972b), the Black Woodpecker (*Dryocopus martius*) and the Green Woodpecker (*Picus viridis*) (Pernkopf

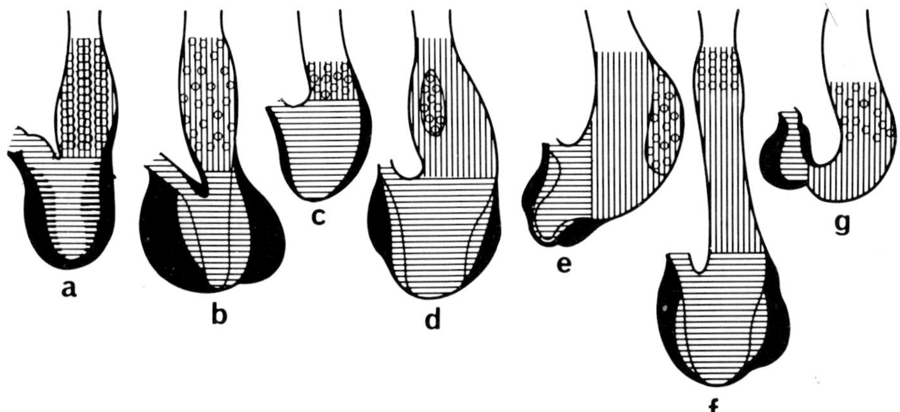

Fig. 3.13. Some variations in the distribution of the deep glands of the proventriculus. The proventriculus and intermediate zones are represented by vertical lines, the gizzard by horizontal lines. (a) *Gyps*, glands arranged in longitudinal tracts. (b) *Pavo*, glands uniformly distributed. (c) *Dacelo*, reduced form of proventriculus. (d) *Picus*, (e) *Struthio*, (f) *Psittacus* and (g) *Procellaria*, glands aggregated. The intermediate zone which lies between the proventriculus and gizzard and in which compound glands are absent is least developed in (a), (b) and (c) and most extensive in (f) and (g). From Pernkopf (1937).

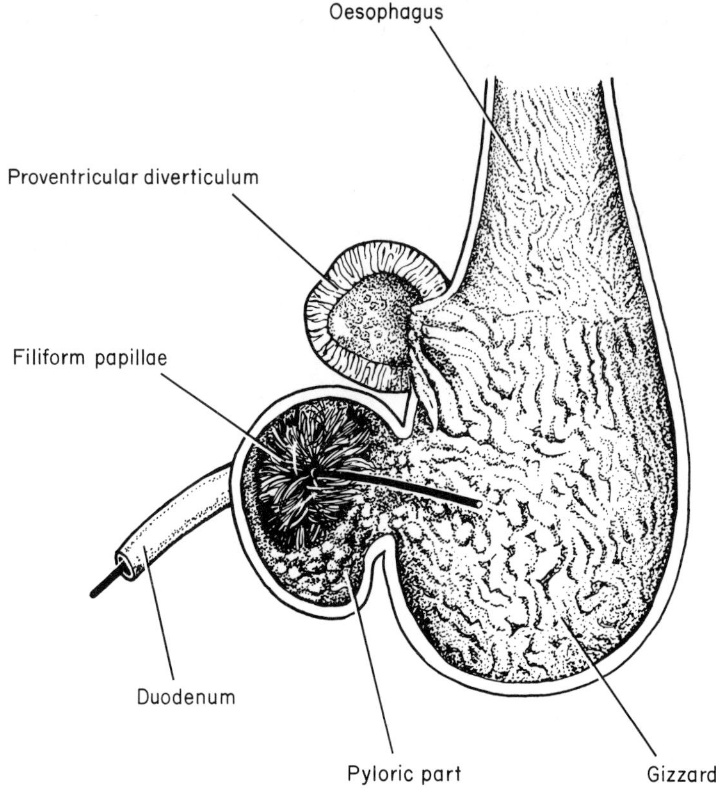

Fig. 3.14. Stomach of the Anhinga (*Anhinga anhinga*) which is characterized by a proventricular diverticulum containing the deep proventricular glands and a well-developed pyloric part. The probe in the pyloric part is inserted into the pyloric orifice which is guarded by hair-like papillae. From Garrod (1876b).

and Lehner, 1937), some screamers (Anhimidae) (Garrod, 1876a; Mitchell, 1895b) and a number of penguins (Spheniscidae) (Watson, 1883). In the screamers (*Chauna* spp.) the glandular patch extends caudally from an annular belt of glands lying close to the junction with the oesophagus. By far the greatest form of glandular concentration exists in the Anhinga (*Anhinga anhinga*) (Fig. 3.14) in which the glands are situated in a diverticulum (*diverticulum proventriculare*) arising from the right side of the organ and which communicates with the main cavity via a narrow opening (Garrod, 1876b). As observed by Steinbacher, G. (1934), all forms of aggregation of the glands whether into tracts, patches or a diverticulum, may be adaptations which permit much greater expansion of the proventriculus than is possible if the glands are more uniformly distributed.

The number of deep glands which are present does not seem to be related to either the size of the proventriculus or the diet. The size of the glands, however, may be influenced by the type of food consumed (Rybicki, 1959; Rybicki and Lubánska, 1959; Cymborowski, 1968). In most birds the glands are unilobular, only a few species, e.g. *Spheniscus demersus*, *Meleagris*, *Gallus*, *Anser*, *Struthio* and *Rhea*, having multilobular glands (Cazin, 1887b). Thus there appears to be no relation between the form of the glands and diet. In *Gallus* each lobule is arranged more or less perpendicular to the luminal surface and consists of many straight closely-packed tubular alveoli extending outwards from a central cavity and draining into the cavity via tertiary ducts (Chodnik, 1947; Calhoun, 1954, pp. 49–53; Menzies and Fisk, 1963; Michel, 1971; Hodges, 1974, p. 51). The central cavity is drained by a secondary duct, all the secondary ducts of a gland uniting to form the primary duct which opens into the lumen. The primary, secondary and tertiary ducts and the central cavity are lined by a simple columnar epithelium almost identical to that lining the lumen of the organ. However, in *Gallus* only the epithelial cells of the primary duct contain PAS-positive mucous granules, the granules being discharged after feeding (Chodnik, 1947). The simple glandular alveolar epithelium consists mainly of one exocrine cell type, the oxynticopeptic cell, which secretes both hydrochloric acid and pepsinogen. The deep proventricular glands of birds therefore resemble the gastric glands of other vertebrates except mammals in which the acid and zymogenic constituents of the gastric juice are secreted by two cell types. An excellent description of the cytology of the oxynticopeptic cell during all phases of digestion is provided in *Gallus* by Chodnik (1947). The fine structure of the cell has been studied by Selander (1963), Toner (1963) and Horváth (1974). One of the characteristic features of the oxynticopeptic cell is that unlike other epithelial cells of the gastrointestinal tract, it is not joined to adjacent cells by apical terminal bars and desmosomes, the terminal bars instead lying near the base of the cell. As Toner pointed out, the absence of apical terminal bars and desmosomes augments six-fold the secretory surface area of the cell, the intercellular clefts then being analogous to the branching canaliculi which increase three-fold the surface area available for secretion by the mammalian oxyntic cell. Other features of the oxynticopeptic cell are typical of the zymogen and oxyntic cells of mammals, the detailed appearance, however, varying with the functional state of the cell. During both fasting and digestion, most of the oxynticopeptic cells are packed with secretory granules. In fasting, the number of secretory granules increases in those cells which have recently been depleted of granules, the maximum accumulation occurring after twenty-four hours. Approximately thirty minutes after a meal the number of secretory granules in the cell starts to decrease through evacuation. Three hours after a meal the number of secretory granules begins to increase and six hours after feeding they are as numerous as in birds which feed *ad libitum*. Chodnik (1947) concluded that the secretory cells act as autonomous units, since they are not all in the same state of functional activity. He observed that the smallest accumulation of secretory granules always occurred in the cells at the bottom of the alveoli. In the fasting birds, the secretory cells contain large amounts of lipids (Wight, 1975). Lipid staining of the surface epithelium and the cells of the deep glands was also observed in procellariiform species by Matthews (1949).

Contrary to an earlier suggestion (Matthews, 1949), the deep proventricular glands in procellariiform species are not the source of the stomach oil which is regurgitated by these birds as food for the chick or in defense. The bulk of this oil apparently is of dietary origin (Lewis, 1969; Cheah and Hansen, 1970a,b; Clark and Prince, 1976).

In addition to the oxynticopeptic cell, the epithelium of the tubular alveoli contains a number of granular endocrine cells, which are dealt with in the chapter on the Endrocrine System in Vol. 2.

For further data on the microanatomy of the deep proventricular glands see Bergmann (1862), Cazin (1887a,b), Schreiner (1900), Swenander (1902), Groebbels (1932, pp. 481–482), Pernkopf and Lehner (1937), Luppa (1962), Ziswiler (1967a,b), Feder (1969, 1972b,c) and Ziswiler and Farner (1972).

5. *Gastric juice*

Gastric juice is composed principally of hydrochloric acid, mucus and the proteolytic enzyme, pepsin. The basal secretory volume in *Gallus* recorded by Long was 15·4 ml/h and contained an acid concentration of 93 mEq/l and a pepsin concentration of 247 Pu/ml. According to Joyner and Kokas (1971) the pH of the gastric juice in fasted chickens is 2·6. As in mammals, stimulation of the vagus provokes the secretion of juice (Friedmann, 1939). A cephalic phase of secretion appears to be present in ducks (Anatidae) (Walter, 1939), and the Barn Owl (*Tyto alba*) (Smith and Richmond, 1972), but it has not been demonstrated in *Gallus* (Ziswiler and Farner, 1972). A gastrin mechanism also appears to exist since using bioassay, radioimmunoassay and immunohistochemistry gastrin has been localized in the stomach and small intestine (Ketterer *et al.*, 1973; Olowo-Okorun and Amure, 1973; Larsson *et al.*, 1974; Polak *et al.*, 1974).

D. *Intermediate zone of stomach*

In most species, the region of the stomach immediately caudal to the proventriculus has a structure between that of the proventriculus and gizzard and is therefore called the intermediate zone (*zona intermedia gastris*) (Figs 3.11e,f,h,i,j, 13, 16). The intermediate zone has no compound glands so that its wall is thinner and less rigid than in the proventriculus. This suggests that the intermediate zone is a region of the stomach which undergoes relatively large changes in calibre, and certainly in *Gallus* the zone is reported to be rich in elastic tissue (Calhoun, 1954, p. 53). Since compound glands are absent, the muscularis mucosae is arranged as a single layer. There is an internal lining, secreted by the inner tubular glands which resembles that of the gizzard in being thicker and firmer than the mucous secretion of the proventriculus although its internal organization is usually less developed than in the gizzard secretion. Accordingly Hodges (1974, p. 55) described this layer in *Gallus*, as a mixture of mucoid secretion and the secretion of the gizzard glands. Externally, as in *Gallus*, the intermediate zone can often be distinguished from the rest of the stomach by its lighter colour. Frequently the zone is separated from the proventriculus by a constriction or isthmus (*isthmus gastris*) (Figs 3.11e,j, 15). Internally the zone usually differs from other parts of the stomach by the relative smoothness of its mucosal surface, the plicae and sulci being much less developed or absent and there being no papillae. An intermediate zone occurs in birds with widely different feeding habits (Groebbels, 1932, p. 458) but between species the extent of the zone varies considerably (Cazin, 1887b;

Swenander, 1902; Groebbels, 1932, p. 458). In *Gallus*, the relatively small intermediate zone (Fig. 3.16) measures only about 0·75 cm in length when the total length of the proventriculus is 4·5 cm (Hodges, 1974, pp. 48, 55). Sometimes, however, the zone is very large as for example in the Black Woodpecker (*Dryocopus martius*) and the Orange-winged Parrot (*Amazona amazonica*) (Fig. 3.11h) in which it is longer than the proventriculus. A well-developed type of intermediate zone like that of *Drocopus martius* is often strongly folded on its inner surface so that the capacity of the zone is greatly increased. In the nectar and pollen feeding Loriidae such as the Red-flanked Lorikeet (*Charmosyna placentis*) the large intermediate zone replaces the relatively small gizzard as the site of gastric proteolysis (Steinbacher, G., 1934). In many birds, however, especially those in which the proventriculus and gizzard are separated by an isthmus, the intermediate zone probably functions mainly when contracted as a barrier separating the two chambers. Unfortunately there appears to be little information on the degree of development of the muscle of the intermediate zone, and whilst thickening has been described in a few species, e.g. *Anser* (Schepelmann, 1906) and the Greater Rhea (*Rhea americana*) (Feder, 1972b), it is not a widespread observation. Amongst the species in which an intermediate zone is reported to be absent are *Phalacrocorax carbo*, *Anas crecca*, *Anas acuta*, *Numenius arquata*, *Haematopus ostralegus*, *Charadrius hiaticula*, *Scolopax rusticola*, *Rallus aquaticus* (Swenander, 1902), and *Cuculus*, *Picoides*, *Caprimulgus*, *Apus*, *Corvus* and *Regulus* (Groebbels, 1932, p. 458).

E. Gizzard

1. Surface features of gizzard

The data on the weight of the stomach provided by Magnan (1911a, pp. 58–72) show that the gizzard (*ventriculus*) is much less developed in carnivorous and piscivorous species which live on relatively soft food than in birds like insectivores and granivores which feed on hard items requiring mechanical breakdown before being acted on by the gastric juice. Since the weight of the gizzard varies with the thickness of the muscle tunic, the weight is therefore an expression of the development of the muscle (Magnan, 1911c, 1912). The development of the muscle tunic determines to a large extent the external appearance of the gizzard. The literature on this aspect of gastric anatomy is particularly detailed, the following description being based mainly on the extensive accounts by Garrod (1872), Cazin (1887b,c), Gadow (1891a, pp. 676–677), Newton and Gadow (1896, pp. 917–918), Swenander (1902), Schepelmann (1906), Magnan (1912), Cornselius (1925), Groebbels (1932, pp. 471–472), Pernkopf and Lehner (1937), Grau (1943, pp. 1088–1089), Lucas and Stettenheim (1965), Ziswiler (1967a,b), Schummer (1973, pp. 50–51) and McLelland (1975).

The gizzard appears in its least differentiated form in the stomach of most fish and meat eaters as a relatively round or oval sac-like structure (Fig. 3.11a, b,c,d,e,i). Opening into the cranial part of it is the proventriculus, the boundary between the two organs often being externally indistinct. Ventrally and to the right is the opening to the duodenum. The short portion of the gizzard between

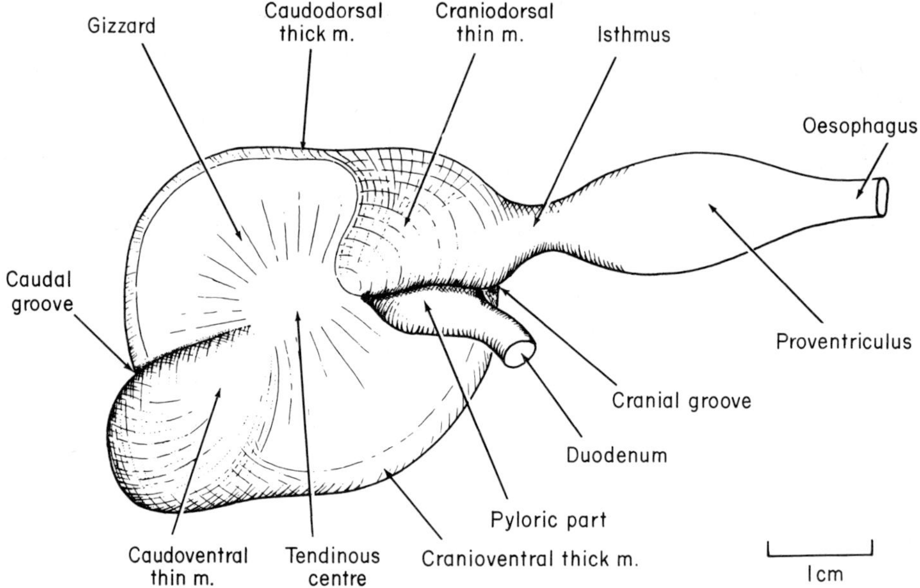

Fig. 3.15. Stomach of *Gallus*. From McLelland (1975).

the openings to the proventriculus and duodenum is the lesser curvature (*curvatura minor*) (Fig. 3.11c) which in some species, e.g. the Gray Heron (*Ardea cinerea*) (Fig. 3.11b) and the Great Cormorant (*Phalacrocorax carbo*) (Fig. 3.11a), is strongly angled to form the angular notch (*incisura angularis*) (Pernkopf and Lehner, 1937). The remaining long margin of the organ makes up the greater curvature (*curvatura major*) (Fig. 3.11c). The two relatively flat surfaces, the tendinous surfaces (*facies tendineae*) (Groebbels, 1932, p. 472), each contains a round tendinous centre (*centrum tendineum*) (Fig. 3.11b). The light-coloured muscle tunic is extremely poorly developed and of uniform thickness, its fibres radiating out in all directions from the tendinous centres.

The best differentiated form of gizzard occurs in the second main type of stomach, i.e. that which is characteristic of insectivores, granivores, herbivores and omnivores; here the gizzard has to deal with relatively hard items of food. The gizzard in these birds is shaped like a biconvex lens, with an oval or rhomboidal circumference and its greater diameter lying along either the longitudinal or transverse axis (Figs 3.11j, 15). The extensive right and left tendinous surfaces (Fig. 3.11j) are usually united dorsally and ventrally by much narrower annular surfaces (*facies annulares*) (Groebbels, 1932, p. 472). The main part or body (*corpus*) of the gizzard separates a small cranial sac (*saccus cranialis*) which protrudes from the cranial extremity of the organ, from a small caudal sac (*saccus caudalis*) protruding from the caudal extremity (Fig. 3.16). The intermediate zone opens into the cranial sac, the junction between the proventriculus and gizzard usually being marked externally by an isthmus. The ventriculopyloric opening (*ostium ventriculopyloricum*) lies on the right tendinous surface immediately ventral to the cranial sac (Fig. 3.16). The muscle tunic is massively developed and can be separated into four semi-autonomous masses radiating out from powerful fan-shaped tendinous centres (Figs 3.15, 16). These tendinous centres in *Anser* appear to be unequally developed (Schepelmann, 1906). The muscular masses are asymmetrically arranged relative to the longitudinal axis and include caudodorsal and cranioventral thick muscles (*m. crassus caudodorsalis*, *m. crassus cranioventralis*) and craniodorsal and caudoventral thin muscles (*m. tenuis craniodorsalis*, *m. tenuis caudoventralis*). The dark coloured thick muscles are composed of outer and inner portions and extend transversely between the tendinous centres and across the dorsal and ventral parts of the gizzard. These thick muscles form the basis of both pairs of tendinous and annular surfaces. In many species, e.g. *Anser*, the cranioventral thick muscle is better developed than the caudodorsal thick muscle (Swenander, 1902; Schepelmann, 1906). The much lighter-coloured, more pliable, thin muscles are unlayered structures corresponding to the inner portions of the thick muscles, and extend between the tendinous centres and across the sacs. If the outer portions of the thick muscles are removed, the muscle tunic of the gizzard is then almost of uniform thickness so that the organ has more the appearance of the gizzard in the undifferentiated type of stomach (Pernkopf and Lehner, 1937). The caudodorsal thick muscle is continuous with the craniodorsal thin muscle, whereas the cranioventral thick muscle is continuous with the caudoventral thin muscle. In summary, the gizzard is bordered to the right and left by the tendinous

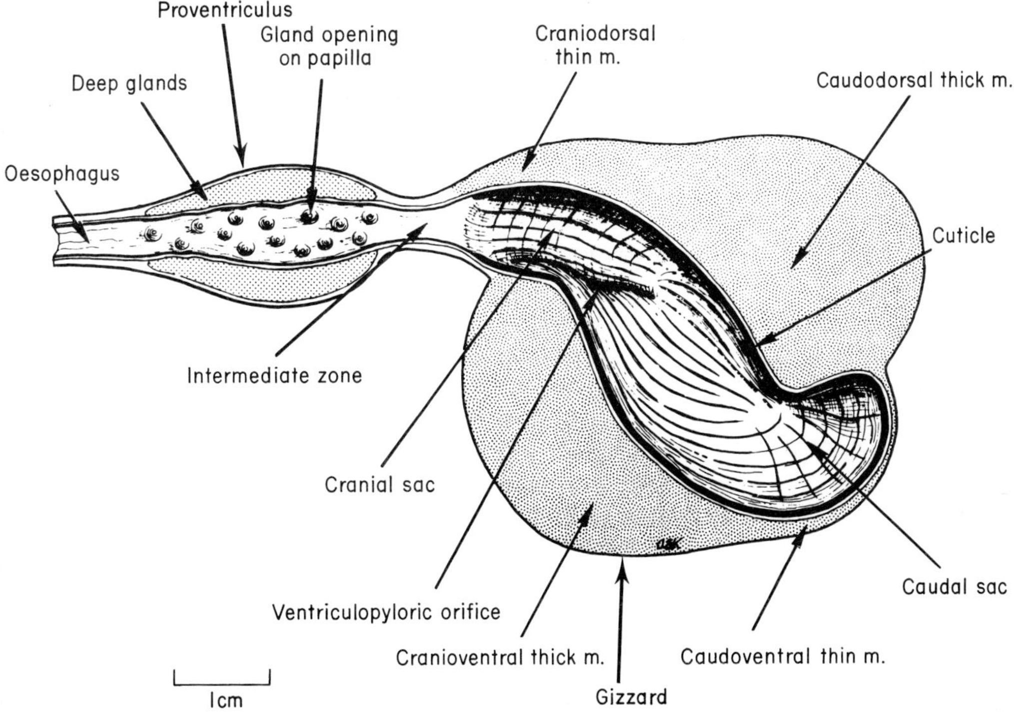

Fig. 3.16. Longitudinal section of stomach of *Gallus*. From McLelland (1975). The caudodorsal thick muscle lies adjacent to the dorsal grinding plate whilst the cranioventral thick muscle lies adjacent to the ventral grinding plate.

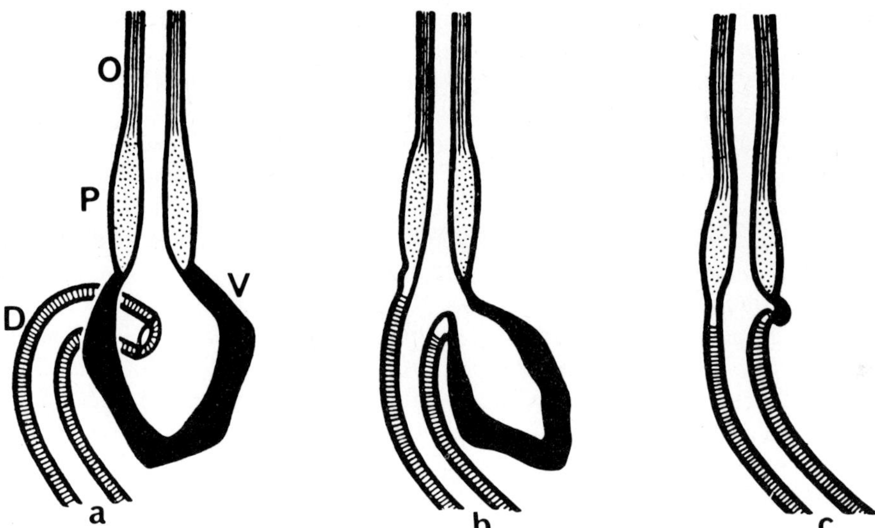

Fig. 3.17. Schematic representation to show the variation in the development of the gizzard in relation to diet in fruit-eating species. (a) Stomach of a primitive flowerpecker (*Dicaeidae*), gizzard unmodified. (b) Stomach of a specialized flowerpecker, e.g. the Black-sided Flowerpecker (*Dicaeum celebicum*), which feeds on fruit and insects. Within the stomach the fruit and insects are separated, the fruit passing directly from the proventriculus to the duodenum, the insects entering the gizzard which is a blind sac suspended from the junction of the proventriculus and duodenum. (c) Stomach of a highly specialized tanager, e.g. the Violaceous Euphonia (*Euphonia violacea*), which feeds entirely on fruit. Since the fruit seeds pass through the gastrointestinal tract undigested the gizzard is rudimentary. In the diagram the tanager gizzard is enlarged for the sake of clarity. D, duodenum; O, oesophagus; P, proventriculus; V, gizzard. From Desselberger (1931).

centres, cranially and caudally by the thin muscles, and dorsally and ventrally by the thick muscles. The junction of the caudodorsal thick muscle and the cranioventral thin muscle is marked on the external surface of the gizzard by a transverse caudal groove (*sulcus caudalis*), a similar cranial groove (*sulcus cranialis*) being present where the cranioventral thick muscle meets the craniodorsal thin muscle (Fig. 3.15). The asymmetrical arrangement of the muscle in this form of gizzard explains its rotatory as well as crushing movements when the organ contracts. Amongst species possessing the highly differentiated second type of stomach the development of the muscle of the gizzard is related to the hardness of the food, it being possible to modify the thickness by changing the diet (Broussy, 1936). The thickness of the muscle is also related to the extent to which physical preparation of the food occurs in the oral cavity, as happens for example in parrots (Psittacidae) (Ziswiler and Farner, 1972).

The gizzard of many species, e.g. most testacivores, granivorous insectivores and frugivores, is intermediate in appearance between the simple gizzard of fish and meat eaters and the highly differentiated gizzard of the second main type of stomach. Examples of these intermediate forms of gizzard are shown in Fig. 3.11f,h.

In certain nectivorous species including hummingbirds (Trochilidae), honeyeaters (Meliphagidae) and sunbirds (Nectariniidae) which feed on both nectar and insects, the gizzard is highly specialized (Desselberger, 1932). The proventricular and pyloric openings of the gizzard in these species lie next to one another so that the easily digestible nectar can pass directly from the proventriculus to the duodenum even although the gizzard is filled with insects undergoing physical breakdown. A basically similar, but more extensive specialization of the gizzard occurs in certain fruit-eaters of the Dicaeidae, e.g. the Golden-edged Flowerpecker (*Dicaeum aureolimbatum*), the Black-sided Flowerpecker (*Dicaeum celebicum*) and Nehrkorn's Flowerpecker (*Dicaeum nehrkorni*), which feed mainly on *Loranthus* berries and to a lesser extent insects (Desselberger, 1931). In these species, the gizzard is suspended from the junction of the proventriculus and duodenum as a blind sac (Fig. 3.17). In the region of the junction is a well-developed, obliquely-placed sphincter, the part of the sphincter on the same side as the gizzard lying below the gizzard opening, whilst the part lying opposite the gizzard is at the level of the opening. Since the gizzard is set obliquely to the longitudinal axis of the proventriculus and duodenum and since its entrance is relatively narrow, the moderately large berries tend to pass directly from the proventriculus to the duodenum. The insect food, however, is much harder and requires to be broken up in the gizzard, its passage into the gizzard being aided by closure of the sphincter which tends to pull the gizzard orifice more into the longitudinal axis of the alimentary tract. This type of specialization of the gizzard is carried still further in certain tanagers, e.g. the Violaceous Euphonia (*Euphonia violacea*), which feed only on mistletoe berries (Forbes, 1880; Wetmore, 1914). In these birds the gizzard is strikingly reduced to a narrow transparent zone extending directly between the proventriculus and duodenum (Fig. 3.17). As noted by Wetmore, the tanagers receive their nutrition from the easily digestible adhesive pulp of the berries, the hard seed passing through the gastrointestinal tract undigested. Since the food therefore does not need to be held in the stomach for any length of time, the gizzard has become incorporated into a straight passageway which allows the direct and rapid movement of the berries from the proventriculus to the duodenum.

2. *Interior of gizzard*

The gizzard in its least differentiated form, as in most fish and meat eaters, is internally undivided (Fig. 3.11c,d,e). The inner surface is lined by a thin continuous layer of secretion, the cuticle (*cuticula gastrica*), produced by the mucosal glands. Usually the cuticle is uniformly thick and of semi-firm consistency, although in some species including the Great Cormorant (*Phalacrocorax carbo*), the Gray Heron (*Ardea cinerea*) and the Eurasian Bittern (*Botaurus stellaris*), it is more jelly-like (Swenander, 1902). In the poorly muscled type of gizzard like this, the function of the cuticle is probably mainly to protect the underlying mucosa from the effects of the gastric juice secreted by the proventriculus. In some species, e.g. kiwis (Apterygidae) and the Great-crested Grebe (*Podiceps cristatus*) (Fig. 3.11e), the entrances to the proventriculus and duodenum form funnel-shaped protrusions into the gizzard so that they appear to be separated by a powerful angular fold (*plica angularis*) (Pernkopf and Lehner, 1937). This fold frequently corresponds in position to the angular notch which is visible externally. When contracted, the inner surface of the less differentiated form of gizzard, is often strongly folded (*plicae ventriculares*) (Fig. 3.12), these folds usually disappearing when the gizzard dilates with food (Swenander, 1902).

In the highly differentiated muscular gizzard characteristic of granivores, herbivores and insectivores, the interior is subdivided into three portions including a cranial sac, a caudal sac and a body (Fig. 3.16) in which can be distinguished dorsal and ventral parts. The cranial sac opens into the dorsal part of the body, whereas the caudal sac opens into the ventral part. The slit-like ventriculopyloric opening lies between the cranial sac and the cranioventral thick muscle. Unlike in the gizzard of fish and meat eaters, the cuticle varies extensively in thickness between different regions of the organ and is of extremely hard consistency. It is best developed in the dorsal and ventral parts of the body and thinnest in the cranial and caudal sacs and over the tendinous centres (Fig. 3.18). In many species, e.g. *Clangula*, *Pavo cristatus*, *Fulica atra*, *Gallinula chloropus*, *Gallus*, *Perdix perdix*, *Columba*, *Tetrao urogallus*, *Gallinago gallinago* and *Crex crex*, the cuticle covering the dorsal and ventral parts of the body of the gizzard is especially thick and forms the so-called dorsal and ventral grinding plates (Flower, 1860; Cazin, 1887b; Swenander, 1902; Schepelmann, 1906; Magnan, 1912; Cornselius, 1925). Each grinding plate is a roundish or elliptical semi-discrete, coarse elevation in which can often be distinguished a flat or concave central portion and a horse-shoe shaped wall. As with the muscle of the gizzard, the thickness of the grinding plate is also asymmetrical. Thus in *Gallus*, the dorsal grinding plate is best developed caudally opposite the caudodorsal thick muscle, whereas the ventral grinding plate is thickest cranially opposite the cranioventral thick muscle (Fig. 3.16). This arrangement enables the grinding plates to fit very closely together and reduce the lumen of the gizzard to a cleft (Schepelmann, 1906). In the fruit-eating pigeon *Ptilinopus*, the cuticle is arranged into four grinding plates, the powerful organ in this species being cruciform in transverse section and not oval as in most other birds with a similarly developed gizzard (Newton and Gadow, 1896, p. 918; Beddard, 1898, pp. 306–307). By far the best developed grinding plates are reported to belong to the Nicobar Pigeon (*Caloenas nicobarica*) in which they form extremely hard, hemispherical, concave masses, 1 cm thick, and lodged in deep depressions of the mucosa (Flower, 1860; Cazin, 1887b). The inner surface of the cuticle over most parts of the gizzard has a system of ring-like elevations and grooves (Fig. 3.18), most of the ridges being based on a system of rugae and grooves of the mucosa (*rugae ventriculares, sulci ventriculares*). The arrangement of the ridges in *Meleagris* appears to be typical of the highly muscular form of gizzard and has been correlated with gastric motility by Dzuik and Duke (1972). Thus the ridges on the grinding plates are arranged only in a longitudinal direction since contraction of the thick muscles results in side to side shortening of the muscles and bulging of the dorsal and ventral walls of the gizzard body into the lumen. The ridges in the sacs, however, are arranged both longitudinally and transversely since during contraction of the thin muscles shortening of the muscles occurs in several directions. The ridges and grooves in the dorsal grinding plate pass without interruption into the cranial sac, whereas those of the ventral grinding plate run directly into the caudal sac (Fig. 3.16). The cuticle is smooth over the tendinous centres and opposite to the externally visible cranial and caudal grooves where presumably there is relatively little distortion of the wall. The outer surface of the cuticle is coloured various shades of yellow, brown or green which is generally agreed to be due mainly to the regurgitation of bile (Groebbels, 1929; Norris, 1961).

The shape of the lumen of the well-muscled form of gizzard varies considerably, depending on the state of contraction of the organ. In *Meleagris*, Dziuk

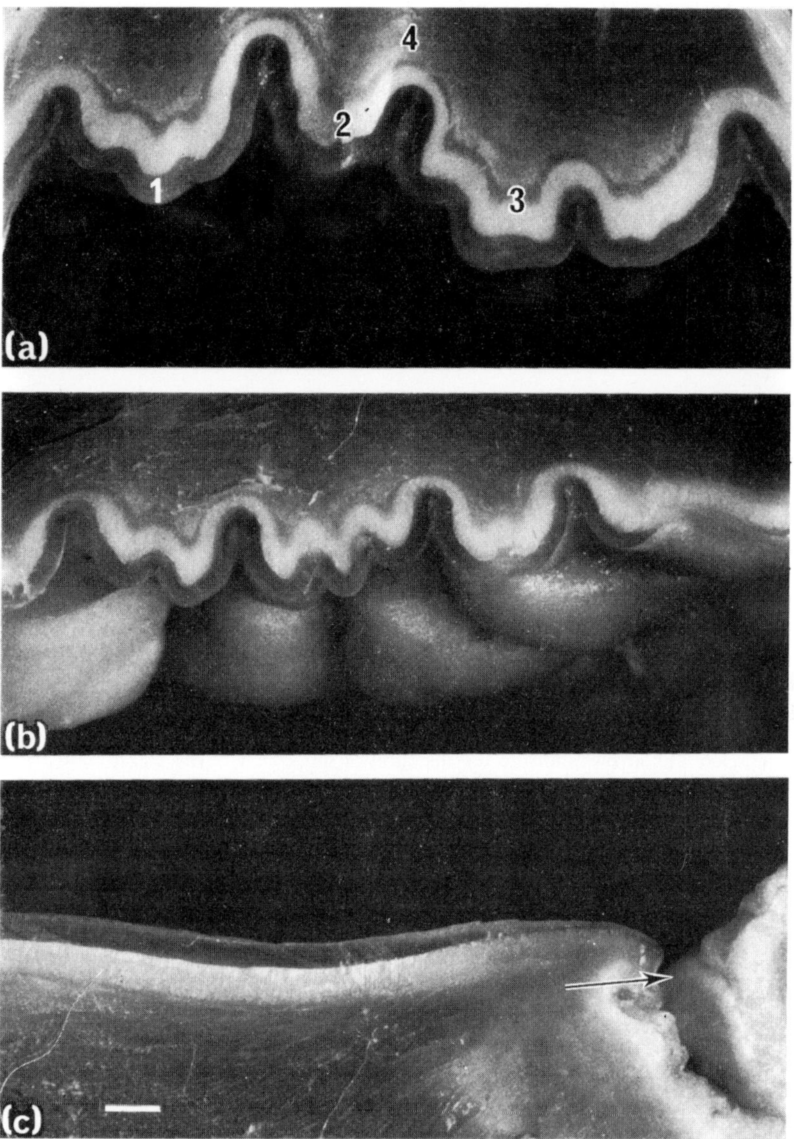

Fig. 3.18. Caudal views of transverse sections through the gizzard wall of *Gallus*, (a) through the roof of the body, (b) through the roof of the cranial sac and (c) through the floor of the cranial sac. 1, cuticle; 2, mucosa; 3, submucosa; 4, muscle. Arrow indicates ventriculopyloric orifice. Scale = 1 mm.

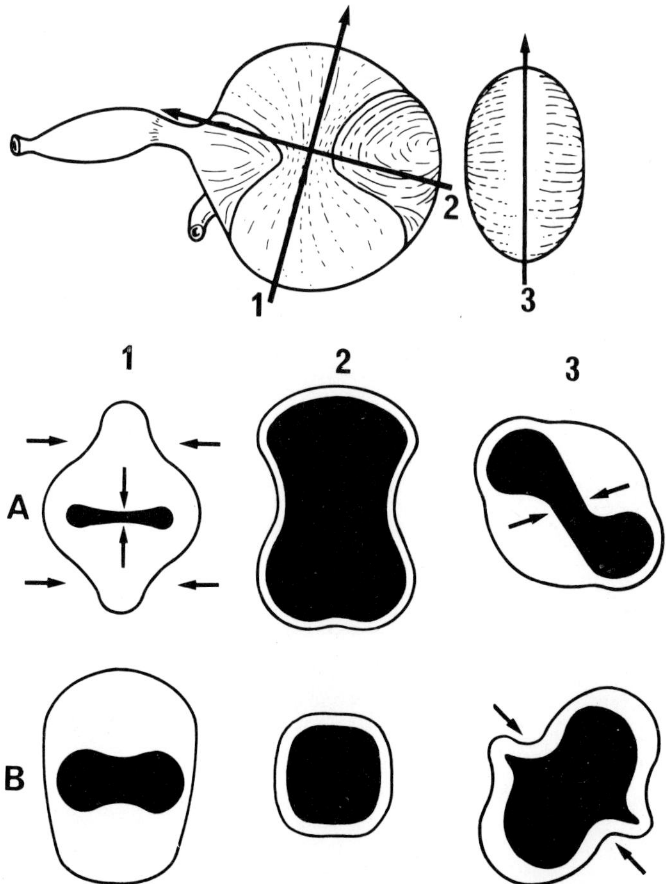

Fig. 3.19. Sections of the gizzard of *Meleagris* during the contraction cycle. Sections (1) and (2) are as indicated in the left-lateral view of the stomach, section (3) as indicated in the caudal view of the stomach. The wall of the gizzard is unshaded, the lumen shaded. In (A) there is contraction of the thick muscles of the gizzard, in (B) contraction of the thin muscles. The arrows indicate the direction of movement of the gizzard wall. From Dziuk and Duke (1972).

and Duke (1972) observed (Fig. 3.19) that contraction of the thick muscles narrows the lumen of the gizzard body in its dorsoventral axis, bringing closer together the caudodorsal and cranioventral thick muscles and their thickened portions of cuticle, and causes distension of the cranial and caudal sacs. Contraction of the thin muscles, in contrast, causes the sacs to disappear whilst the cavity of the gizzard body is shortened in its craniocaudal axis and lengthened dorsoventrally. The ventriculopyloric opening appears to be closed by contraction of the cranioventral thick muscle and the craniodorsal thin muscle.

The appearance of the interior of the intermediate forms of gizzard like those of frugivores, granivorous, insectivores and testacivores depends to a great extent upon whether its primary function is storage or the physical digestion of food. When the gizzard is primarily a container, its cavity tends to be undivided and lined with a cuticle which is soft and of uniform thickness. A gizzard adapted more for the mechanical preparation of food tends to have a subdivided lumen and a hardish cuticle of unequal thickness. Usually in intermediate forms the inner surface of the cuticle has a variably developed system of ridges and grooves, the ridges being formed either by thickening of the cuticle as in the Puffin (*Fratercula artica*) or folding of the mucosa as in the Common Gull (*Larus canus*) (Fig. 3.11f). In some species, e.g. *Mergus*, *Uria* and *Larus*, the cuticular folds are especially well-developed in the dorsal and ventral parts of the body of the gizzard, i.e. in the position of the grinding plates in the highly differentiated type of stomach, and seem to be acting as alternatives to the plates. Compared to the grinding plates the folds, being mucosal in origin, have the advantage of allowing considerable expansion of the organ (Swenander, 1902). An unusual adaptation of the cuticle in an intermediate form of gizzard occurs in certain frugivorous pigeons (*Columbidae*), the secretory layer being raised into a number of hard pointed conical processes (*procc. conicales*) which are used for crushing nutmegs and other hard fruits. In Peale's Pigeon (*Ducula latrans*) Garrod (1878a) observed 23 such cones, the largest of which were 7 mm in width at the base and 4 mm in height. Eighteen cones are distributed in the dorsal and ventral parts of the gizzard close to one another in rows of three counting either longitudinally or transversely, the remaining five cones lying on the tendinous centres. The cuticle between the cones is relatively soft. Wood (1924) found a similar number of conical processes in the Pacific Pigeon (*Ducula pacifica*) but unlike in *Ducula latrans* they are usually irregularly arranged. In the Island Imperial Pigeon (*Ducula myristicivora*) the 22 cones are distributed in coarse, parallel, longitudinal folds and interdigitate with one another like cogs (Cadow, 1933). Conical processes have also been reported in the cuticle of some tubinarine species (Newton and Gadow, 1896, p. 918). Highly detailed accounts of the internal appearance of the intermediate forms of gizzard are available for many species in the descriptions of Cazin (1887b), Swenander (1902), Magnan (1912) and Cornselius (1925).

3. *Structure of gizzard*

The wall of the gizzard consists of four layers: mucous membrane, submucosa, muscle tunic and serosa.

The mucous membrane (*tunica mucosa gastris*) (Fig. 3.18) is lined by a simple columnar epithelium and is invaginated by numerous shallow crypts at the bottom of which open the branched or unbranched tubular glands (*gll. ventriculares*) of the lamina propria. The glands are distributed either regularly or irregularly, the regular pattern of arrangement usually being in rows as in *Athene noctua*, *Passer domesticus*, *Sturnis vulgaris* and *Columba*, or groups as in *Crex crex*, *Gallinago gallinago*, *Fulica atra*, *Gallus* and *Perdix perdix* (Swenander, 1902). In *Gallus* 10–30 glands are present in each group, 5–8 glands opening into a crypt (Eglitis and Knouff, 1962). The glands may open into the crypts separately as in *Gallus* or occasionally, as in the King Eider (*Somateria spectabilis*) and the Sheld Duck (*Tadorna tadorna*), via a common duct (Swenander, 1902). The glands are best developed in the more highly muscular forms of gizzard and then they are largest where the cuticle is thickest. The cells of the surface epithelium and the epithelium of the crypts and glands secrete the cuticle lining the inner surface of the organ. These cells appear to originate in the bases of the glands and migrate towards the mucosal surface where they degenerate and become sloughed off (Chodnik, 1947; Toner, 1964a). The fine structure of the cells was investigated by Toner (1964a,b) in *Gallus*. In the bases of the glands are a small number of cuboidal basal cells which usually occur in pairs and have characteristically long bulbous-tipped microvilli. The overall appearance of the basal cell suggests that it is relatively undifferentiated and therefore likely to be the source of the cells higher up in the glands. In support of this view, mitotic activity was observed here by Chodnik (1947). The principal cell type in the glands is the chief cell which has an appearance typical of a protein-secreting cell. The apical cytoplasm of the chief cells is packed with secretory granules which are especially numerous close to the mouths of the glands. The granules release their contents at the surface of the cells to form a narrow layer of secretion. The surface cells line the crypts and mucosal surface, and are taller than the chief cells. The apex of the surface cell has numerous microvilli and bulges into the lumen. The apical cytoplasm contains secretory granules. A constant feature of the surface cell is degeneration, especially in the Golgi apparatus and the mitochondria. These changes occur first in the cells at the bases of the crypts and become progressively more advanced towards the mouths of the crypts where the cells die and become sloughed off. Chodnik (1947) found no change in the appearance of the cells after fasting or feeding. In addition to the cell types described above, the epithelium also contains endocrine granular cells which are dealt with in the chapter on the Endocrine System in Vol. 2. A muscularis mucosae appears to be usually absent, the mucosal muscle of the intermediate zone joining the muscular tunic of the gizzard. Swenander (1902), however, observed muscle (*lamina muscularis mucosae*) in the gizzard mucosa of raptors which extended caudally to the level of the tendinous centres, as well as in several other species including the Wryneck (*Jynx torquilla*) and the Green Woodpecker (*Picus viridis*), in which he reported it to be well-developed.

The submucosa (*tela submucosa gastris*) (Fig. 3.18) is composed of dense connective tissue which by some authorities has been interpreted as a stratum compactum of the mucosa (Swenander, 1902; Plenk, 1932). According to Schepelmann (1906) the submucosa in *Anser* has three zones depending on the direction of the fibres and is best developed under the grinding plates where the cuticle is thickest. Broad connective tissue bands extend from the submucosa into the muscular tunic. The submucosa functions therefore to hold the muscular tunic tightly to the mucous membrane as well as providing a firm basis for the grinding action of the cuticular lining. A submucosal nerve plexus is absent.

The muscular tunic (*tunica muscularis gastris*) (Fig. 3.18) consists basically of inner circular and outer longitudinal layers (*stratum circulare, stratum longitudinale*) (Bauer, 1901; Swenander, 1902; Cornselius, 1925; Pernkopf, 1930, 1937; Plenk, 1932; Pernkopf and Lehner, 1937). The circular layer (Fig. 3.20) in all species is relatively well-developed. The longitudinal layer in contrast is weakly developed and usually appears to be absent over most of the gizzard except the lesser curvature (Pernkopf, 1930, 1937). In *Gallus* a longitudinal layer of muscle is present in the embryo but is lost about the 10th day of incubation (Bennett and Cobb, 1969a). A thin longitudinal layer, presumably extending over a large part of the gizzard was reported by Groebbels (1932, p. 471) to be present in hummingbirds (Trochilidae). The differentiation in the well-developed type of stomach of the circular layer into semi-autonomous masses has already been described. The muscle of the gizzard attaches to the tendinous centres on the tendinous surfaces of the organ (Figs 3.11b, 15). In *Gallus* and *Meleagris* the junction between the muscle and tendon has some fibrocartilage which has a strengthening effect (Calhoun, 1954, p. 56; Malewitz and Calhoun, 1958). Muscle fibres are absent opposite the centres of the tendons (Calhoun, 1954, p. 56).

The fine structure of the muscle of the gizzard was investigated in *Gallus* by Choi (1962), Bennett and Cobb (1969a,b) and Cobb and Bennett (1969). A characteristic feature of the muscle is its extensive arrangement into interlocking bundles separated by connective tissue, the bundles in the thick muscles being large and closely packed together, whilst those in the thin muscles are smaller and more loosely arranged. Anastomoses between adjacent bundles are

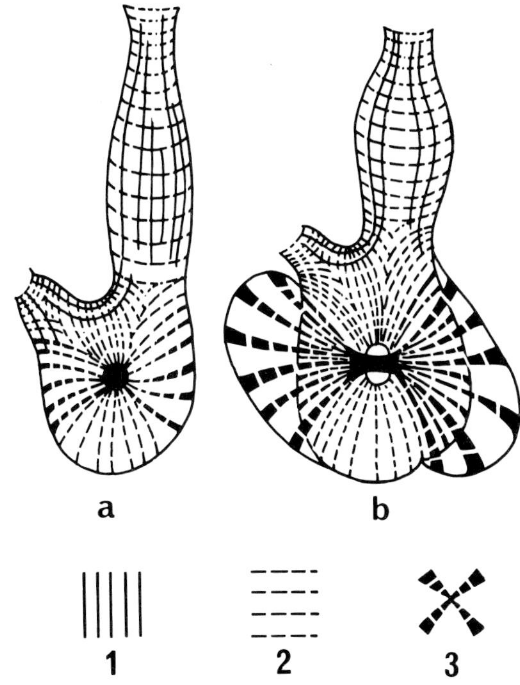

Fig. 3.20. The musculature of the stomach (a) in *Gyps* in which the gizzard is relatively poorly differentiated, and (b) in *Gallus* in which the gizzard is highly differentiated. 1, longitudinal muscle layer; 2, circular muscle layer; 3, thickened part of circular layer. In the proventriculus inner circular and outer longitudinal layers are present. In the gizzard the longitudinal layer occurs only at the lesser curvature. The fibres of the circular layer radiate out from the tendinous centres. In *Gallus* the circular layer is best developed caudodorsally and cranioventrally to form semiautonomous masses. From Pernkopf (1937).

numerous. The muscle cells contain both thick and thin myofilaments and extremely large mitochondria, and nexuses between adjacent cells are more numerous than in the smooth muscle in other regions. Bennett and Cobb (1969a,b) concluded that the mechanism of contraction of the smooth muscle of the gizzard is much more active than in other viscera. Certainly the structure of the muscle allows for a rapid spread of impulses throughout the organ and an unusually powerful form of contraction.

On the outer surface of the muscle, immediately below the serosa, lies the myenteric nerve plexus.

4. *Gizzard cuticle*

Much of our present knowledge of the formation and structure of the gizzard cuticle (*cuticula gastrica*) has come from recent studies in *Gallus* by Eglitis and Knouff (1962), Toner (1964a), Hill (1971a) and Michel (1971) on which the following account is mainly based. The cuticle of *Gallus* (Fig. 3.18) is a thick, pliable membrane typical of species with a highly muscular form of gizzard and is composed of a scaffolding of interconnecting vertical rods (*columnae verticales*) in which is distributed horizontal matrix (*matrix horizontalis*). The vertical rods are secreted by the mucosal glands, the secretion of each gland hardening within the lumen in the form of a filamentous bundle. The filaments of all the glands opening into a crypt unite to form a vertical rod. The vertical rod projects slightly beyond the surface of the cuticle as a dentate process (*proc. dentatus*). Lateral branches of the rods unite with those of neighbouring rods to confer great mechanical strength on the cuticle. The distribution of the vertical rods varies between species in a manner similar to that of the mucosal glands (Cazin, 1887b). The horizontal matrix is secreted by the cells of the crypts and surface epithelium. Unlike the vertical rods it does not harden immediately, but only after it has spread over the surface of the epithelium and around the rods. The hardening of the horizontal matrix is believed to be due to a fall in its pH as a result of diffusion through the cuticle of hydrochloric acid from the proventriculus. The cyclic nature of the production of the horizontal matrix is indicated by variations in its staining reaction, the dark stripes representing periods of rest. Trapped within the horizontal matrix are clusters of surface epithelial cells which have degenerated and become sloughed off. In contrast, the thin, relatively soft cuticle characteristic of the poorly muscled type of gizzard has much less internal organization, the division into vertical rods and horizontal matrix frequently being unclear (Cornselius, 1925), and in some species, e.g. *Larus*, *Podiceps* and *Alauda*, it may have an almost homogenous appearance. Such a cuticle was described by Groebbels (1932, p. 492) in *Falco tinnunculus*, *Athene* and *Cuculus canorus* as being composed of rods of secretion lying in a very small amount of horizontal matrix and distributed on top of one another almost parallel to the surface of the cuticle, the more superficial rods lying so close together that the cuticle here appears almost unstructured. For further comparative details of the cuticle consult Curschmann (1866), Cazin (1886, 1887b), Swenander (1902) and Cornselius (1925).

Recent histochemical and biochemical analyses have demonstrated the cuticle to be a carbohydrate–protein complex (Luppa, 1959; Eglitis and Knouff, 1962; Webb and Colvin, 1964; Glerean and Katchbarian, 1965; Michel, 1971).

The surface of the cuticle is usually being continuously worn away and replaced. In some birds, however, the whole cuticle at intervals is suddenly shed. Amongst the species listed by McAtee (1906, 1917) and Smith (1913) in which massive shedding like this is known to occur are *Cuculus canorus*, *Numenius arquata*, *Colinus virginianus*, *Falco sparverius*, *Coccyzus americanus*, *Coccyzus erythropthalmus*, *Pica pica*, *Corvus ossifragus*, *Sturnus vulgaris*, *Turdus viscivorus*, *Sturnella magna*, *Toxostoma redivivum*, *Mimus polyglottos*, *Buceros* and many Anatidae. Generally the cast off cuticle is ground down and excreted. In certain hornbills (Bucerotidae), however, the shed cuticle is regurgitated in the form of a sac-like structure containing seeds, which in the breeding season the male bird feeds to the female on the nest (Bartlett, 1869; Flower, 1869; Murie, 1874).

5. *Grit*

Most birds feeding on coarse material such as vegetable matter ingest grit which assists in the grinding down of the food in the gizzard to a digestible form. Extensive surveys of this phenomenon are provided by Meinertzhagen (1954, 1964). Most of the grit appears to be in the form of quartz, the size of the particles being proportional to the coarseness of the food. Thus the Lesser Flamingo (*Phoeniconaias minor*) which feeds on algae and diatoms ingests a fine form of grit, whereas the Ostrich (*Struthio camelus*) which feeds on coarse vegetation takes in pebbles. The intake of grit is sometimes also closely governed by seasonal or regional variations in diet as in the Scandanavian Willow Ptarmigan (*Lagopus lagopus*) and the Corn Bunting (*Emberiza calandra*). Some birds which do not take in grit utilize other materials for the same purpose. The Dipper (*Cinclus cinclus*), for example, ingests small shells, whilst diving ducks (Anatidae) use the hard outer coverings of the molluscs and crustaceans which form their diet. In the gizzard of certain species, e.g. divers (Gaviidae) there is grit of unknown function.

F. Pyloric part of stomach

The pyloric part (*pars pylorica*) of the stomach is the variably developed portion between the gizzard and duodenum. It communicates with the gizzard at the ventriculopyloric orifice and with the duodenum at the pyloric orifice (*ostium pyloricum*). Unfortunately, the precise form of this pyloric part in different species is often difficult to establish with certainty, because of the lack of sufficient detail in most accounts, the conflicting nature of the accounts, and the varied terminology.

In its least developed form as in the domestic birds (Fig. 3.15), the pyloric part is an extremely short, narrow tube of almost uniform calibre (Zeitzschmann, 1908; Aitken, 1958; Hodges, 1974, pp. 63–64; Larsson *et al.*, 1974). In *Gallus*, for example, the pyloric part is a lighter-coloured portion of the gut, only 0·5 cm in length, and separated from the duodenum by a fold of the mucosa.

Its mucous membrane is intermediate in structure to that of the gizzard and duodenum. The mucosal glands (*gll. pyloricae*) are basically similar to those of the gizzard, but secrete mucus instead of a cuticle. In about half of the specimens examined by Hodges (1974, p. 64), the tubular glands immediately proximal to the mucosal fold were replaced by branched acinar glands lined by intensely eosinophilic cells. The epithelium of the pyloric part has been reported to contain a large concentration of granular endocrine cells (see the chapter on the Endocrine System in Vol. 2.

In a number of species (Figs 3.11a,b,c,e,g, 14) the pyloric part of the stomach is a distinct expansion of the digestive tract which externally can usually be distinguished without difficulty from both the gizzard and duodenum (Cazin, 1887b; Gadow, 1891a, p. 679; Swenander, 1902; Cornselius, 1925; Groebbels, 1932, p. 459; Pernkopf and Lehner, 1937). This form of pyloric part reaches its greatest development in *Phalacrocorax carbo*, *Ardea cinerea*, *Botauris stellaris*, the Anhingidae, *Ciconia nigra*, *Podiceps cristatus*, *Eudyptula minor*, *Apteryx*, *Struthio*, *Pelecanus*, *Numenius arquata*, *Vanellus*, *Haematopus*, *Procellaria parkinsoni*, *Scolopax rusticola*, *Rallus*, *Gallinula chloropus*, *Fulica*, *Tadorna*, *Clangula*, *Somateria spectabilis*, *Somateria mollissima*, *Anas acuta*, *Anas crecca*, *Grus* and *Cygnus olor*. In the Gray Heron (*Ardea cinerea*) (Fig. 3.11b) the expansion is divided into proximal and distal portions, the smaller distal portion forming the pyloric bulb (*bulbus pyloricus*) of Pernkopf and Lehner (1937). A moderately developed expansion (Fig. 3.11d,f), only slightly larger than the duodenum occurs in *Aquila chrysaetos*, *Haliaeetus albicilla*, *Pandion haliaetus*, *Buteo buteo*, *Pernis apivorus*, *Falco peregrinus*, *Falco subbuteo*, *Falco tinnunculus*, *Athene noctua*, *Larus* and *Cuculus canorus*.

From the very limited information available, the general structure of the enlarged forms of the pyloric part appears to be basically similar to that of the gizzard (Cazin, 1887b; Swenander, 1902). The mucosa is often folded (*plicae pyloricae*) as in the Gray Heron (*Ardea cinerea*) (Cazin, 1887b), and the Little Blue Penguin (*Eudyptula minor*) and many birds of prey (Pernkopf and Lehner, 1937). Similar folds occur in *Ardea* at the boundaries of the pyloric bulb. Internally the mucosa is lined by a secretory layer or cuticle (*cuticula gastrica*) which is generally reported to be somewhat softer than in the gizzard. This cuticle in the Great-crested Grebe (*Podiceps cristatus*) and the Black Stork (*Ciconia nigra*) is raised into a number of dentate ridges (Swenander, 1902). In grebes (Podicipedidae) the muscle tunic, where it crosses the dorsal wall, is five to six times thicker than in other regions of the stomach. Wetmore (1920) suggested that contraction of this thickened muscle closes off the pyloric part from the gizzard.

The extremely well-developed pyloric part of the stomach of darters (Anhingidae) (Fig. 3.14) has a relatively complex structure compared to that of other species, and consequently has received most attention (Garrod, 1876b, 1878b; Forbes, 1882b; Cazin, 1884, 1887b; Beddard, 1898, p. 415). In the African Darter (*Anhinga rufa*) the expansion is characterized by a well-developed conical process (*torus pyloricus*) protruding from its distal wall into the lumen. The enlarged apex of the process is free and directed towards the body of the gizzard, whilst the narrow base is attached to the stomach wall close to the pyloric opening. Fibres from the muscular tunic extend into the process and act to retract it so that it occludes the opening. A less developed form of pyloric process is also present in the pyloric part of the stomach of the Asian Darter (*Anhinga melanogaster*). In *Anhinga rufa*, *Anhinga melanogaster* and *Anhinga anhinga* the cuticle covering the distal region of the pyloric part is raised into a dense mat of hair-like papillae (*papillae filiformes pyloricae*) (Fig. 3.14). The papillae in *Anhinga anhinga* are 1·27 cm in length and usually fill the chamber. The histological structure of the papillae was investigated by Cazin (1884, 1887b) in *Anhinga melanogaster*. Each papilla represents the secretion of all the simple tubular glands which open into a crypt and which in the distal region of the pyloric part are especially well-developed. Cazin described the papillae as corresponding to the vertical rods of secretion in the cuticle lining the main part of the gizzard, but unlike the rods they are not joined together by either lateral branches or horizontal matrix.

What the function is of the various enlarged forms of the pyloric part does not appear to have been established. However, most investigators have commented on the association between the better developed forms of expansion and the intake of food which is accompanied by much water. Possibly as suggested by Gadow (1891a, p. 679) and Swenander (1902), a large pyloric part has the effect of lengthening the stay of food in the stomach thereby prolonging the period of digestion. This would be especially important when the diet contains relatively indigestible items such as fish bones and shells which might damage the intestines. Certainly the large conical process and hair-like papillae characteristic of the pyloric part of darters (Anhingidae) would seem ideally suited to act as a valvular apparatus or sieve preventing the passage of hard material into the intestines. The role which the cuticle would play in this is uncertain. As well as probably being involved in the physical breakdown of the retained food it may also help to trap the food, since Swenander (1902) frequently observed that the cuticle in the Great-crested Grebe (*Podiceps cristatus*) contained many fish bones which he presumed were held there until worn away or chemically digested.

G. Blood supply of stomach

The following account is based mainly on the observations in *Gallus* by Nishida *et al.* (1969). The arteries arise from the coeliac artery, the branches to the proventriculus forming a network in the deep part of the wall, whereas those to the gizzard form a network on the surface of the organ. Extensions of the gizzard network in *Gallus* perforate the wall of the organ in the region of the tendinous centres, whereas in *Anser* they enter via the annular surfaces (Nishida *et al.*, 1976). The veins of the proventriculus form a dense network superficial to that of the arteries. Approximately half of the venous blood from the proventriculus is drained by right and left hepatic portal veins and half by the cranial vena cava. In anseriform species, however, most of the proventricular blood is drained by the cranial vena cava since the proventricular branch to the right hepatic portal vein is either rudimentary as in *Anser* (Nishida *et al.*, 1976) or absent as in *Anas* (Nishida and Mochizuki, 1976). The venous drainage of the proventriculus of *Columba* appears to be similar to that of *Anas* (Malinovský,

1965a). In the highly differentiated stomach therefore, the venous network of the proventriculus directly connects the hepatic portal and systemic circulations. In *Buteo buteo*, however, which unlike the other researched species has a relatively simple type of stomach, the proventriculus is drained solely by the portal veins (Malinovský, 1965a). The venous network which is formed on the surface of the gizzard by vessels perforating the annular surfaces of the organ is drained by the portal veins.

H. Nerve supply of stomach

The stomach is innervated by branches of the vagi and by perivascular nerve fibres from the coeliac and mesenteric plexuses (Malinovský, 1963). According to Bennett (1969), the vagal branches consist mainly of cholinergic nerve fibres, a few noradrenergic fibres being present below the level of the vagosympathetic anastomosis. The perivascular nerves consist of both cholinergic and noradrenergic fibres. The microanatomy of the enteric nervous system of the gizzard of *Gallus* and *Columba* has recently been studied in considerable detail by Malinovský (1964), Bennett (1969), Bennett and Cobb (1969a,b,c), Bennett and Malmfors (1970) and Bennett *et al.* (1971). The myenteric nerve plexus lies just beneath the serosa and on the surface of the circular layer of the muscle tunic. Over the thin muscles the plexus is wide-meshed, whilst over the thick muscles it is condensed forming on the dorsal and ventral annular surfaces a narrow-meshed plexus in *Gallus*, and a single nerve trunk in *Columba*. The plexus is composed mainly of cholinergic ganglion cells and noradrenergic fibres, some catecholamine cell bodies also being present. Branches of the plexus consisting of cholinergic cells associated with cholinergic and noradrenergic fibres extend into the muscle tunic. The density of the innervation of the muscle is relatively constant (1 axon bundle per 670 μm^2 of tissue), the majority of the axon bundles lying within the muscle bundles. The cholinergic fibres innervate the muscle cells whilst the noradrenergic fibres appear to be related mainly to the blood vessels and the ganglion cells. A submucosal plexus is lacking. The histology of the myenteric and submucosal plexuses of the proventriculus has been described by Iwanow (1930), Okamura (1934) and Ábrahám (1936) but information on the histochemistry of this region of the enteric nervous system does not appear to be available.

It is generally agreed that the neural control of the muscular activity of the stomach is exceedingly complex. The myenteric plexus is responsible for the basic rhythmic pattern of motility of the stomach. This basic pattern is modified by discharges from the vagal and perivascular nerve pathways both of which contain excitatory and inhibitory fibres. For a detailed analysis of the available structural and physiological data on the innervation see Farner (1970), Ziswiler and Farner (1972) and Bennett (1974).

I. Stomach motility

1. Motility of the strigiform stomach. Pellet egestion

Investigations on the motility of the relatively undifferentiated type of stomach appear to be restricted to raptors (strigiform and falconiform species) in which the primary interest has usually been in the arrangement of the indigestible portion of the stomach contents into pellets which are regurgitated and egested via the mouth. In the following account therefore, the motility of this type of stomach and pellet formation are considered side by side. Gastric pellets are usually composed of hard items such as bones, teeth, beaks, claws and the chitinous remains of insects which are bound together by softer material such as feathers, fur and vegetable matter. Apart from raptors, pellet formation is also known to occur in many other groups of birds which are associated with a variety of feeding habits, and which in many instances possess a highly differentiated type of stomach. Amongst the non-raptorial species listed by Tucker (1944), for example in which pellet formation has definitely been established are the Ardeidae, *Corvus corax*, *Motacilla alba*, *Muscicapa striata*, *Locustella naevia*, *Acrocephalus scirpaceus*, *Sylvia communis*, *Turdus merula*, *Oenanthe oenanthe*, *Phoenicurus phoenicurus*, *Erithacus rubecula*, *Hirundo rustica*, *Delichon urbica*, *Apus apus*, *Caprimulgus europaeus*, *Coracias garrulus*, *Columba palumbus*, *Numenius arquata*, *Tringa totanus*, *Tetrao urogallus* and *Larus argentatus*.

Most of the following description of stomach motility and pellet formation is based on the extensive observations by Kostuch and Duke (1975), Duke *et al.* (1976c) and Rhoades and Duke (1977) in the Great Horned Owl (*Bubo virginianus*). In these studies intragastric pressure changes were measured either by strain gauge transducers implanted in the gizzard or by an intragastric pressure telemetry method, whilst gastroduodenal contraction sequences were recorded using radiographic techniques. Captured prey is moved to the gizzard by oesophageal and proventricular peristalsis aided by extension of the neck and jerking movements of the head. Within the gizzard seven phases in gastric digestion can be distinguished. In each phase cycles of contraction occur which arise near the isthmus and proceed in a clockwise manner over the greater curvature to the region of the pylorus. *Phase I, filling*. Lasts 50–60 min and characterized by vigorous, rapidly moving contractions of high frequency. *Phase II, early chemical digestion*. Lasts 4·5–5 h and characterized by moderately forceful contractions of long duration and low frequency. During this phase most of the flesh of the prey is digested. *Phase III, late chemical digestion*. Lasts 3–3·5 h and characterized by large contractions which are similar to those in phase II but of lower frequency and longer duration. During phases II and III the more fluid parts of the ingesta are expelled from the gizzard. At the completion of phase III the gizzard contents consist only of indigestible material. *Phase IV*, fluid evacuation. Lasts 30–45 min and characterized by vigorous "paired" contractions of high frequency separated from each other by a distance of about half the length of the greater curvature. In this phase fluid is squeezed out of the ingesta. *Phases V and VI, early and late pellet compaction*. Lasts 5–6 h and characterized by large, slow contractions similar to those in phase

III. In phase IV paired contractions occur. Phases V and VI are often reversed in their order. Late in phase VI gizzard activity is usually much reduced. At the end of phase VI the gizzard contents are tightly compacted and are beginning to take on the shape of a pellet. *Phase VII, final pellet formation and egestion.* Lasts 12–20 min and characterized by two types of contraction, one type involving the caudal part of the gizzard, the other type involving the left side. These two types of contraction complete the compaction of the pellet and 2–3 min before egestion expel it into the proventriculus. Ten to twelve seconds before egestion, the pellet enters the oesophagus where it is moved orally by reverse peristalsis. Egestion is frequently accompanied by gaping and shaking of the head. Gastric motility restarts 30–60 s after egestion. The mechanism of pellet egestion seems to be different from both the vomiting and regurgitation processes observed in mammals.

A cephalic phase of gastric motility was demonstrated in raptors by Duke *et al.* (1976b). Following ingestion in raptors, stomach movements cease for a short while, which probably allows the stomach to accommodate the large portions of food characteristic of these species. Duke *et al.* (1976b) suggested that this inhibition of gastric motility is analogous to the "receptive relaxation" of the stomach of mammals. Kostuch and Duke (1975) found the mean meal to pellet interval in the Great Horned Owl, when 20–40 g of food are eaten, was 12·8 h. Small raptors have a shorter meal to pellet interval than larger raptors (Duke *et al.*, 1976a). The meal to pellet interval in owls increases when a larger quantity of food is eaten (Chitty, 1938; Kostuch and Duke, 1975; Duke *et al.*, 1976a; Duke and Rhoades, 1977) and usually in owls one pellet is egested after each meal (Chitty, 1938; Duke *et al.*, 1976a). The egestion of pellets in hawks (Accipitridae), however, is not related to the weight of the meal but appears to be associated with the occurrence of dawn, one pellet being egested after two or three meals (Duke *et al.*, 1976a). Other factors which have been shown to influence the meal to pellet interval in owls are the composition of the meal, the time of feeding and the availability of the next meal (Chitty, 1938; Smith and Richmond, 1972; Duke and Rhoades, 1977).

2. *Motility of the galliform stomach*

The motility of the highly differentiated type of stomach has been investigated by Mangold (1906), Ashcraft (1930), Nolf (1938a,b), Vonk and Postma (1949), Otani (1965), Pastea *et al.* (1966, 1968a, 1969), Matuura *et al.* (1967), Duke *et al.* (1972), Duke and Evanson (1972). Dziuk and Duke (1972), Oguro and Ikeda (1974a,b) and Hill and Strachan (1975). The following account is based mainly on the cineradiographic studies in young turkeys by Dziuk and Duke (1972) who demonstrated a gastric contraction cycle in which five phases can usually be recognized. In *phase I* (4 s), the thin muscles of the gizzard begin to contract and the isthmus closes. In *phase II* (4 s), the thin muscles are maximally contracted. The pylorus opens and ingesta passes from the stomach to the duodenum. In *phase III* (2 s), the thick muscles of the gizzard begin to contract whilst the thin muscles start to relax. The pylorus closes and there is contraction of the duodenum. Opening of the isthmus allows refluxing of ingesta from the

intestinal artery. Blood is drained to the right hepatic portal vein from the duodenum, pancreas, ileum and caeca by the gastropancreaticoduodenal vein; from the jejunum, ileum and caeca by the cranial mesenteric vein; and from the rectum by the caudal mesenteric vein.

H. Nerve supply of intestines

The intestinal tract is innervated by a number of prevertebral nerve plexuses formed by the splanchnic nerves from the thoracic and synsacral sympathetic nerve trunks and by branches of the vagus nerve and the pudendal plexus (Hsieh, 1951). Most of the available histochemical and pharmacological data on the innervation summarized by Bennett (1974) suggests a cholinergic and adrenergic innervation. However, other evidence has been presented which raises the possibility that contraction of the muscle of the rectum in *Gallus* is mediated through the excitation of fibres other than cholinergic and adrenergic nerves (Bartlet and Hassan, 1971; Bartlet, 1974; Takewaki and Ohashi, 1977; Takewaki *et al.*, 1977). Histochemical data on the enteric plexuses in *Gallus* have recently been provided by Ali and McLelland (1978). At all levels of the intestinal tract the intramural nerves are distributed as myenteric, submucosal, muscle and mucosal plexuses, as well as a perivascular plexus which extends throughout the thickness of the gut wall. All plexuses contain cholinesterase-positive and fluorescent nerve fibres. Ganglia are restricted to the myenteric and submucosal plexuses and are generally cholinesterase positive. Nerve cell bodies showing specific fluorescence, however, have not been observed. The ganglion cells in both the myenteric and submucosal plexuses are surrounded by many varicose terminal fluorescent fibres. Except in the rectum, the longitudinal layer of muscle is sparsely innervated.

I. Intestinal function

The function of the intestines has been extensively reviewed recently by Hill (1971b), Ziswiler and Farner (1972) and Sturkie (1976a,b). The intestines are concerned with the digestion of food and the absorption of the products of digestion. Food is moved down the tract by peristaltic waves of contraction. In many galliform species, the large intestine has also been shown to exhibit retroperistaltic waves which appear to originate in the cloaca and move the intestinal contents, including both ingesta and urine, from the cloaca to the rectum and caeca (Browne, 1922; Yasukawa, 1959; Akester *et al.*, 1967; Nechay *et al.*, 1968; Fenna and Boag, 1974a). Tindall (1976), however, was unable to demonstrate coordination of the mechanical activities of the rectum and caeca in the domestic fowl, and therefore suggested that material is moved along the rectum towards the caeca by haphazard contractions. On the basis of their investigations on Japanese Quail (*Coturnix japonica*), Fenna and Boag (1974a) have suggested the following sequence of intestinal contractions. Ingesta are moved caudally along the intestinal tract until sufficient filling of the cloaca occurs to initiate, by some as yet unknown mechanism, waves of retroperistalsis in the rectum. The opposing retroperistaltic contractions of the rectum and the peristaltic contractions of the small intestines meet at the ileo-rectal junction. The resulting increase in pressure forces the intestinal contents into the caeca. The constricted openings of the caeca and their relatively long villi probably act as filters allowing only fluid and fine food particles, i.e. the liquid fraction of the intestinal contents, to enter. Within the caeca the contractions serve to both mix the intestinal contents and move them distally. Retroperistalsis continues in the rectum until the caeca are filled. Emptying of the caeca seems to be brought about by massive retroperistaltic waves which occur simultaneously in both caeca (Yasukawa, 1959). These waves continue distally along the rectum and are responsible for the viscous semiliquid "caecal" droppings which are interposed with the well-formed cylindrical droppings solely of rectal origin. Fenna and Boag (1974a) pointed out that these alternating waves of contraction in the large intestine permit the caeca to be maintained in a full state whilst still allowing movement of the ingesta distally along the tract.

The segmenting movements of the intestinal tract serve only to break up the food and to mix it with the digestive juices. The small intestine seems to be the principal site for chemical digestion. Nearly all the enzymes in the intestinal juice are of gastric and pancreatic origin as well as coming from intestinal epithelial cells sloughed from the villi. The enzyme contribution of the intestinal glands to the intestinal juice is difficult to establish because of the contamination of the juice by other sources. However, there is some evidence in *Gallus* for enzyme secretion, probably of amylase, by the crypts of the duodenum (Kokas *et al.*, 1967). For excellent analyses of the present state of knowledge of enzyme activity in the avian small intestine consult Hill (1971b), Ziswiler and Farner (1972) and Sturkie (1976b). The precise role of the caeca in digestion has still to be established. Most studies have been in galliform species in which the caeca are well-developed. According to Fenna and Boag (1974b) the major role of the caeca in wild galliforms is to separate the contents of the intestines into a nutrient rich fluid fraction which enters the caeca for further digestion and absorption, and a relatively indigestible solid fibrous fraction which remains for a short while in the rectum before being extruded as "rectal" droppings. This mechanism they argued would allow birds to make efficient use of large amounts of green vegetation which is usually readily available but of relatively low nutritional value without at the same time having to accommodate large masses of cellulose. The stored food in the caeca undergoes bacterial fermentation, although the fermentation appears to supply only a small portion of the total energy requirement of the birds (Thompson and Boag, 1975; Gasaway, 1976a,b).

For the efficient absorption of the products of digestion, the intestines have a number of specializations which serve to increase the surface area available for absorption. These include lengthening of the small intestine, development of the complex surface relief of folds and villi, and the presence of microvilli on the absorptive epithelial cells. Evidence of the absorption mechanisms by which specific nutrients pass across the intestinal wall and of the preferential sites of absorption for the nutrients appears to be still relatively limited. The available information suggests that in *Gallus* the most active absorption of dietary nitrogen occurs in the proximal half of the jejunum (Immondi and Bird, 1965) whilst the most active absorption of fat takes place in the middle part of the small intestine (Renner, 1961). During fat absorption in birds chylomicra pass directly into the network of blood capillaries in the folds and villi (Graney, 1967) and not as in mammals into the lymphatic system. In the rectum there is absorption of water and also under certain conditions, of electrolytes (Hill and Lumijarvi, 1968). The rectum and coprodaeum also appear to be important in the resorption of salt and water from ureteral urine (Bindslev and Skadhauge, 1971; Skadhauge, 1973, 1974) (see also the Urinary system and Cloaca).

V. Pancreas

The pancreas is a long narrow gland most of which lies longitudinally in the caudal part of the body cavity, extending for a variable distance between the two limbs of the duodenal loop (Figs 3.7, 21, 22, 27). The cranial part of the gland may reach forward to the level of the spleen where characteristically as in *Gallus*, it is very attenuated and frequently hidden by fat (Schummer, 1973, p. 61). Part of the pancreas in the Budgerigar (*Melopsittacus undulatus*) lies along the outside of the duodenal loop (Fig. 3.21) (Evans, 1969; Feder, 1969), and in the Hoopoe (*Upupa epops*) according to Beddard (1911) the pancreas extends into the mesentery of the first two jejunal loops.

The avian pancreas like that of other vertebrates combines both exocrine and endocrine functions. The tubulo-acinar exocrine portion of the gland drains by ducts into the duodenum (Fig. 3.21). The endocrine pancreas consists of the islets and is discussed in detail in the chapter on the Endocrine System in Vol. 2.

A. Pancreas size

An excellent source of data on the size of the pancreas in many species of bird is the account of Magnan (1911a, pp. 73–79) which demonstrates that in general the gland is relatively large in piscivorous and insectivorous species and relatively small in carnivores and granivores. Another useful set of measurements of the gland is that recently obtained for domestic birds by Fehér and Fáncsi (1971) and which, unlike most of the other data, covers more than one parameter. In *Gallus* the pancreas weighs 3·44 g and makes up 0·17% of the total body weight. The relative weight of the gland in *Gallus* reaches a maximum approximately 20 days after hatching and then gradually declines (Oakberg, 1949). The absolute weight increases 214 times between hatching and maturity (Latimer, 1925).

B. External appearance of pancreas

Most of the following account of the gross morphology of the pancreas is based on descriptions by Gadow (1891a, pp. 684–685), Clara (1924), Hill (1926), Groebbels (1932, p. 487), Fehér and Fáncsi (1971), Ziswiler and Farner (1972), Schummer (1973, p. 61) and Paik *et al.* (1974). The pancreas is a pale yellow or red organ with a finely lobulated surface, and in the majority of species seems to be composed of three lobes although there has not been complete agreement in the literature on the precise nature of these lobes or their nomenclature (for a summary of the different interpretations of the organization of the gland in *Gallus* see Paik *et al.*, 1974). Unfortunately although full descriptions of the lobes exist for domestic species, much of the information which is available for wild birds is fragmentary and incomplete. In the present description the subdivisions of the gland follow the recent accounts in domestic species by Fehér and Fáncsi (1971), Schummer (1973, p. 61) and Paik *et al.* (1974), all of which describe similar dorsal, ventral and splenic lobes.

The dorsal and ventral lobes (*lobus pancreatis dorsalis*, *lobus pancreatis ventralis*) make up most of the gland, the dorsal lobe accompanying the descending limb of the duodenum, the ventral lobe lying alongside the ascending limb (Fig. 3.27). Generally, the two lobes are joined together by bridges of parenchymatous tissue which vary considerably in extent both interspecifically as well as between individuals of the same species. Especially well-developed interlobar connections are reported to be present in *Casuarius*, *Pelecanus*, *Ardea*, *Phoenicopterus* and *Otis*, the Accipitres, and several insectivorous passerines, so much so that in these birds it is difficult to distinguish two lobes (Gadow, 1891a, p. 684). In contrast, a total absence of parenchymatous bridging tissue seems to be the rule in a number of birds including *Anas* and *Columba*. Subdivisions of one or both of the lobes can usually be made out macroscopically, each lobe in *Gallus* for example, having dorsal and ventral parts (*pars dorsalis*, *pars ventralis*) which are partially separated from one another by a groove (Fig. 3.27) (Paik *et al.*, 1974). The relatively small splenic lobe (*lobus pancreatis splenalis*) lies cranial to the dorsal and ventral lobes and is generally continuous with one or both of these lobes (Fig. 3.27), although like other aspects of the gross anatomy of the gland this also appears to be subject to individual variation. A splenic lobe which is completely separate from the rest of the pancreas has been described in a number of birds including *Columba* and the Common Kestrel (*Falco tinnunculus*).

C. Exocrine pancreas and pancreatic ducts

The exocrine part of the pancreas is a compound tubulo-acinar gland which is divided into lobules by indistinct connective tissue septa. The lobules consist mainly of acini formed by typical zymogenic columnar cells similar to those of the mammalian pancreas. The formation and the release of the granules of the acinar cells in relation to feeding has been described in *Gallus* by Chodnik (1948). Following a period of starvation, the supranuclear portions of the cells are packed with granules. Soon after feeding many of the granules are released, although no cell has its store of granules totally depleted. The appearance of the acini, half an hour after feeding suggests that they act as autonomous units since the amount of granules they contain varies considerably from one acinus to another. Evacuation of the granules proceeds for approximately two hours after the intake of food. The formation of new granules starts immediately secretion commences and continues for about five or six hours after feeding. For details of the ultrastructure of the acinar cells consult the accounts of Zeigel (1962), Braun-Blanquet (1969) and Hodges (1974, pp. 104–105).

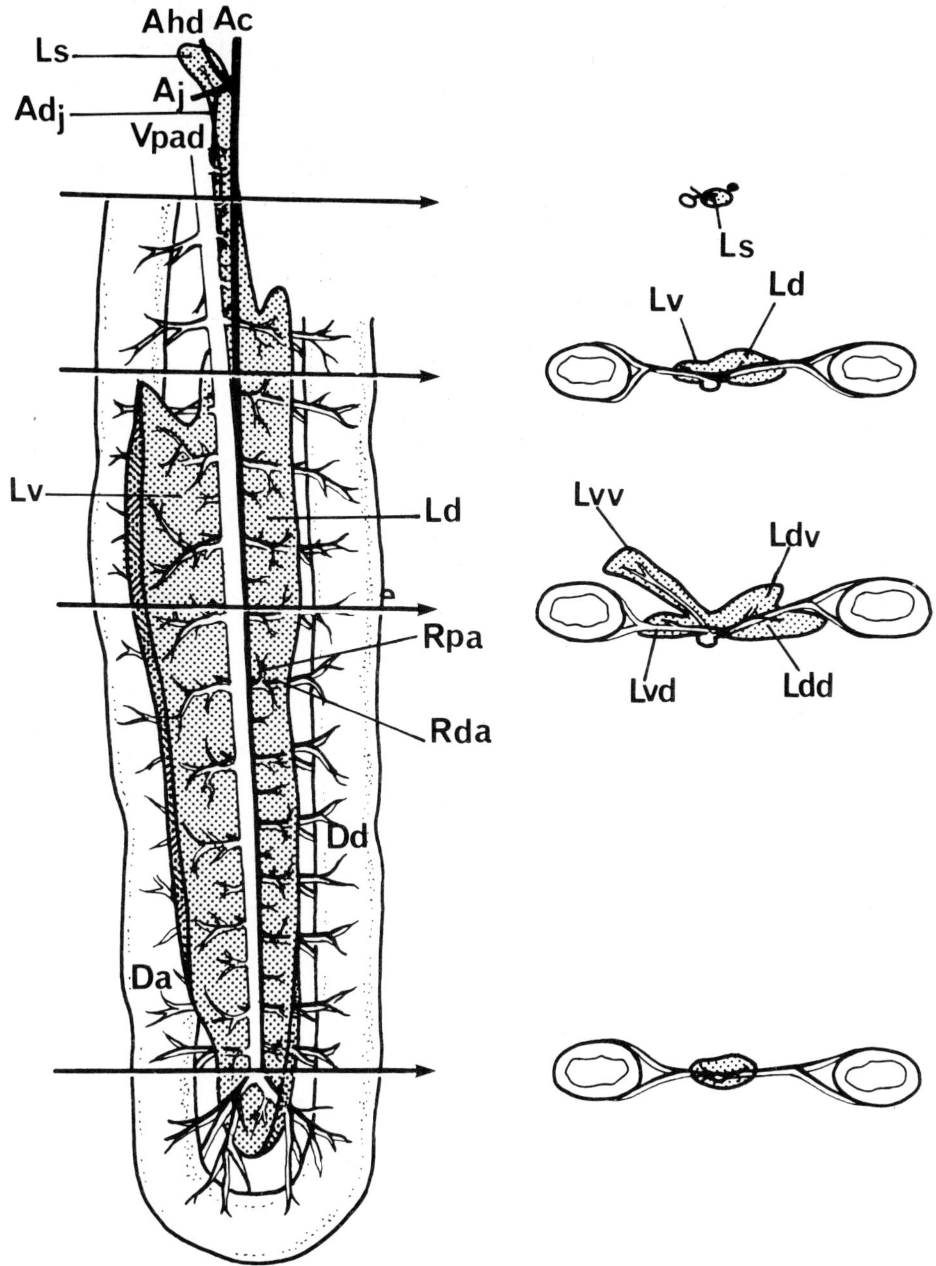

Fig. 3.27. Semischematic drawing of pancreas of *Gallus*, ventral view. Ac, coeliac artery; Adj, duodenojejunal artery; Ahd, right hepatic artery; Aj, jejunal artery; Da, ascending duodenum; Dd, descending duodenum; Ld, dorsal lobe of pancreas; Ldd, dorsal part of dorsal lobe; Ldv, ventral part of dorsal lobe; Ls, splenic lobe of pancreas; Lv, ventral lobe of pancreas; Lvd, dorsal part of ventral lobe; Lvv, ventral part of ventral lobe; Rda, duodenal arterial ramus; Rpa, pancreatic arterial ramus; Vpad, pancreaticoduodenal vein. From Paik *et al.* (1969).

The pancreatic ducts vary in number from one to three depending on the species, one duct being present for example in *Pelecanus* and *Caprimulgus*, two ducts in *Apteryx*, *Ciconia*, *Threskiornis*, *Phoenicopterus*, *Otis*, *Rallus*, *Phasianus*, *Meleagris*, *Psittacus*, *Corvus* and many Anatidae, and three ducts in *Larus*, several Anatidae, *Ardea*, *Burhinus*, *Gallus*, *Columba*, *Aquila*, Striges, *Cuculus* and Pici (Groebbels, 1932, p. 487). The manner in which the ducts are distributed amongst the lobes also varies considerably and has been well-documented for a number of species by Hill (1926). However, detailed information on the amount of parenchymatous tissue drained by each of the ducts seems to exist only for *Gallus*. Thus in the majority of individuals examined by Paik *et al.* (1974) most of the gland was drained by the ventral duct (*ductus pancreaticus ventralis*); the dorsal and accessory ducts (*ductus pancreaticus dorsalis*, *ductus pancreaticus accessorius*) together were restricted to the splenic lobe, to the cranial portion of the dorsal lobe and to the cranial portion of the dorsal part of the ventral lobe. In most birds, the pancreatic ducts open into the ascending limb of the duodenal loop (Fig. 3.21), the precise positions of the openings, however, varying with the species. Widely separated openings are a frequent occurrence as in *Turdus merula*, *Gallinula chloropus* and *Larus argentatus* (Hill, 1926). In *Larus argentatus*, for example, two of the ducts open into the duodenum near the apex of the loop whilst a third duct joins the distal part of the ascending limb. Frequently, all the pancreatic ducts open into the distal part of the ascending duodenum and often, as in the domestic species, the openings lie close to those of the bile ducts (Fig. 3.21) (Fehér and Fáncsi, 1971; Paik *et al.*, 1974). Apparently the pancreatic ducts rarely open into the descending limb of the duodenum as they do in the hornbill (Bucerotidae) described by Gadow (1891a, p. 684). Whilst the pancreatic ducts like the bile ducts, generally open into the duodenal loop on its inner surface, in the Budgerigar (*Melopsittacus undulatus*) the duct which drains the lobe of the pancreas lying outside the duodenal loop terminates on the outer surface of the loop and opposite to the other ducts (Fig. 3.21) (Evans, 1969). Both the pancreatic ducts and bile ducts in domestic birds open into the duodenum by a common papilla (*papilla duodenalis*) (Batojeva and Batojev, 1972; Paik *et al.*, 1974), the order in which the ducts penetrate the wall of the duodenum being highly variable (Paik *et al.*, 1974). This part of the duodenum in the domestic species was found by Fehér and Fáncsi (1971) to form an ampulla-shaped swelling which in *Gallus* is about 5 mm long and 3 mm wide. The walls of the pancreatic ducts are composed of three main layers; the mucous membrane which is strongly folded and lined by a simple columnar epithelium, the muscle tunic which consists of outer circular and inner longitudinal layers, and the adventitia (Hodges, 1974, pp. 105–106). The fine structure of the epithelium of the ducts has recently been described by Weyrauch and Schnorr (1978). At the duct openings the muscle has a sphincter-like arrangement (Fehér and Fáncsi, 1971).

D. *Blood supply of pancreas*

The blood supply of the pancreas (Fig. 3.27) has been established for the domestic fowl by Paik *et al.* (1969, 1970). The gland receives its arterial supply from the pancreaticoduodenal, jejunal and duodenojejunal branches of the coeliac artery. The pancreaticoduodenal artery extends longitudinally through the gland giving off numerous duodenal and pancreatic rami to the dorsal and ventral lobes. The duodenal rami terminate eventually outside the gland in the ascending and descending limbs of the duodenal loop. The jejunal and duodenojejunal arteries give off pancreatic rami to the splenic lobe. Interlobular arteries arising from the pancreaticoduodenal and pancreatic rami form in turn intralobular arteries which open into a blood capillary network in the interacinar tissue. From this network afferent vessels extend to the islets where they form in small A and B islets, a convoluted glomerulus-like arrangement of sinusoids, and in the large A islets of the splenic lobe, a rete mirabile-like structure. The islets are drained by efferent vessels back into the interacinar capillary network. The remaining veins including the intralobular and interlobular veins, and the duodenal and pancreatic rami of the pancreaticoduodenal veins (Fig. 3.27) are mostly satellites of the arteries and carry blood to the right hepatic portal vein.

E. Nerve supply of pancreas

Recent descriptions of the innervation of the avian pancreas are provided by Watari (1968), Kobayashi and Fujika (1969), Kudo (1971a), Dahl (1973), Smith (1974), Trandaburu (1974) and Watanabe and Yasuda (1977). The pancreatic nerves arise from the pancreaticoduodenal nerve plexus, a subsidiary of the coeliac plexus (Baumel, 1975). The intrapancreatic nerve bundles are ganglionated and consist of unmyelinated axons. They are distributed with the blood vessels in the interlobular and interacinar connective tissue. Some of the axons terminate close to the acinar cells. The innervation of the islets is described in the chapter on the Endocrine System in Vol. 2.

F. Exocrine pancreatic secretion

Pancreatic juice is a clear fluid secreted by the acinar cells, its composition and flow rate being the subject of many investigations which have recently been reviewed by Hill (1971b), Ziswiler and Farner (1972) and Sturkie (1976b). The enzyme content of the juice (amylase, lipase and proteinases) appears to be basically similar to that of mammals and is important in chemical digestion. The high pH of the juice acts to neutralize the acid chyme which enters the duodenum from the stomach and therefore ensures the presence of an alkaline medium which is essential for the optimum activity of the pancreatic enzymes. The rate of flow of the juice seems to be controlled by nervous and hormonal mechanisms as in mammals. Whilst the precise nature of these mechanisms does not appear to have been established, much of the existing evidence points to the control being mediated by the vagi and by the hormones pancreozymin and secretin.

VI. Liver

The liver (*hepar*) is the largest gland in the body and fills most of the ventral parts of the cranial and middle regions of the body cavity where it lies ventral to the lungs, stomach, spleen, intestines and gonads (Fig. 3.7). As in other vertebrates, it functions both as an exocrine gland secreting bile via a system of ducts and as an endocrine gland secreting substances directly into the blood stream.

A. *Physical characteristics of liver*

Most data in the literature on the size of the liver in birds are for domestic species only (Latimer, 1925; Latimer and Rosenbaum, 1926; Marsden, 1940; Hafez, 1955; Simić and Janković, 1959, 1960; Yasuda, 1962; Al-Dabagh and Abdulla, 1963; Boldizsár and Kozma, 1968; Kapp and Balázs, 1970; Schummer, 1973, p. 58). In *Gallus* the liver weighs between 31 and 51 g and forms 1·7–2·3% of the total body weight (Schummer, 1973, p. 58). Unfortunately corresponding figures to these do not seem to be available for the livers of wild birds, the best data available for these birds probably being the short series of body to liver weight ratios cited by Gadow (1891a, p. 681). The type of food consumed appears to have an important correlation with the size of the liver and in general the largest livers, relative to the size of the body, are found in piscivorous and insectivorous species (Magnan, 1910b, 1911d, 1912b). The influence of age on liver size has been studied only in *Gallus*, and although much of the evidence suggests that in this species the relative weight of the liver reaches a maximum sometime between hatching and maturity and then declines (Yasuda, 1962; Al-Dabagh and Abdulla, 1963; Boldizsár and Kozma, 1968), little change was observed by Hafez (1955) during the first two years of life.

The liver at hatching has a bright yellow colour due to pigments which are absorbed by it along with the lipids of the yolk (Hanes, 1912; Calhoun, 1954, p. 76; Cook and Bletner, 1955; Kingsbury *et al.*, 1956; Kapp and Balázs, 1970). Later the colour gradually changes to the red-brown of the adult liver, the transition occurring in *Gallus* between 8 and 14 days of age (Cook and Bletner, 1955) and in *Anas* between 5 and 10 days of age (Kapp and Balázs, 1970). These colour changes in the liver can be correlated with changes in its histological appearance.

B. *Liver form*

The following account of the gross morphology of the liver is based mainly on descriptions in domestic birds by Georgescu (1910), Grau (1943), Lucas and Denington (1956), Simić and Janković (1959, 1960), Kern (1963), Wildfeuer (1963, pp. 4–7), Flechsig (1964), Wissdorf *et al.* (1971), Schummer (1973, pp. 58–61) and McLelland (1975), the few details that it has been possible to add for wild species being obtained from the limited reports by Gadow (1891a, pp. 680–683) and Beddard (1898, pp. 31–32). Whilst the general form of the avian liver seems to be relatively constant within a species, considerable variations in detail may occur as has been demonstrated statistically by Lucas and Denington (1956) in *Gallus*. In all birds the liver is divided by cranial and caudal longitudinal incisures (*incisura interlobaris cranialis*, *incisura interlobaris caudalis*) into right and left lobes (*lobus hepaticus dexter*, *lobus hepaticus sinister*) which are joined together cranially in the midline between the incisures by an interlobar part (*pars interlobaris*) (Fig. 3.28). The relative proportions of these two lobes greatly vary both interspecifically (Gadow, 1891a, p. 681; Beddard, 1898, pp. 31–32) and between individuals of the same species (Lucas and Denington, 1956). In most birds examined by Gadow (1891a) and Beddard (1898) the right lobe was larger than the left. Much less commonly the two lobes were equal in size, as for example in *Rhea*, *Buceros rhinoceros*, *Scolopax rusticola* and *Synthliboramphus antiquus*, and in the Tinamidae and Cathartidae, or the left lobe was larger as in *Grus leucogeranus*, *Brachyramphus marmoratus* and *Alca torda*. The greatest discrepancy in size which was observed between the lobes by Gadow (1891a, p. 681) occurred in a shearwater (Procellariidae) in which the right lobe was six times larger than the left. In many species, at least one of the lobes is subdivided. Amongst domestic birds (Fig. 3.28) the left lobe is subdivided in *Gallus* and *Meleagris* into caudodorsal and caudoventral parts (*pars caudodorsalis*, *pars caudoventralis*). Gadow (1891a) found the right lobe subdivided in most of the Caprimulgidae, Apodidae and Trochilidae as well as in the majority of passerine species.

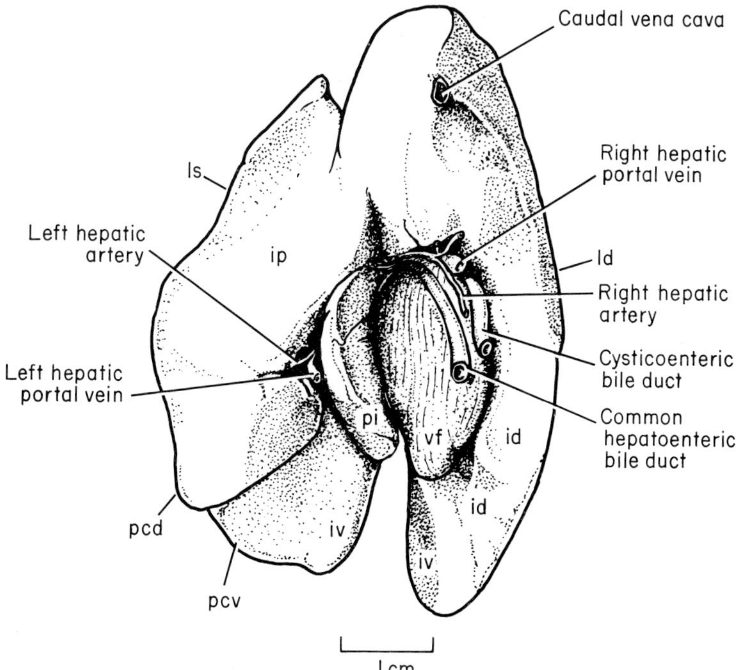

Fig. 3.28. Liver of *Gallus*, visceral surface. id, impression of duodenum; ip, impression of proventriculus; iv, impression of gizzard; ld, right hepatic lobe; ls, left hepatic lobe; pcd, pcv, caudodorsal and caudoventral parts of left hepatic lobe; pi, intermediate process; vf, gallbladder.

The ventrally directed parietal surface (*facies parietalis*) of the liver is convex and smooth except on its cranioventral part where there is a deep impression formed by the heart (*impressio cardiaca*) (Fig. 3.7). The dorsally directed visceral surface (*facies visceralis*) in contrast is concave and quite irregular since it is closely moulded to the adjacent viscera which make permanent and well-defined impressions on it. The visceral surface of the liver in *Gallus* is shown in Fig. 3.28. In this species there are impressions on the visceral surface of the left lobe formed by the proventriculus (*impressio proventricularis*), the gizzard (*impressio ventricularis*) and the spleen (*impressio splenalis*), and on the visceral surface of the right lobe by the gizzard, spleen, the ascending and descending limbs of the duodenum (*impressio duodenalis*), the umbilical vein and the right testis (*impressio testicularis*). In *Anas* and *Anser* the right lobe has impressions formed by the jejunum (*impressio jejunalis*). The gallbladder (*vesica fellea*) when present, lies in a pit-like depression (*fossa vesicae felleae*) on the visceral surface of the right lobe between the impressions formed by the two limbs of the duodenal loop (Fig. 3.28). Stretching across the visceral surfaces of both the lobes is a transverse groove, the hilus of the liver (*porta hepatis*), where the hepatic arteries and portal veins enter the organ and the bile ducts emerge. Arising from the visceral surface ventral to the hilus, in the domestic species at least, are one (*Gallus* and *Meleagris*) or two (*Anas* and *Anser*) intermediate processes (*procc. intermedii*). Simić and Janković (1959, 1960) have also described in the domestic birds, except *Columba*, a papillary process (*proc. papillaris*) arising from the visceral surface of the craniodorsal part of the right lobe. Totally embedded within the craniodorsal part of the right lobe is a small length of the caudal vena cava (Fig. 3.28).

C. Hepatic blood supply

As in mammals, the avian liver receives blood through hepatic portal veins and hepatic arteries and is drained by hepatic veins. The intrahepatic vessels have recently been described in the Adelie Penguin (*Pygoscelis adeliae*) by Andrews and Andrews (1976). The following short account in *Gallus* is based mainly on Jablan-Pantić and Antonijević (1961), Pavaux and Jolly (1968), Purton (1969b), Baumel (1975) and Miyaki (1978).

(1) *Hepatic portal veins.* Blood from the gastrointestinal tract is carried to the liver by the right and left hepatic portal veins (Fig. 3.28). According to Miyaki (1978) the territory of liver tissue supplied by each of the portal veins varies between birds. The larger right vein divides on the medial surface of the right lobe and supplies the right lobe, the interlobar tissue, and frequently a part of the left lobe. The smaller left hepatic portal vein enters the medial surface of the left lobe to which it is restricted, an anastomosis often being present between it and the right hepatic portal vein. In 48% of the domestic fowl examined by Miyaki (1978) small afferent veins entered the lateral or ventrocaudal borders of the liver from the abdominal air sacs and the peritoneum.

(2) *Hepatic arteries.* The arterial blood supply of the liver consists of right and left hepatic arteries which arise from branches of bifurcation of the coeliac artery (Figs 3.21, 28). The right hepatic artery gives off branches to the inter-

lobar part of the liver, the gallbladder and the proximal ends of the extrahepatic bile ducts before entering the medial surface of the right lobe of the liver. The left hepatic artery enters the medial surface of the left lobe. The two hepatic arteries are joined together in the hilus by an artery which extends transversely alongside the left branch of the right hepatic portal vein and from which a number of small arteries enter the liver directly. The intrahepatic tributaries of the arteries are distributed with branches of the portal vein.

(3) *Hepatic veins*. The liver is drained by two relatively large right and left hepatic veins and by one or more smaller middle hepatic veins. The right hepatic vein drains the right lobe of the liver via cranial dorsal, caudal dorsal and ventral intralobar branches. The left hepatic vein receives blood from the left lobe via similar intralobar branches to the right vein. Both right and left hepatic veins empty into the caudal vena cava close to the point where the vena cava leaves the cranial surface of the liver. The middle hepatic veins drain the interlobar part of the liver and open into the embedded portion of the vena cava.

D. *Internal structure of liver*

The internal organization of the liver of birds closely adheres to the typical vertebrate pattern (Elias and Bengelsdorf, 1952; Hickey and Elias, 1954; Purton, 1969a,b). The portal veins and hepatic veins extend through the liver in opposite directions to each other between the hilus and the periphery of the organ. Arising from these vessels are two approximately parallel systems of repeatedly branching portal and hepatic veins. The terminal portal veins interdigitate with the terminal hepatic veins and are linked to the hepatic veins by a short network of sinusoids. Coursing beside each of the terminal portal veins is a branch of the hepatic artery and a bile duct. These three structures are supported by a small amount of connective tissue (*capsula fibrosa perivascularis*) continuous with the capsule of the organ and form a portal tract. The hepatic artery connects with the sinusoidal network either directly through an arteriole or indirectly via a peribiliary capillary plexus which drains into the terminal portal vein.

The parenchyma of the liver consists, as in other vertebrates, of a continuous system of fenestrated, branching and anastomosing sheets of hepatocytes, the muralium of Elias (1955). Enclosed by the muralium is a network of lacunae containing the sinusoids which connect the portal and hepatic venous systems. Depending on the species, the sheets of liver cells are two cells thick as they predominantly are in lower vertebrates or one cell thick as in all mammals. Amongst the species examined by Elias and Bengelsdorf (1952) and Hickey and Elias (1954) the parenchymal sheets were two cells thick in *Gallus*, the Wood Duck (*Aix sponsa*) and the American Coot (*Fulica americana*) and one cell thick in the Red-necked Grebe (*Podiceps grisegena*) and the Prairie Chicken (*Tympanuchus cupido*) and in all sixteen of the passerine species which they investigated. Parenchymal sheets which are either one or two cells thick occurred in the Pintail (*Anas acuta*) and the Ruffed Grouse (*Bonasa umbellus*). Within the sheets of hepatocytes lie the bile canaliculi. In *Gallus* which has a 2-cell-thick muralium, the bile canaliculi run between the two layers of cells and therefore have walls formed by 3, 4 or 5 adjacent hepatocytes. This arrangement differs therefore from the 1-cell-thick muralium of mammals in which the canaliculi form a network around the opposing cell surfaces of the hepatocytes so that the walls are composed of only 2 or 3 hepatocytes. Elias and Bengelsdorf (1952) suggested that a 2-cell-thick muralium like that of *Gallus*, is less stable anatomically and less efficient physiologically than the 1-cell-thick muralium of mammals. The bile canaliculi drain into interlobular ducts (*ductuli interlobulares*) in the portal tracts.

The classic hepatic lobule (*lobulus hepaticus*) in birds, which is centred on the terminal branch of the hepatic vein, i.e. the central vein (*vena centralis*) and is surrounded by portal tracts, is not sharply separated off from neighbouring hepatic lobules as in the pig and some other mammals. Consequently hepatic lobules are difficult to identify histologically except close to the hilus. However, as Purton (1969b) emphasized, the hepatic lobules are not discrete vascular units anyway, since adjacent lobules are interconnected by sinusoids. A much more valuable unit of liver tissue from a structural and functional standpoint is the liver acinus which is the region of liver parenchyma supplied by the terminal branch of the portal vein and drained by the accompanying bile duct. Differences in the physiological activity of the hepatic cells in different regions of the acinus were suggested by the histochemical studies of Shah *et al.* (1972, 1975), Pilo *et al.* (1973), Ratzlaff and Tyler (1973) and Asnani *et al.* (1975).

The remainder of this account of the microanatomy of the liver is based on the ultrastructural observations by David (1961), Yasuda (1962), Theron (1965), Allen and Carstens (1966) Köhler and Schumacher (1967), Schildmacher *et al.* (1968), Adamiker (1969), Duncan *et al.* (1969), Purton (1969a,b), Kapp and Bálzs (1970) and Hodges (1972, 1974, pp. 88–100). The sinusoids in the lacunae between the sheets of hepatocytes are lined by two kinds of cells, a flattened endothelial cell and a protuberant irregular phagocytic cell (of Kupffer). Whilst the endothelium of the sinusoids is fenestrated as in mammals, a number of the fenestrations are closed by a diaphragm. The basal lamina of the sinusoids is discontinuous. The sinusoids are separated from the sheets of hepatocytes by the perisinusoidal space (of Disse). This space is continuous with the sinusoidal lumen through the fenestrations in the wall of the sinusoids and consequently is filled with blood plasma. In addition, Purton (1976) demonstrated the presence in the space in *Gallus* of a fat-storing cell and a free mesenchymal cell. The phagocytic cells of the sinusoids are an important component of the reticuloendothelial system and as such remove red blood cells from the circulation. The breakdown products of haemoglobin are returned by the phagocytic cells into the blood stream from which they are taken up again by the hepatocytes and excreted as bile pigments.

The hepatocyte is a polygonal cell containing a large spherical to oval nucleus which is generally situated close to the vascular pole of the cell. The cell membrane at both the vascular and biliary poles has numerous microvilli which greatly increase the surface areas available at these poles for absorption and secretion. At the biliary pole, the hepatocyte is firmly bound to adjacent cells by tight junctions which prevent the bile canaliculus from rupturing when high pressures develop within its lumen. The wide range of cytoplasmic organelles possessed by the hepatocyte reflects the large number of activities in which the cell is involved. Amongst these are roles in the metabolism of carbohydrates, proteins and fat, the synthesis of plasma proteins and proteins concerned with blood coagulation, the storage of glycogen, fats and fat-soluble vitamins, and the secretion of bile. In support of these many functions the mitochondria of the cell are especially numerous, and there is abundant rough endoplasmic reticulum, some smooth endoplasmic reticulum, a Golgi complex, lysosomes and microtubules (peroxisomes), which have themselves been the subject of investigations by Shnitka (1966), Scott *et al.* (1969) and Essner (1970). Unlike in mammals, glycogen granules are relatively scarce in the hepatocytes of most birds, large accumulations only being observed in newly-hatched domestic fowl (Allen and Carstens, 1966; Stephens and Bils, 1967; Kapp and Balázs, 1970) and in mature birds in lay (Hodges, 1972).

Differences in the structure of the liver in early and late postembryonic life were demonstrated in ducklings by Kapp and Balázs (1970) using a combination of histochemical methods and electron microscopy. In the early postembryonic period up to 8–10 days of age the parenchymal cells are relatively large and unlike those of adult birds contain large quantities of glycogen and lipids, the latter consisting in the first few days of cholesterol. During this period two types of hepatocytes, pale and dark, can be distinguished with the electron microscope. The pale cell is characterized by the relatively low electron density of its cytoplasm which contains lipid droplets and glycogen granules but few organelles. In contrast, the dark hepatocyte has a more electron dense cytoplasm with many mitochondria and free ribosomes, and large quantities of rough endoplasmic reticulum, but few lipid droplets and no glycogen granules. Between days 9 and 20 after hatching the number of pale cells decreases so that between the 20th and 42nd day most of the hepatocytes resemble the parenchymal cells of the adult liver. Kapp and Balázs have suggested that since the histochemical changes in the hepatocytes parallel the absorption from the yolk sac of yolk which is also rich in glycogen and cholesterol, the large quantities of glycogen and lipids in the hepatocytes during the first few days of life are probably derived from the yolk sac and act as an immediate source of energy to the bird.

E. Biliary system

The bile canaliculi drain into interlobular ducts (*ductuli interlobulares*) which extend through the liver in close association with the branches of the portal veins and hepatic arteries. The lobar bile ducts (*ductuli biliferi*) unite finally at the hilus to form right and left hepatic ducts (*ductuli hepatici*) (Fig. 3.29). For details of the larger lobar bile ducts in *Gallus* and some variations in the formation of the hepatic ducts see Miyaki (1973). Bile is usually transported from the liver to the small intestine in birds by two ducts unlike in mammals which have a single common bile duct. For detailed descriptions in domestic species see Simić and Janković (1959, 1960), Crompton and Nesheim (1972) and Miyaki (1973). In *Gallus*, as described by Miyaki (1973), and shown in Fig. 3.29, and probably also in birds generally, the right and left hepatic ducts unite on the medial surface of the right lobe of the liver to form a common hepatoenteric duct (*ductus hepatoentericus communis*) which opens into the duodenum. A

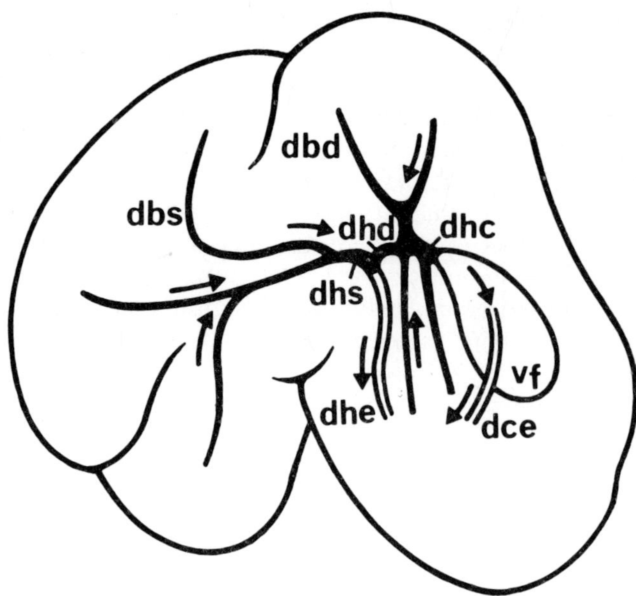

Fig. 3.29. Schematic drawing of the bile ducts of *Gallus*. The black ducts are intrahepatic, the white ducts extrahepatic. The direction of bile flow is shown by arrows. dbd, lobar bile duct of the right lobe; dbs, lobar bile duct of left lobe; dce, cysticoenteric duct; dhc, hepatocystic duct; dhd, right hepatic duct; dhe, common hepatoenteric duct; dhs, left hepatic duct; vf, gallbladder. From Miyaki (1973).

hepatocystic duct (*ductus hepatocysticus*) branches from the right hepatic duct and enters the gallbladder which is drained in turn by the cysticoenteric duct (*ductus cysticoentericus*) into the duodenum. Two hepatoenteric ducts, in addition to the cysticoenteric duct, were reported in certain Cracidae by Gadow (1891a, p. 683) and in *Anser* by Simić and Janković (1959, 1960). In species without a gallbladder the branch from the right hepatic duct opens directly into the duodenum and is referred to as the "right hepatoenteric duct" (*ductus hepatoentericus dexter*) to distinguish it from the common hepatoenteric duct which is formed by the union of the left and right hepatic ducts. A right hepatoenteric duct is absent in the Ostrich (*Struthio camelus*) (Newton and Gadow, 1896, p. 299). The level of the duodenum into which the bile ducts open seems to vary in different species although there is little information on this for most birds. According to Gadow (1891a, p. 681) the ducts usually join the duodenum somewhere on its middle portion or on the ascending limb (Fig. 3.21). However, in some birds including the Ostrich (*Struthio camelus*) and *Columba*, the common hepatoenteric duct empties into the duodenum in the proximal part of the descending limb. Gadow (1891a, p. 683) found that the openings of the ducts on the inner surface of the duodenum were marked by small elevations, the lumen of each opening being guarded by a flap-like valve. In most domestic species, the bile ducts open into a diverticulum on the distal part of the ascending limb of the duodenum in close association with the pancreatic ducts.

By far the most wide-ranging investigation into the occurrence of a gallbladder (*vesica fellea*) in birds was by Gorham and Ivy (1938) who collated most of the available literature as well as adding themselves much new data for a wide range of species. The result showed that a gallbladder is present in most birds (Fig. 3.28) and that amongst the relatively few species in which its absence has been noted are most of the Columbidae, many psittaciforms, *Struthio camelus*, *Rhea*, *Opisthocomus hoazin*, *Turnix tanki*, *Podica senegalensis*, *Falco peregrinus*, *Chordeiles*, *Archilochus colubris*, *Campylopterus ensipennis*, *Halcyon smyrnensis*, *Picumnus squamulatus*, *Pitta oatesi*, *Philepitta*, *Alcippe nipalensis*, *Ficedula parva*, *Bombycilla garrulus* and *Aethopyga saturata*. However, the presence of a gallbladder in some species appears to be variable as observed, for example, by Gadow (1891a, p. 683) in *Mergus merganser*, *Numida meleagris*, *Falco peregrinus*, *Nymphicus hollandicus*, *Cuculus* and *Rhea*. The presence or absence of a gallbladder Gadow concluded, is of no taxonomic value. The gall-bladder in the small number of species for which details of it are available, is partly embedded in the medial surface of the right lobe of the liver and varies greatly in size and shape between different species. In *Gallus*, for example, it is a pear-shaped sac, 2–5 cm in length, with its blind distal extremity lying opposite the caudal edge of the liver (Simić and Janković, 1959). An unusually long and tubular gallbladder occurs in some Picidae, Capitonidae and Ramphastidae, reaching in one aracari species at least almost to the level of the cloaca (Forbes, 1882a). The gallbladder is generally agreed to be especially well-developed in carnivorous species.

The microanatomy of the biliary system was investigated by Calhoun (1954, p. 76), Bader (1966), Purton (1969b), Yamada (1969, 1973), Hodges (1974, pp. 100–101) and Weyrauch and Schnorr (1978). The interlobular ducts in the portal tracts are lined by a simple cuboidal or columnar epithelium with microvilli. The walls of the extrahepatic ducts are composed of three main layers: mucous membrane, muscle tunic and adventitia. The mucous membrane is folded when the ducts are empty, and has a simple columnar epithelium. The epithelial cells are characterized by apically-situated secretory granules which are PAS positive and probably mucus in nature. The presence of microvilli at the luminal surface of the cells and pinocytotic vesicles just below the surface indicate a resorptive function, although Yamada (1969) concluded, from the general ultrastructure of the cells in the Japanese Quail (*Coturnix japonica*) that their main activity is secretory rather than absorptive. In *Gallus* the epithelium of the ducts consists of secretory cells, undifferentiated cells, ciliated cells and migrating cells (Weyrauch and Schnorr, 1978). The muscle tissue is arranged into outer longitudinal, middle circular and inner oblique or longitudinal layers.

The wall of the gallbladder appears to consist of the same three layers as the bile ducts. The mucosal folds may be high or low depending on whether the organ is distended or empty and contracted. The simple columnar epithelium was investigated by Yamada in the Japanese Quail (*Coturnix japonica*). The ultrastructural characteristics of the epithelial cells resemble those of the bile ducts and indicate the existence of resorptive and secretory functions. However, both roles do not seem to be shared equally by the same epithelial cell since the cells lining the crests of the folds are specialized more for absorption than secretion, whilst those in the troughs between the folds appear to have mainly a secretory function. The greater capacity for resorption possessed by the gall-bladder compared with the bile ducts, is reflected in the ultrastructure of some of the gallbladder epithelial cells, especially those on the crests of the folds. In these cells the microvilli are relatively numerous and well-developed, and there are more pinocytotic vesicles and larger amounts of smooth endoplasmic reticulum than in the cells of the bile ducts. Furthermore, dilated intercellular spaces occur at the bases of the cells. Yamada (1973) has suggested that the pinocytotic vesicles and the smooth endoplasmic reticulum are involved in the absorption of lipids from the bile and might explain the occurrence of relatively large numbers of lipid droplets in the basal parts of the cytoplasm. The microvilli and dilated intercellular spaces are probably concerned with the absorption of water, electrolytes and other substances of low molecular weight. As in the bile ducts, the secretion of the gallbladder epithelium appears to be mucous in nature. The muscle coat (*tunica muscularis*) is complex and the arrangement of its fibres appears to vary with the degree of distension of the organ. Hodges (1974) found it to be composed basically of two layers, an outer circular or oblique layer from which fibres extend into the mucosal folds, and an irregular inner longitudinal layer.

F. Bile

Bile is the exocrine secretion of the liver and consists principally of bile pigments, bile salts, cholesterol, amylase and electrolytes, further details of its composition being obtainable from the accounts of Farner (1943), Anderson *et al.* (1957), Clarkson *et al.* (1957), Haslewood (1964) and Lind *et al.* (1967). The two bile pigments of birds, biliverdin and bilirubin, are formed in the reticulo-endothelial system from the breakdown of haemoglobin. The pigments are carried to the hepatocytes where they are made water-soluble by conjugation with glucuronic acid before being excreted in the bile as waste products. Biliverdin is responsible for the green colour of the bile. The bile salts and cholesterol in contrast are both synthesized by the hepatocytes. Bile salts are emulsifying agents and are involved in the intestine in the digestion and absorption of fats. The secretion of bile by the hepatocyte is continuous and in *Gallus* there is a bile flow rate varying between 0·5 and 1 ml/hour (Lind *et al.*, 1967). As in mammals, the main stimulus to bile secretion appears to be the bile salts. Bile is stored in the gallbladder where its volume is reduced by the resorption of water and inorganic salts (Schmidt and Ivy, 1937). For further data on the formation and function of bile see Clarkson and Richards (1971), Hill (1971b), Ziswiler and Farner (1972) and Sturkie (1976b).

References

Ábrahám, A. (1936). Beiträge zur Kenntnis der Innervation des Vogeldarmes, *Z. Zellforsch. mikrosk. Anat.* **23**, 737–745.

Adamiker, D. (1969). Elektronenmikroskopischen Untersuchungen zur Virushepatitis der Entenküken. Zentbl, *VetMed. Reihe* B **16**, 620–636.

Aitken, R. N. C. (1958). A histochemical study of the stomach and intestine of the chicken, *J. Anat., Lond.* **92**, 453–466.

Akester, A. R., Anderson, R. S., Hill, K. J. and Osbaldiston, G. W. (1967). A radiographic study of urine flow in the domestic fowl, *Br. Poult. Sci.* **8**, 209–212.

Al-Dabagh, M. A. and Abdulla, M. (1963). Correlation of sizes and weights of livers and spleens to the ages and body weights of normal chicks with a note on the histology of these organs in chicks, *Vet. Rec.* **75**, 397–400.

Ali, H. A. and McLelland, J. (1978). Avian enteric nerve plexuses. A histochemical study, *Cell Tiss. Res.* **189**, 537–548.

Allen, J. R. and Carstens, L. A. (1966). Electron microscopic alterations in the liver of chickens fed toxic fat, *Lab. Invest.* **15**, 970–979.

Anderson, I. G., Haslewood, G. A. D. and Wooton, I. D. P. (1957). Comparative studies of bile salts. X. Bile salts of the King Penguin, *Aptenodytes patagonica*, *Biochem. J.* **67**, 323–328.

Andrews, C. J. H. and Andrews, W. H. H. (1976). Casts of hepatic blood vessels: a comparison of the microcirculation of the penguin, *Pygoscelis adeliae*, with some common laboratory animals, *J. Anat., Lond.* **122**, 283–292.

Antony, M. (1920). Über die Speicheldrüsen der Vögel, *Zool. Jb., Abt. Anat. Ontog. Tiere* **41**, 547–660.

Ashcraft, D. W. (1930). The correlative activities of the alimentary canal of the fowl, *Am. J. Physiol.* **93**, 105–110.

Asnani, M. V., Yadav, P. L., Pilo, B. and Shah, R. V. (1975). Comparative histochemical studies on avian liver. 6. Distribution patterns of lipids and some enzymes in the liver of nectar feeder, sunbird (*Nectarinia asiatica*), *PAVO* **11**, 85–91.

Bader, G. (1966). Die submikroskopische Struktur des Gallenblasenepithels und seiner Regeneration 2. Mitteilung: Huhn und verschiedene Säugetiere, *Z. mikrosk.-anat. Forsch.* **74**, 303–320.

Barthels, P. (1895). Beitrag zur Histologie des Ösophagus der Vögel, *Z. wiss. Zool.* **59**, 655–689.

Bartlett, A. L. (1974). Actions of putative transmitters in the chicken vagus nerve/oesophagus and Remak nerve/rectum preparations, *Br. J. Pharmac. Chemother.* **51**, 549–558.

Bartlett, A. L. and Hassan, T. (1971). Contraction of chicken rectum to nerve stimulation after blockade of sympathetic and parasympathetic transmission, *Q. Jl exp. Physiol.* **56**, 178–183.

Bartlett, A. D. (1869). Remarks upon the habits of the hornbills (*Buceros*), *Proc. zool. Soc. Lond.* **1869**, 142–146.

Bartram, E. (1901). Anatomische, histologische und embryologische Untersuchungen über den Verdauungstraktus von *Eudyptes chrysocome*, *Z. Naturw.* **74**, 173–236.

Bath, W. (1906). Die Geschmacksorgane der Vögel und Krokodile, *Arch. Biontologie* **1**, 5–47.

Batojeva, S. Ts. and Batojev, Ts. Zh. (1972). On the anatomy of the pancreas of domestic birds, *Arkh. Anat. Gistol. Embriol.* **63**, 105–108.

Bauer, M. (1901). Beiträge zur Histologie des Muskelmagens der Vögel, *Arch. mikrosk. anat. EntwMech.* **57**, 653–676.

Baumel, J. J. (1975). Aves heart and blood vessels. *In* "Sisson and Grossman's The Anatomy of the Domestic Animals" (R. Getty, Ed.), Vol. 2, 5th edn. Saunders, Philadelphia.

Bayer, R. C., Chawan, C. B., Bird, F. H. and Musgrave, S. D. (1975). Characteristics of the absorptive surface of the small intestine of the chicken from 1 day to 14 weeks of age, *Poult. Sci.* **54**, 155–169.

Beams, H. W. and Meyer, R. K. (1931). The function of pigeon "milk", *Physiol. Zoöl.* **4**, 486–500.

Beddard, F. E. (1886). On some points in the anatomy of *Chauna chavaria*, *Proc. zool. Soc. Lond.* **1886**, 178–181.

Beddard, F. E. (1890). On the alimentary canal of the Martineta Tinamou (*Calidromas elegans*), *Ibis* **1890**. 61–66.

Beddard, F. E. (1891). On the tongue of *Zosterops*, *Ibis* **1891**, 510–512.

Beddard, F. E. (1898). "The Structure and Classification of Birds." Longmans Green and Co, London.

Beddard, F. E. (1911). On the alimentary tract of certain birds and on the mesenteric relations of the intestinal loops, *Proc. zool. Soc. Lond.* **1**, 47–93.

Beddard, F. E. and Mitchell, P. C. (1894). On the anatomy of *Palamedea cornuta*, *Proc. zool. Soc. Lond.* **1894**, 536–557.

Beeker, W. J. (1953). Feeding adaptations and systematics in the avian order Piciformes, *J. Wash. Acad. Sci.* **43**, 293–299.

Bego, U. and Rapić, S. (1964). Topografski odnosi vratnog dijelà dušnika i jednjaka u peradi, *Vet. Arh.* **34**, 243–253.

Bego, U. and Rapić, S. (1965). Die topographischen Verhältnisse des Halsteiles der Luft- und der Speirseöhre beim Geflügel, *Acta anat.* **61**, 456.

Bennett, T (1969). Studies on the avian gizzard: histochemical analysis of the extrinsic and intrinsic innervation, *Z. Zellforsch. mikrosk. Anat.* **98**, 188–201.

Bennett, T. (1974). Peripheral and autonomic nervous systems. *In* "Avian Biology" (D. S. Farner and J. R. King, Eds). Academic Press, New York and London.

Bennett, T. and Cobb, J. L. S. (1969a). Studies on the avian gizzard: morphology and innervation of the smooth muscle, *Z. Zellforsch. mikrosk. Anat.* **96**, 173–185.

Bennett, T. and Cobb, J. L. S. (1969b). Studies on the avian gizzard: the development of the gizzard and its innervation, *Z. Zellforsch. mikrosk. Anat.* **98**, 599–621.

Bennett, T. and Cobb, J. L. S. (1969c). Studies on the avian gizzard: Auerbach's plexus, *Z. Zellforsch. mikrosk. Anat.* **99**, 109–120.

Bennett, T., Cobb, J. L. S. and Malmfors, T. (1971). Fluorescence histochemical observations on Auerbach's plexus and the problem of the inhibitory innervation of the gut, *J. Physiol., Lond.* **218**, 77–78P.

Bennett, T. and Malmfors, T. (1970). The adrenergic nervous system of the domestic fowl (*Gallus domesticus* (L.), *Z. Zellforsch. mikrosk. Anat.* **106**, 22–50.

vanden Berge, J. C. (1978). Myologia. *In* "Nomina Anatomica Avium" (J. J. Baumel, A. S. King and A. M. Lucas, Eds). Academic Press, London and New York.

Bergmann, C. (1862). Einiges über den Drüsenmagen der Vögel, *Arch. Anat. Physiol.* **1862**, 581–587.

Berkhoudt, H. (1976). The epidermal structure of the bill tip organ in ducks, *Neth. J. Zool.* **26**, 561–566.

Berkhoudt, H. (1977). Taste buds in the bill of the Mallard (*Anas platyrhynchos* L.), *Neth. J. Zool.* **27**, 310–331.

Bhaduri, J. L., Biswas, B. and Das, S. K. (1957). The arterial system of the domestic pigeon (*Columba livia* Gmelin), *Anat. Anz.* **104**, 1–14.

Bindslev, N. and Skadhauge, E. (1971). Salt and water permeability of the epithelium of the coprodeum and large intestine in the normal and dehydrated fowl (*Gallus domesticus*). *In vivo* perfusion studies, *J. Physiol., Lond.* **216**, 735–751.

Bock, W. J. (1961). Salivary glands in the Gray Jays (*Perisoreus*), *Auk* **78**, 355–365.

Bock, W. J., Balda, R. P. and Vander Wall, S. B. (1973). Morphology of the sublingual pouch and tongue musculature in Clark's Nutcracker, *Auk* **90**, 491–519.

Böker, H. (1929). Flugvermogen unk Kopf bei *Opisthocomus cristatus* und *Stringops habroptilus*. *Morph. Jb.* **63**, 152–207.

Boldizsár, H. and Kozma, M. (1968). Correlation between the body weight, liver weight and hepatic ribonucleic acid level in chickens of various ages, *Acta vet. hung.* **18**, 57–62.

Bolton, T. B. (1971). The structure of the nervous system. *In* "Physiology and Biochemistry of the Domestic Fowl" (D. J. Bell and B. M. Freeman, Eds), Vol. 2. Academic Press, London and New York.

Botezat, E. (1904). Geschmacksorgane und andere nervöse Endapparate im Schnabel der Vögel, *Biol. Zentralbl.* **24**, 722–736.

Botezat, E. (1906). Die Nervenendapparate in den Mundteilen der Vögel und die einheitliche Endigungsweise der peripheren Nerven bei den Wirbeltieren, *Z. Wiss. Zool.* **84**, 205–360.

Botezat, E. (1910). Morphologie, Physiologie und phylogenetische Bedeutung der Geschmacksorgane der Vögel, *Anat. Anz.* **36**, 428–461.

Braun-Blanquet, M. (1969). Examen du pancréas de canard normal au microscope électronique précédé de sou observation macroscopique et microscopique. I. Glande exocrine, *Acta anat.* **72**, 161–194.

Broussy, J. (1936). Sur un point particulier de l'histophysiologie de la muqueuse de gésier des granivores, *C.r. Séanc. Soc. Biol.* **121**, 1298–1300.

Browne, T. G. (1922). Some observations on the digestive system of the fowl, *J. comp. Path. Ther.* **35**, 12–32.

Bujard, E. (1906). Sur les villosités intestinales. Quelques types chez les oiseaux, *C. R. Assoc. Anat.* **8** Réunion Bordeaux 128–132.

Bujard, E. (1909). Etude des types appendiciels de la muqueuse intestinale, en rapport avec les régimes alimentaires. Morphologie comparée. Sitiomophoses naturelles et expérimentales, *Int. Mschr. Anat. Physiol.* **26**, 101–196.

Cade, T. J. and Greenwald, L. I. (1966). Drinking behaviour of mouse birds in the Namib Desert, southern Africa, *Auk* **83**, 126–128.

Cadow, G. (1933). Magen und Darm der Fruchttauben, *J. Orn., Lpz.* **81**, 236–252.

Calhoun, M. H. (1954). "Microscopic Anatomy of the Digestive System." Iowa State Coll. Press, Ames.

Cattaneo, G. (1884). Istologia e sviluppo dell' apparato gastrico degli uccelli, *Atti Soc. ital. Sci. nat.* **27**, 90–175.

Cazin, M. (1884). Note sur la structure de l'estomac du *Plotus melanogaster*, *Annls Sci. nat. Zool.* **18**, 128.

Cazin, M. (1886). Recherches sur la structure de l'estomac des oiseaux, *C.r. hebd. Séanc. Acad. Sci., Paris* **102**, 1031–1033.

Cazin, M. (1887a). Glandes gastriques à mucus et à ferment chez les oiseaux. *C.r. hebd. Séanc. Acad. Sci., Paris* **104**, 590–592.

Cazin, M. (1887b). Recherches anatomiques histologiques et embryologiques sur l'appareil gastrique des oiseaux, *Annls Sci. nat. Zool.* **4**, 177–323.

Cazin, M. (1887c). Structure et le mécanisme du gésier des oiseaux, *Bull. Soc. Philomath., Paris* **12**, 19–22.

Chapin, J. P. (1922). The function of the oesophagus in the bittern's booming, *Auk* **39**, 196–202.

Cheah, C. C. and Hansen, I. A. (1970a). Wax esters in the stomach oil of petrels, *Int. J. Biochem.* **1**, 198–202.

Cheah, C. C. and Hansen, I. A. (1970b). Stomach oil and tissue lipids of the petrels *Puffinus pacificus* and *Pterodroma macroptera*, *Int. J. Biochem.* **1**, 203–208.

Chitty, D. (1938). A laboratory study of pellet formation in the Short-eared Owl (*Asio flammeus*), *Proc. zool. Soc., Lond.* **108**, 267–287.

Chodnik, K. S. (1947). A cytological study of the alimentary tract of the domestic fowl (*Gallus domesticus*), *Q. Jl microsc. Sci.* **88**, 419–443.

Chodnik, K. S. (1948). Cytology of the glands associated with the alimentary tract of domestic fowl (*Gallus domesticus*), *Q. Jl microsc. Sci.* **89**, 75–87.

Choi, J. K. (1962). Fine structure of the smooth muscle of the chicken's gizzard. Electron Microsc., Proc. Int. Congr., 5th, 1962 Vol. 2, Art. M-9.

Clara, M. (1924). Das Pankreas der Vögel, *Anat. Anz.* **57**, 257–265.

Clara, M. (1925). Beiträge zur Kenntnis des Vogeldarmes. I. Teil. Mikroskopische Anatomie, *Z. mikrosk. anat. Forsch.* **4**, 346–416.

Clara, M. (1926a). Beiträge zur Kenntnis des Vogeldarmes. II. Teil. Die Hauptzellen des Darmepithels, *Z. mikrosk. anat. Forsch.* **6**, 1–27.

Clara, M. (1962b). Beiträge zur Kenntnis des Vogeldarmes. III. Teil. Die basalgekörnten Zellen im Darmepithel, *Z. mikrosk. anat. Forsch.* **6**, 28–54.

Clara, M. (1926c). Beiträge zur Kenntnis des Vogeldarmes. IV. Teil. Über das Vorkommen von Körnerzellen vom Typus der Panethschen Zellen bei den Vögeln, *Z. mikrosk. anat. Forsch.* **6**, 55–75.

Clara, M. (1926d). Beiträge zur Kenntnis des Vogeldarmes. V. Teil. Die Schleimbildung im Darmepithel mit besonderer Berücksichtigung der Becherzellenfrage, *Z. mikrosk. anat. Forsch.* **6**, 256–304.

Clara, M. (1926e). Beiträge zur Kenntnis des Vogeldarmes. VI. Teil. Das lymphoretikuläre Gewebe im Darmrohre mit besonderer Berücksichtigung der leukozytären Zellen, *Z. mikrosk. anat. Forsch.* **6**, 305–350.

Clara, M. (1927a). Beiträge zur Kenntnis des Vogeldarmes. VII. Teil. Die Lieberkühn'schen Krypten, *Z. mikrosk. anat. Forsch.* **8**, 22–72.

Clara, M. (1927b). Beiträge zur Kenntnis des Vogeldarmes. VIII und letzter Teil. Das Problem des Rumpfdarmschleimhautreliefs, *Z. mikrosk. anat. Forsch.* **9**, 1–48.

Clara, M. (1934). Über den Bau des Magendarmkanals bei dem Amseln (Turdidae). *Z. Anat.* **102**, 718–771.

Clark, G. A. Jr. (1961). Occurrence and timing of egg teeth in birds, *Wilson Bull.* **73**, 268–278.

Clark, H. L. (1918). Notes on the anatomy of the Cuban Trogon, *Auk* **35**, 286–289.

Clarke, A. and Prince, P. A. (1976). The origin of stomach oil in marine birds: analyses of the stomach oil from six species of subantarctic procellariiform birds, *J. exp. mar. Biol. Ecol.* **23**, 15–30.

Clarke, L. F., Rahn, H. and Martin, M. D. (1942). Seasonal and sexual dimorphic variations in the so-called "air sac" region of the Sage Grouse, *Bull. Wyo Game Fish Dep.* **2**, 13–27.

Clarke, R. (1967). On the constancy of the number of villi in the duodenum of the post-embryonic domestic fowl, *J. Embryol. exp. Morph.* **17**, 131–138.

Clarke, P. L. (1978). The structure of the ileo-caeco-colic junction of the domestic fowl (*Gallus gallus* L.), *Br. Poult. Sci.* **19**, 595–600.

Clarkson, M. J. and Richards, T. G. (1971). The liver with special reference to bile formation. *In* "Physiology and Biochemistry of the Domestic Fowl" (D. J. Bell and B. M. Freeman, Eds), Vol. 2. Academic Press, London and New York.

Clarkson, T. B., King, J. S. and Warnock, N. H. (1957). A comparison of the effect of gallogen and sulfarlem on the normal bile flow of the cockerel, *Am. J. vet. Res.* **18**, 187–190.

Cobb, J. L. S. and Bennett, T. (1969). A study of nexuses in visceral smooth muscle, *J. cell. Biol.* **41**, 287–297.

Cook, R. E. and Bletner, J. K. (1955). Normal appearance of the internal organs of chicks from hatching through six weeks of age, *Poult. Sci.* **34**, 1188–1189.

Cornselius, C. (1925). Morphologie, Histologie, und Embryologie des Muskelmagens der Vögel, *Gegenbaurs morph. Jb.* **54**, 507–559.

Corti, A. (1923). Contributo alla migliore conoscenza dei diverticuli ciechi dell'intestino posteriore degli uccelli, *Ric. Morf.* **3**, 211–295.

Crompton, D. W. T. and Nesheim, M. C. (1972). A note on the biliary system of the domestic duck and a method for collecting bile, *J. exp. Biol.* **56**, 545–550.

Curschmann, H. (1866). Zur Histologie des Muskelmagens der Vögel, *Z. wiss. Zool.* **16**, 224–235.

Cymborowski, B. (1968). Influence of diet on the histological structure of the gullet and glandular stomach of the Common Tern (*Sterna hirundo* L.), *Zoologica Pol.* **18**, 451–468.

Dahl, E. (1973). The fine structure of the pancreatic nerves of the domestic fowl, *Z. Zellforsch.* **136**, 501–510.

David, H. (1961). Zur Morphologie der Leberzellmembran, *Z. Zellforsch. mikrosk. Anat.* **55**, 220–234.

Demke, D. D. (1954). A brief histology of the intestine of the turkey poult, *Am. J. vet Res.* **15**, 447–449.

Desselberger, H. (1931). Der Verdauungskanal der Dicaeiden nach Gestalt und Funktion, *J. Orn., Lpz.* **79**, 353–370.

Desselberger, H. (1932). Ueber den Verdauungskanal nektarfressender Vögel, *J. Orn., Lpz.* **80**, 309–318.

Dow, D. D. (1965). The role of saliva in food storage by the Gray Jay, *Auk* **82**, 139–154.

Duerdon, J. E. (1912). Experiments with ostriches—XX. The anatomy and physiology of the Ostrich. C—The internal organs, *Agric. J. Un. S. Afr.* **III**, 492–507.

Duke, G. E., Dzuik, H. E. and Evanson, O. A. (1972). Gastric pressure and smooth muscle electrical potential changes in turkeys, *Am. J. Physiol.* **222**, 167–173.

Duke, G. E. and Evanson, O. A. (1972). Inhibition of gastric motility by duodenal contents in turkeys. *Poult. Sci.* **51**, 1625–1636.

Duke, G. E., Evanson, O. A. and Jegers, A. A. (1976a). Meal to pellet intervals in 14 species of captive raptors, *Comp. Biochem. Physiol.* **53A**, 1–6.

Duke, G. E., Evanson, O. A. and Redig, P. T. (1976b). A cephalic influence on gastric motility upon seeing food in domestic turkeys (*Meleagris gallopavo*), Great-horned Owls (*Bubo virginianus*) and Red-tailed Hawks (*Buteo jamaicensis*), *Poult. Sci.* **55**, 2155–2165.

Duke, G. E., Evanson, O. A. and Redig, P. T. and Rhoades, D. D. (1976c). Mechanism of pellet egestion in Great-horned Owls (*Bubo virginianus*), *Am. J. Physiol.* **231**, 1824–1829.

Duke, G. E. and Rhoades, D. D. (1977). Factors affecting meal to pellet intervals in Great-horned Owls (*Bubo virginianus*), *Comp. Biochem. Physiol.* **56A**, 282–286.

Dumont, J. N. (1965). Prolactin-induced cytologic changes in the mucosa of the pigeon crop during crop-"milk" formation, *Z. Zellforsch. mikrosk. Anat.* **68**, 755–782.

Duncan, D., Rigdon, R. H. and Morales, R. (1969). Fine structure of amyloid containing livers from White Pekin ducks, *Tex. Rep. Biol. Med.* **27**, 969–984.

Dzuik, H. E. and Duke, G. E. (1972). Cineradiographic studies of gastric motility in turkeys, *Am. J. Physiol.* **222**, 159–166.

Eber, G. (1956). Vergleichende Untersuchungen uber die Ernährung einiger Finkenvögel, *Biol. Abh.* **13**, 1–60.

Eglitis, I. and Knouff, R. A. (1962). An histological and histochemical analysis of the inner lining and glandular epithelium of the chicken gizzard, *Am. J. Anat.* **111**, 49–66.

Elias, H. (1955). Liver morphology, *Biol. Rev.* **30**, 263–310.

Elias, H. and Bengelsdorf, H. (1952). The structure of the liver of vertebrates, *Acta anat.* **14**, 297–337.

Essner, E. (1970). Observations on hepatic and renal peroxisomes (microbodies) in the developing chick, *J. Histochem. Cytochem.* **18**, 80–92.

Evans, H. E. (1969). Anatomy of the Budgerigar. *In* "Diseases of Cage and Aviary Birds" (M. L. Petrak, Ed.). Lea and Febiger, Philadelphia.

Fahrenholz, C. (1937). Drüsen der Mundhöhle. *In* "Handbuch der vergleichenden Anatomie der Wirbeltiere" (L. Bolk, E. Göppert, E. Kallius and W. Lubosch, Eds), Vol. 3. Urban and Schwarzenberg, Berlin.

Farner, D. S. (1943). Biliary amylase in the domestic fowl, *Biol. Bull., Woods Hole* **84**, 240–253.

Farner, D. S. (1970). Some glimpses of comparative avian physiology, *Fedn Proc. Fedn Am. exp. Biol.* **29**, 1649–1663.

Feder, F.-H. (1969). Beitrag zur makroskopischen und mikroskopischen Anatomie des Verdauungsapparates beim Wellensittich (*Melopsittacus undulatus*), *Anat. Anz.* **125**, 233–255.

Feder, F.-H. (1972a). Strukturuntersuchungen am Oesophagus verschiedener Vogelarten, *Zentbl. VetMed.* **1**, 201–211.

Feder, F.-H. (1972b). Zur mikroskopischen Anatomie des Verdauungsapparates beim Nandu (*Rhea americana*), *Anat. Anz.* **132**, 250–265.

Feder, F.-H. (1972c). Zur Lage der tiefen Drüsen im Proventriculus verschiedener Vogelarten, *Zentbl. VetMed.* **1C**, 266.

Fehér, G. (1975). Die ontogenetische Entwicklung des Dottersackstiels der Hausvögel und seine Rolle bei der Absorption des Dotters, *Anat. Histol. Embryol.* **4**, 113–126.

Fehér, Gy. and Fáncsi, T. (1971). Vergleichende Morphologie der Bauchspeicheldrüse von Hausvögeln, *Acta vet. hung.* **21**, 141–164.

Fehér, G. and Gyürü, F. (1971). Adatok a házimadarak sziktömlojének posztembrionális változásaihoz I. A sziktömlö posztembrionális változasai csirkében, *Magy. Állatorv. Lap.* 353–360.

Fehér, G. and Gyürü, F. (1972a). Adatok a házimadarak sziktömlojének posztembrionális változásaihoz II. A sziktömlö posztembrionális változasai kacsában és ludban, *Magy. Állatorv. Lap.* 297–306.

Fehér, G. and Gyürü, F. (1972b). Postnatale Stukturveränderungen des Dottersackes bei Ente und Gans, *Zentbl. VetMed.* **1C**, 266–267.

Fenna, L. and Boag, D. A. (1974a). Filling and emptying of the galliform caecum, *Can. J. Zool.* **52**, 537–540.

Fenna, L. and Boag, D. A. (1974b). Adaptive significance of the caeca in Japanese Quail and Spruce Grouse (Galliformes), *Can. J. Zool.* **52**, 1577–1584.

Fernando, M. A. and McCraw, B. M. (1973). Mucosal morphology and cellular renewal in the intestine of chickens following a single infection of *Eimeria acervulina*, *J. Parasit.* **59**, 493–501.

Fisher, H. I. and Dater, E. E. (1961). Esophageal diverticula in the Redpoll, *Acanthis flammea*, *Auk* **78**, 528–531.

Flechsig, G. (1964). "Makroskopische und Mikroskopische Anatomie der Leber und das Pankreas bei Huhn, Truthuhn, Ente, Gans, und Taube." Diss. Leipzig.

Flower, W. H. (1860). On the structure of the gizzard of the Nicobar Pigeon, and other granivorous birds, *Proc. zool Soc. Lond.* **1860**, 330–334.

Flower, W. H. (1869). Note on a substance ejected from the stomach of a hornbill (*Buceros corrugatus*), *Proc. zool. Soc. Lond.* **1869**, 150.

Forbes, W. A. (1880). Contributions to the anatomy of passerine birds—1. On the structure of the stomach in certain genera of tanagers, *Proc. zool. Soc. Lond.* **1880**, 143–147.

Forbes, W. A. (1882a). Report on the anatomy of the petrels (Tubinares), collected during the voyage of H.M.S. Challenger. Report on the Scientific Results of the Voyage of H.M.S. Challenger during the years 1873–76, *Zoology* **4**, 1–64.

Forbes, W. A. (1882b). On some points in the anatomy of the Indian Darter (*Plotus melanogaster*), and on the mechanism of the neck in the darters (*Plotus*), in connexion with their habits, *Proc. zool. Soc. Lond.* **1882**, 208–212.

French, G. H. (1898). The glandular stomach of birds, *J. appl. Microsc. Lab. Meth.* **1**, 107–108.

French, N. R. (1954). Notes on breeding activities and on gular sacs in the Pine Grosbeak, *Condor* **56**, 83–85.

Friedman, M. H. F. (1939). Gastric secretion in birds, *J. cell. comp. Physiol.* **13**, 219–233.

Fritz, E. (1961). Dotterresorption und histologische Veränderungen des Dottersacks nach den Schlüpfen des Kükens. Wilhelm Roux Arch, *EntwMech. Org.* **153**, 93–119.

Fujii, S. and Tamura, T. (1966). Histochemical studies on the mucin of the chicken salivary glands, *J. Fac. Fish. Anim. Husb. Hiroshima Univ.* **6**, 345–355.

Gadhoke, J. S., Lindsay, R. T. and Desmond, R. K. (1975). Comparative study of the major arterial branches of the descending aorta and their supply to the abdominal viscera in the domestic turkey (*Meleagris gallopavo*), *Anat. Anz.* **138**, 438–443.

Gadow, H. (1883). On the suctorial apparatus of the Tenuirostres, *Proc. zool. Soc. Lond.* **1883**, 62–69.

Gadow, H. (1889). On the taxonomic value of the intestinal convolutions in birds, *Proc. zool. Soc. Lond.* **1889**, 303–316.

Gadow, H. (1891a). Vogel. *In* "Bronn's Klassen und Ordnungen des Thierreichs Anat. Theil", Vol. 6. C. F. Winter'sche, Leipzig.

Gadow, H. (1891b). Crop and sternum of *Opisthocomus cristatus*: a contribution to the question of the correlation of organs and the inheritance of acquired characters. Proc. R. Ir. Acad. Series III, Vol. II, 147–154.

Gardner, L. L. (1926). The adaptive modifications and the taxonomic value of the tongue in birds. Proc. U.S. natn. Mus. **67**, Art. 19.

Gardner, L. L. (1927). On the tongue in birds, *Ibis* **3**, 185–196.

Garrod, A. H. (1872). On the mechanism of the gizzard in birds, *Proc. zool. Soc. Lond.* **1872**, 525–529.

Garrod, A. H. (1874a). On the "showing-off" of the Australian Bustard (*Eupodotis australis*), *Proc. zool. Soc. Lond.* **1874**, 471–473.

Garrod, A. H. (1874b). Further note on the mechanism of the "show-off" in bustards, *Proc. zool. Soc. Lond.* **1874**, 673–674.

Garrod, A. H. (1876a). On the anatomy of *Chauna derbiana*, and on the systematic position of the screamers (Palamedeidae), *Proc. zool. Soc. Lond.* **1876**, 189–200.

Garrod, A. H. (1876b). Notes on the anatomy of *Plotus anhinga*, *Proc. zool. Soc. Lond.* **1876**, 335–345.

Garrod, A. H. (1878a). Note on the gizzard and other organs of *Carpophaga latrans*, *Proc. zool. Soc. Lond.* **1878**, 102–105.

Garrod, A. H. (1978b). Note on points in the anatomy of Levaillant's Darter (*Plotus levaillanti*), *Proc. zool. Soc. Lond.* **1878**, 679–681.

Gasaway, W. C. (1976a). Seasonal variation in diet, volatile fatty acid production and size of the cecum of Rock Ptarmigan, *Comp. Biochem. Physiol.* **53A**, 109–114.

Gasaway, W. C. (1976b). Volatile fatty acids and metabolizable energy derived from cecal fermentation in the Willow Ptarmigan, *Comp. Biochem. Physiol.* **53A**, 115–121.

Georgescu, P. (1910). Leber bei Geflügel, *Jb. vet. Med.* **30**, 284.

Glerean, A. and Katchbarian, E. (1965). Estudo histologico e histoquimico da moela de *Gallus (gallus) domesticus*, *Rec. Fac. Farm. Bioquim. Univ. Sao Paulo* **2**, 73–84.

Goodman, D. C. and Fisher, H. I. (1962). "Functional Anatomy of the Feeding Apparatus in Waterfowl. Aves: Anatidae." Southern Illinois University Press, Carbondale.

Göppert, E. (1903). Die Bedeutung der Zunge für den sekundären Gaumen und den Ductus nasopharyngeus, *Morph. Jb.* **31**, 311–359.

Gorham, F. W. and Ivy, A. C. (1938). General function of the gall bladder from the evolutionary standpoint, *Field Mus. Natur. Hist. Publ., Zool. Ser.* **22**, 159–213.

Gottschaldt, K.-M. and Lausmann, S. (1974). The peripheral morphological basis of tactile sensibility in the beak of geese, *Cell Tiss. Res.* **153**, 477–496.

Graney, D. O. (1967). Electron microscopic observations on the morphology of intestinal capillaries in the chicken and the transcapillary passage of chylomicra during fat absorption, *Anat. Rec.* **157**, 250.

Grau, H. (1943). Anatomie der Hausvögel. *In* "Ellenberger-Baum's Handbuch der vergleichenden Anatomie der Haustiere", 18th edn. Springer-Verlag, Berlin.

Greschik, E. (1912). Mikroskopische Anatomie des Endarmes der Vögel, *Aquila* **19**, 210–269.

Greschik, E. (1913). Histologische Untersuchungen der Unterkieferdrüse (*Glandula mandibularis*) der Vögel, *Aquila* **20**, 331–374.

Greschik, E. (1914). Histologie des Darmkanales der Saatkrähe (*Corvus frugilegus* L.), *Aquila* **21**, 121–136.

Greschik, E. (1922). Über Paneth'schen Zellen und basalgekörnte Zellen im Dünndarm der Vögel, *Aquila* **29**, 149–155.

Greulich, R. C. (1949). Intestinal Schollenleukozyten in the chicken, *Anat. Rec.* **103**, 571.

Grey, R. D. (1972). Morphogenesis of intestinal villi. I. Scanning electron microscopy of the duodenal epithelium of the developing chick embryo, *J. Morph.* **137**, 193–214.

Grimm, R. J. and Whitehouse, W. M. (1963). Pellet formation in a Great Horned Owl: a roentgenographic study, *Auk* **80**, 301–306.

Groebbels, F. (1924). Beiträge zur histologischen Physiologie der Verdauungsdrüsen. I. Untersuchungen über die histologische Physiologie der Magenschleimhaut einiger Säugetiere und Vögel, *Z. Biol.* **80**, 1–22.

Groebbels, F. (1929). Über die Farbe der Cuticula im Muskelmagen der Vögel, *Z. vergl. Physiol.* **10**, 20–25.

Groebbels, F. (1932). "Der Vogel. Bau, Funktion, Lebenserscheinung, Einpassung", Vol. 1. Borntraeger, Berlin.

Hafez, E. S. E. (1955). Differential growth of organs and edible meat in the domestic fowl. *Poult. Sci.* **34**, 745–753.

Halnan, E. T. (1949). The architecture of the avian gut and tolerance of crude fibre, *Br. J. Nutr.* **3**, 245–253.

Hanes, F. M. (1912). Lipid metabolism in the developing chick and its relation to calcification, *J. exp. Med.* **16**, 512–526.

Hanke, B. (1957). Zur Histologie des Ösophagus der Tinamidae, *Bonn. zool. Beitr.* **8**, 1–4.

Hanke, B. and Niethammer, G. (1955). Zur Morphologie und Histologie des Oesophagus von *Thinocorus orbignyanus*, *Bonn. zool. Beitr.* **6**, 207–211.

Harrison, J. G. (1964). Tongue. *In* "A New Dictionary of Birds" (A. L. Thomson, Ed.). Nelson, London.

Haslewood, G. A. D. (1964). The biological significance of chemical differences in bile salts, *Biol. Rev.* **39**, 537–574.

Heidrich, K. (1908). Die Mund-Schlundkopfhöhle der Vögel und ihr Drüsen, *Morph. Jb.* **37**, 10–69.

Henry, K. M., MacDonald, A. J. and Magee, H. E. (1933). Observations on the functions of the alimentary canal in fowls, *J. exp. Biol.* **10**, 153–171.

Herpol, C. (1966). Is de voedingswijze bij vogels een determinerende faktor voor de darmlengte? *Gerfaut* **56**, 79–99.

Herpol, C. (1967). Zuurtegraad en vertering in de maag von vogels, *Naturw. Tijdschr. (Ghent)*. **49**, 201–215.

Hickey, J. J. and Elias, H. (1954). The structure of the liver of birds, *Auk* **71**, 458–462.

Hill, F. W. and Lumijarvi, D. H. (1968). Evidence for an electrolytic-conserving function of the colon in chickens, *Fedn Proc. Fedn Am. Socs. exp. Biol.* **27**, 421.

Hill, K. J. (1971a). The structure of the alimentary tract. *In* "Physiology and Biochemistry of the Domestic Fowl" (D. J. Bell and B. M. Freeman, Eds), Vol. 1. Academic Press, London and New York.

Hill, K. J. (1971b). The physiology of digestion. *In* "Physiology and Biochemistry of the Domestic Fowl" (D. J. Bell and B. M. Freeman, Eds), Vol. 1. Academic Press, London and New York.

Hill, K. J. and Strachan, P. J. (1975). Recent advances in digestive physiology of the fowl. *In* "Symp. zool. Soc. Lond. No 35" (M. Peaker, Ed.). Academic Press, London and New York.

Hill, W. C. O. (1926). A comparative study of the pancreas, *Proc. zool. Soc. Lond.* **1**, 581–631.

Hilton, W. A. (1902). The morphology and development of intestinal folds and villi in vertebrates, *Am. J. Anat.* **1**, 459–505.

Hodges, R. D. (1972). The ultrastructure of the liver parenchyma of the immature fowl (*Gallus domesticus*), *Z. Zellforsch. mikrosk. Anat.* **133**, 35–46.

Hodges, R. D. (1974). "The Histology of the Fowl." Academic Press, London and New York.

Hodges, R. D. and Michael, E. (1975). Structure and histochemistry of the normal intestine of the fowl. III. The fine structure of the duodenal crypt, *Cell Tiss. Res.* **160**, 125–138.

Hoffmann-Fezer, G. (1973). Histologische Untersuchungen an lymphatischen Organen des Huhnes (*Gallus domesticus*) während des ersten Lebenjahres, *Z. Zellforsch. mikrosk. Anat.* **136**, 45–58.

Holman, J. (1968). Ultrastructural localization of acid phosphatase in the specific granules of globule leucocytes of the chick intestine, *Acta histochem.* **31**, 212–214.

Honess, R. F. and Allred, W. J. (1942). Structure and function of the neck muscles in inflation and deflation of the esophagus in the Sage Grouse, *Bull. Wyo. Game Fish Dep.* **2**, 5–12.

Hooper, P. A. and Schneider, R. (1970). The effects of "autonomic drugs" on villous movement in the small intestine of the pigeon, *Br. J. Pharmac. Chemother.* **40**, 426–436.

Hooper, P. A. and Schneider, R. (1971). The mechanism of small intestinal villous movement in the pigeon, *Life Sci.* **10**, Part I, 61–66.

Horváth, I. (1974). Electron microscope study of chicken proventriculus, *Acta Vet. Hung.* **24**, 85–97.

Hsieh, T. M. (1951). "The Sympathetic and Parasympathetic Nervous Systems of the Fowl." Ph.D. Thesis, University of Edinburgh.

Hugon, J. S. and Borgers, M. (1969). Localization of acid and alkaline phosphatase activities in the duodenum of the chick, *Acta histochem.* **34**, 349–359.

Humphrey, C. D. and Turk, D. E. (1970). An electron microscope investigation of chick intestinal epithelial cells, *Poult. Sci.* **49**, 1399.

Humphrey, C. D. and Turk, D. E. (1974). The ultrastructure of normal chick intestinal epithelium, *Poult. Sci.* **53**, 990–1000.

Ihnen, K. (1928). Beiträge zur Physiologie des Kropfes bei Huhn und Taube. I. Mitteilung. Bewegung und Innervation des Kropfes, *Pflügers Arch. ges. Physiol.* **18**, 767–782.

Imondi, A. R. and Bird, F. H. (1965). The sites of nitrogen absorption from the alimentary tract of the chicken, *Poult. Sci.* **44**, 916–920.

Imondi, A. R. and Bird, F. H. (1966). The turnover of intestinal epithelium in the chick, *Poult. Sci.* **45**, 142–147.

Iwanow, I. F. (1930). Die sympathische Innervation des Verdauungstraktes einiger Vogelartern, *Z. mikrosk. anat. Forsch.* **22**, 469–492.

Jablan-Pantić, O. and Antonijević, N. (1961). Blutgefässe der Hühnerleber, *Acta vet., Beogr.* **11**, 17–27.

Jabobshagen, E. (1937). Mittel- und Enddarm. *In* "Handbuch der vergleichenden Anatomie der Wirbeltiere" (L. Bolk, E. Göppert, E. Kallius and W. Lubosch, Eds), Vol. 3. Urban-Schwarzenberg, Berlin and Vienna.

Jenkin, P. M. (1951). The filter-feeding and food of flamingoes (Phoenicopteri), *Phil. Trans. R. Soc.* **240B**, 401–493.

Jerrett, S. A. and Goodge, W. R. (1973). Evidence for amylase in avian salivary glands, *J. Morph.* **139**, 27–46.

Johnson, O. W. and Skadhauge, E. (1975). Structural—functional correlations in the kidneys and observations of colon and cloacal morphology in certain Australian birds, *J. Anat.* **120**, 495–505.

Johnston, D. W. (1958). Sex and age characters and salivary glands of the Chimney Swift, *Condor* **60**, 73–84.

Joyner, W. L. and Kokas, E. (1971). Action of serotonin on gastric (proventriculus) secretion in chickens. *Comp. gen. Pharmac.* **2**, 145–150.

Kaden, L. (1936). Über Epithel und Drüsen des Vogelschlundes, *Zool. Jb., Abt. Anat. Ontog. Tiere* **61**, 421–466.

Kaiser, H. (1925). Beiträge zur makro- und mikroskopischen Anatomie des Gänse- und Taubendarms, *Dt. tierärztl. Wschr.* **33**, 729–731.

Kapp, P. and Balázs, M. (1970). Postembryonic histomorphology and histochemistry of liver cells in ducklings, *Acta vet. Hung.* **20**, 309–323.

Kappelhoff, W. (1959). "Zum mikroskopischen Bau der Blinddärme des Huhnes (*Gallus domesticus* L.) unter besonderer Berücksichtigung ihrer postembryonelen Entwicklung". Inaug. Diss., Univ. Giessen.

Kare, M. R. and Rogers, J. G. (1976). Sense organs. *In* "Avian Physiology (P. D. Sturkie, Ed.). Springer-Verlag, New York.

Kern, D. (1963). "Die Topographie der Eingeweide der Körperhöhle des Haushuhnes (*Gallus domesticus*) unter besonderer Berücksichtigung der Serosa- und Gekröserhältnisse." Inaug. Diss., Univ. Giessen.

Ketterer, H., Ruoff, H.-J. and Sewing, K.-Fr. (1973). Do chickens have gastrin-like compounds? *Experientia* **29**, 1096.

King, A. S. (1975). Aves lymphatic system. *In* "Sisson and Grossman's The Anatomy of the Domestic Animals" (R. Getty, Ed.), Vol. 2. Saunders, Philadelphia.

Kingsbury, J. W., Alexanderson, M. and Kornstein, E. S. (1956). The development of the liver in the chick, *Anat. Rec.* **124**, 165–187.

Kobayashi, S. and Fujita, T. (1969). Fine structure of mammalian and avian pancreatic islets with special reference to D cells and nervous elements, *Z. Zellforsch.* **100**, 340, 363.

Köhler, H. and Schumacher, A. (1967). Licht- und elektronenoptische Untersuchungen zur Leberzirrhose nach experimenteller Aflatoxinvergiftung bei Entenküken, *Zentbl. VetMed.* **14A**, 395–415.

Kokas, E., Phillips, J. L. and Brunson, W. D. (1967). The secretory activity of the duodenum in chickens, *Comp. Biochem. Physiol.* **22**, 81–90.

Kolossow, N. G., Sabussow, G. H. and Iwanow, J. F. (1932). Zur Innervation des Verdauungskanales der Vögel: eine experimentell-morphologische Untersuchung, *Z. mikrosk. anat. Forsch.* **30**, 257–294.

Kostuch, T. E. and Duke, G. E. (1975). Gastric motility in Great Horned Owls (*Bubo virginianus*), *Comp. Biochem. Physiol.* **51A**, 201–205.

Krüger, A. (1926). Beiträge zur makro- und mikroskopischen Anatomie des Darmes von *Gallus domesticus* mit besonderer Berüchsichtigung der Darmzotten, *Dt. tierärztl. Wschr.* **34**, 112–113.

Krulis, V. (1978). Struktur und Verteilung von Tastrezeptoren im Schnabel-Zungenbereich von Singvögeln, im besonderen der Fringillidae, *Revue suisse Zool.* **85**, 385–447.

Kudo, S. (1971a). Fine structure of autonomic ganglion in the chicken pancreas, *Arch. histol. jap.* **32**, 455–497.

Kudo, S. (1971b). Electron microscopic observations on avian esophageal epithelium, *Arch. histol. jap.* **33**, 1–30.

Lang, E. M. (1963). Flamingos raise their young on a liquid containing blood, *Experientia* **19**, 532–533.

Lang, E. M., Thiersch, A., Thommen, H. and Wackernagel, H. (1962). Was füttern die Flamingos (*Phoenicopterus ruber*) ihren Jungen? *Orn. Beob.* **59**, 173–176.

Larsson, L.-I., Sundler, F., Håkanson, R., Rehfeld, J. F. and Stadil, F. (1974). Distribution and properties of gastrin cells in the gastrointestinal tract of chicken, *Cell Tiss. Res.* **154**, 409–421.

Latimer, H. B. (1925). The relative postnatal growth of the systems and organs of the chicken, *Anat. Rec.* **31**, 233–253.

Latimer, H. B. and Rosenbaum, J. A. (1926). A quantitative study of the anatomy of the turkey hen, *Anat. Rec.* **34**, 15–23.

Leasure, E. E. and Link, R. P. (1940). Studies on the saliva of the hen, *Poult. Sci.* **193**, 131–134.

Lehmann, V. W. (1941). Altwater's Prairie Chicken, *N. Am. Fauna* **57**, 1–63.

Leiber, A. (1907). Vergleichende Anatomie der Spechtzunge, *Zoologica, N.Y.* **20**, 1–79.

Leibovitz, L. (1968). Wenyonella Philiplevinei, N.SP., a coccidial organism of the White Pekin Duck, *Avian Dis.* **12**, 670–681.

Leopold, A. S. (1953). Intestinal morphology of gallinaceous birds in relation to food habits, *J. Wildl. Mgmt* **17**, 197–203.

Lewin, V. (1963). Reproduction and development of young in a population of California Quail, *Condor* **65**, 249–278.

Lewis, R. W. (1969). Studies on the stomach oils of marine animals—II. Oils of some procellariiform birds, *Comp. Biochem. Physiol.* **31**, 725–731.

Lienhart, R. (1953). Recherches sur le rôle des cailloux contenus dans le gésier des oiseaux granivores, *Bull. Soc Sci. Nancy* **12**, 5–9.

Lind, G. W., Gronwall, R. R. and Cornelius, C. E. (1967). Bile pigments in the chicken, *Res. vet. Sci.* **8**, 280–282.

Litwer, G. (1926). Die histologischen Veränderungen der Kropfwandung bei Tauben, zur Zeit der Bebrütung und Ausfütterung ihrer Jungen, *Z. Zellforsch. mikrosk. Anat.* **3**, 695–722.

Long, J. F. (1967). Gastric secretion in unanesthetized chickens, *Am. J. Physiol.* **212**, 1303–1307.

Lucas, A. M. and Denington, E. M. (1956). Morphology of the chicken liver, *Poult. Sci.* **35**, 793–806.

Lucas, A. M. and Stettenheim, P. R. (1965). Avian anatomy. *In* "Diseases of Poultry" (H. E. Biester and L. H. Schwarte, Eds), 5th edn. Iowa State University, Press, Ames.

Lucas, A. M. and Stettenheim, P. R. (1972). Avian Anatomy. Integument. Part I. Agricultural Handbook 362. United States Department of Agriculture, Washington, D.C.

Lucas, F. A. (1896). The taxonomic value of the tongue in birds, *Auk* **13**, 109–115.

Lucas, F. A. (1897). The tongues of birds, *Rep. U.S. natn. Mus.* **1895**, 1003–1020.

Luppa, H. (1959). Histogenetische und histochemische Untersuchungen am Epithel des Embryonalen Hühnermagens, *Acta anat.* **39**, 51–81.

Luppa, H. (1962). Histologie, Histogenese und Topochemie der Drüsen des Sauropsidenmagens II. Aves, *Acta histochem.* **13**, 233–300.

Macowan, M. M. and Magee, H. E. (1932). Observations on digestion and absorption in fowls, *Q. Jl exp. Physiol.* **21**, 275–280.

Magnan, A. (1910a). Influence du régime alimentaire sur l'intestin chez les oiseaux, *C.r. hebd. Séanc. Acad. Sci., Paris* **150**, 1706–1707.

Magnan, A. (1910b). Sur une certaine loi de variation du foie et du pancréas chez les oiseaux, *C.r. hebd. Séanc. Acad. Sci., Paris* **151**, 159–160.

Magnan, A. (1911a). "Le Tube Digestif et la Régime Alimentaire des Oiseaux." Thèses, Paris.

Magnan, A. (1911b). Sur la variation inverse du ventricule succenturié et du gésier chez les oiseaux, *C.r. hebd. Séanc. Acad. Sci., Paris* **152**, 1705–1707.

Magnan, A. (1911c). La surface digestive du ventricule succenturié et la musculature du gésier chez les oiseaux, *C.r. hebd. Séanc. Acad. Sci., Paris* **153**, 295–297.

Magnan, A. (1911d). Le foie et sa variation en poids chez les oiseaux, *Bull. Mus. Hist. nat., Paris* **17**, 492–493.

Magnan, A. (1912). Essai de morphologie stomacale en fonction du régime alimentaire chez les oiseaux, *Ann. Sci. nat. Zool.* **15**, 1–41.

Malewitz, T. D. and Calhoun, M. L. (1958). The gross and microscopic anatomy of the digestive tract, spleen, kidney, lungs and heart of the turkey, *Poult. Sci.* **37**, 388–398.

Malinovský, L. (1963). The nerve supply of the stomach in the domestic pigeon (*Columba domestica*), *Čslká Morf.* **11**, 16–27.

Malinovský, L. (1964). Mikroskopická struktura nervových pleteni svalnatého žaludku holuba domáciho (*Columba domestica*), *Cslká Morf.* **12**, 30–39.

Malinovský, L. (1965a). A contribution to the comparative anatomy of vessels in the abdominal part of the body cavity in birds. II. A comparison of the vascular supply to the stomachs and adjacent organs of the Buzzard (*Buteo buteo* L.) and domestic pigeon (*Columba livia* L., *f. domestica*), *Folia morph.* **13**, 202–211.

Malinovský, L. (1965b). Contribution to the comparative anatomy of the vessels in the abdominal part of the body cavity in birds. III. Nomenclature of branches of *a coeliaca* and of tributaries of the *v. portae*, *Folia morph.* **13**, 252–264.

Malinovský, L. and Novotná, M. (1977). Branching of the coeliac artery in some domestic birds. III. A comparison of the pattern of the coeliac artery in three breeds of the domestic fowl (*Gallus gallus f. domestica*), *Anat. Anz.* **141**, 136–146.

Malinovský, L., Roubal, P. and Višňanská, M. (1973a). Vascularization of viscera in abdominal part of body cavity in some domestic birds, *Folia morph.* **21**, 292–295.

Malinovský, L. and Višňanská, M. (1975). Branching of the coeliac artery in some domestic birds. II. The domestic goose, *Folia morph.* **23**, 128–135.

Malinovský, L., Višňanská, M. and Roubal, P. (1973b). Branching of *a. coeliaca* in some domestic birds. I. Domestic duck (*Anas platyrhynchos f. domestica*). Spisy lék ták Masaryk. Univ. 325–336.

Mangold, E. (1906). Der Muskelmaqen der körnerfressenden Vögel, seine motorischen Funktionen und ihre Abhängigkeit vom Nervensystem, *Pflügers Arch. ges. Physiol.* **111**, 163–240.

Marsden, S. J. (1940). Weights and measurements of parts and organs of turkeys, *Poult. Sci.* **19**, 23–28.

Marshall, A. J. and Folley, S. J. (1956). The origin of nest-cement in edible-nest swiftlets (*Collocalia* spp.), *Proc. zool. Soc. Lond.* **126**, 383–389.

Matthews, L. H. (1949). The origin of stomach oil in the petrels, with comparative observations on the avian proventriculus, *Ibis* **91**, 373–392.

Matuura, M., Akahori, F. and Arai, K. (1967). Studies on the movement of the alimentary canal in fowls. I. Movement of crop and gizzard, *Bull. Azabu Vet. Coll.* **16**, 55–65.

Maumus, J. (1902). Les caecums des oiseaux, *Annls Sci. nat. Zool.* **15**, 1–148.

McAtee, W. L. (1906). The shedding of the stomach lining by birds, *Auk* **23**, 346.

McAtee, W. L. (1917). The shedding of the stomach lining by birds, particularly as exemplified by the Anatidae, *Auk* **4**, 415–421.

McCallion, D. J. and Aitken, M. E. (1953). A cytological study of the anterior submaxillary glands of the fowl, *Gallus domesticus*, *Can. J. Zool.* **31**, 173–178.

McLelland, J. (1975). Aves digestive system. *In* "Sisson and Grossman's The Anatomy of the Domestic Animals" (R. Getty, Ed.), Vol. 2. Saunders, Philadelphia.

McLelland, J. (1978). Systema digestorium. *In* "Nomina Anatomica Avium" (J. J. Baumel, A. S. King, A. M. Lucas, J. F. Breazile and H. E. Evans, Eds). Academic Press, London and New York.

Medway, Lord (1962). The relation between the reproductive cycle, moult and changes in the sublingual salivary glands of the swiftlet *Collocalia maxima* Hume, *Proc. zool. Soc. Lond.* **138**, 305–315.

Meinertzhagen, R. (1954). Grit, *Bull. Br. Orn. Club* **74**, 97–102.

Meinertzhagen, R. (1964). Grit. *In* "A New Dictionary of Birds" (A. L. Thomson, Ed.). Nelson, London.

Menzies, G. and Fisk, A. (1963). Observations on the oxyntico-peptic cells in the proventricular mucosa of *Gallus domesticus*, *Q. Jl microsc. Sci.* **104**, 207–215.

Michael, E. and Hodges, R. D. (1973). Structure and histochemistry of the normal intestine of the fowl. I. The mature absorptive cell, *Histochem. J.* **5**, 313–333.

Michel, G. (1971). Zur Histologie und Histochemie der Schleimhaut des Drüsen- und Muskelmagens von Huhn und Ente, *Mb. Vet. med.* **23**, 907–911.

Michel, G. and Gutte, G. (1971). Zur mikroskopischen Anatomie und Histochemie des Darmkanals von Huhn und Ente, *Arch. exp. VetMed.* **25**, 601–613.

Miller, A. H. (1941). The buccal food-carrying pouches of the Rosy Finch, *Condor* **43**, 72–73.

Mitchell, P. C. (1895a). On the proventricular crypts of *Pseudotantalus ibis*, *Proc. zool. Soc. Lond.* **1895**, 271–273.

Mitchell, P. C. (1895b). On the anatomy of *Chauna chavaria*, *Proc. zool. Soc. Lond.* **1895**, 350–358.

Mitchell, P. C. (1896). On the intestinal tract of birds, *Proc. zool. Soc. Lond.* **1896**, 136–159.

Mitchell, P. C. (1901). On the intestinal tract of birds, with remarks on the valuation and nomenclature of zoological characters, *Trans. Linn. Soc. Lond. Zoology* **8**, 173–275.

Miyaki, T. (1973). The hepatic lobule and its relation to the distribution of blood vessels and bile ducts in the fowl, *Jap. J. vet. Sci.* **35**, 403–410.

Miyaki, T. (1978). The afferent venous vessels to the liver and the intrahepatic portal distribution in the fowl, *Zbl. Vet.Med. C. Anat. Histol. Embryol.* **7**, 129–139.

Moncrieff, R. W. (1951). "The Chemical Senses." Hill, London.

Moon, H. W. and Skartvedt, S. M. (1975). Effect of age on epithelial cell migration in small intestine of chickens, *Am. J. vet. Res.* **36**, 213–215.

Mountfort, G. R. (1964). Bill. *In* "A New Dictionary of Birds" (A. L. Thomson, Ed.). Nelson, London.

Moyano, T., Blanco, A., Jover, A. and Vaamonde, R. (1974). Ultraestructura de las células de absorción del intestino ciego de pollo (*Gallus domesticus*), *Archos Zootecnia* **23**, 49–55.

Müller, S. (1922). Zur Morphologie des Oberflächenreliefs der Rumpfdarmsschleimhaut bei den Vögeln, *Jena Z. Naturw.* **58**, 533–606.

Murie, J. (1868). Observations concerning the presence and function of the gular pouch in *Otis kori* and *Otis australis*, *Proc. zool. Soc. Lond.* **1868**, 471–477.

Murie, J. (1869). Note on the sublingual aperture and sphincter of the gular pouch in *Otis tarda*, *Proc. zool. Soc. Lond.* **1869**, 140–142.

Murie, J. (1874). On the nature of the sacs vomited by the hornbills, *Proc. zool. Soc. Lond.* **1874**, 420–425.

Murton, R. K. (1964). Milk, pigeon. *In* "A New Dictionary of Birds" (A. L. Thomson, Ed.). Nelson, London.

Naik, D. R. and Dominic, C. J. (1962). The intestinal caeca of some Indian birds in relation to food habits, *Naturwissenschaften* **49**, 287.

Naik, D. R. and Dominic, C. J. (1963). The intestinal caeca as a criterion in avian taxonomy. Proc. 50th Ind. Sci. Congr., 1962, Part III, 533.

Naik, D. R. and Dominic, C. J. (1968). Intestinal caeca of owls. Proc. 55th Ind. Sci. Congr., 1968, Part III, 522.

Naik, D. R. and Dominic, C. J. (1969). A study of the intestinal caeca of some Indian birds. Proc. 56th Ind. Sci. Congr., 1969, Part III, 473–474.

Nechay, B. R., Boyarsky, S. and Catacutan-Labay, P. (1968). Rapid migration of urine into intestine of chickens, *Comp. Biochem. Physiol.* **26**, 369–370.

Newton, A. and Gadow, H. (1896). "A Dictionary of Birds." Black, London.

Newton, I. (1967). The adaptive radiation and feeding ecology of some British finches, *Ibis* **109**, 33–98.

Niethammer, G. (1931). Zur Histologie und Physiologie des Taubenkropfes, *Zool. Anz.* **97**, 93–103.

Niethammer, G. (1933). Anatomisch-histologische und physiologische Untersuchungen über die Kropfbildungen der Vögel, *Z. wiss. Zool. Abt.* A **144**, 12–101.

Niethammer, G. (1937). Ueber den Kropf der männlichen Grosstrappe, *Orn. Mber.* **45**, 189–192.

Niethammer, G. (1961). Sonderbildungen an Ösophagus und Trachea beim Weibchen von *Turnix sylvatica lepurana*, *J. Orn., Lpz.* **102**, 75–79.

Niethammer, G. (1966). Sexualdimorphismus am Ösophagus von *Rostratula*, *J. Orn., Lpz.* **107**, 201–204.

Nishida, T. and Mochizuki, K. (1976). The venous system of the proventriculus of duck (*Anas domesticus*), *Jap. J. vet. Sci.* **38**, 255–262.

Nishida, T., Paik, Y. K. and Yasuda, M. (1969). Comparative and topographical anatomy of the fowl. LVIII. Blood vascular supply of the glandular stomach (*ventriculus glandularis*) and the muscular stomach (*ventriculus muscularis*), *Jap. J. vet. Sci.* **31**, 51–70.

Nishida, T., Tsugiyama, I. and Mochizuki, K. (1976). The gastric venous system of the domestic goose, *Jap. J. vet. Sci.* **38**, 595–610.

Nolf, P. (1938a). L'appareil nerveux de l'automatisme gastrique de l'oiseau. I. Essai d'analyse par la nicotine, *Archs int. Physiol.* **46**, 1–85.

Nolf, P. (1938b). L'appareil nerveux de l'automatisme gastrique de l'oiseau. II. Étude des effets causés par une ou plusieurs sections de l'anneau nerveux du gésier, *Archs int. Physiol.* **46**, 441–559.

Norris, R. A. (1961). Colors of stomach linings of certain passerines, *Wilson Bull.* **73**, 380–383.

Oakberg, E. F. (1949). Quantitative studies of pancreas and islands of Langerhans in relation to age, sex and body weight in White Leghorn chickens, *Am. J. Anat.* **84**, 279–310.

Oguro, K. and Ikeda, M. (1974a). Studies on the transit of the content in the chicken gastro-intestine. I. Regurgitation of the content of the small intestine into the gizzard, *Jap. J. vet. Sci.* **36**, 291–298.

Oguro, K. and Ikeda, M. (1974b). Studies on the transit of the content in the chicken gastro-intestine. II. Relationship between movements of gizzard and duodenum and the transit of contents in the small intestine, *Jap. J. vet. Sci.* **36**, 513–523.

Ohashi, H. (1971). An electrophysiological study of transmission from intramural excitory nerve to smooth muscle cells of the chicken oesophagus, *Jap. J. Pharmac.* **21**, 585–596.

Okamura, C. (1934). Über die Darstellung des Nervenapparates in der Magen-Darmwand mittels der Vergoldungsmethode, *Z. mikrosk. anat. Forsch.* **35**, 218–253.

Oliveira, A. (1958). Contribuicao para o estudo anatômico da artéria celiaca e sua distribuicaoni no *Gallus domesticus*, *Veterinaria* **12**, 1–22.

Oliveira, A. (1959). Contribuicáo para o estudo anatômico das affluentes e confluentes do distrito venoso portal no *Gallus gallus domesticus*, *Veterinaria* **13**, 43–78.

Olson, L. D., Garrett, P. D. and Bond, R. E. (1974). Anatomic study of the auditory tube in the turkey head, *Am. J. vet. Res.* **35**, 811–815.

Olowo-Okorun, M. O. and Amure, B. O. (1973). Gastrin activity in the chicken proventriculus, *Nature, Lond.* **246**, 424–425.

Otani, I. (1965). Fundamental studies on the digestion in the domestic fowl. I. Observations on the place of deposition of the bolus and the movements of gizzard, *J. Fac. Fish. Anim. Husb. Hiroshima Univ.* **6**, 281–295.

Overton, J. and Shoup, J. (1964). Fine structure of cell surface specializations in the maturing duodenal mucosa of the chick, *J. cell Biol.* **21**, 75–85.

Owen, R. (1866). "Anatomy of Vertebrates", Vol. II. Longmans, Green, London.

Paik, Y. K., Fujioka, T. and Yasuda, M. (1974). Comparative and topographical anatomy of the fowl. LXXVIII. Division of pancreatic lobes and distribution of pancreatic ducts, *Jap. J. vet. Sci.* **36**, 213–229.

Paik, Y. K., Nishida, T. and Yasuda, M. (1969). Comparative and topographical anatomy of the fowl. LVII. The blood vascular system of the pancreas in the fowl, *Jap. J. vet. Sci.* **31**, 241–251.

Paik, Y. K., Nishida, T. and Yasuda, M. (1970). Comparative and topographical anatomy of the fowl. LXII. Distribution of the fine blood vascular system in the pancreas, *Jap. J. vet. Sci.* **32**, 177–183.

Paştea, E., Nicolau, A., Popa, V., May, I. and Roşca, I. (1969). Untersuchungen zur Morpho-Physiologie des Magen-Komplexes der Ente, *Zentbl. VetMed.* **16A**, 450–459.

Paştea, E., Nicolau, A., Popa, V. and Roşca, I. (1966). Contribution à l'étude de la dynamique des premières portions du tube digestif et des estomacs chez la poule, *Recl Méd. vét. Éc Alfort.* **142**, 185–200.

Paştea, E., Nicolau, A., Popa, V. and Roşca, I. (1968a). Dynamics of the digestive tract in hens and ducks, *Acta physiol. Hung.* **33**, 305–310.

Paştea, E., Nicolau, A., Popa, V., Roşca, I., May, I. and Căprārin, A. (1968b). Morphologisch-physiologische Beobachtungen über den Mund-Pharynx-Ösophagus der Ente, *Zentbl. VetMed.* **15**, 572–580.

Patzelt, V. (1936). Der Darm. *In* "Handbuch der mikroskopischen Anatomie des Menschen" (W. von Mollendorf, Ed.), Vol. 5, Part III. Springer-Verlag, Berlin and New York.

Pavaux, Cl. and Jolly, A. (1968). Note sur la structure vasculo-canaliculaire du foie des oiseaux domestiques, *Revue méd. vét.* **119**, 445–466.

Payne, L. N. (1971). The lymphoid system. *In* "Physiology and Biochemistry of the Domestic Fowl" (D. J. Bell and B. M. Freeman, Eds), Vol. 2. Academic Press, London and New York.

Pendergast, B. A. and Boag, D. A. (1973). Seasonal changes in the internal anatomy of Spruce Grouse in Alberta, *Auk* **90**, 307–317.

Pernkopf, E. (1930). Beiträge zur vergleichenden Anatomie des Vertebraten-Magens, *Z. Anat. EntwGesch.* **91**, 329–390.

Pernkopf, E. (1937). Die Vergleichung der verschiedenen Formtypen des Vorderdarmes der Kranioten. *In* "Handbuch der vergleichenden Anatomie der Wirbeltiere" (L. Bolk, E. Göppert, E. Kallius and W. Lubosch, Eds), Vol. III. Urban and Schwarzenberg, Berlin and Vienna.

Pernkopf, E. and Lehner, J. (1937). Vorderdarm. Vergleichende Beschreibung des Vorderdarm bei den einzelnen Klassen der Kranioten. *In* "Handbuch der vergleichende Anatomie der Wirbeltiere" (L. Bolk, E. Göppert, E. Kallius and W. Lubosch, Eds), Vol. III. Urban and Schwarzenberg, Berlin and Vienna.

Pilo, B., Shah, R. V., Asnani, M. V. and Yadav, P. L. (1973). Comparative histochemical studies on avian liver. 4. Relationship of dietary preference of various representative birds with the fat and glycogen contents and distribution pattern of histochemically demonstrable lipids in their livers, *PAVO* **11**, 12–20.

Pilz, H. (1937). Artmerkmale am Darmkanal des Hausgeflügels (Gans, Ente, Huhn, Taube), *Morph. Jb.* **79**, 275–304.

Pintea, V., Jarubescu, V. and Cotrut, M. (1957). Contributiuni la studiul esofagului de gaina, *Lucr. stiint.* **1**, 297–310.

Plenk, H. (1932). Der Magen. *In* "Handbuch der mikroskopischen Anatomie des Menschen" (W. von Möllendorf, Ed.), Vol. 5. Springer-Verlag, Berlin and New York.

Polak, J. M., Pearse, A. G. E., Garaud, J.-C. and Bloom, S. R. (1974). Cellular localization of a vasoactive intestinal peptide in the mammalian and avian gastro-intestinal tract, *Gut* **15**, 720–724.

Portmann, A. (1950). Le tube digestif. *In* "Traite de Zoologie" (P.-P. Grassé, Ed.), Vol. 15. Masson, Paris.

Preuss, F., Donat, K. and Luckhaus, G. (1969). Funktionelle Studie über die Zunge der Hausvögel, *Berl. tierärztl. Wschr.* **82**, 45–48.

Prévost, J. (1961). "Écologie du Manchot Empereur (*Aptenodytes forsteri Gray*)." Hermann, Paris.

Prévost, J. and Vilter, V. (1963). Histologie de la sécrétion oesophagienne du Manchot Empereur. Proc. XIII Int. Ornithol. Congr. 1963, Vol. 2, 1085–1094.

Pulliainen, E. (1976). Small intestine and caeca lengths in the Willow Grouse (*Lagopus lagopus*) in Finnish Lapland, *Ann. zool. Fennici* **13**, 195–199.

Purton, M. D. (1969a). The structure and ultrastructure of the liver in *Gallus domesticus*, *J. Anat.* **105**, 212.

Purton, M. D. (1969b). Structure and ultrastructure of the liver in the domestic fowl, *Gallus gallus*, *J. Zool., Lond.* **159**, 273–282.

Purton, M. D. (1976). Extravascular cells within the perisinusoidal space of the avian liver, *Experientia* **32**, 737–740.

Pycraft, W. P. (1910). "A History of Birds." Methuen, London.

Ratzlaff, M. H. and Tyler, W. S. (1973). A histochemical study of the avian liver, *Poult. Sci.* **52**, 1419–1428.

Reed, C. I. and Reed, B. P. (1928). The mechanism of pellet formation in the Great Horned Owl (*Bubo virginianus*), *Science, N.Y.* **68**, 359–360.

Remouchamps, E. (1880). Sur la glande gastrique du Nandu d'Amérique (*Rhea americana*), *Archs Biol., Paris* **1**, 583–594.

Renner, R. (1961). Site of fat absorption in the chick. *Poult. Sci.* **40**, 1447.

Rhoades, D. D. and Duke, G. E. (1977). Cineradiographic studies of gastric motility in the Great Horned Owl (*Bubo virginianus*), *Condor* **79**, 328–334.

Robbins, C. A. (1932). The advantage of crossed mandibles; a note on the American Red Crossbill, *Auk* **49**, 159–165.

Romanoff, A. L. (1960). "The Avian Embryo." Macmillan, New York.

Rouff, H.-J. and Sewing, K.-Fr. (1971). Die Rolle der Kropfs bei der Steurung der Magensaftsekretion von Hühnern. *Naunyn-Schmiedebergs, Arch. exp. Path. Pharmak.* **271**, 142–148.

Rybicki, M. (1959). Differences in the histological structure of the ingluvies and the glandular stomach of domestic and wild geese, *Acta Biol. Exper.* **19**, 33–40.

Rybicki, M. and Lubańska, L. (1959). The digestive mechanism of green plants in the ingluvies and glandular stomach of *Anser anser* L, *Acta Biol. Exper.* **19**, 5–32.

Saito, I. (1966). Comparative anatomical studies of the oral organs of the poultry. IV. Macroscopical observations of the salivary glands, *Bull. Fac. Agric. Univ. Miyazaki* **12**, 110–120.

Savory, C. J. and Gentle, M. J. (1976a). Effects of dietary dilution with fibre on the food intake and gut dimensions of Japanese Quail, *Br. Poult. Sci.* **17**, 561–570.

Savory, C. J. and Gentle, M. J. (1976b). Changes in food intake and gut size in Japanese Quail in response to manipulation of dietary fibre content, *Br. Poult. Sci.* **17**, 571–580.

Scharnke, H. (1930). Über den Bau der Kolibrizunge, *Orn. Mber.* **38**, 150–151.

Scharnke, H. (1931a). Beiträge zur Morphologie und Entwicklungsgeschichte der Zunge der Trochilidae, Meliphagidae und Picidae, *J. Orn. Lpz.* **79**, 425–491.,

Scharnke, H. (1931b). Die Nektaraufnahme mit der Kalibrizunge, *Orn. Mber.* **39**, 22–23.

Scharnke, H. (1932). Ueber den Bau der Zunge der Nectariniidae, Promeropidae und Drepanididae nebst Bemerkungen zur Systematik der blütenbesuchenden Passeres, *J. Orn. Lpz.* **80**, 114–123.,

Scharnke, H. (1933). Ueber eine rückgebildete Honigfresser-Zunge, *J. Orn., Lpz.* **81**, 355–359.

Schepelmann, E. (1906). Über die gestaltende Wirkung verschiedener Ernährung auf die Organe der Gans, insbesondere über die funktionelle Anpassung an sie Nahrung, *Arch. EntwMech. Org.* **21**, 500–595.

Schildmacher, H., Wohlrab, F. and Cossel, L. (1968). Biochemische, licht- und electronenmikroskopiche Untersuchungen der Leber eines Zugvogels (*Fringilla montifringilla* L.) unter dem Einflus zunehmender Tageslänge, *Z. Zellforsch. mikrosk. Anat.* **91**, 604–616.

Schmidt, C. R. and Ivy, A. C. (1937). The general function of the gall bladder, *J. cell. comp. Physiol.* **10**, 365–383.

Schreiner, K. E. (1900). Beiträge zur Histologie und Embryologie des Vorderdarmes der Vögel, *Z. wiss. Zool.* **68**, 481–580.

Schumacher, S. (1925). Der Bau der Blinddärme und des übrigen Darmrohres von Spielhahn (*Lyrurus tetrix* L.), *Z. Anat. EntwGesch.* **76**, 640–644.

Schummer, A. (1973). Anatomie der Hausvogel. *In* "Lehrbuch der Anatomie der Haustiere" (R. Nickel, A. Schummer and E. Seiferle, Eds), Vol. V. Parey, Berlin.

Scott, P. J., Visentin, L. P. and Allen, J. M. (1969). The enzymatic characteristics of peroxisomes of amphibian and avian liver and kidney, *Ann. N.Y. Acad. Sci.* **168**, 244–264.

Selander, U. (1963). Fine structure of the oxyntic cell in the chicken proventriculus, *Acta anat.* **55**, 299–310.

Shah, R. V., Pilo, B., Asnani, M. V. and Yadav, P. L. (1972). Comparative histochemical studies on avian liver. 1. Relationship of dietary peculiarities with the distribution pattern of histochemically demonstrable alkaline and acid phosphatases in livers of certain representative birds, *PAVO* **10**, 58–70.

Shah, R. V., Yadav, P. L., Asnani, M. V. and Pilo, B. (1975). Comparative histochemical studies on avian liver. 3. The relationship of dietary preferences of various representative birds with the distribution of adenosine triphosphatase in their livers, *J. Anim. Morph. Physiol.* **22**, 60–64.

Shnitka, T. K. (1966). Comparative ultrastructure of hepatic microbodies in some mammals and birds in relation to species differences in uricase activity, *J. Ultrastruct. Res.* **16**, 598–625.

Sick, H. (1954). Zur Biologie des amazonischen Schirmvogels, *Cephalopterus ornatus*, *J. Orn., Lpz.* **95**, 233–244.

Sick, H. (1964). Hoatzin. *In* "A New Dictionary of Birds" (A. L. Thomson, Ed.). Nelson, London.

Simić, V. and Janković, N. (1959). Ein Beitrag zur Kenntnis der Morphologie und Topographie der Leber beim Hausgeflügel und der Taube, *Acta Vet., Beogr.* **9**, 7–34.

Simić, V. and Janković, N. (1960). Ein Beitrag zur Morphologie und Topographie der Leber beim Hausgeflügel und der Taube, *Wien. tierärztl. Mschr.* **154**, 175.

Sitna, B. (1965). Effect of diet on caeca structure in *Fulica atra* L., *Zoologica Pol.* **15**, 213–230.

Skadhauge, E. (1973). Renal and cloacal salt and water transport in the fowl (*Gallus domesticus*), *Dan. med. Bull.* **20**, suppl. 1, 1–82.

Skadhauge, E. (1974). Cloacal resorption of salt and water in the Galah (*Cacatua roseicapilla*), *J. Physiol., Lond.* **240**, 763–773.

Smith, H. H. (1913). Note on the ejection of the lining-membrane of the gizzard by the Curlew, *Br. Birds*, **6**, 334–336.

Smith, P. H. (1974). Pancreatic islets of the *Coturnix* quail. A light and microscopic study with special reference to the islet organ of the splenic lobe, *Anat. Rec.* **178**, 567–586.

Smith, C. R. and Richmond, M. E. (1972). Factors influencing pellet egestion and gastric pH in the Barn Owl, *Wilson Bull.* **84**, 179–186.

Spielvogel, A. M., Farley, R. D. and Norman, A. W. (1972). Studies on the mechanism of action of calciferol. V. Turnover time of chick intestinal epithelial cells in relation to the intestinal action of vitamin D, *Expl. Cell Res.* **74**, 359–366.

Stadtmüller, F. (1938). Zunge, Mundehöhlenboden. *In* "Handbuch der vergleichenden Anatomie der Wirbeltiere" (L. Bolk, E. Göppert, E. Kallius and W. Lubosch, Eds), Vol. 5. Urban and Schwarzenberg, Berlin and Vienna.

Steinbacher, G. (1934). Zur Kenntnis des Magens blütenbesuchender Papageien, *Orn. Mber.* **42**, 80–83.

Steinbacher, G. (1951). Die Zungenborsten der Loris, *Zool. Anz.* **146**, 57–65.

Steinbacher, J. (1934). Untersuchungen über den Zungenapparat indischer Spechte, *J. Orn. Lpz.* **8**, 399–408.

Steinbacher, J. (1935). Ueber den Zungenapparat südafrikanischer Spechte, *Orn. Mber.* **43**, 85–89.

Steinbacher, J. (1941). Weitere Untersuchungen über den Zungenapparat afrikanischer Spechte, *Orn. Mber.* **49**, 126–136.

Steinbacher, J. (1955). Zur Morphologie und Anatomie des Zungenapparates brasilianscher Spechte, *Senckenberg. biol.* **36**, 1–8.

Steinbacher, J. (1957). Ueber den Zungenapparat einiger neotropischer Spechte, *Seneckenberg. biol.* **36**, 259–270.

Steinbacher, J. (1964). Woodpecker. *In* "A New Dictionary of Birds" (A. L. Thomson, Ed.). Nelson, London.

Stephens, R. J. and Bils, R. F. (1967). Ultrastructural changes in the developing chick liver. 1. General cytology, *J. Ultrastruct. Res.* **18**, 456–474.

Stevenson, J. (1933). Experiments on the digestion of food by birds, *Wilson Bull.* **45**, 155–167.

Storer, R. W. (1971). Adaptive radiation of birds. *In* "Avian Biology" (D. S. Farner and J. R. King, Eds), Vol. I. Academic Press, New York and London.

Studer-Thiersch, A. (1967). Beiträge zur Brutbiologie der Flamingos, (Gattung *Phoenicopterus*), *Zool. Gart., Lpz.* **34**, 159–229.

Sturkie, P. D. (1976a). Alimentary canal: anatomy, prehension, deglutition, feeding, drinking, passage of ingesta and motility. *In* "Avian Physiology" (P. D. Sturkie, Ed.). Springer-Verlag, New York.

Sturkie, P. D. (1976b). Secretion of gastric and pancreatic juice, pH of tract, digestion in alimentary canal, liver and bile, and absorption. *In* "Avian Physiology" (P. D. Sturkie, Ed.). Springer-Verlag, New York.

Suzuki, M. and Nomura, S. (1975). Electromyographic studies on the deglutition movements in the fowl, *Jap. J. vet. Sci.* **37**, 289–293.

Swenander, G. (1899). Beiträge zur Kenntnis des Kropfes der Vögel, *Zool. Anz.* **22**, 140–142.

Swenander, G. (1902). Studien über den Bau des Schlundes und des Magens der Vögel. K. norske Vidensk, *Selsk. Skr.* **6**, 1–240.

Takewaki, T. and Ohashi, O. (1977). Non-cholinergic excitatory transmission to intestinal smooth muscle cells, *Nature, Lond.* **268**, 749–750.

Takewaki, T., Ohashi, H. and Okada, T. (1977). Non-cholinergic and non-adrenergic mechanisms in the contraction and relaxation of the chicken rectum, *Jap. J. Pharmac.* **27**, 105–115.

Theron, J. J. (1965). Acute liver injury in ducklings as a result of Aflatoxin poisoning, *Lab. Invest.* **14**, 1586–1603.

Thompson, D. C. and Boag, D. A. (1975). Role of the caeca in Japanese Quail energetics, *Can. J. Zool.* **53**, 166–170.

Thomson, A. L. (1964). "A New Dictionary of Birds" (A. L. Thomson, Ed.), Nelson, London.

Tindall, A. (1976). Hind gut function in the domestic fowl, *Comp. Biochem. Physiol.* **53C**, 83–89.

Toner, P. G. (1963). The fine structure of resting and active cells in the submucosal glands of the fowl proventriculus, *J. Anat.* **97**, 575–583.

Toner, P. G. (1964a). The fine structure of gizzard gland cells in the domestic fowl, *J. Anat.* **98**, 77–86.

Toner, P. G. (1964b). Fine structure of the argyrophil and argentaffin cells in the gastro-intestinal tract of the fowl, *Z. Zellforsch. mikrosk. Anat.* **63**, 830–839.

Toner, P. G. (1965). The fine structure of the globule leucocyte in the fowl intestine, *Acta anat.* **61**, 321–330.

Tordoff, H. B. (1954). Social organization and behavior in a flock of captive nonbreeding Red Crossbills, *Condor* **56**, 346–358.

Trandaburu, T. (1974). Ultrastructural and acetylcholinesterase investigations on the pancreas intrinsic innervation of two birds species (*Columba livia domestica* Gnn and *Eucdice cantans* Gm.), *Gegenbaurs morph. Jb.* **120**, 888–904.

Tucker, B. W. (1944). The ejection of pellets by passerine and other birds, *Br. Birds* **38**, 50–52.

Vandeputte-Poma, J. (1968). Quelques données sur la composition du "lait de pigeon", *Z. vergl. Physiol.* **58**, 356–363.

Vonk, H. J. and Postma, N. (1949). X-ray studies on the movements of the hen's intestine, *Physiologia comp. Oecol.* **1**, 15–23.

Wackernagel, H. (1964). Was füttern die Flamingos ihren Jungen? *Int. Z. VitamForsch.* **34**, 141–143.

Walter, W. G. (1939). Bedingte Magensaftsekretion bei der Ente, *Acta brev. neerl. Physiol.* **9**, 56–57.

Watanabe, T. and Yasuda, M. (1977). Electron microscopic study on the innervation of the pancreas of the domestic fowl, *Cell Tiss. Res.* **180**, 453–465.

Watari, N. (1968). Fine structure of nervous elements in the pancreas of some vertebrates, *Z. Zellforsch. mikrosk. Anat.* **85**, 291–314.

Watson, M. (1883). Report on the anatomy of the Spheniscidae collected during the voyage of H.M.S. Challenger during the years 1873–76. Report on the Scientific Results, Zoology (Sect. 5), 7.

Webb, T. E. and Colvin, J. R. (1964). The composition, structure, and mechanism of formation of the lining of the gizzard of the chicken, *Can. J. Biochem. Physiol.* **42**, 59–70.

Weber, W. (1962). Zur Histologie und Cytologie der Kropfmilchbildung der Taube, *Z. Zellforsch. mikrosk. Anat.* **56**, 247–276.

Wenzel, B. M. (1973). Chemoreception. *In* "Avian Biology" (D. S. Farner and J. R. King, Eds), Vol. III. Academic Press, New York and London.

Westpfahl, U. (1961). Das Arteriensystem des Haushuhnes (*Gallus domesticus*), *Wiss. Humboldt-Univ. Berl.* **10**, 93–124.

Wetherbee, D. K. (1959). Egg teeth and hatched shells of various bird species, *Bird-Banding* **30**, 119–121.

Wetmore, A. (1914). The development of the stomach in the euphonias, *Auk* **31**, 458–461.

Wetmore, A. (1920). A peculiar feeding habit of grebes, *Condor* **22**, 18–20.

Weymouth, R. D., Lasiewski, R. C. and Berger, A. J. (1964). The tongue apparatus in hummingbirds, *Acta anat.* **58**, 252–270.

Weyrauch, K. D. and Schnorr, B. (1978). Die Feinstruktur des Epithels der Hauptausführungsgänge der Leber und des Pankreas vom Haushuhn, *Anat. Anz.* **143**, 37–49.

White, S. S. (1968). Mechanisms involved in deglutition in *Gallus domesticus*, *J. Anat.* **104**, 177.

White, S. S. (1970). "The Larynx of *Gallus domesticus*." Ph.D. Thesis, University of Liverpool.

Wight, P. A. L. (1975). The occurrence of lipid in the oxyntico-peptic cells of the proventriculus of the fasting domestic fowl, *J. Anat.* **120**, 485–494.

Wildfeuer, A. (1963). "Ein Beitrag zur Morphologie der Leber des Huhnes." Diss. Univ. Giessen.

Wissdorf, von H., Geyer, H. and Lutz, H. (1971). Anatomische Grundlagen und Operationsbeschreibung zur Gewinnung einer funktionsfähigen Hühnleber für Perfusionsversuche, *Dt. tierärztl. Wachr.* **78**, 365–371.

Wood, C. A. (1924). The Polynesian Fruit Pigeon, *Globicera pacifica*, its food and digestive apparatus, *Auk* **41**, 433–438.

Yamada, K. (1969). On the fine structure of the quail common bile duct epithelium. *Acta anat.* **74,** 76–87.

Yamada, K. (1973). The ultrastructure of quail gall bladder epithelium, *Acta anat.* **84**, 282–301.

Yasuda, M. (1962). Histological, histochemical and electron microscopic studies on the development of the liver of the domestic fowl, *Arch. histol. jap.* **23**, 79–112.

Yasukawa, M. (1959). Studies on the movements of the large intestine. VII. Movements of the large intestine of fowls, *Jap. J. vet. Sci.* **21**, 1–8.

Zeigel, R. F. (1962). Cytogenic study of embryonic chick pancreas. 1. Exocrine tissue, *J. natn. Cancer Inst.* **28**, 269–303.

Zeitzschmann, O. (1908). Über eine eigenartige Grenzzone in der Schleimhaut zwischen Muskelmagen und Duodenum beim Vogel, *Anat. Anz.* **33**, 456–460.

Ziswiler, V. (1964). Neue Aspekte zur Systematik körnerfressender Singvögel, *Ver. Schweiz. Naturforsch. Ges.* 133–134.

Ziswiler, V. (1965). Zur Kenntnis des Samenöffnens und der Struktur des hörnernen Gaumens bei körnerfressenden Oscines, *J. Orn., Lpz.* **106**, 1–48.

Ziswiler, V. (1967a). Vergleichend morphologische Untersuchungen am Verdauungstrakt körnerfressender Singvögel zur Abklärung ihrer systematischen Stellung, *Zool. Jb. Syst.* **94**, 427–520.

Ziswiler, V. (1967b). Der Verdauungstrakt körnerfressender Singvögel als taxonomischer Merkmalskomplex, *Revue Suisse Zool.* **74**, 620–628.

Ziswiler, V. and Farner, D. S. (1972). Digestion and the digestive system. *In* "Avian Biology" (D. S. Farner and J. R. King, Eds), Vol. II. Academic Press, New York and London.

Zweers, G. A., Gerritsen, A. F. Ch. and van Kranenburg-Voogd, P. J. (1977). Mechanics of feeding of the Mallard (*Anas platyrhynchos* L.; Aves, Anseriformes). *In* "Contributions to Vertebrate Evolution" (M. K. Hecht and F. S. Szalay, Eds), Vol. 3. Karger, Basel.

4
Urinary organs

OSCAR W. JOHNSON

Department of Biology, Moorhead State University, Moorhead, Minnesota 56560, U.S.A.

CONTENTS

I. Introduction and historical perspective 184

II. Gross anatomy of the kidney 185
A. Structural features 185
B. Weight relationships 186

III. Microscopic anatomy of the kidney 188
A. Lobes and lobules 188
B. The nephron 199
C. Organization of the medulla 209

IV. Vascular system of the kidney 211
A. General plan 211
B. Arteries 211
C. Veins and portal valve 213
D. Vasa recta 217

V. Associated organs 217
A. Ureter 217
B. Rectum and cloaca 221

VI. Physiological considerations 224

Acknowledgements 229

References 230

I. Introduction and historical perspective

The urinary organs consist of the paired kidneys, ureters and associated blood vessels. Since post-renal modification of urine occurs in the rectum and cloaca, certain features of these organs will also be considered. The fundamental component of the system is the kidney (*ren*), and most of the following discussion will be devoted to it.

The basic anatomy and histology of the avian kidney were first studied many years ago by classic descriptive anatomists using a wide array of the techniques available at the time. The principal sources for this information are: Hyrtl (1846, 1863), Hüfner (1866), Lindgren (1868), Policard and Lacassagne (1910), Huber (1917), Tchang (1923), Spanner (1925), Feldotto (1929), Waldeyer (1931) and Vilter (1935). Probably because much of the early research was inconclusive, certain features of the anatomy of the avian kidney remained poorly understood for a surprisingly long time. Thus as recently as 1963, Bartholomew and Cade, in their important review of water economy in birds, noted the virtual absence of microanatomical data needed to correlate renal structure and function. Over the past decade, substantial progress has been made in elucidating many of these problems.

The literature on the mammalian kidney far surpasses in its volume that devoted to kidneys of other vertebrates. Hence, present understanding of avian renal structure and function often rests upon concepts related to the kidneys of mammals. Unfortunately, the terminology applied to the mammalian kidney combined with *a priori* reasoning have often distorted descriptions of the structure of the avian kidney. For example, some authors have spoken of the renal papilla in birds, although this is strictly a mammalian feature.

The presence of a renal portal system (*systema portale renale*) in birds is probably the most fundamental difference relative to the mammalian kidney, and furthermore is basic to the pattern of lobular construction which will be described later. However, for well over a century there was disagreement (summarized by Spanner, 1925 and Sperber, 1949) as to whether or not birds even possessed a renal portal system. Interestingly, some illustrious biologists were involved in the argument. Meckel, Cuvier, Neugebauer, Hyrtl, Gadow and others either denied the presence or functional significance of avian renal portal circulation; while Jacobson, Gratiolet and Jourdain were the chief proponents of this venous pathway. Although the extensive work of Spanner (1925) satisfied most investigators that a renal portal flow existed in birds, the controversy was not completely resolved until the experiments of Sperber (1949).

Prior to more recent investigations, cortico-medullary organization and vascular organization in bird kidneys had received considerable attention (Spanner, 1925; Feldotto, 1929; Möllendorf, 1930; Sperber, 1949; Mouchette and Cuypers, 1959; Poulson, 1965; Siller and Hindle, 1969). Although several of these authors suggested the essence of three-dimensional kidney construction, such information did not fully emerge in subsequent accounts. The major sources in the literature have been based on diagrams and descriptions which provide mainly a two-dimensional perspective (histological sections) or are otherwise somewhat difficult to follow (Sperber, 1960; Siller, 1971; Shoemaker, 1972; Hodges, 1974). Hopefully, this chapter will resolve most of these problems.

II. Gross anatomy of the kidney

A. Structural features

The paired kidneys are prominent, flattened, retroperitoneal organs on the ventral surface of the synsacrum, and in most birds are deeply recessed into bony depressions of the synsacrum, termed synsacral fossae (Fig. 4.1). Various morphological and developmental features of the fossae are described by Radu (1975). Large nerve trunks and blood vessels traverse the kidney substance as they pass through the synsacral region. The combined effect of fossae, vessels and nerves is to secure the kidneys in place rather intricately. Thus it is generally difficult to excise them intact, especially without prior fixation.

In the past, anatomists described the avian kidney as consisting of three prominent lobes. Goodchild (1956) correctly pointed out that this was a misapplication of the term "lobe", and suggested the term "division" as an alternative. The three divisions are the cranial renal division (*divisio renalis*

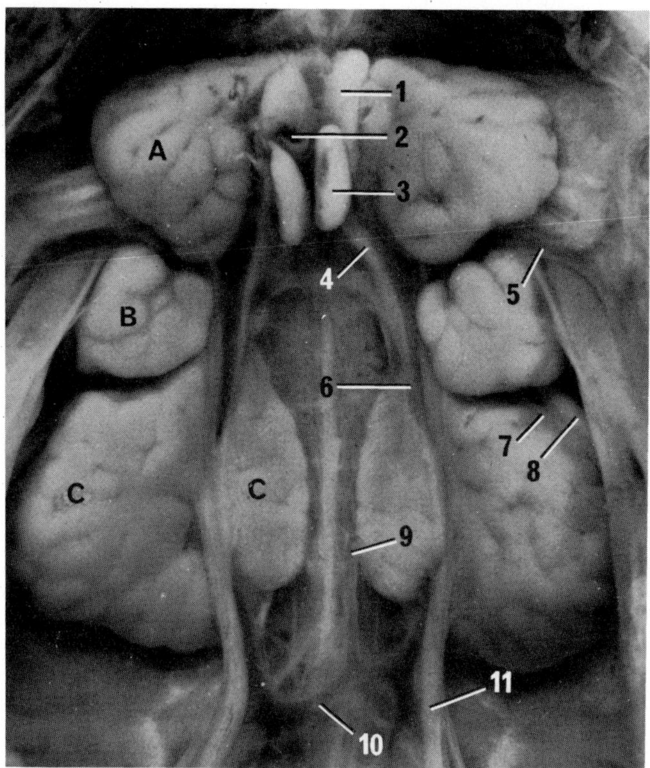

Fig. 4.1. Ventral view of the synsacral region of a Mourning Dove (*Zenaida macroura*) with the kidneys *in situ*. In contrast to most birds, the Columbidae have shallow synsacral fossae. Since the kidneys are not recessed deeply, their anatomical features can be seen without extensive dissection. A, cranial division; B, middle division; C, caudal division; 1, adrenal gland; 2, caudal vena cava; 3, testis; 4, common iliac vein; 5, external iliac vein; 6, caudal renal vein; 7, ischiadic artery; 8, ischiadic nerve; 9, caudal mesenteric vein; 10, venous arch; 11, ureter. ×4.

cranialis), the middle renal division (*divisio renalis media*) and the caudal renal division (*divisio renalis caudalis*). Each division is in turn multilobar. In many birds the divisions are very evident superficially, but in others they are less conspicuous. However, the position of the four prominent blood vessels which pass through the kidney can generally be used to identify the three divisions (Sperber, 1960). Thus the external iliac artery and vein mark the boundary between the cranial and middle divisions, while the ischiadic artery and vein form the boundary between the middle and caudal divisions (Fig. 4.1).

The middle division of the kidney is typically the smallest, but the relative development of the other two divisions varies considerably. Based upon 181 species from 20 taxonomic orders, Johnson (1968) defined five categories of kidney shape which reflect the varying development of the three divisions. In some birds the cranial division is much larger than the caudal and vice versa. Kidneys from nonpasserine birds almost invariably display three obvious divisions. In passerine kidneys, on the other hand, the middle division often seems to be lacking. This results primarily from an intimate association between the middle and caudal divisions which obscures their junction; nonetheless, close scrutiny often reveals evidence of a boundary between them (Fig. 4.2).

Kuroda (1963) and Francis (1964) suggest that in some birds the kidney consists of more than three divisions. For example, Francis indicates that there are four divisions in Ciconiiformes and Charadrii, and five in *Apteryx*. However, this may be illusory since the ventral surface of the kidney is often indented by structures which cross it, particularly the common iliac vein, caudal renal vein and ureter. The resultant grooves are filled with connective tissue and then seem to divide the kidney into more than three divisions. The kidneys of *Apteryx* are unusual in that the cranial and middle divisions are both much larger than the caudal division. In hornbills (Bucerotidae) each kidney consists of a separate cranial and caudal division (Das, 1924; Van Tyne and Berger, 1959; Feinstein, 1962), a point confirmed by my own dissections. Since these two divisions are separated by the iliac and ischiadic vessels it would appear that the middle division has been lost altogether in this group. Feinstein (1962) reported separation between divisions (but with the middle division retained and fused to either the cranial or caudal division) in certain species of old world Columbidae, Coraciidae, Cuculidae and Strigidae. In my opinion, based on dissections of these species, attenuation between divisions is more common than total separation. Such attenuation is related to the contours of the synsacral fossae and the resultant fit of the kidneys within these bony recesses.

B. Weight relationships

Several workers provide information on kidney–body weight relationships in birds (Riboisiere, 1910; Crile and Quiring, 1940; Rensch, 1948; Quiring, 1962). The earlier data plus additional measurements (representing a grand total of 356 species from 23 taxonomic orders) were analysed by Johnson (1968). The results indicate that for body weights of 100–1000 g, the kidneys are approximately 1% of the body weight; at lower body weights they generally exceed 1%, whereas at higher body weights they tend to fall below 1%. Because smaller

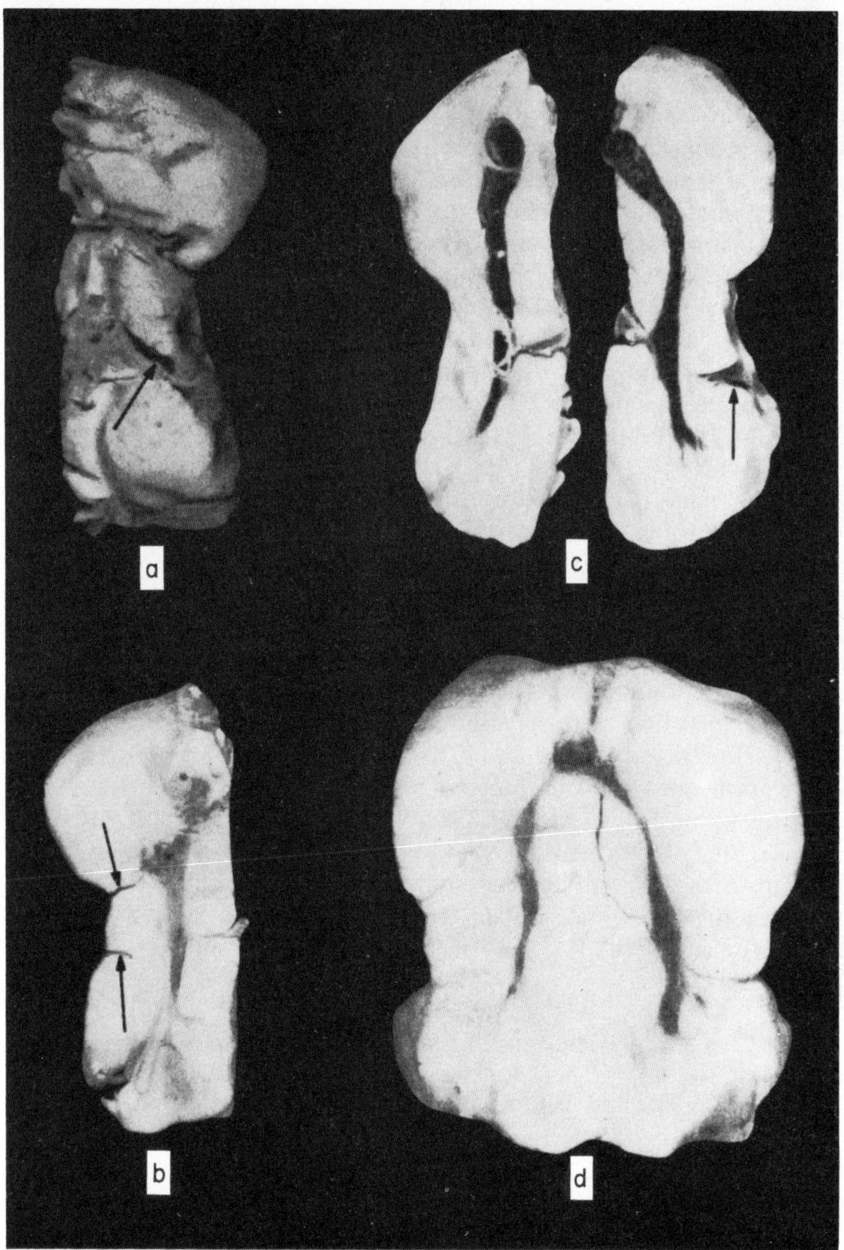

Fig. 4.2. Anatomy of passerine kidneys. (a) Dorsal surface of the right kidney from a Western Bluebird (*Sialia mexicana*). The arrow points to a shallow groove for the ischiadic blood vessels forming the boundary between the middle and the caudal divisions. ×4. (b) Ventral surface of the right kidney from *Sialia mexicana*. Arrows indicate delicate connective tissue-filled septa at the boundaries of the middle division. ×4. (c) Ventral surface of the kidneys from a Brown-headed Cowbird (*Molothrus ater*). The arrow indicates a cleft at the junction of the middle and caudal divisions. Note the absence of this separation in the other kidney. ×4. (d) Ventral surface of the kidneys from a Bridled Titmouse (*Parus wollweberi*) showing mid-line fusion between the middle and caudal divisions. Such fusion produces a bridge of continuous cortex, and is especially common in passerines. ×7. (From Johnson, 1968).

birds metabolize more actively per unit body weight than do larger forms (King and Farner, 1961; Calder and King, 1974), the inverse kidney to body weight relationship probably reflects relative excretory demands. Furthermore, birds below 100 g are mostly passerines which have a higher metabolic rate in relation to body weight than do non-passerines (Lasiewski and Dawson, 1967). In this regard it is notable that the largest kidneys which I have found among small (less than 100 g) non-passerines were in certain Charadriiformes (particularly shorebirds or waders).

From the available data Johnson (1968) calculated that the relationship between kidney and body weight was: log kidney wt. in g = 0·9127 (log body wt. in g)—1·7994. A similar relationship was observed by Hughes (1970) in non-passerines with salt glands. Hughes also made the interesting observation that the kidneys are relatively larger in birds with functional salt glands than in those without.

When renal mass is expressed as mg of kidney per g of body weight, a high degree of intraspecific constancy is usually found. However, there are variations without apparent explanation (Johnson, 1968). For example, two adult male Common Magpies (*Pica pica*) of similar body weight (199 and 210 g, respectively) had kidneys of entirely different sizes (1·8 and 3·2 g, respectively); comparable differences were also noted in several other species. Among birds of similar body weight, both Riboisiere (1910) and Rensch (1948) noted a tendency for insectivores to have larger kidneys than herbivores. Neither my own studies nor those of Hughes (1970) agree with this generality, and indeed such a relationship would be difficult to support amidst other variables. For example, the kidneys of swallows (Hirundinidae) are larger than those of similar size finches (Fringillidae), but this may relate to the intense activity of swallows rather than to their food habits. Rensch (1948) reported an intraspecific trend for larger kidneys in female birds, but this was not substantiated by Johnson (1968) or Hughes (1970).

III. Microscopic anatomy of the kidney

A. Lobes and lobules

1. Basic concepts

The following features of the mammalian kidney are of particular relevance in describing the construction of the bird kidney. (1) Each lobe in the mammalian kidney consists of a medullary pyramid together with its associated cap of cortical substance. In some mammals (such as the rat or rabbit) the entire kidney consists of one such unit and hence is unilobar. In other mammals a variable number of lobes contribute to form a multilobar kidney. (2) Each lobe contains a series of lobules, the outlines of which are not clearly marked by connective tissue partitions. The core of a lobule is formed by a single branched collecting duct which ascends the medulla and continues into the cortex as a component of a medullary ray. One branched collecting duct together with the nephrons that drain into it constitute a lobule. Since medullary rays are conspicuous against a background of cortex and since interlobular arteries furnish an occasional landmark, it is easy to visualize the cortical portion of a lobule. In the medulla, however, the lobule merges with neighbouring lobules and cannot be delineated. (3) The collecting duct system coalesces to form large papillary ducts which drain into the ureter at the tip (papilla) of a medullary pyramid. The papilla protrudes into the expanded upper end or pelvis of the ureter much as a finger is pushed into an inflated balloon.

To a worker familiar with the basic mammalian plan outlined above, the avian kidney is initially confusing. The cortex (*cortex renalis*) and medulla (*medulla renalis*) are not sharply regionalized but rather are intermingled throughout the organ. Furthermore, there is no ureteral pelvis. Instead the ureter branches to form a continuous dendritic system of collecting ducts and collecting tubules. When a renal portal system is added, the structural plan which results is even more complex. The first accurate textbook accounts of lobular organization in avian kidneys were by King (1975) and King and McLelland (1975), and the concept is further analysed in the Nomina Anatomica Avium (Baumel *et al.*, 1979).

2. *Medullary cones*

Medullary tissue in the avian kidney is distributed as a series of cone-shaped masses. Each cone can be likened to a tiny funnel delicately wrapped in connective tissue which is continuous with the connective tissue coat of the ureter (Fig. 4.3). A cone is packed with nephronal loops (of Henle) (*ansae nephronorum*), medullary collecting tubules (*tubuli colligentes medullare*) and vasa recta (*arteriolae rectae* and *venulae rectae*) all of which enter at the wide cortical end of the cone. The size and complexity of a medullary cone gradually decline towards its narrow ureteral end. This is caused by variation in the length of the nephronal loops such that their number steadily decreases, and by gradual dendritic fusion of the collecting tubules. At the narrow extremity of the cone a single large collecting duct (*ductus colligens*) remains, and this is generally one and the same as the tertiary level of ureteral branching. The collecting ducts from several cones fuse to form a secondary ureteral branch (*r. uretericus secundarius*) which drains via a primary branch (*r. uretericus primarius*) into the ureter proper (Fig. 4.3).

At the wide end of the cone, the emerging collecting tubules produce a flower-like effect (Fig. 4.4). The petals of the flower are perilobular collecting tubules (*tubuli colligentes perilobulares*), which wrap themselves about a given mass of cortex like a series of barrel staves. Smaller tributaries of the perilobular collecting tubules penetrate the cortical substance to unite with distal convoluted tubules (*tubuli convoluti distales*).

Diameters of medullary cones at their wide cortical ends have been measured in numerous species (Johnson and Mugaas, 1970a). There is considerable variation, but typical values in passerines range from around 0·2–0·4 mm, and in non-passerines from about 0·3–0·5 mm. The number of collecting tubules entering a cone is variable, but is usually about 20 to 30. The overall length of cones (between their wide and narrow ends) varies greatly among species

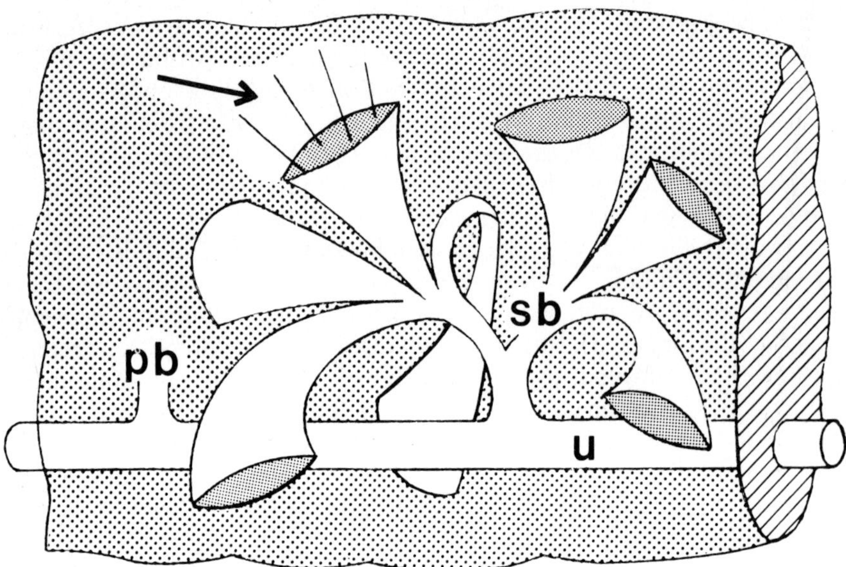

Fig. 4.3. Part of a kidney drawn as though transparent to show cortex (stippled) and medullary cones. Five of the cones have their open ends toward the viewer. Cones may be straight or curved to varying degrees. Curving is often more pronounced than shown here, with cones assuming S- or W-shaped configurations. Several cones converge to form a secondary branch of the ureter (sb), the latter in turn joins a primary branch (pb) which enters the ureter (u). The ureter, utereral branches and cones all share a common investment of connective tissue. The arrow indicates the general plan whereby collecting tubules and other intramedullary components extend from the cortex into the cones.

(Johnson and Mugaas, 1970b; Johnson and Ohmart, 1973a,b; Johnson, 1974; Johnson and Skadhauge, 1975). Mean values range from 0·94 mm in the American Coot (*Fulica americana*) to 4·17 mm in the Emu (*Dromaius novaehollandiae*).

3. *Cortical organization*

The avian renal cortex with its associated afferent and efferent venous circulation (renal portal and renal veins, respectively), resembles the lobular pattern of the liver. The efferent renal veins (*vv. renales craniales* and *v. renalis caudalis*) branch repeatedly within the kidney eventually forming many smaller vessels called intralobular (or central) veins (*vv. intralobulares*). Each intralobular vein constitutes a central axis around which is arranged a concentric mass of cortical tissue. The cortical wrapping is bounded by renal portal vessels from which blood passes to the intralobular vein through a peritubular capillary plexus (*rete capillare peritubulare corticale*) (Figs 4.4, 5). By its close association with a group of intralobular veins, a given area of cortex becomes correspondingly dendritic, as suggested by Spanner (1925) and Sperber (1949).

Cortical organization is based upon areas containing the peritubular capillary plexus (subsequently referred to as cortical units) which are separated by zones

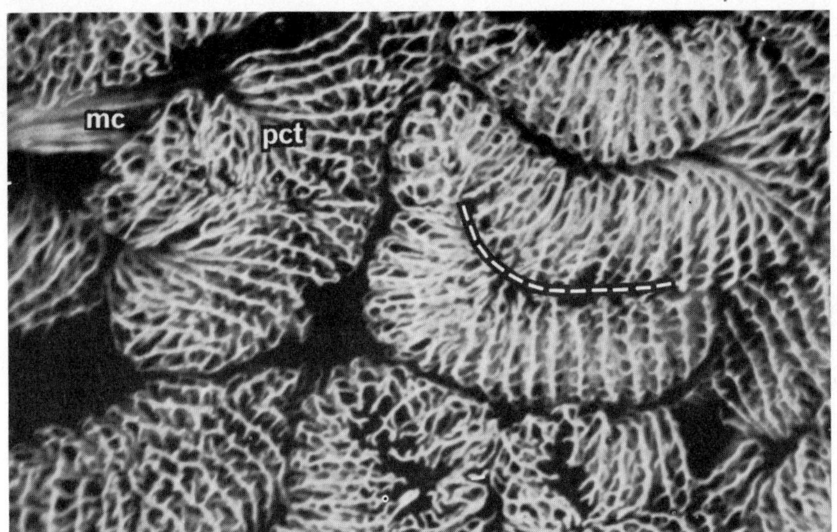

Fig. 4.4. Surface view of cleared kidney tissue from a Sooty Tern (*Sterna fuscata*) showing the arrangement of the collecting tubules as revealed through retrograde ureteral injection of "Microfil". A single medullary cone (mc) is delimited by the cluster of medullary collecting tubules contained within it. A series of perilobular collecting tubules (pct) sprays out from the cone's opening. Additional medullary cones (not visible) are associated with the other perilobular collecting tubules in the photograph. Numerous lateral tributaries join the perilobular tubules; each forms a link with the distal tubule of a nephron. The underlying position of one intralobular vein is represented by the broken line. ×26. (From Johnson *et al.*, 1972).

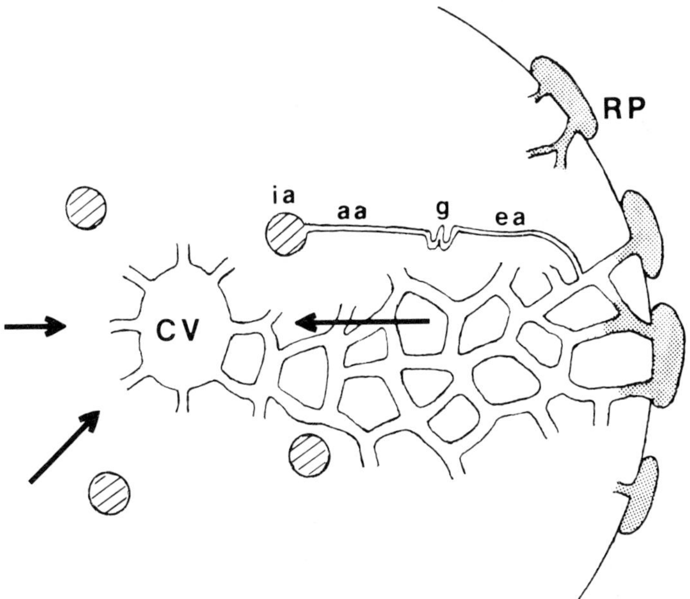

Fig. 4.5. Diagram of cortical organization. The elaborate peritubular capillary plexus lies amidst the uriniferous tubules and surrounds the intralobular (central) vein (CV); all blood flow is centripetal (arrows). Branches (RP) of the renal portal veins are perilobular. Four intralobular arteries (ia) are shown. From the latter, afferent glomerular arterioles (aa) lead to glomeruli (g); and efferent glomerular arterioles (ea) thence drain to the capillaries of the peritubular network.

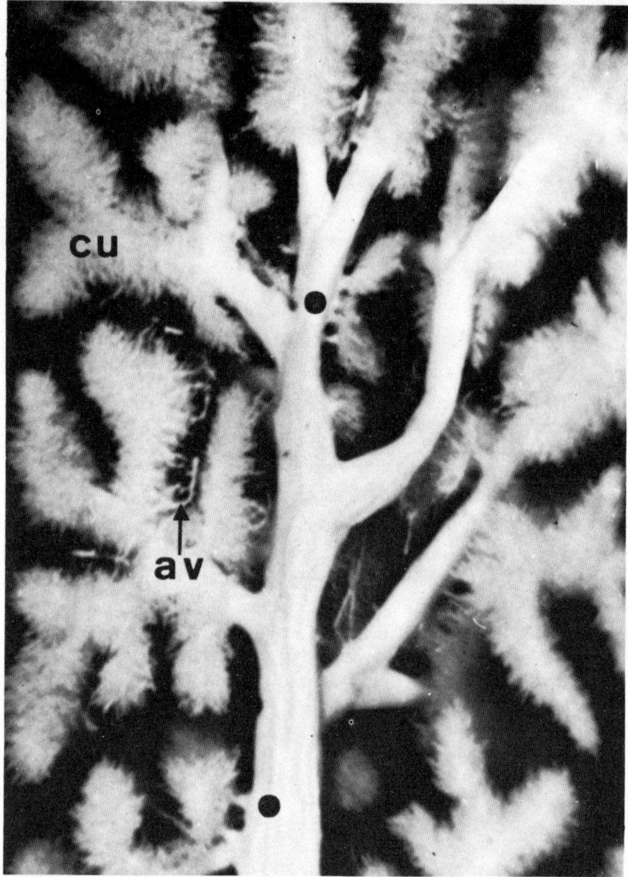

Fig. 4.6. Cleared caudal division of the kidney of Harris' Sparrow (*Zonotrichia querula*) in which the peritubular capillary network is revealed by injection of "Microfil" via the caudal renal vein. The zones containing the peritubular network are termed cortical units (cu) and are separated from each other by areas with no capillaries. Small cortical units can be seen adjacent to the black dots. Some injection medium has crossed the peritubular capillary plexus to reveal interlobular afferent vessels (av) of the portal system. ×24. (Modified from Johnson *et al.*, 1972).

containing no capillary plexus (Figs 4.6, 7). The peritubular capillary plexus can be demonstrated by retrograde injection, as in Fig. 4.6; the injectant successively fills the major efferent venous pathways, the intralobular veins, and finally the peritubular capillary plexus. A schematic representation of a preparation produced in this manner is shown in Fig. 4.7.

Usually six to seven orders of branching occur between the main trunk of an efferent renal vein and its terminal intralobular veins. Branching is typically dichotomous throughout, although trichotomy is not uncommon (e.g. the three *B* branches on the right side of Fig. 4.7). Exceptions occur in very small kidneys such as those of the Zebra Finch (*Poephila guttata*) where branching is less extensive and terminates at the fourth and fifth orders (*D* and *E*). The terminal intralobular veins in units of all sizes are generally paired or in a group of three.

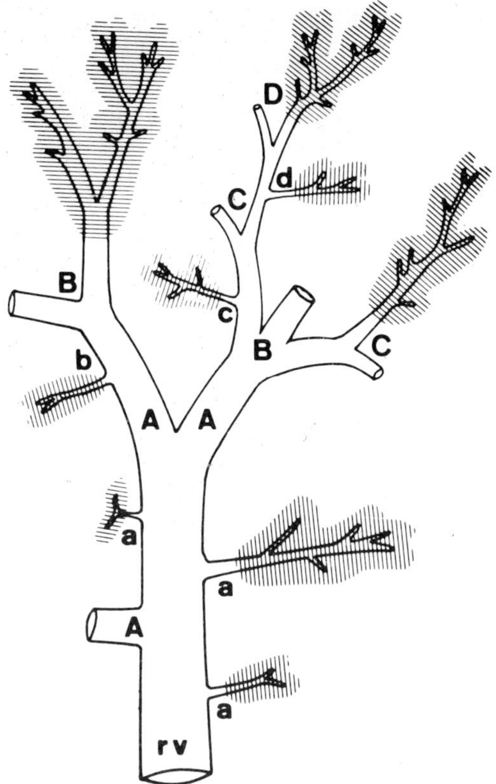

Fig. 4.7. Generalized pattern of branching of the efferent caudal renal vein (rv) in the caudal division of the kidney. Capital letters represent branches of equal or nearly equal diameter. When branching is unequal, the smaller tributary is designated by a lower case letter. The smaller tributaries are often simple with few branches, but occasionally they exhibit several orders of branching. Crosshatched areas represent zones of continuous portal capillaries which define the boundaries of individual cortical units. (From Johnson *et al.*, 1972).

The levels at which individual cortical units arise from the efferent venous system are extremely varied. In the species studied by Johnson *et al.* (1972) (Table I), the cortical units typically begin at some point on the first four orders of branching of the renal vein (Fig. 4.7). The Zebra Finch is an exception in that cortical units were not observed to arise beyond the third level of branching (*C* and *c*). This reflects the diminutive kidney of this bird with its simpler pattern of branching.

There is no apparent microstructural basis for subdividing a cortical unit. Hence, cortical organization is dendritic and conforms to the branching pattern of the efferent renal veins. Individual cortical units vary greatly in size. A given unit can be measured by determining the combined length of all its branches. Such measurements are shown in Table I, and demonstrate broad intraspecific ranges extending from less than 1 mm to over 18 mm. Between species, there is major variation in the maximum size of cortical units. Small passerines such as the Zebra Finch (*Poephila guttata*) and Harris' Sparrow (*Zonotrichia querula*)

TABLE I. **Sizes of cortical units from the kidneys of various birds.**[a]

Species	Distribution of values (mm)					Range (mm)	Mean kidney weight(g)[b]
	<1·0	1·0–5·0	5·1–9·0	9·1–13·0	>13·0		
American Golden Plover (*Pluvialis dominica*) 46, 3, 3[c]	8	14	12	5	7	0·2–16·7	0.68 (6)
Ruddy Turnstone (*Arenaria interpres*) 44, 2, 2	15	11	12	4	2	0·3–17·2	0.59 (5)
Wandering Tattler (*Heteroscelus incanus*) 42, 3, 3	4	15	7	7	9	0·5–16·6	0·64 (3)
Sooty Tern (*Sterna fuscata*) 45, 3, 2	4	17	11	7	6	0·5–16·2	1·54 (8)
Brown Noddy (*Anous stolidus*) 45, 4, 4	5	18	6	7	9	0·3–18.9	1·29 (15)
White Tern (*Gygis alba*) 47, 3, 3	11	17	10	4	5	0·3–16·0	0·94 (13)
Mourning Dove (*Zenaida macroura*) 35, 3, 3	9	9	3	7	7	0·5–16·2	0·35 (13)
Budgerigar (*Melopsittacus undulatus*) 52, 5, 3	12	20	11	4	5	0·2–15·4	0·12 (12)
Common Flicker (*Colaptes auratus*) 54, 2, 1	18	14	11	7	4	0·3–18·5	0·49 (1)
Zebra Finch (*Poephila guttata*) 49, 4, 2	19	25	5	0	0	0·2–9·0	0·05 (16)
Common Grackle (*Quiscalus quiscula*) 57, 4, 2	15	20	10	11	1	0·1–15·7	0·52 (3)
Harris' Sparrow (*Zonotrichia querula*) 39, 3, 2	6	20	11	2	0	0·4–9·6	0·19 (6)

Size of each unit measured by determining the combined length of all its branches.

[a] All data from Johnson *et al.* (1972).
[b] Weights represent individual kidneys after hardening in alcohol, or fixation in AFA; numbers of birds sampled shown in parentheses.
[c] Number of cortical units measured, number of kidneys evaluated, number of donor birds.

have correspondingly smaller units than those found in other species. It is common to find very small cortical units (those less than 1 mm in length in Table I) originating along the main trunk of the efferent vein or from its *A* branches. Their locations represent areas of deep-lying cortex adjacent to large vessels (Figs 4.6, 7). Each species also has substantial numbers of cortical units in the 1–5 mm range (Table I), which typically arose from small vessels (*a*, *b*, *c* and *d*) along the first four orders of branching of the renal vein. Units larger than 5 mm generally originated from larger vessels at the second to fourth levels of branching (*B*, *C*, *D*) with the largest units arising at the second and third levels (*B*, *C*).

The size of a cortical unit is correlated with its relative position within the kidney, as can be shown in a schematic cross-section (Fig. 4.8). In some species the larger efferent vessels are located a considerable distance beneath the ventral surface of the kidney. This allows space for the development of short cortical units (cu_1 in Fig. 4.8). In other birds, the major efferent channels are more

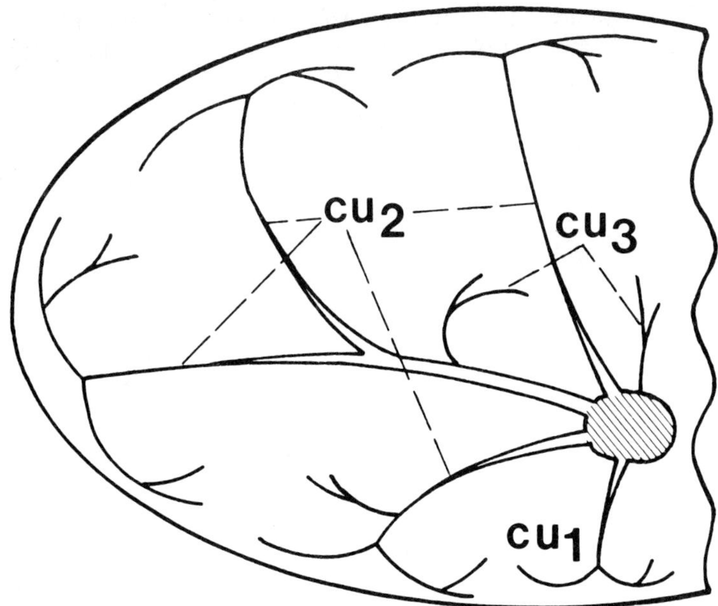

Fig. 4.8. A schematic transverse section of the left half of a kidney showing a large first order efferent vein (crosshatched) and its tributaries. Cortical units ($cu_{1,2,3}$) vary in size relative to their position within the kidney. (Modified from Johnson et al., 1972).

superficial and the development of short cortical units is less evident. In all birds, regardless of the size of the kidney, the largest units appear to be associated with efferent drainages which are directed laterally and dorsally (cu_2 in Fig. 4.8). Many of the branches of these large units ramify with an umbrella-like effect near the kidney's surface, producing there an irregular form of relief pattern. Deep-lying renal tissue typically contains small cortical units (cu_3 in Fig. 4.8). Overall, the efferent network is one in which cortical units of varying sizes are economically packed within the available intrarenal space. Siller (1971) and Hodges (1974) equate each polyhedral irregularity on the kidney surface (Fig. 4.1) with the base of a lobule, but this is incorrect since the pattern is produced by the ramified cortex as described above.

Regarding the efferent veins as the cortical axis obviously places the collecting tubules and afferent veins in an interlobular position. Most workers have followed this approach, but Goodchild (1956) argued for a reversal of the foregoing, proposing that collecting tubules and afferent veins should comprise the intralobular axis with central (efferent) veins in the interlobular position. This has the advantage of corresponding to the mammalian pattern, wherein medullary rays form the axes of the renal lobules. However, it is extremely difficult to equate the elaborately arranged collecting tubules of birds (Fig. 4.4) with medullary rays. Furthermore, the arterial distribution in the avian kidney (though it needs further study) is less orderly than in the mammal (Siller and Hindle, 1969). Hence, it is much more convenient to regard the efferent drainage as intralobular despite the lack of direct homology with the mammalian renal lobule; this takes advantage of an organizational pattern which can be readily appreciated, and furthermore it emphasizes the presence of a renal portal system.

4. *Cortico-medullary relationships*

Unfortunately, investigations of the organization of the cortex and medulla have often involved an essentially two-dimensional histologic approach, the term renal lobule being applied in a manner which tended to be confusing (Fig. 4.9). Recently, however, injection-clearing techniques have demonstrated the lobular organization of the avian kidney in three dimensions (Johnson and Mugaas, 1970a,b; Johnson *et al.*, 1972; Johnson, 1974). Injected specimens have

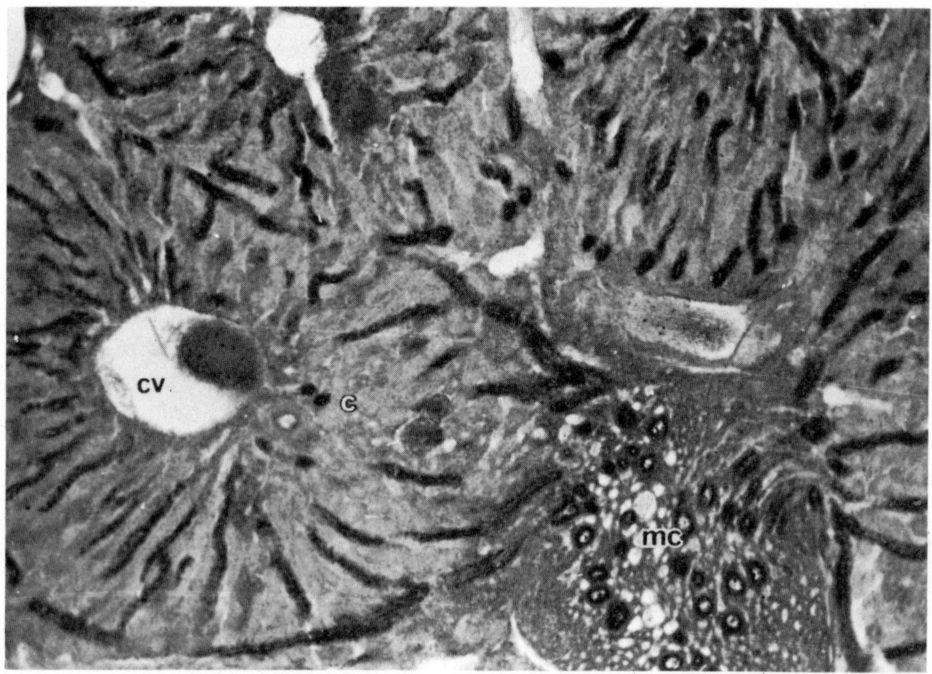

Fig. 4.9. Representative section of the kidney of the Budgerigar (*Melopsittacus undulatus*) with cortex (c) surrounding intralobular (central) veins (cv) and deeply stained collecting tubules passing into a medullary cone (mc). Some collecting tubules are entering the cone in the plane of the section; others entered at a more distal level. In the past, the term "lobule" was applied to sections like this one. Such an approach is misleading since it fails to emphasize important three-dimensional perspectives. Alcian blue, ×94. From Johnson *et al.*, 1972).

clearly shown that a typical cortical unit is associated with several medullary cones (Fig. 4.10a,b), and that each cone receives perilobular collecting tubules from a particular portion of the unit. Individual medullary cones may be either straight or curved to varying degrees (Johnson and Mugaas, 1970b). Straight cones often extend peripherally where they are associated with larger cortical units (cu_2 in Fig. 4.8); curved cones (Fig. 4.10b) typically drain deeper and smaller cortical units (cu_1 and cu_3 in Fig. 4.8).

Throughout the cortico-medullary relationship there is considerable overlap in the organizational pattern. Although a given medullary cone may receive collecting tubules from a portion of only one cortical unit, the same medullary cone is often associated with parts of separate cortical units (Fig. 4.10a). The latter may originate from the same major venous branch (an *A* or *B*, for example), or they may occur on separate but adjoining major branches.

Obviously, the structural relationship between the cortex and medulla in the avian kidney is relatively complex. The size and shape of a cortical unit is variable, as is the manner by which its collecting tubules relate to the medullary cones. This variability is reflected in the work of Kurihara and Yasuda (1975a) who described two distinctly different lobular arrangements. Despite these variations, it is possible to visualize a renal lobe (*lobus renalis*). Groups or clusters of medullary cones (usually 2–5 per group) arise from ureteral branches (generally at the secondary level). Each group of cones tends to remain separate from other similar groups, and can be teased away as a relatively distinct entity (Fig. 4.10c). A particular group is usually associated with a region of cortex along the same major venous branch. However, one or more cones from the same cluster may diverge away and associate with regions of cortex having separate venous drainages. Furthermore, a cortical unit can provide collecting tubules to cones originating from different ureteral branches (Fig. 4.10b).

Because of the lack of clear-cut boundaries in the pattern described above, the avian renal lobe can be most readily and usefully defined as being composed of a group of medullary cones which typically drains into a secondary branch of the ureter, and the cortex that is associated with these cones (Fig. 4.11). Such an approach emphasizes the discrete nature of individual clusters of cones. This distinction is lost if the renal lobe is considered to originate only at the level of the primary ureteral branch, as implied by Atterbury (1923) and by Siller and Hindle (1969). A lobule can be defined as one medullary cone together with the cortex that contributes collecting tubules to it.

Homologies with the mammalian kidney are still very indefinite and await further research. It is possible that the medullary cones in a lobe are *in toto* homologous to a mammalian medullary pyramid. Furthermore, the avian kidney resembles the extreme renculus type of kidney possessed by cetaceans (Sperber, 1944). In both birds and cetaceans there are many lobes at varying depths throughout the kidney. However, in cetaceans the lobes are delimited by connective tissue, whereas in birds they are more or less fused into a continuous renal mass. While the connective tissue of the avian medullary cones (peripheral wrapping as well as internal components) can be readily demonstrated histologically, it has unfortunately proved impossible to detect any connective tissue partitions between the cortical portions of the lobes and lobules.

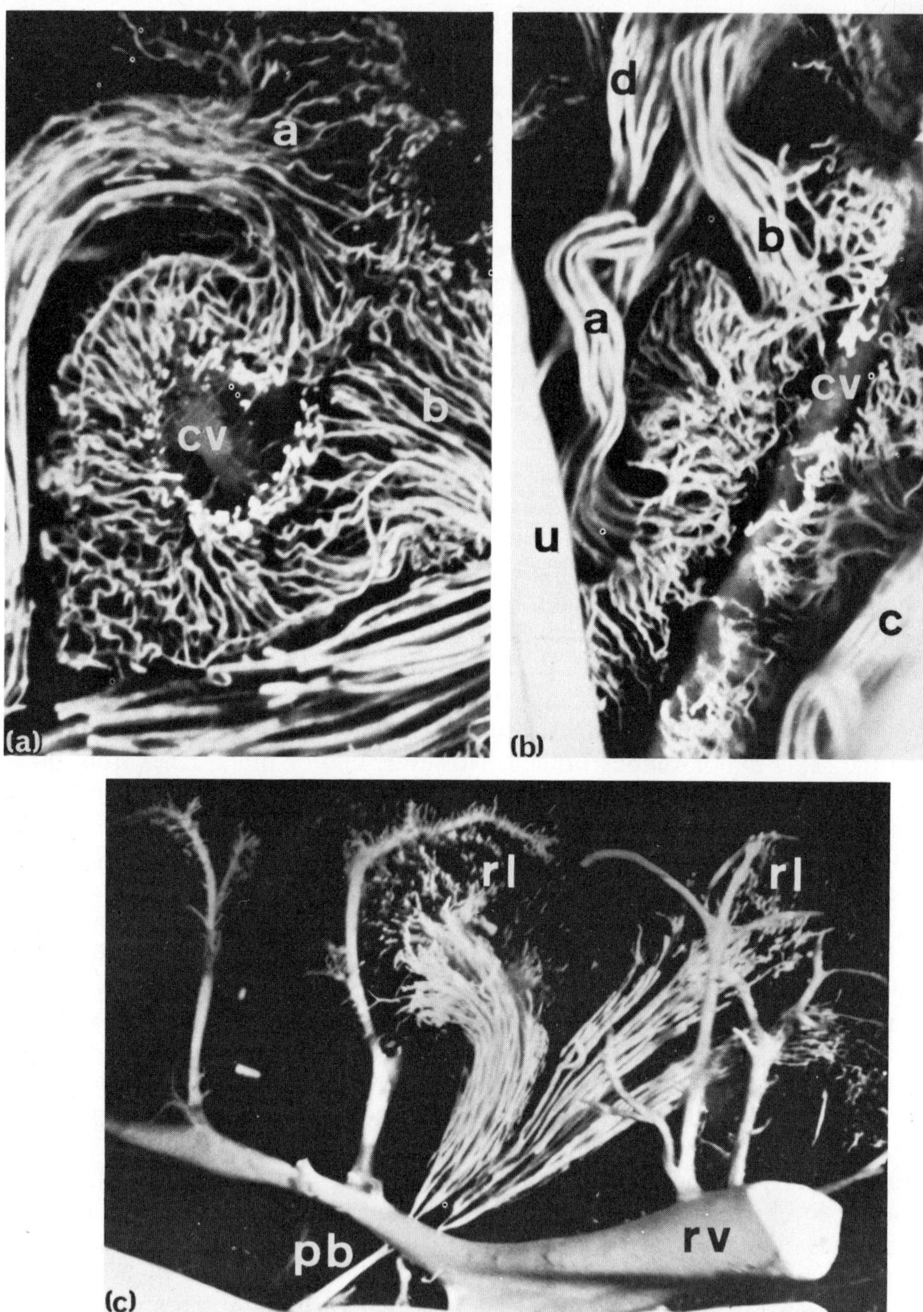

Fig. 4.10. Relationship between medullary cones and the cortex as revealed through injection and clearing. (a) A portion of one branch of a cortical unit with the intralobular (central) vein (cv) surrounded by collecting tubules which enter two medullary cones (a and b). The collecting tubules entering (a) from the top of the picture could be coming from either another branch of the same cortical unit, or from a separate cortical unit. Sooty Tern (*Sterna fuscata*) kidney. $\times 52$. (b) Two strongly recurved medullary cones (a and b) associated with one side of a cortical branch the axis of which is revealed by an

B. The nephron

1. General features

The orderly cortico-medullary plan in mammalian kidneys encourages precise studies of the intrarenal course of individual nephrons and the exact arrangement of the medullary components (Beeuwkes, 1971; Kriz *et al.*, 1972; Kriz and Koepsell, 1974; Kaissling *et al.*, 1975). There is no research on the avian kidney which is as detailed as these mammalian studies. Nevertheless, a substantial body of information is available.

As in mammals, the basic structural and functional unit is the nephron (*nephronum*). The latter begins with the renal corpuscle and ends where the distal convoluted tubule joins a collecting tubule. Some avian nephrons are medullary (*nephronum medullare*) and these are mammalian-type nephrons since they possess nephronal loops (loops of Henle); the majority are cortical (*nephronum corticale*), and these are reptilian-type nephrons devoid of loops (Figs 4.11, 12). Thus, the kidneys of birds are an evolutionary composite exhibiting both mammalian and reptilian features. The distinction between mammalian and reptilian-type nephrons is not always clear-cut. Various authors (Huber, 1917; Feldotto, 1929; Sperber, 1960; Braun and Dantzler, 1972) have described transitional nephrons with very short loops.

As described earlier, the medullary portion of a lobule (i.e. the medullary cone) resembles a funnel in its relationship to its associated cortex (Figs 4.3, 11). Those nephrons originating from glomeruli near the mouth of the funnel are thus juxtamedullary in position, and are mammalian-type nephrons. Their nephronal loops extend down into the medullary cone to varying depths. Many of these loops are short, some are intermediate, and a few continue all the way to the apex of the cone (Figs 4.12, 13).

In contrast, the reptilian-type nephrons are restricted to cortex which is peripheral to a medullary cone (Fig. 4.13). Indeed the superficial part of the cortex consists entirely of reptilian-type nephrons. These reptilian-type nephrons lie closely packed with their long axes at right angles to the perilobular collecting tubules (Fig. 4.13). Their arrangement is very orderly, and in macerated tissue a section of cortex parallel to an intralobular vein can be followed by teasing away each consecutive nephron like opening the pages of a book.

The large perilobular collecting tubules receive smaller tributaries from within the cortical substance (Fig. 4.4). Only the general features of the embryonic development of the distal collecting tubules appear to be documented (Romanoff, 1960), and unfortunately, as pointed out by Rouiller (1969) and Ericsson

---

intralobular vein (cv). Additional collecting tubules, representing a separate medullary cluster (c) arising elsewhere along the ureter (u), drain the opposite side of the branch. The origins of straight medullary cones (d) which extend to peripheral cortex are evident. Sooty Tern kidney. ×40. (c) A primary ureteral branch (pb) forms two secondary branches which are partially hidden beneath the caudal renal vein (rv). From each secondary branch, a group of medullary cones extends to associate with venous tributaries (central veins of cortex). Each of these complexes is a renal lobe (rl). Bristle-thighed Curlew (*Numenius tahitiensis*) kidney. ×18. (All photographs from Johnson *et al.*, 1972).

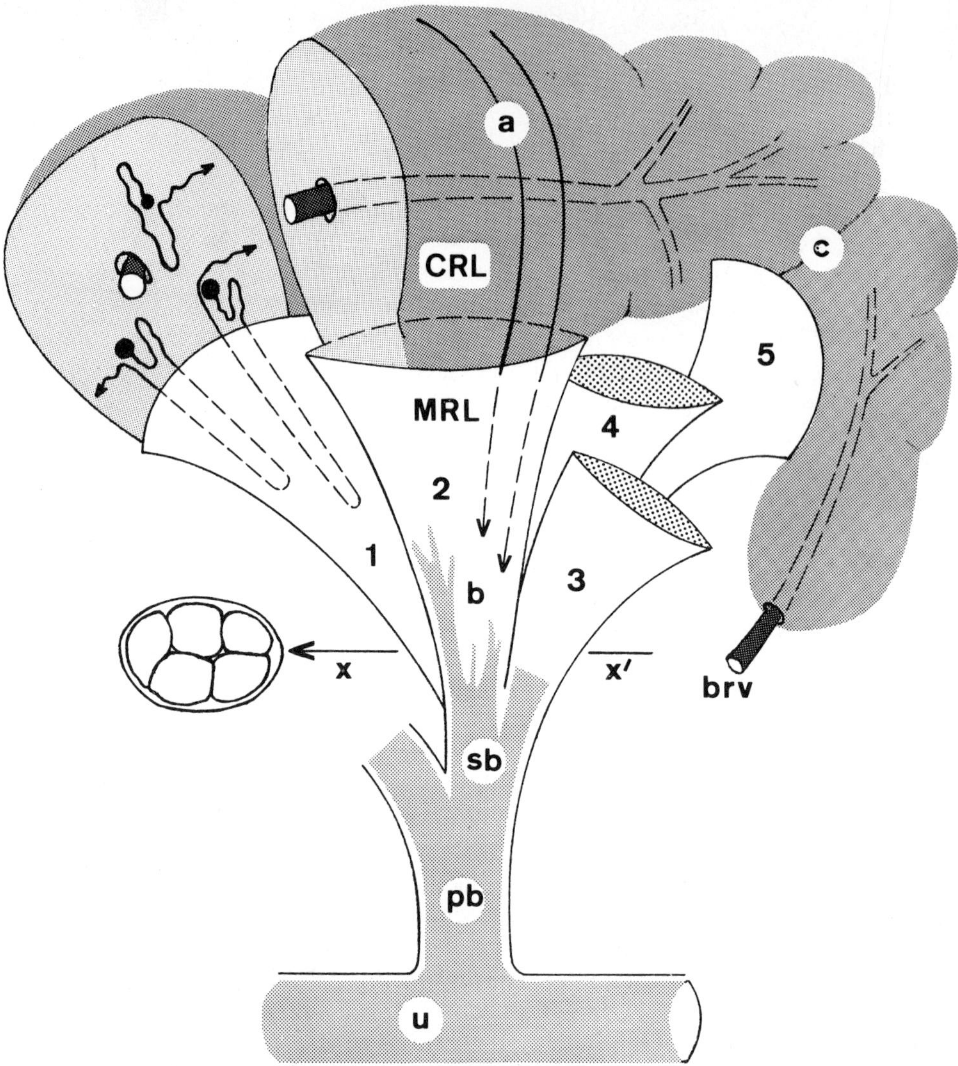

Fig. 4.11. Plan of the renal lobe and lobule. One lobe consisting of five lobules is shown. Each lobule has a cortical region (CRL) and a medullary region or cone (MRL); the cortical regions of lobules 3 and 4 are omitted. The cortical region is not sharply delimited, being simply that region of cortex which contributes collecting tubules and nephronal loops to a medullary region or cone. A given cone is often associated with cortex from different branches of the renal vein (brv). For example, at "c" two cortical units lie close to each other and cone 5 would receive collecting tubules, etc. from both units. Representative perilobular collecting tubules at "a", together with other such tubules, enter each cone and undergo gradual dendritic fusion (in the direction of the arrows) to form increasingly larger ducts (b). The latter are directly continuous with the lumen of the ureter (u). The dark outlining shows the continuous connective tissue coat of the ureter, primary and secondary ureteral branches (pb, sb) and medullary cones. When a group of cones converges upon a secondary branch, the group acquires a common connective tissue investment (as in the transverse section at level X–X′), whereupon the connective tissue surrounding each individual cone largely disappears. One reptilian-type nephron (without a nephronal loop), and two mammalian-type nephrons (with nephronal loops) are visible in the cut surface of cortex at the upper left. The arrows indicate the direction by which their distal convoluted tubules will eventually link with perilobular collecting tubules. (From Johnson, 1974).

and Trump (1969), knowledge of this region of the uriniferous tubules is inexact and the terminology somewhat arbitrary. For instance, these small tributaries have been variously referred to as "connecting" or "junctional tubules" (Huber, 1917), "initial collecting ducts" (Sperber, 1960), and "collecting tubules" (Siller, 1971; Hodges, 1974). Their junctions with the distal convoluted tubules can be easily demonstrated histochemically since the collecting tubule produces an abundance of mucus (Fig. 4.14) whereas the lining of the nephron does not. Nevertheless, the exact nature of the initial collecting tubule (as it will be termed here) still seems to be in question. Siller (1971) equated this region with the terminal part of the nephron (*pars conjungens* of the *tubulus convolutus distalis*) regardless of its content of mucus.

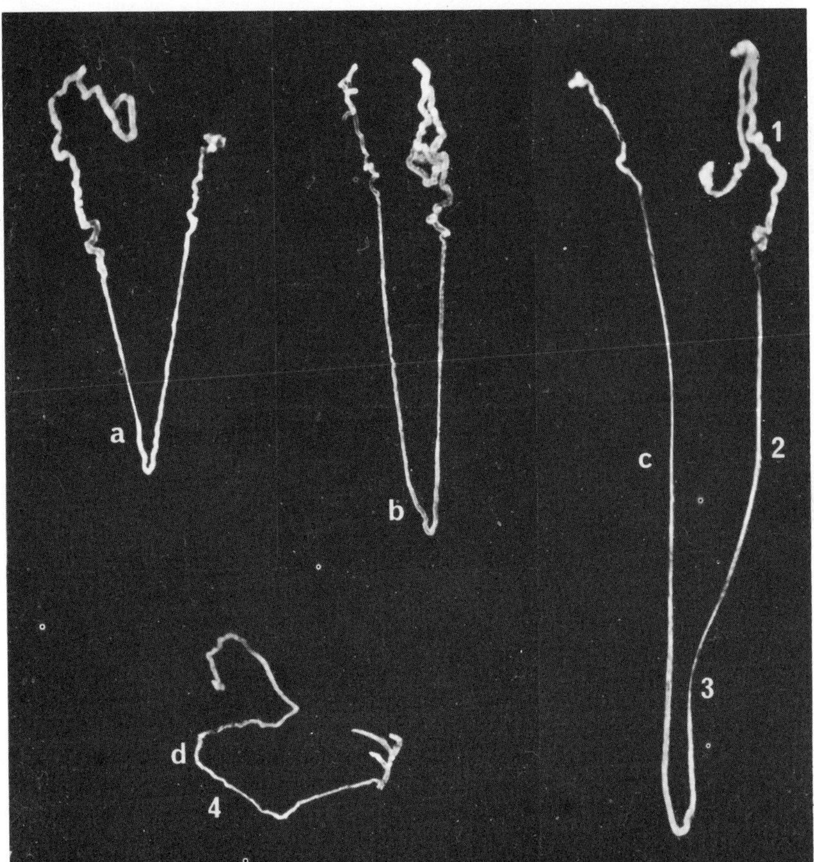

Fig. 4.12. Nephrons teased from the macerated kidney tissue of Gambel's Quail (*Lophortyx gambelii*). The mammalian-type nephrons (a, b and c) show variation in the length of the nephronal loop. Only (c) would reach the apex of a medullary cone. The reptilian-type nephron (d) joins a segment of collecting tubule on which the broken ends of three other nephrons are evident. 1, convoluted part of the proximal convoluted tubule; 2, straight part of the proximal convoluted tubule; 3, thin limb of the nephronal loop; 4, intermediate segment. ×25.

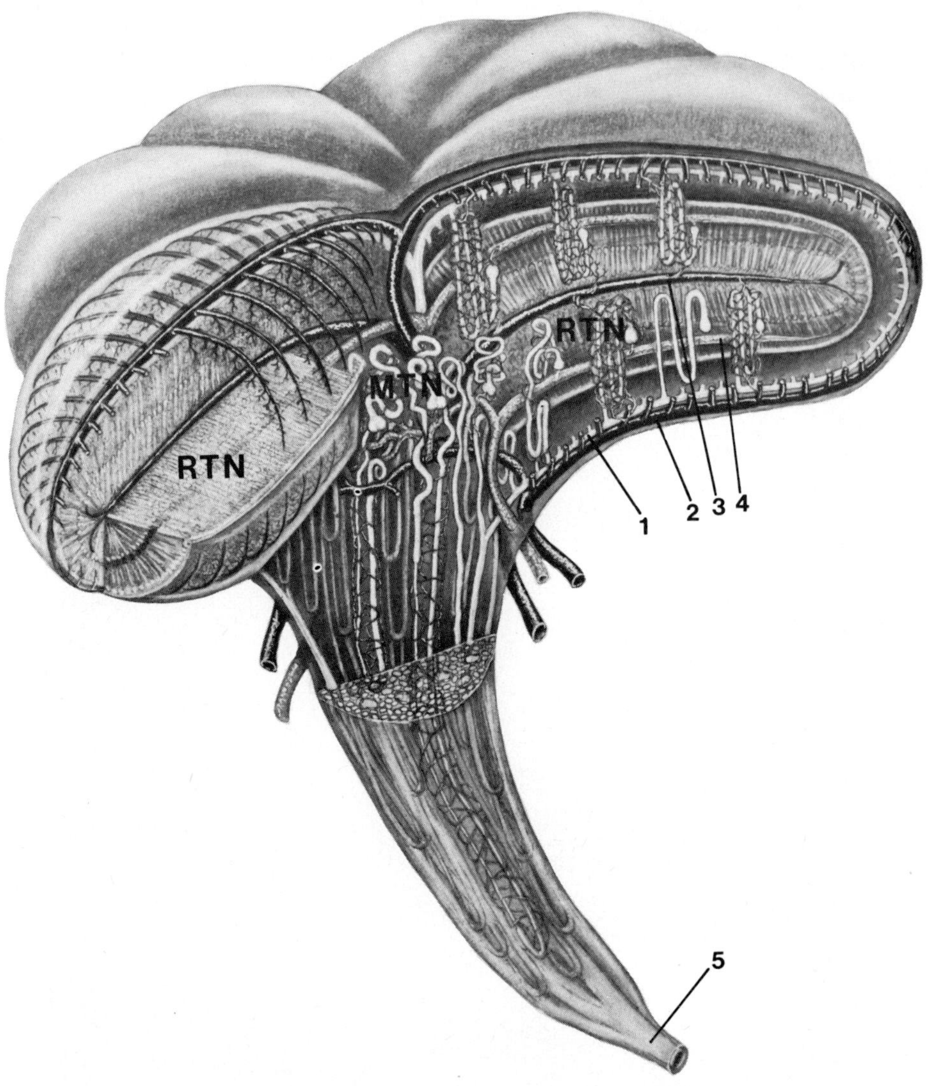

Fig. 4.13. A diagrammatic representation of one renal lobule. Mammalian-type nephrons (MTN) originate in a juxtamedullary position and their nephronal loops extend to varying depths within the cone. Reptilian-type nephrons (RTN) are located in cortex peripheral to the juxtamedullary zone. 1, perilobular collecting tubule; 2, afferent interlobular vein; 3, efferent intralobular vein; 4, intralobular artery; 5, single collecting duct, or tertiary branch of ureter. ×18. (Modified from Braun and Dantzler, 1972.)

Distal convoluted tubules of mammalian-type nephrons join the initial collecting tubules in the region of the mouth of a medullary cone; reptilian-type nephrons typically form this link peripheral to the mouth of a cone (Fig. 4.13).

2. *Renal corpuscle*

The avian renal corpuscle (*corpusculum renale*) is similar to its mammalian counterpart in some respects and dissimilar in others. Of the glomerular capsule (*capsula glomerularis*), which was previously known as Bowman's capsule, the external or parietal part (*pars externa*) consists of a layer of flattened epithelial cells; the internal or visceral part (*pars interna*) is reflected at the vascular pole (*polus vascularis*) to form a membrane which invests the glomerular capillaries. The latter membrane is composed of podocytes which in electron-micrographs appear comparable to those of mammals (Pak Roy and Robertson, 1957; Siller, 1971).

The glomerulus (*glomerulus corpusculi renalis*) consists of capillary loops, but in the bird the loops are fewer and simpler in arrangement than in the mammal. Tchang (1923) was convinced that the avian glomerulus consists of a single,

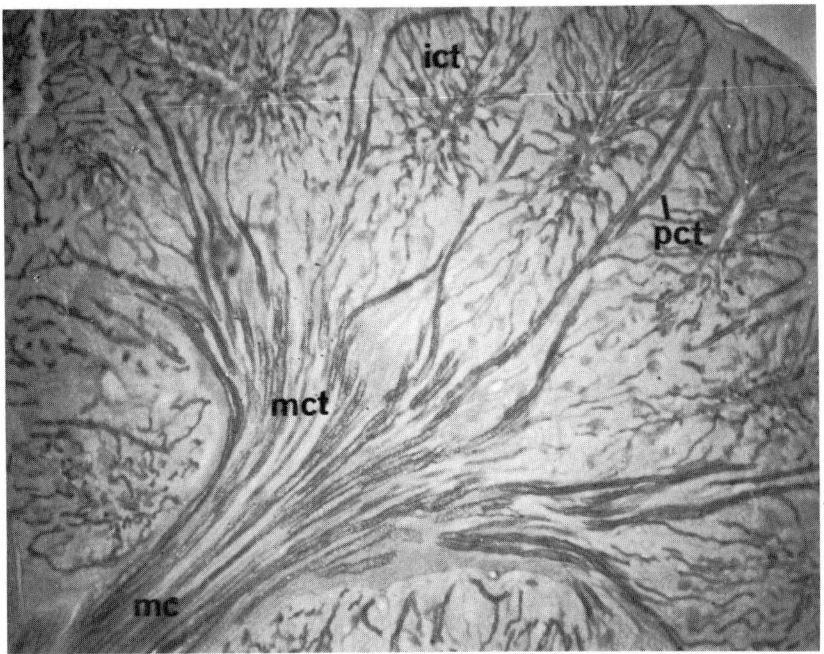

Fig. 4.14. Section of the kidney of a Mourning Dove (*Zenaida macroura*) stained to show mucus in the collecting duct system. Numerous initial collecting tubules (ict) are evident throughout the photograph. These in turn link to perilobular collecting tubules (pct). The latter fuse dendritically to form medullary collecting tubules (mct) which descend the medullary cone (mc). Alcian blue, $\times 39$. (Modified from Johnson and Mugaas, 1970a).

unbranched, convoluted capillary and similar findings were noted in Gambel's Quail (*Lophortyx gambelii*) by E. J. Braun (personal communication). Other investigators reported either multiple or branched capillaries (Hyrtl, 1863; Möllendorf, 1930; Vilter, 1935; Siller and Hindle, 1969). Sperber (1960) and Kudo *et al.* (1971) indicate that some glomeruli have one capillary loop, others several. The exact nature of capillary organization in the avian glomerulus needs further study, particularly the variation between the glomeruli of reptilian-type and mammalian-type nephrons. The avian glomerulus contains a central core of mesangial cells (*mesangium*) (Vilter, 1935; Pak Roy and Robertson, 1957), while in the mammal mesangiocytes are dispersed as intercapillary elements throughout the glomerulus. The endothelium of avian glomerular capillaries is very thin with fenestrations 50–100 nm in diameter (Siller, 1971).

Avian glomeruli are relatively small. Those of reptilian-type nephrons are often not much wider than their associated proximal convoluted tubules, but the glomeruli of mammalian-type nephrons are substantially bigger. Kudo *et al.* (1971) provided comparative data indicating that glomerular diameters in birds are only about 0·2–0·5 those of mammals. As a rough guide based upon numerous species of birds, glomeruli of reptilian-type nephrons range from about 25–35 μm, and mammalian-type glomeruli range from about 75–125 μm. According to Marshall (1934) birds have about twice as many glomeruli as mammals per unit of kidney cortex. Romanoff (1960), however, believed the difference to be several orders of magnitude greater than this. The small size of the glomerulus coupled with the central core of mesangium, led some investigators to regard avian glomeruli as "degenerate" structures reflecting a trend towards an aglomerular condition (Marshall and Smith, 1930; Vilter, 1935; Smith, 1951). However, this point of view is unacceptable since avian glomeruli function efficiently despite their size (Sperber, 1960; Skadhauge, 1972) and also appear fundamental to certain water conservation mechanisms (see below, p. 229; Braun and Dantzler, 1972).

3. *Proximal convoluted tubule*

The proximal convoluted tubule (*tubulus convolutus proximalis*) is a continuation of the glomerular capsule. Sperber (1960) noted that a short, somewhat constricted neck, often connects the tubule and the capsule. With the possible exception of nephrons possessing a lengthy nephronal loop, the proximal tubule is the longest segment of the nephron and is the predominant element in the cortex. The epithelium of the proximal tubule in birds resembles that of mammals in being highly eosinophilic, with a well-developed brush border. It varies from cuboidal to low columnar, and Siller (1971) noted that mitochondria are concentrated in the basal two-thirds of each cell. Portions of the proximal tubule lying relatively close to the glomerulus are slightly narrower than those situated further away. On the basis of diameter, it is convenient to divide the proximal tubule of the reptilian-type nephron into a thin part (*pars tenuis*) and a thick part (*pars crassa*). In contrast, the proximal tubule of a mammalian-type nephron is more elaborate, and consists of a convoluted part and a straight part (*pars convoluta* and *pars recta*) (Fig. 4.12).

4. *Nephronal loop and distal convoluted tubule*

In a mammalian-type nephron the transition between the convoluted and straight parts of the proximal tubule occurs in the juxtamedullary cortex. Studies in the Gambel's Quail (*Lophortyx gambelii*) (Braun and Johnson, unpublished) indicate that as the straight part passes into the medulla its cells decrease in size, so that when the tubule is cut in cross-section its nuclei appear to be crowded together; furthermore, the cytoplasm becomes basophilic (Fig. 4.15), and the nuclei tend to be elongated parallel to the long axis of the duct. Although observations comparable to these were made by Berger (1966) in *Columba*, Michel and Junge (1972) were unable to find similar histological features in *Gallus* and *Anas*. Crowding and elongation of the nuclei continue on into the thin limb of the nephronal loop (*pars descendens ansae*), where they become particularly conspicuous. However, the apparent decrease in cell size and the altered shape of the nucleus could perhaps be artifacts caused by post-mortem collapse of this relatively thin walled part of the nephron. It is conceivable that the thin limb of the nephronal loop (and to a lesser degree the straight part of the proximal convoluted tubule) are distended in the functional kidney. Such an interpretation seems likely from photographs of functioning thin limbs (Imai and Kokko, 1974), and measurements obtained from tissues frozen during

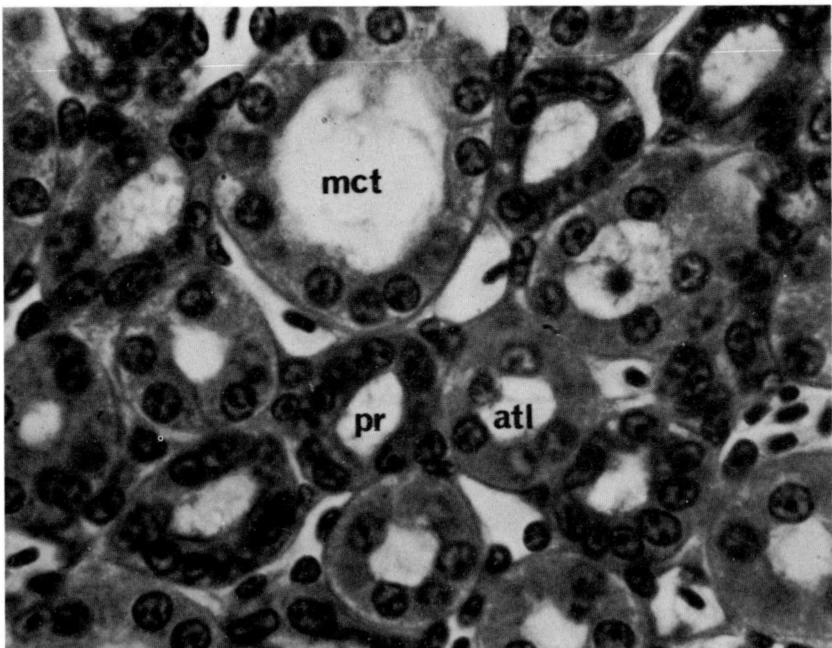

Fig. 4.15. Medulla sectioned near the opening of a medullary cone. The major constituents are: medullary collecting tubules (mct), straight parts of proximal convoluted tubules (pr), and ascending thick limbs of nephronal loops (atl). The thin limbs of the nephronal loops look very much like the straight parts of the proximal tubules except for reduced diameter. Elf Owl (*Micrathene whitneyi*), haematoxylin and eosin, $\times 900$.

functional activity (Rouiller, 1969). Tisher *et al.* (1971) presented evidence for lability in the morphology of the components of the renal medulla.

The straight part of the proximal convoluted tubule tapers very gradually to produce a thin limb of uniform diameter (Siller, 1971), so that the exact point of transformation is not altogether clear with the light microscope (Fig. 4.12). This is in contrast to the mammal in which the thin limb forms rather abruptly (Ericsson and Trump, 1969). The nature of this gradual transition from proximal tubule to thin limb in the mammalian-type nephron of birds needs clarification through ultrastructural studies.

While it is difficult to be certain exactly where the thin limb begins, its transition into the thick limb is marked by a relatively abrupt change in diameter (Fig. 4.12). The thick limb invariably begins proximal to the turning-point of the loop so that the turn is in the thick segment, as it is in the short-type of mammalian nephron. At the origin of a thick limb the size of the cells rapidly increases so that the nuclei are much less crowded, and the cytoplasm becomes and remains weakly eosinophilic. Thick limbs are appreciably wider in the turn

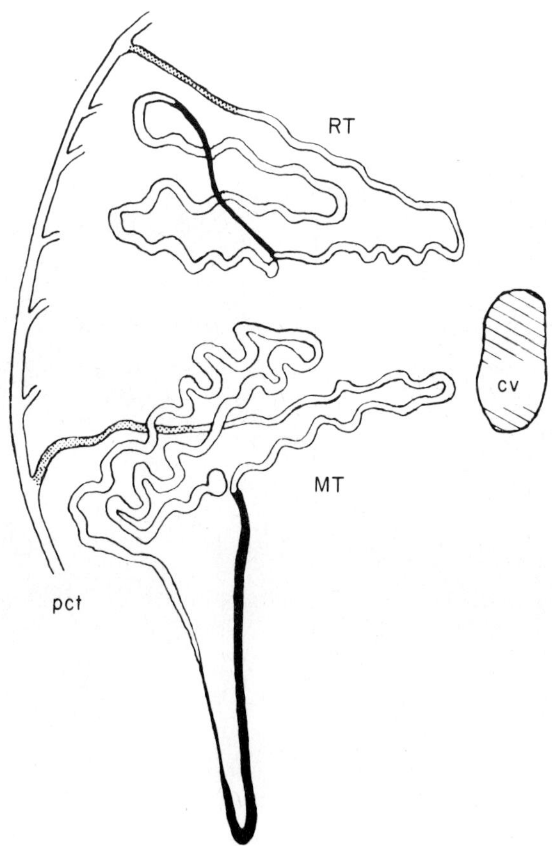

Fig. 4.16. The course of reptilian-type (RT) and mammalian-type (MT) nephrons relative to an intralobular (central) vein (cv) and a perilobular collecting tubule (pct). Black represents the intermediate segment (reptilian-type nephron), and nephronal loop (mammalian-type nephron). The finely stippled areas are initial collecting tubules.

of the loop and narrower as they ascend the medulla (Fig. 4.12). For example, the respective diameters in Gambel's Quail (*Lophortyx gambelii*) are 24–26 μm and 18–22 μm (Braun and Johnson, unpublished observations). In comparison to mammals, the epithelial lining throughout the whole of the nephronal loop is much thicker. While the mammalian thin limb is lined by cells resembling an endothelium, such is not the case in birds (Fig. 4.15).

The ascending part of the thick limb (*pars ascendens ansae*) proceeds into the cortex and comes into contact with its parent renal corpuscle. In accordance with mammalian terminology, the distal convoluted tubule begins at this point of contact and ends when it joins an initial collecting tubule. In both mammalian-type and reptilian-type nephrons, the distal tubule as compared to the proximal tubule is narrower in diameter, shorter in total length (Fig. 4.12), and composed of cells which are smaller, weakly eosinophilic, and devoid of a brush border. Distal tubules course toward the intralobular vein, make a few convolutions and then abruptly turn to join initial collecting tubules in the peripheral part of the cortex (Fig. 4.16). The cortex thus becomes subdivided into two zones, an inner zone enclosed by an outer zone: the inner zone consists of the intralobular vein surrounded by distal tubules, and the outer zone is formed mainly by proximal tubules (Fig. 4.17).

Several investigators have described a rather ill-defined short intermediate segment (*segmentum intermedium*) (Figs 4.12, 16) in most but not all reptilian-type

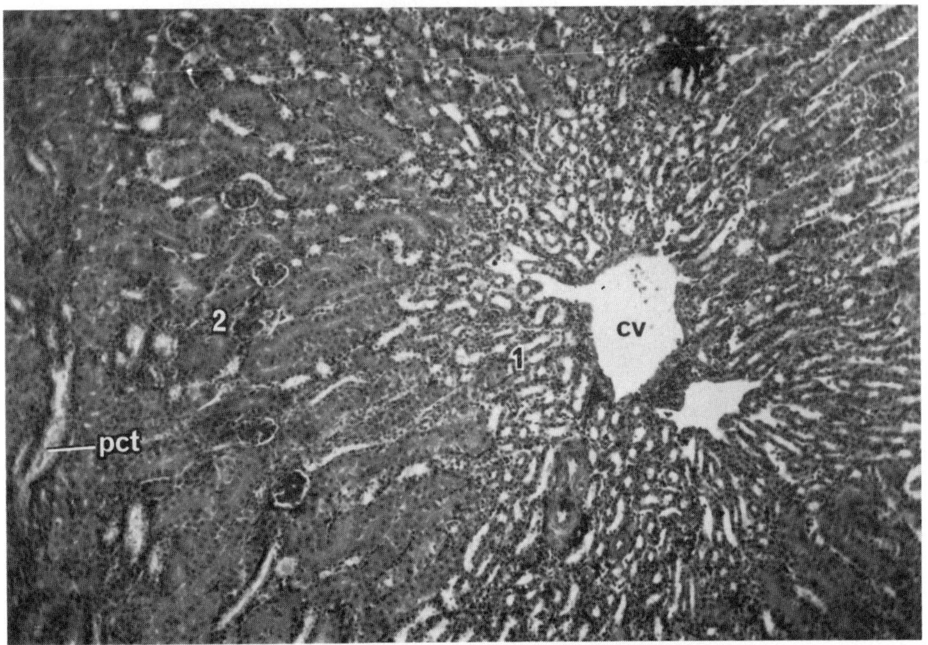

Fig. 4.17. Distal convoluted tubules (1) are particularly abundant in a zone surrounding the intralobular (central) vein (cv), whereas proximal convoluted tubules (2) are localized more peripherally. One boundary of this particular area of cortex is marked by a perilobular collecting tubule (pct). Pied-billed Grebe (*Podilymbus podiceps*), haematoxylin and eosin, ×93.

nephrons (Huber, 1917; Feldotto, 1929; Sperber, 1960; Siller, 1971). The intermediate segment forms a link between proximal and distal convoluted tubules and probably represents a primitive nephronal loop. However, Siller (1971) has pointed out that the exact relationship of the intermediate segment to the nephronal loop awaits resolution through ultrastructural studies.

5. *Juxtaglomerular complex*

A complete juxtaglomerular complex (*complexus juxtaglomerularis*), consisting of the macula densa, extraglomerular mesangium (*insula juxtavascularis*) and juxtaglomerular cells (*cellulae juxtaglomerulares*) of the tunica media of the afferent glomerular arteriole, is present in birds. The macula densa occurs where the ascending limb contacts its renal corpuscle (Fig. 4.18), and is especially well developed in mammalian-type nephrons. Sokabe and Ogawa (1974) argue that the epithelial cells of the avian macula densa are structurally intermediate between those of the typical macula densa cells in mammals and the ordinary

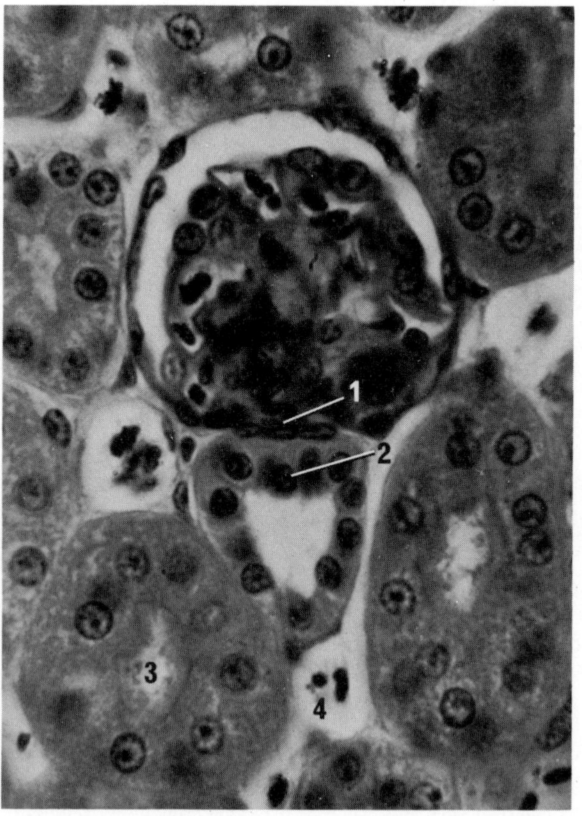

Fig. 4.18. Glomerulus of a mammalian-type nephron in the kidney of a Common Nighthawk (*Chordeiles minor*). 1, extraglomerular mesangium; 2, macula densa; 3, proximal convoluted tubule; 4, peritubular capillary. Haematoxylin and eosin, ×980.

cells of the distal tubules. They concluded that the macula densa (based on its structural characteristics in the mammal) does not occur in birds. However, personal observations and those of other workers (Edwards, 1940; McKelvey, 1963; Berger, 1966; Sutherland, 1966; Kurihara and Yasuda, 1973) suggest that such a conclusion is unwarranted and that a macula densa is almost certainly present in the avian kidney.

There is also some disagreement about the extraglomerular mesangium. Sokabe and Ogawa (1974) claim that this structure is not found in the avian kidney. However, its presence was demonstrated in photographs by Siller (1971), and it can also be seen in Fig. 4.18. According to Siller (1971) the granules (renin) within avian juxtaglomerular cells are difficult to stain. Nonetheless, numerous investigators have described them in the kidneys of birds and renin has been demonstrated in a number of avian species (Sokabe and Ogawa, 1974). The functional significance of renin in birds, and the extent to which it may be related to the roles of angiotensin and aldosterone (as in mammals) is not altogether clear (Sokabe, 1974).

For additional cytological and ultrastructural data, and a series of electron-micrographs covering all regions of the avian nephron, consult Siller (1971) and Hodges (1974).

C. Organization of the medulla

As described earlier (see p. 189 and Figs 4.11, 13), the medullary collecting tubules (*tubuli colligentes medullares*) descend the medullary cones and undergo gradual dendritic fusion to produce large ureteral branches. In cross-sections of the medulla variations in the organization of its components are obvious (Johnson and Mugaas, 1970a). The cones of the Passeriformes demonstrate the most highly organized pattern, with a central core composed of the thin limbs and capillaries, next to which is a ring of collecting tubules, and at the periphery thick limbs arranged in several layers (Fig. 4.19a). When compared to the highly organized intramedullary plan of passerines, the majority of non-passerines display two simpler patterns.

In one pattern there is a general intermingling of all elements, the thin limbs of the loops often being peripheral and the collecting tubules occasionally central (Fig. 4.19b). This pattern occurs in various Podicipediformes, Pelecaniformes, Ciconiiformes, Anseriformes, Galliformes, Gruiformes and Columbiformes. The second pattern is similar to that of passerines with the collecting tubules centralized and often arranged in rings, although unlike passerines (see below) transitions from thin to thick limbs are present within the area bounded by collecting tubules. This pattern occurs in various Charadriiformes, Psittaciformes, Cuculiformes, Strigiformes, Caprimulgiformes and Piciformes. Exceptions to these two patterns are the American Kestrel (*Falco sparverius*) and the hummingbirds (*Trochilidae*). In the former, intramedullary organization is passerine-like while hummingbirds show poorly developed medullary cones consisting almost entirely of collecting tubules with only a few nephronal loops.

Although the general disposition of the parts of the nephron within the cortex and medulla is known, there have been no detailed studies of the exact course

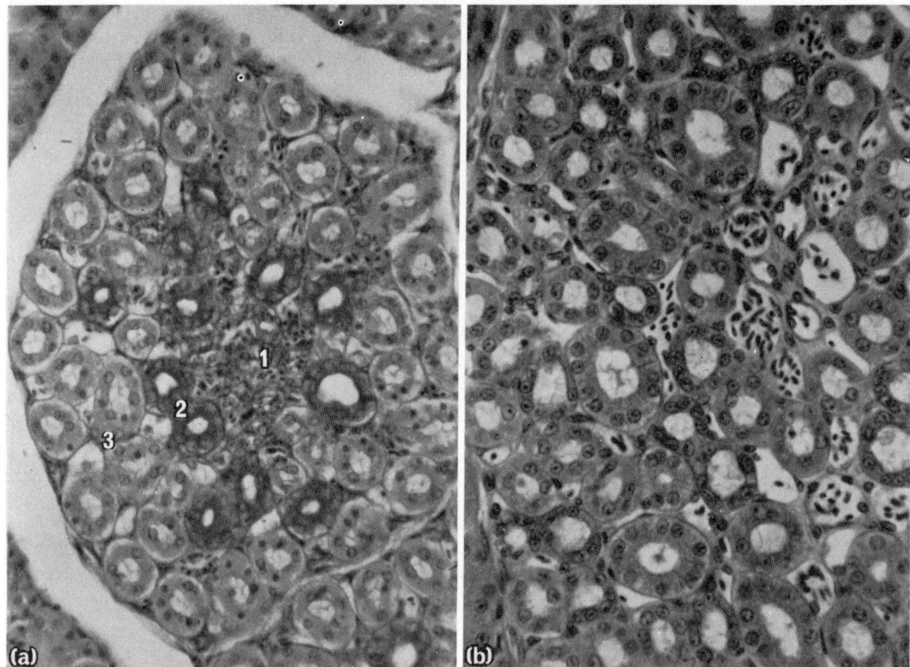

Fig. 4.19. Variations in intramedullary organization. (a) Typical passerine medullary cone with a core of thin limbs (1), a prominent ring of darkly stained medullary collecting tubules (2) and a peripheral zone of thick limbs (3). Savannah Sparrow (*Ammodramus sandwichensis beldingi*), haematoxylin and eosin plus alcian blue, × 235. (b) Non-passerine medullary tissue in which there is a relatively disorderly intermingling of the constituents. Scaled Quail (*Callipepla squamata*), haematoxylin and eosin, × 340. (Both photographs from Johnson and Mugaas, 1970a).

taken by individual tubules of either reptilian-type or mammalian-type nephrons. This is particularly important for mammalian-type nephrons, since precise information on intramedullary organization would probably clarify the functions of the avian renal medulla. Ideally, individual nephrons within blocks of cleared tissue should be injected and the course of the injectant traced by means of microcinematography (Beeuwkes, 1971).

By following serial cross sections of passerine medullary cones, commencing at their wide cortical ends and progressing towards their narrow ureteral ends, Johnson and Mugaas (1970a) interpreted the arrangement of the loops in the medulla to be as follows. Over most of their length all the thin limbs of the loops are centralized within the ring of collecting tubules. However, each thin limb leaves this position after a variable distance, the latter depending upon the length of the loop. Just prior to loop formation, the thin limb courses peripherally and becomes localized immediately outside the ring of collecting tubules. Once peripheral to the collecting tubules the thin limb runs parallel to them, its cells quickly becoming cuboidal and eosinophilic, thus producing the initial narrowed part of the thick limb. The thick limb rapidly expands in diameter and then turns to form the ascending limb of the loop.

IV. Vascular system of the kidney

A. General plan

Blood flow to the kidney is via two systems, arterial and portal. A very extensive peritubular capillary plexus pervades the cortical portion of the lobule. Hence, reptilian-type nephrons (lying entirely in the cortex) and cortical segments of the mammalian-type nephrons are literally bathed in blood. Post-glomerular arterial blood plus renal portal blood flow together into the capillaries of the peritubular plexus, and all subsequent drainage is centripetal, i.e. towards the centre of the lobule into an efferent intralobular (central) vein (Fig. 4.5). Several intralobular veins become confluent to form an efferent renal radix (*rdx. renalis efferens*); the many efferent renal radices are the tributaries of the cranial and caudal renal veins. The cranial and caudal renal veins drain into the common iliac veins, and from there the blood flows into the caudal vena cava (Fig. 4.20).

B. Arteries

Studies of the arterial blood supply to the avian kidney are relatively few and are limited mainly to those on *Gallus* by Mouchette and Cuypers (1959), Siller and Hindle (1969) and Kurihara and Yasuda (1975a).

There are three pairs of renal arteries, cranial, middle and caudal. Only the left and right cranial renal arteries (*a. renalis cranialis*) arise directly from the aorta (Fig. 4.20). The left and right middle and caudal renal arteries (*a. renalis media, a. renalis caudalis*) both originate from the left and right ischiadic arteries as the latter pass through the kidneys. The branches of each cranial renal artery are limited to the cranial division of the kidney. The branches of the middle and caudal renal arteries supply blood to the middle and caudal divisions of the kidney, respectively. The three renal arteries also give rise to a delicate arterial network which supplies the ureter and its intrarenal branches (Siller and Hindle, 1969). The external iliac artery typically remains unbranched as it traverses the kidney.

Possibly, more extensive comparative research would reveal greater inter-specific variation in the pattern of arterial arrangement. On the other hand, Siller and Hindle (1969) reviewed the very old literature and found that of 21 species (including *Gallus*) examined by Barkow (1829), only the kidney of the Gray Heron (*Ardea cinerea*) received branches from the external iliac artery. The other species all followed the pattern described above for *Gallus*. Thus both Spanner (1925) and Sperber (1960) were incorrect in their generalization that the external iliac artery supplies blood to the kidney.

Siller and Hindle (1969) were unable to correlate major branches (extra-lobular) of the renal artery with those of the ureter. Hence, they felt that it would not be valid to use terms like "interlobar" or "arcuate" in an attempt to establish homologies with the mammalian arterial plan. Comparing mammals and birds, Braun (1976) concluded that in birds the initial ramifications of the renal arteries within the kidneys are less clearly defined. Nonetheless, Kurihara

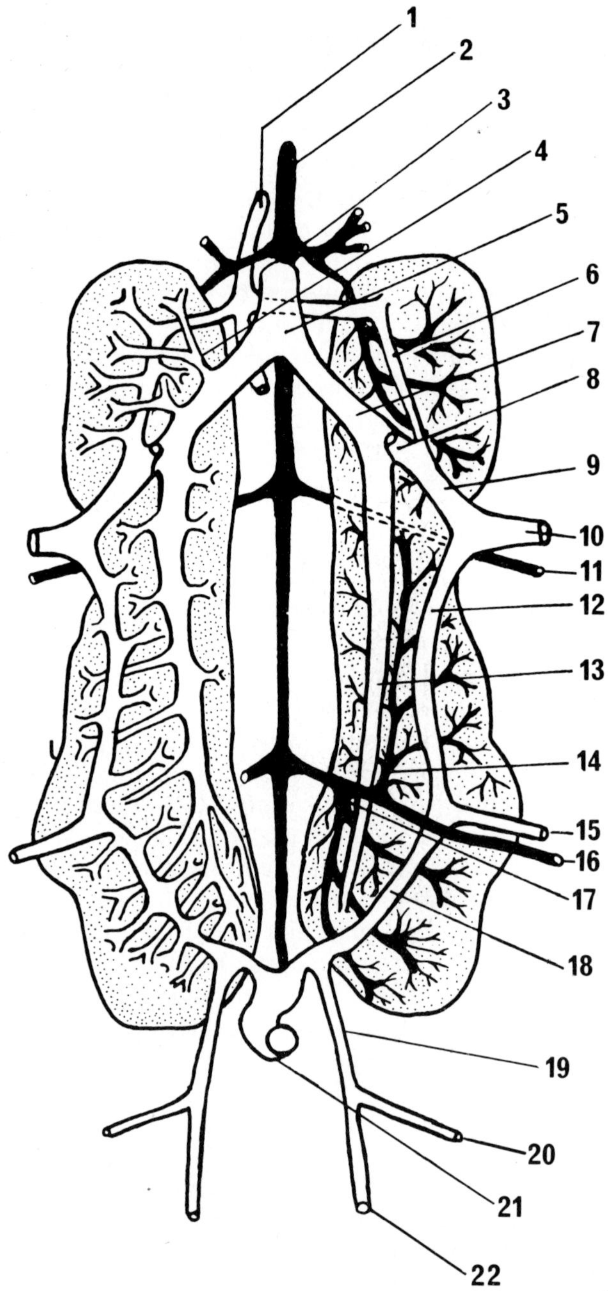

Fig. 4.20. Ventral view of the kidneys of *Gallus* drawn as though transparent to reveal the blood vessels within. The left side of the diagram shows renal portal and efferent veins; the right side shows arteries. 1, internal vertebral venous sinus; 2, aorta; 3, cranial renal artery; 4, one of several cranial renal veins; 5, caudal vena cava; 6, cranial renal portal vein; 7, common iliac vein; 8, portal valve; 9, common iliac vein; 10, external iliac vein; 11, external iliac artery; 12, caudal renal portal vein; 13, caudal renal vein; 14, middle renal artery; 15, ischiadic vein; 16, ischiadic artery; 17, caudal renal artery; 18, caudal renal portal vein; 19, internal iliac vein; 20, lateral caudal vein; 21, caudal mesenteric vein; 22, pudendal vein. (Modified from King and McLelland, 1975).

and Yasuda (1975a) do employ the terms interlobar and intralobar arteries, although their concept of the lobe and lobule may not correspond with the definitions given earlier (p. 197). A completely satisfactory nomenclature for the extralobular arteries awaits further study.

Once the arterial branches have entered the cortical portion of a lobule there are no problems with terminology. Such vessels are correctly called intralobular arteries (*aa. intralobulares*), and their pattern of distribution has been well documented (Siller and Hindle, 1969; Kurihara and Yasuda, 1975a; Braun, 1976). Several intralobular arteries run parallel to an intralobular vein approximately midway between the intralobular vein and the interlobular (portal) veins (Fig. 4.5). As it courses toward the extremity of a cortical unit each intralobular artery gives off a series of afferent glomerular arterioles (*arteriolae glomerulares afferentes*). Afferent arterioles are remarkably constant in their arrangement, since they invariably arise from the side of the intralobular artery opposite the intralobular (central) vein. Thus the glomeruli are all positioned towards the periphery of the cortex (Fig. 4.5).

The efferent glomerular arteriole (*arteriola glomerularis efferens*) leaves the glomerulus at the vascular pole in close proximity to the afferent arteriole. Attempts to follow its subsequent course by means of injection preparations have not been altogether satisfactory (Siller and Hindle, 1969). However, it appears that efferent arterioles typically extend peripherally to join the peritubular capillary plexus in the outer cortex (Fig. 4.5). The combined flow of portal and arterial blood then proceeds towards the intralobular vein. Both Siller and Hindle (1969) and Kurihara and Yasuda (1975a) have shown that occasional branches of the intralobular artery do not lead to glomeruli, but instead discharge directly into the capillaries. The significance of such glomerular bypass vessels is not understood, although (as suggested by Braun and Dantzler, 1974) they possibly function as shunts when reptilian-type nephrons are inactive (p. 229).

C. *Veins and portal valve*

The arrangement of the veins is shown in Fig. 4.20. The major afferent (portal) channel is the external iliac vein bringing blood from the hind limb. On entering the kidney, this vessel branches to form the caudal renal portal vein (*v. portalis renalis caudalis*) and common iliac vein. The latter gives off the cranial renal portal vein (*v. portalis renalis cranialis*). The relatively small cranial renal veins and the large caudal renal vein drain the kidney into the common iliac vein. The right and left common iliac veins unite forming the caudal vena cava.

Numerous branches leave the major afferent veins (cranial and caudal renal portal veins) and eventually drain into the peritubular capillary plexus described on p. 190. The terminal parts of these branches are interlobular veins (*vv. interlobulares*), which are interlobular or perilobular in relation to the cortical portion of a lobule (Fig. 4.5). The exact pattern of the afferent venous branching relative to the kidney lobe is unclear. After traversing the capillaries, blood enters the efferent intralobular (central) vein which forms the axis of cortical organization (see p. 190 and Figs 4.5, 11). Intralobular veins fuse into a series of efferent renal branches (p. 211) which drain into the cranial and caudal renal veins (Fig. 4.20).

The renal portal valve (*valva portalis renalis*), is a smooth muscle sphincter lying just peripheral to the junction of the caudal renal vein and the common iliac vein. Depending on how much the valve is contracted, more or less blood is prevented from flowing directly to the caudal vena cava. Diverted blood is routed into the renal portal system via the cranial and caudal renal portal veins. To what extent this blood will actually traverse the peritubular capillary plexus of the kidney cortex is subject to further vasoconstrictor control (see below). In appearance, the renal portal valve is funnel-shaped with the apex of the funnel directed toward the caudal vena cava. Sperber (1949) described a number of interspecific variations in valve morphology. For example, in some birds the valve has a single orifice while in others there are multiple perforations. According to Oelofsen (1977) an unusual vascular arrangement occurs in the Ostrich (*Struthio camelus*) in which the caudal renal portal vein and the caudal renal vein join via an elongated dilation housing three separate renal portal valves.

Blood flow to the cranial renal portal vein is from the common iliac vein only. The caudal renal portal vein, however, is more complex. Caudally, it receives the ischiadic vein in the region where the middle and caudal divisions of the kidney adjoin. Further caudally the left and right caudal renal portal veins anastomose in the midline. In the region of the anastomosis, the caudal mesenteric (coccygeomesenteric) vein and the internal iliac veins also join the portal circulation. In effect, the major portal vessels in both kidneys form a vascular ring (Fig. 4.20), for which Kurihara and Yasuda (1975b) have proposed the term renal portal circle.

Studies of the renal portal system in *Gallus* (Mouchette and Cuypers, 1959; Akester, 1967; Baumel, 1975; Kurihara and Yasuda, 1975b) and in *Columba* (Sperber, 1960) demonstrate anastomoses linking the cranial and caudal renal portal veins with the internal vertebral venous sinus. Also, there is evidence for an anastomosis between the caudal renal portal vein and the caudal renal vein in *Gallus* (Akester, 1967; Kurihara and Yasuda, 1975b). Since Akester found this anastomosis only once in a sample of about 100 birds, he speculated that it might be simply a developmental abnormality.

Portal blood flow can enter the kidney via the external iliac vein, ischiadic vein, internal iliac vein, and under certain physiological conditions the caudal mesenteric vein (Fig. 4.20). The radiographic studies of blood flow in the kidney of *Gallus* by Akester (1964, 1967) demonstrated a highly labile system. Various flow patterns occur according to the degree of contraction of the portal valve coupled with additional sphincter effects (Akester, 1967) at points where venous branches leave the cranial and caudal renal portal veins. If these latter sphincters are mostly closed, much of the blood entering the portal system will be shunted through the kidney without traversing the peritubular capillary network of the cortex and vice versa. Akester (1967) demonstrated that the direction of blood flow in the caudal mesenteric vein of *Gallus* is variable. At times, the caudal mesenteric flow was toward the kidneys, as suggested by both Spanner (1925) and Sperber (1949); at other times the flow was in the opposite direction toward the liver. Comparable findings have been reported in the Jackass Penguin

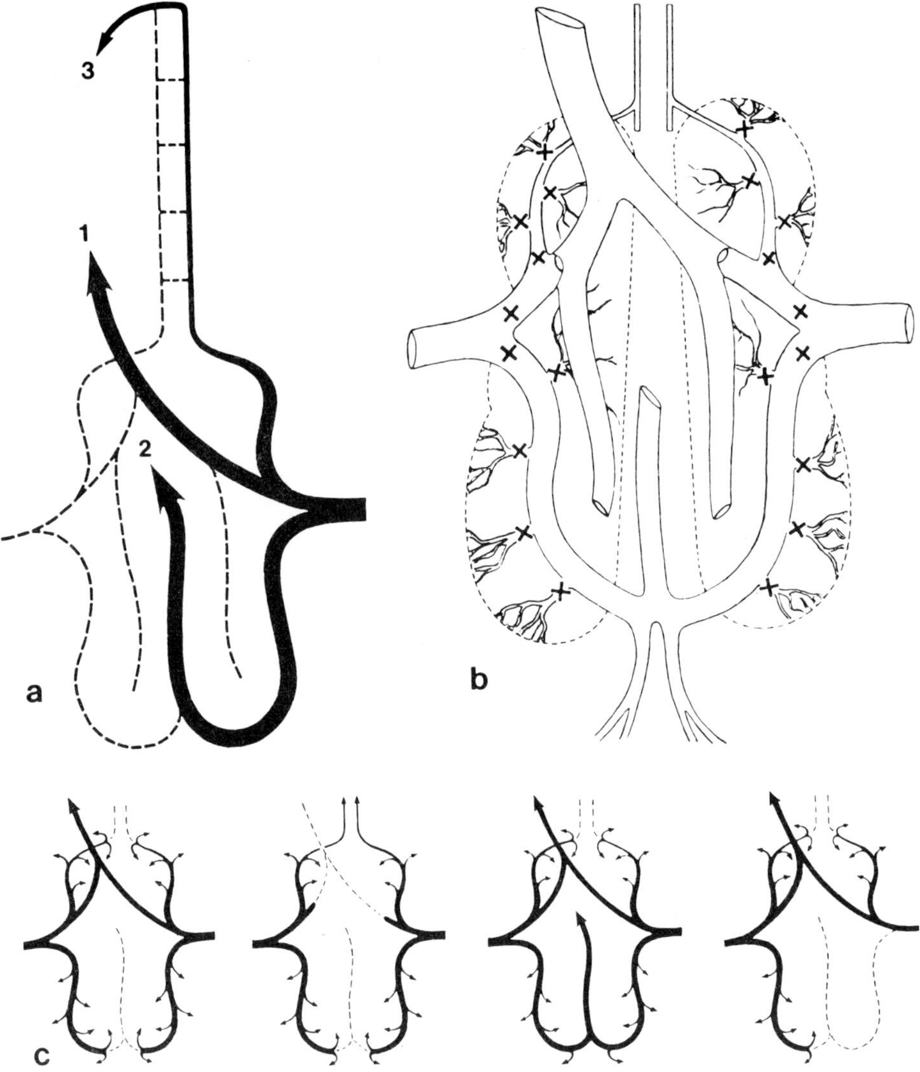

Fig. 4.21. Renal portal shunts, portal sphincters and variations in portal blood flow. All diagrams represent ventral views; the exact sequence of veins can be ascertained from Fig. 4.20. (a) The three renal portal shunts as marked on the left kidney: 1, through the portal valve to the caudal vena cava; 2, caudal renal portal vein, to the caudal mesenteric vein; 3, cranial renal portal vein to the internal vertebral venous sinuses. Shunts 1 and 2 can function individually, or all three collectively, to provide a complete by-pass of the kidney. However, partial shunting is more common with at least some blood entering the peri-tubular capillary network. (b) Points of vasoconstriction marked by crosses. Numerous combinations of constriction result in highly varied patterns of blood flow. (c) Selected examples of routes of flow. From left to right: *portal valves open*, partial shunting to caudal vena cava, all portions of kidneys receiving blood from the cranial and caudal renal portal veins, no flow to the caudal mesenteric vein; *portal valves closed*, no flow to caudal vena cava, partial shunt of cranial divisions, middle and caudal divisions receiving blood from the caudal renal portal veins, no flow to caudal mesenteric vein; *portal valves open*, partial shunting to caudal vena cava, all portions of kidneys receiving blood from the cranial and caudal renal portal veins, partial shunting to caudal mesenteric vein; *portal valves open*, partial shunting to caudal vena cava, all divisions in right kidney receiving blood from the cranial and caudal renal portal veins, the left kidney has flow in the cranial renal portal vein but not in the caudal renal portal vein, no flow in the caudal mesenteric vein. (From Akester, 1967; with permission of Cambridge University Press).

(*Spheniscus demersus*) by Oelofsen (1973). The factors that control such variability are not understood. There are three renal portal shunts, numerous muscular sphincters, and many temporal variations in the pattern of venous flow (Fig. 4.21).

Physiologically, the renal portal system is of fundamental importance in the excretion of uric acid in all birds, and possibly urea in Spheniscidae (Oelofsen, 1973). Uric acid is the primary avian nitrogenous waste product and its removal largely depends upon tubular secretion. The latter process occurs at the tubule–blood interface within the vast peritubular capillary plexus. Nonetheless, the system is functionally labile as evidenced by the extreme variability in flow patterns described above. Possibly, the shutting down of portal flow is a physiological response which temporarily provides extra blood to the general circulation at times of emergencies (Siller and Hindle, 1969). It is likely that adjustments in portal blood flow are mediated through neurohumoral effects similar to those demonstrated with the renal portal valve. The latter has a rich nerve supply (Gilbert, 1961; Akester and Mann, 1969; Dolezel and Zlabek, 1969) and is constricted by histamine or acetylcholine and dilated by adrenaline (Rennick and Gandia, 1954; Oelofsen, 1973). Radu *et al.* (1973) demonstrated that intrarenal blood vessels have an adrenergic and cholinergic innervation.

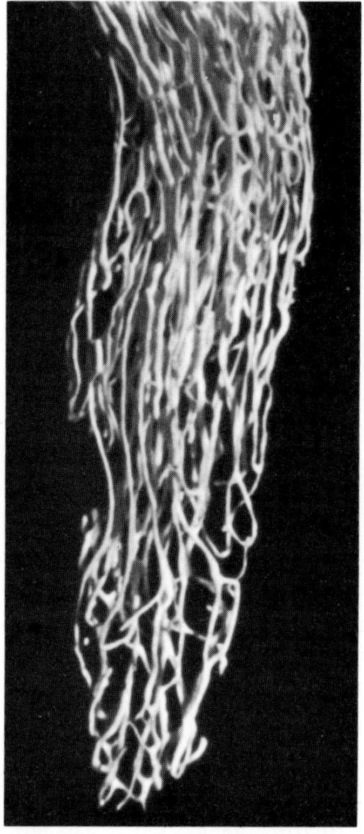

Fig. 4. 22. Vasa recta and associated capillary plexus in a cleared medullary cone from a Gambel's Quail (*Lophortyx gambelii*). "Microfil" injection via the ischiadic artery, ×49.

D. *Vasa recta*

There are a number of detailed studies of the vascular architecture of the mammalian renal medulla (e.g. Plakke and Pfeiffer, 1964; Prong *et al.*, 1969; Kriz and Dieterich, 1970; Fourman and Moffat, 1971; Kriz and Koepsell, 1974), but unfortunately comparable work has not yet been done on the avian kidney. However, it is apparent that the efferent glomerular arterioles of juxtamedullary glomeruli supply blood to the descending arteriolae rectae in the same manner as in mammals (Siller and Hindle, 1969). The arteriolae rectae open into a capillary network extending throughout the medullary cone (Fig. 4.22). Presumably, there is a system of venulae rectae through which blood ascends towards the cortex. However, the arteriolae rectae and venulae rectae of birds do not appear to be as histologically distinct as they are in mammals (Fourman and Moffat, 1971), and in histological sections they are difficult to distinguish with certainty. Venulae rectae could drain into the peritubular capillary network at the cortico–medullary junction or directly into efferent veins. Siller and Hindle (1969) suggested, though without solid evidence, that the vasa recta system links with arterial elements near the apex of the medullary cone.

V. Associated organs

A. *Ureter*

1. General structure

A system of ureteral branches extends throughout the kidney and links the ureter proper with the collecting ducts. As described earlier (p. 189) the arrangement is a dendritic one, devoid of a renal pelvis or papillae. Injection preparations reveal conspicuous primary ureteric branches (*rr. ureterici primarii*) which subdivide dendritically into secondary branches (*rr. ureterici secundarii*). Each of the latter receive large collecting ducts (*ductuli colligentes*), one from each of several medullary cones (p. 189 and Figs 4.3, 11, 13). The length of the branches is directly related to size of the kidney. Large kidneys require more extensive ureteral branching to drain medullary cones which are remote from the ureter (Fig. 4.23a). The renal part of the ureter (*pars renalis*) arises from the coalescence of several primary branches within the cranial end of the cranial division of the kidney. At first, the ureter is embedded within the kidney, but at or slightly caudal to the junction between the middle and caudal divisions, it emerges and thereafter occupies a prominent superficial position on the ventral surface of the kidney (Fig. 4.1).

The number of primary branches and the pattern by which they join the ureter vary considerably between species (Johnson and Mugaas, 1970a). The simplest pattern occurs in small birds (e.g. various sparrows, the Budgerigar *Melopsittacus undulatus*, etc.) in which the relatively small size of the kidney precludes numerous primary branches spaced evenly along the ureter. In these species short primary branches enter the ureter as clusters in each kidney division. Typically, there are 2–3 clusters per division, each cluster having up

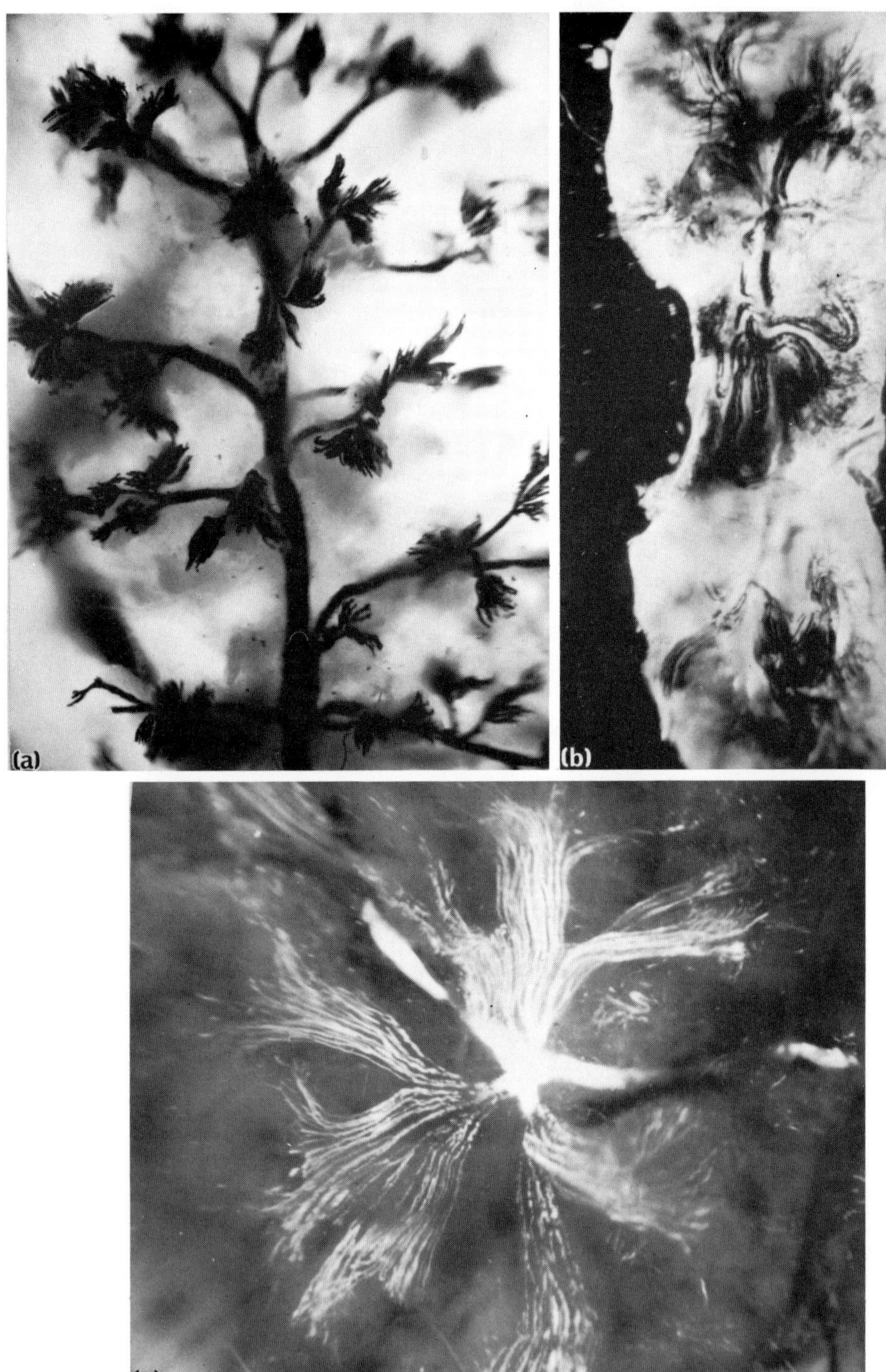

Fig. 4.23. Various patterns of ureteral branching as revealed in cleared kidney tissue. (a) Cranial division of the kidney of the American Coot (*Fulica americana*) showing long ureteral branches. Tuft-like clusters of collecting tubules define the individual medullary cones. India ink injection via ureter, ×8. (b) Kidney of a House Finch (*Carpodacus mexicanus*) in which the medullary cones attach to short ureteral branches. The latter are clustered in each kidney division, and not spaced out along the ureter. India ink injection

to four primary branches (Fig. 4.23b). In the kidneys of larger birds, the more numerous primary branches generally are not clustered, but tend to be single and evenly spaced along the ureter. The kidney of the Mourning Dove (*Zenaida macroura*) has ureteral branches which form a palmate pattern in each division (Fig. 4.23c).

The pelvic portion of the ureter (*pars pelvina*) extends from the point where the ureter leaves the kidney to its junction with the urodeum. In this region, the ureters are prominent as they traverse the dorsal body wall (Fig. 4.1). In male birds each ureter lies parallel to the ductus deferens to which it is closely joined by connective tissue. In immature females the left ureter is similarly related to the oviduct, but in older females the effects of sexual recrudescence largely obliterate this relationship.

According to Hinman *et al.* (1939) and Inoue (1953) there is no functional sphincter at the cloacal orifice of the ureter (*ostium ureteris*). However, Inoue further indicated that the ureter does not simply pierce the cloacal wall, but empties into a deep recess of the urodeal mucosa which he termed the "mucosal duct". Inoue concluded that the oblique angle at which the ureter joins the cloaca combines with the "mucosal duct" to produce a sphincter-like effect. This problem, however, needs further investigation. There are irregular folds in the urodeal mucosa at the point where the ureter opens, which may perhaps constitute a mucosal flap rather than a mucosal duct (Fig. 4.24). Furthermore, the presence of a muscular sphincter cannot be completely discounted; for example, in Fig. 4.24 a circular band of muscle is present which could well have a sphincter action.

2. *Histology*

As collecting ducts merge to form ureteral branches, there is a relatively abrupt change in the epithelial lining. The simple cuboidal epithelium characteristic of collecting ducts first becomes stratified cuboidal and then, closer to the ureter, changes into a pseudostratified epithelium. In the walls of the ureteral branches there is a stroma of connective tissue and mainly circular smooth muscle. The connective tissue-muscle coat becomes gradually thicker down the ureteral tree, attaining its maximum development in the pelvic portion of the ureter. Mucin-filled vacuoles are present in the epithelial cells of the ureteral branches and become progressively more abundant towards the ureter. Longley *et al.* (1963) and McNabb *et al.* (1973) showed that mucus varies histochemically within the collecting tubules, ducts, ureteral branches and ureter. It also appears that the secretion of mucus varies with diet and consequent excretory demands; for example, a high protein intake probably triggers greater production of mucus by the urinary tract (McNabb *et al.*, 1973). In birds, mucus appears to have an important function in lubricating the movement of the viscous urine through

via ureter, ×8. (c) Caudal division of the kidney of a Mourning Dove (*Zenaida macroura*) in which the collecting tubules and ureteral complex are filled with uric acid precipitates. ×13. (From Johnson and Mugaas, 1970a,b).

the ureteral system. Furthermore, McNabb (1974) demonstrated that urinary mucus may serve a protective role by converting lyophobic urate colloids into lyophilic colloids having increased stability. This would prevent excessive precipitation which might result in clogging of the urinary tract.

In sections of the ureter taken at or near its pelvic region, the lumen generally is stellate (presumably a fixation artifact), and frequently contains mucous strands and irregularly shaped precipitates of uric acid (Fig. 4.25a,b). It is lined by a pseudostratified epithelium in which are numerous mucin-filled vacuoles. The epithelium rests upon a gland-free lamina propria containing many dense aggregations of lymphocytes which may reflect protective mechanisms against bacterial infection (Inoue, 1953). There is no submucosa. The muscle tunic consists of inner longitudinal, middle circular and outer longitudinal layers. The middle circular layer is by far the best developed. The longitudinal layers (especially the outer one) are usually discontinuous and form strands of muscle rather than complete coats. Much connective tissue pervades the muscle layers, and often produces a fibromuscular stroma in which muscle orientation is disorderly. Hence, the muscle tunic lacks the crisply defined layers which characterize organs such as the small intestine. An adventitia surrounds the muscle tunic. The innervation of the avian ureter and its branches has received only limited attention, but autonomic fibres and endings were demonstrated by Radu *et al.* (1973), and Ivanov (1974).

The above description of ureteral structure (based largely upon unpublished personal observations) applies to most of the avian orders with little variation, and generally agrees with the findings of others (Hinman *et al.*, 1939; Inoue,

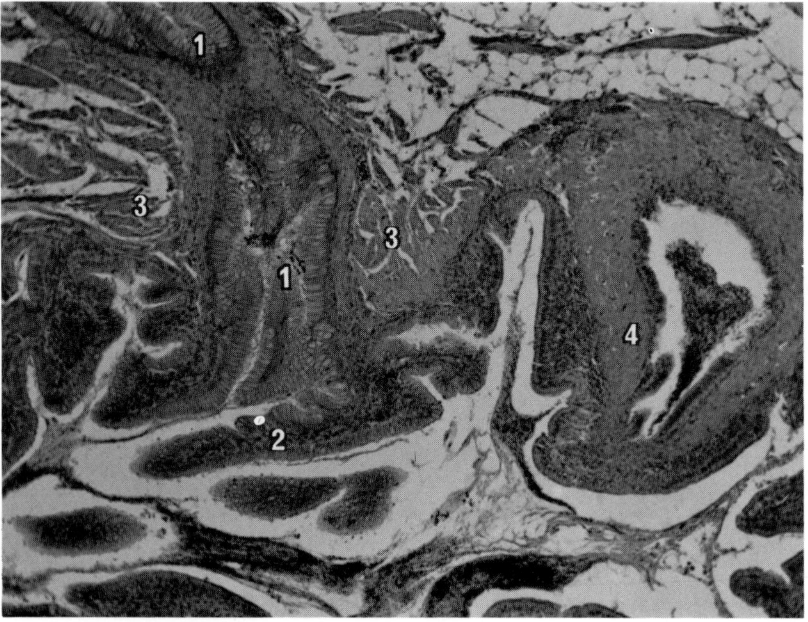

Fig. 4.24. Frontal section showing the left half of the urodeum in a Zebra Finch (*Poephila guttata*). 1, ureter; 2, apparent urodeal mucosal flap over the ureteral orifice; 3, possible ureteral sphincter muscle; 4, ductus deferens. Haematoxylin and eosin, ×92.

1953; Goodchild, 1956; Bortolami and Palmiera, 1962; Liu, 1962; Oelofsen, 1973; Hodges, 1974). Occasionally, the epithelium of the ureteral tree is not pseudostratified but rather transitional and possibly stratified squamous in some areas, as in the American White Pelican (*Pelecanus erythrorhynchos*) and the Marbled Godwit (*Limosa fedoa*). It is apparent from Oelofsen's (1973) photographs that penguins (Spheniscidae) also have these variations, and Goodchild (1956) reported transitional epithelium in some sections of the ureter of *Gallus*.

B. Rectum and cloaca

The role of the rectum (the portion of gut extending between the ileum and cloaca and sometimes referred to as "colon") and cloaca in resorbtion of salt

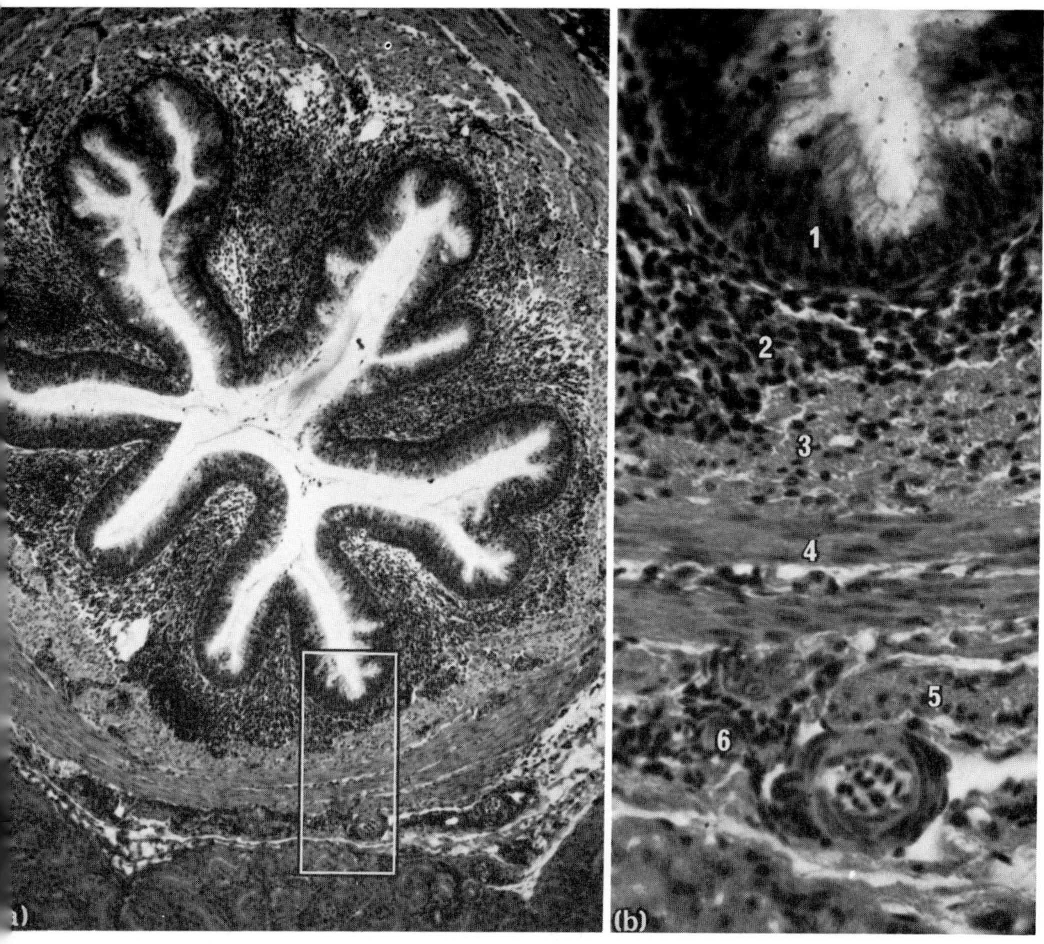

Fig. 4.25. Ureter of Gambel's Quail (*Lophortyx gambelii*). (a) A transverse section of the ureter near the caudal edge of the caudal division of the kidney. Many lymphocytes are evident in the lamina propria. ×92. (b) Detail of the area enclosed by the rectangle. 1, epithelial lining; 2, lamina propria; 3, inner longitudinal muscle; 4, middle circular muscle; 5, outer longitudinal muscle; 6, adventitia. ×417. Haematoxylin and eosin.

and water from ureteral urine appears to be significant. Although these organs are not parts of the urinary system *per se*, it is appropriate to describe certain of their basic features. For an extensive review of cloacal water and salt transport consult Skadhauge's (1973) monograph and his briefer papers (Skadhauge, 1972, 1975, 1976a,b).

Skadhauge (1973) considered the coprodeum and rectum to function essentially as one chamber in their role of modifying the urine, the urodeum and proctodeum having little if any role in this function. Such an interpretation is consistent with the mucosal histology of these organs. The proctodeal and urodeal portions of the cloaca are lined by multilayered epithelia of a non-resorbtive nature. Stratified squamous epithelium predominates in the proctodeum; it is also found in the urodeum along with areas of pseudostratified and transitional epithelium (Johnson and Skadhauge, 1975). The coprodeum and rectum, on the other hand, are lined by simple columnar epithelium suited to resorbtive processes. Furthermore, the surface area in the coprodeum and rectum is greatly increased by the presence of villi. In many (probably most) birds, the surface structure of the mucosa resembles that of the mammalian small intestine (Fig. 4.26a). However, some species (the Galah *Eolophus roseicapillus* and Laughing Kookaburra *Dacelo novaeguineae*, for example) lack villi in the coprodeum and the rectum, and the resultant histology is like that of the mammal colon (Fig. 4.26b). For further details see the chapters on the Digestive System and the Cloaca in Vol. 2.

Retrograde movement of urine and faeces has been demonstrated radiographically with radio-opaque material (Koike and McFarland, 1966; Akester *et al.*, 1967; Nechay *et al.*, 1968; Ohmart *et al.*, 1970). Often the retrograde movement carries the material into the caeca. However, there is no apparent correlation between xerophilia and the relative development of the caeca. Caecal function is much more clearly related to the digestive processes associated with herbivorous diets than to any excretory role (Skadhauge, 1973). The phenomenon of the retrograde movement of urine and faeces has been studied almost exclusively in *Gallus*, and studies of other species are needed. Nonetheless, the physiological basis for the potential modification of urine by this mechanism has been well established.

Skadhauge (1974a) noted little correlation between the capacity for renal concentration and xerophilia in a variety of Australian birds. This emphasizes the point that xerophilic adaptation is not simply a matter of altered kidney function but includes other factors as well. It is attractive to speculate that the coprodeum and rectum furnish a site for resorbtive mechanisms in some xerophilic species. The pattern of the villi seen in the Zebra Finch (*Poephila guttata*) (Fig. 4.26a) may reflect such a relationship, especially in view of the rather unimpressive U/P osmolar ratio (2·8) in this species (Skadhauge, 1974a). It seems likely that the elaborately organized coprodeum and rectum in the Emu (*Dromaius novaehollandiae*) (Fig. 4.26c) resorb water during retrograde movement of the urine. Skadhauge (1974a) reported that the Emu kidney was inefficient (U/P osmolar ratio of 1·4) and that post-renal water resorbtion was probable.

Most of the experimental work on cloacal salt and water transport in birds has been on *Gallus*. The main conclusion from these studies is that the coprodeum and rectum resorb quantities of sodium chloride and water in dehydrated individuals (Skadhauge, 1973). More recently, Skadhauge (1974b) studied these parameters in the Galah (*Eolophus roseicapillus*), a xerophilic parrot from Australia. His results indicate that post-renal water conservation is substantially more effective in the Galah than in *Gallus*. Additional studies of desert birds are needed to clarify fully the role of the coprodeum and rectum in xerophilic adaptation.

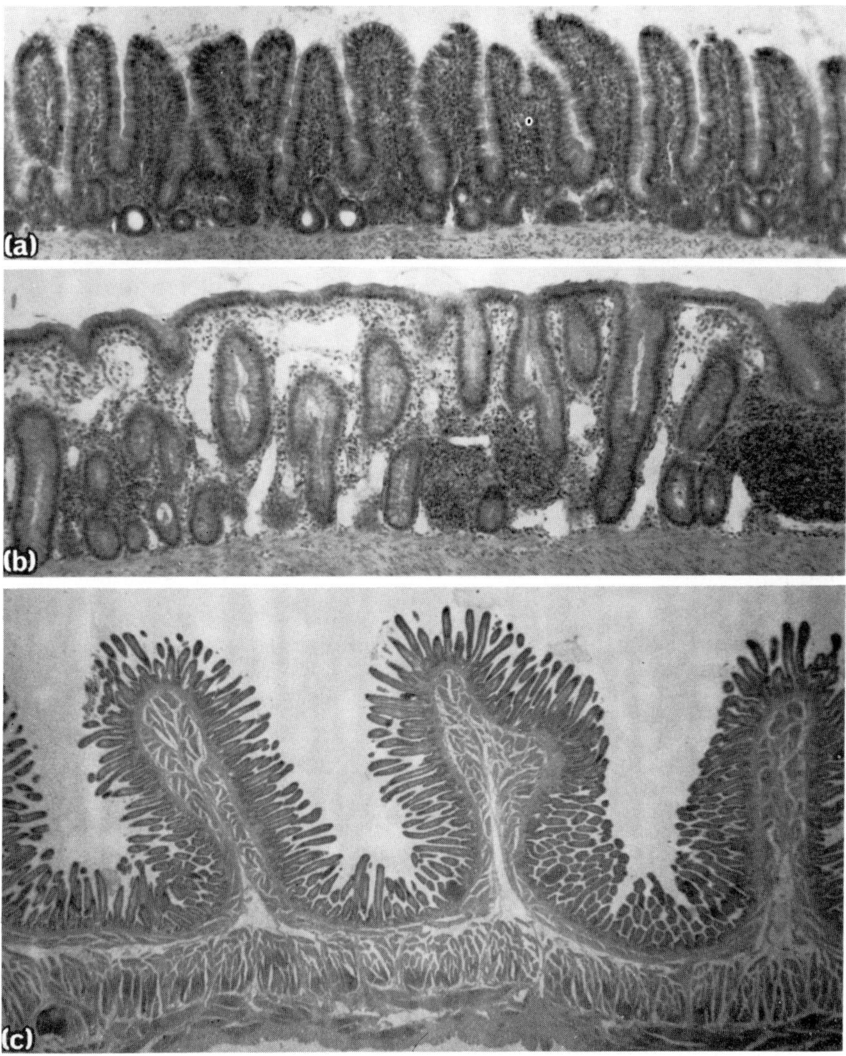

Fig. 4.26. Mucosa of the rectum. (a) Zebra Finch (*Poephila guttata*) showing the typical avian mucosa with prominent villi. ×100. (b) Galah (*Eolophus roseicapillus*) in which the villi are absent and the mucosa resembles that of the mammalian colon. ×100. (c) Emu (*Dromaius novaehollandiae*) with tall submucosal folds to which numerous villi are anchored. Such a pattern produces extensive surface area in the lower gut. ×8. All sections stained with haematoxylin and eosin. (From Johnson and Skadhauge, 1975).

VI. Physiological considerations

This section is not intended to be a thorough discussion of osmoregulation in birds. Extensive treatments of this topic have been published by Sykes (1971), Shoemaker (1972), Skadhauge (1973) and Sturkie (1976). Rather, the emphasis here will be upon possible correlations between structure and function in the kidney.

Numerous physiological studies concerned with avian water economy under varying conditions of saline stress or water deprivation have been published over the past few years (e.g. McNabb, 1969a,b; Moldenhauer and Wiens, 1970; Ohmart and Smith, 1970; Carey and Morton, 1971; Braun and Dantzler, 1972; Johnson and Ohmart, 1973a,b; Skadhauge, 1974a). Ecological adaptation is a major theme in these studies, since investigators have examined species (or subspecies) representing environments ranging from mesic to stressful desert or salt marsh habitats.

Relatively efficient osmoregulation has been demonstrated conclusively in birds from dry environments. Nonetheless, various findings have emerged in recent research which add complexity to the physiology of water economy in birds. Ward *et al.* (1975a,b) demonstrated that changes in diet, water availability, and feeding time cause *Gallus* to exhibit diurnal variations in the rate of urine flow and in urinary uric acid and ammonia concentrations. Moldenhauer and Taylor (1973) note that a diversity of diets has been employed in studies of avian water economy, including millet, mash, or combinations of these. Moldenhauer and Taylor showed that the Chipping Sparrow (*Spizella passerina passerina*), a bird of relatively moist environments, can mimic xerophilic species by surviving lengthy periods without water. However, this was possible only when the animals were on a millet diet; other individuals fed mash or mixed diets (which contain more nitrogen and salt than millet) invariably died. Such findings have obvious implications for the design of future experiments, and complicate comparisons of past research. Finally, there is the question of whether "laboratory xerophiles" truly reflect eco-physiological adaptations fundamental to the birds in nature. Such a point is implicit in the extensive field observations reported by Fisher *et al.* (1972). These investigators documented the natural daily drinking habits of 118 avian species found in the Australian desert. Many species were independent of water, but others which we tend to think of as classic xerophiles were not. For example, the authors state "It appears paradoxical that the Zebra Finch is one of the most widespread birds in arid parts of Australia since its drinking habits reveal it to be highly dependent on water, despite its ability to survive for long periods at moderate temperatures in the laboratory with little or no water". Careful consideration should be given to field data such as these in further studies of water economy in birds, particularly in the selection of appropriate experimental xerophiles.

Especially acute problems of water conservation exist in small birds which display an inherently high rate of pulmocutaneous water loss but lack nasal glands (Bartholomew and Cade, 1963). Certain physiological parameters of such taxa are summarized in Table II. As might be expected, the data represented by this wide range of studies are not uniform. For example, the ability to concentrate urine, as expressed in the relative osmolality of urine compared to that of plasma (U:P ratios), has been determined for some species but not for others. Overall it is impossible to compare the various taxa using the same basis throughout. Nonetheless, it is apparent (Table II) that some forms are relatively unspecialized and have essentially no tolerance toward either water deprivation or saline drinking water: examples are the House Finch (*Carpodacus mexicanus*) and Black-rumped Waxbill (*Estrilda troglodytes*). In contrast, other taxa such as the salt marsh subspecies of the Savannah Sparrow (*Ammodramus sandwichensis beldingi* and *rostratus*), the Black-throated Sparrow (*Amphispiza bilineata*), Brewer's Sparrow (*Spizella breweri*), Vesper Sparrow (*Pooecetes gramineus*), Zebra Finch (*Poephila guttata*) and Budgerigar (*Melopsittacus undulatus*) display considerable saline tolerance and/or the ability to withstand long periods without drinking water.

Before discussing kidney adaptations which appear to be characteristic of xerophilic birds, it is very important to recognize that there are other mechanisms which may be involved in meeting osmotic stress. Among these are the ability to tolerate the elevated plasma osmolality which accompanies dehydration, lessened pulmocutaneous evaporation rates during water deprivation and high ambient temperatures, behavioral adaptations to avoid periods of maximum heat stress, resorbtion of water from faeces and urine in the coprodeum and rectum, the presence of functional nasal glands, and unrecognized variations within a species in haemodynamics and osmotic transfer processes of the kidney. It is also important to note that substantial quantities of cations occur as urinary precipitates in birds (McNabb and McNabb, 1975). Unless these precipitates are completely dissolved when determining U:P ratios, the results will not reflect the total renal osmoregulatory ability. Some of the U:P ratios shown in Table II are likely to be too low.

As is apparent from the preceding paragraph, the kidneys of a xerophilic bird will not necessarily demonstrate obvious microstructural adaptations. Nonetheless, such modifications have been clearly defined in a number of taxa (Poulson, 1965; Johnson and Mugaas, 1970b; Johnson and Ohmart, 1973a,b; Johnson and Skadhauge, 1975). Table III is a summary of a large volume of research which relates to the correlation between structure and function in the avian kidney. Space does not permit a discussion of how such parameters as the ratio of mg cortex to medullary cone are estimated, but this information is available through references cited in the table. Two major points emerge from Table III. (a) The kidneys of highly effective water conservers contain two to three times *less* cortex per medullary cone than those of species with low effectiveness. Furthermore, there is no obvious tendency toward an increase in the length of the nephronal loops (as determined by measuring the length of medullary cones). (b) Among relatively inefficient water conservers there are fewer medullary cones per kidney and thus *more* cortex per cone. The cones tend to be slightly shorter than those of efficient species. Hence, we can generalize that there is a tendency for a greater number of medullary cones in species of birds displaying effective water conservation, a feature originally described by Poulson (1965). Concomitantly, each cone in such birds is associated with a relatively small quantity of cortex indicating high populations of mammalian-type nephrons (and thus greater potential for countercurrent hypertonicity in the medulla) and fewer reptilian-type nephrons.

As in mammals, the avian renal medulla contains a countercurrent multiplier system and an associated osmotic gradient (Poulson, 1965; Skadhauge and Schmidt-Nielsen, 1967; Braun and Dantzler, 1972; Emery *et al.*, 1972). Although there are certain exceptions to the general rule, concentrating ability in the mammalian kidney varies directly with the relative (not absolute) thickness of the medulla. This concept is an outgrowth of Sperber's (1944) classic work in which relative thickness of the renal medulla was found to be greatest in mammals adapted to arid environments. Sperber defined relative medullary thickness as the absolute medullary thickness (in mm) × 10/kidney size (in mm), where kidney size is the cube root of the product of the kidney's dimensions. Obviously, the latter product is an approximation of kidney volume. Schmidt-Nielson and O'Dell (1961) extended Sperber's observations and found "a close correlation between the urinary concentrating ability and the relative medullary thickness" in various mammals.

TABLE II. **Water economy in small birds.**

Species[a]	Current information on osmoregulatory abilities		
	Effects of water deprivation[b]	Max. saline regimen allowing weight maintenance (M NaCl)[b]	Urine/plasma ratios[b,c]
Budgerigar (1) (*Melopsittacus undulatus*)	Survival	0·2–0·3	2·3–2·7 (os)
Stark's Short-toed Lark (2) (*Calandrella starki*)	Survival	0·3	—
Grey-backed Finch-Lark (2) (*Eremopterix verticalis*)	Survival	0·3	—
Scaly-fronted Weaver (3) (*Sporopipes squamifrons*)	Survival	—	—
Vesper Sparrow (4) (*Pooecetes gramineus*)	Survival	0·25	—
Black-throated Sparrow (5) (*Amphispiza bilineata*)	Survival	0·4	—
Savannah Sparrow (6) (*Ammodramus sandwichensis brooksi*)	—	0·25	2·2 (os)
Savannah Sparrow (7) (*A. s. anthinus* and *A. s. nevadensis*)	Limited survival	0·4	—
Savannah Sparrow (6) (*A. s. beldingi*)	Limited survival likely	0·6–0·7	4·5 (os)
Savannah Sparrow (8) (*A. s. rostratus*)	Results vary[d]	0·6 and sea water[d]	—
Zebra Finch (9) (*Poephila guttata*)	Results vary[e]	0·2–0·8[e]	2·8 (os)
Brewer's Sparrow (10) (*Spizella breweri*)	Most birds survive	0·55	—
Chipping Sparrow (11) (*Spizella passerina passerina*)	Survival if fed millet	—	—
Seaside Sparrow (12) (*Ammodramus maritimus*)	—	0·3	3·5 (Cl)
Sharp-tailed Sparrow (12) (*Ammodramus caudacutus*)	—	0·3	3·5 (Cl)

Although the data are more difficult to secure in birds, the same parameters as described above can be obtained. The mean length of medullary cones (as shown in Table III) is analogous to the absolute thickness of the mammalian renal medulla. Volumes of irregularly contoured bird kidneys can be measured by fluid displacement and their cube roots calculated. Therefore, using these variables, one can calculate values for the intraspecific "medullary thickness" of birds, which are directly comparable to Sperber's (1944) data for mammals.

If the physiological significance of relative medullary thickness were the same in birds and mammals, one would expect to find such evidence in birds which are effective water conservers. However, results have been ambiguous, and largely the reverse of the mammalian pattern. In one study, Johnson (1974) found no obvious relationship between relative medullary thickness and the ability to conserve water (as reflected by survival without water, the ability to drink relatively concentrated saline solutions, or high U:P ratios) among 12

TABLE II. (*continued*)

Species[a]	Current information on osmoregulatory abilities		
	Effects of water deprivation[b]	Max. saline regimen allowing weight maintenance (M NaCl)[b]	Urine/plasma ratios[b,c]
Brown-headed Cowbird (13) (*Molothrus ater*)	—	0·3–0·4	2·5 (Cl)
Tree Sparrow (10) (*Spizella arborea*)	—	0·2	—
House Sparrow (14) (*Passer domesticus*)	—	0·3	—
Singing Honeyeater (15) *Meliphaga virescens*	—	—	2·4 (os)
House Finch (16) (*Carpodacus mexicanus*)	Death	0·25	2·1 (os)
Sage Sparrow (17) (*Amphispiza belli nevadensis*)	Death	0·25	2·4 (Cl)
Black-rumped Waxbill (18) (*Estrilda troglodytes*)	Death	0·15	—

[a] Water economy data based upon the following sources: (1) Cade and Dybas (1962), Greenwald *et al.* (1967), Krag and Skadhauge (1972); (2) Willoughby (1968); (3) Cade (1965); (4) Ohmart and Smith (1971); (5) Smyth and Bartholomew (1966); (6) Cade and Bartholomew (1959), Poulson and Bartholomew (1962a); (7) Cade and Bartholomew (1959); (8) Cade and Bartholomew (1959), Johnson and Ohmart (1973b); (9) Oksche *et al.* (1963), Calder (1964), Lee (1964), Cade *et al.* (1965), Skadhauge and Bradshaw (1974); (10) Ohmart and Smith (1970); (11) Moldenhauer and Taylor (1973); (12) Poulson (1969); (13) Lustick (1970); (14) Minock (1969); (15) Skadhauge (1974a); (16) Bartholomew and Cade (1956, 1958), Poulson and Bartholomew (1962b); (17) Moldenhauer and Wiens (1970); (18) Cade *et al.* (1965).

[b] Dashes indicate that no data are available.

[c] Expressed as chloride ratio (Cl) or total osmotic ratio (os).

[d] Cade and Bartholomew (1959) reported survival during hydropenia; Johnson and Ohmart (1973b) were unable to substantiate this ability. Only one-half of the test birds survived a sea water regimen (Johnson and Ohmart, 1973b).

[e] There are contradictory findings which probably relate to domesticated strains of birds used by some investigators. Many individuals survive hydropenia for long periods; wild-caught birds can utilize 0·7–0·8 M NaCl drinking solutions (Oksche *et al.*, 1963; Skadhauge and Bradshaw, 1974).

TABLE III. Relative effectiveness of water economy in small birds and some features of their renal microanatomy.

Species	Effectiveness[a]	Range in number of medullary cones/kidney[b]	Approx. mg cortex/medullary cone[b]	Mean length of medullary cones (mm)[b]
Song Sparrow (*Zonotrichia melodia juddi*)	Low (1)[c]	76–83[d]	1·7	1·4
Savannah Sparrow (*Ammodramus sandwichensis nevadensis*)	Low (2)	79–86	1·6	1·5
House Finch (*Carpodacus mexicanus*)	Low (3)	69–79	1·8	1·8
Singing Honeyeater (*Meliphaga virescens*)	Low (4)	63–72	1·7	1·8
Budgerigar (*Melopsittacus undulatus*)	Intermediate (5)	76–86	1·6	2·4
Vesper Sparrow (*Pooecetes gramineus*)	Intermediate (6)	97–110	1·3	1·7
Zebra Finch (*Poephila guttata*)	High (7)	116–130	1·1	1·6
Black-throated Sparrow (*Amphispiza bilineata*)	High (8)	124–134	0·9	2·1
Savannah Sparrow (*A. s. beldingi*)	High (9)	180–195	0·6	1·7
Savannah Sparrow (*A. s. rostratus*)	High (10)	156–190	0·5	2·4

[a] Based upon physiological abilities to tolerate dehydration, hydropenia, saline drinking solutions, etc.

[b] Data from Johnson and Mugaas (1970b), Johnson and Ohmart (1973a, b), Johnson and Skadhauge (1975).

[c] Relative effectiveness based upon information from the following sources: (1) Bartholomew and Cade (1963); (2) Cade and Bartholomew (1959); (3) Bartholomew and Cade (1956, 1958), Poulson and Bartholomew (1962b); (4) Skadhauge (1974a); (5) Cade and Dybas (1962), Greenwald *et al.* (1967), Krag and Skadhauge (1972); (6) Ohmart and Smith (1971); (7) Oksche *et al.* (1963), Calder (1964), Lee (1964), Cade *et al.* (1965), Skadhauge and Bradshaw (1974); (8) Smyth and Bartholomew (1966); (9) Cade and Bartholomew (1959), Poulson and Bartholomew (1962a); (10) Cade and Bartholomew (1959), Johnson and Ohmart (1973b).

[d] All values are proportional since calculations were made relative to a common base. This base was the mean weight of the largest kidneys found among the taxa listed (i.e. 0·15 g/kidney in both *Ammodramus sandwichensis beldingi* and *A. sandwichensis rostratus*).

passerine taxa and the Budgerigar (*Melopsittacus undulatus*). In another study involving several Australian species (Johnson and Skadhauge, 1975) there was an apparent correlation between medullary thickness and U:P ratios. Conflicting data representing the Savannah Sparrows (*Ammodramus sandwichensis*) further complicate the matter. For example, the inland races (*nevadensis* and *anthinus*) and the salt marsh race (*beldingi*) have nearly identical medullary thickness (Johnson, 1974), but the osmoregulatory abilities of *beldingi* far surpass those of either *nevadensis* or *anthinus*. In fact, *beldingi* represents the most efficient avian kidney studied to date, with a mean maximum total osmotic U:P ratio of 4·5 (Table II).

Overall, relative medullary thickness in birds ranges from about 0·6–5·4 (Johnson, 1974). The low value approximates minimal mammalian values (Sperber, 1944), but the high value is unimpressive among mammals where, for example, some desert rodents have relative medullary thickness values of about 12 (Dantzler, 1970). In birds from either desert or rigorous salt marsh environments the values range from around 3–5. It is important to note that the degree to which desert rodents can concentrate their urine is not always consistent with relative medullary thickness. Dantzler (1970) pointed out that some species are better concentrators than others despite reduced medullary thickness, and that two Australian rodent species with approximately the same medullary development differed markedly in their respective abilities to concentrate urine. Obviously, factors other than relative thickness (such as the efficiency of ionic transport) influence the total phenomenon. Current physiological understanding is inadequate to evaluate either the variation in relative medullary thickness found between birds and mammals, or among birds alone.

Thus, of the possible structural–functional relationships which have been studied in the avian kidney, the only satisfactory correlation is that between greater numbers of medullary cones per kidney and effective water conservation. Each cone in an effective conserver is associated with a relatively small quantity of cortex implying a high proportion of mammalian-type nephrons in the kidney.

An intriguing aspect of avian kidney function concerns the relative importance of mammalian-type and reptilian-type nephrons. Braun and Dantzler (1972) found a dramatic reduction in the number of functioning reptilian-type nephrons in the kidneys of Gambel's Quail (*Lophortyx gambelii*) undergoing osmotic stress through salt-loading. At the same time, their mammalian-type nephrons continued to function, thus maintaining countercurrent hypertonicity in the medulla. Comparable results were obtained upon the administration of arginine vasotocin, the naturally occurring antidiuretic hormone of birds (Braun and Dantzler, 1974). Apparently, arginine vasotocin causes constriction of the afferent arterioles of reptilian-type nephrons (Braun, 1976) which significantly reduces total glomerular filtration rate. Similar intermittent functioning of nephrons may exist in other avian species and be of fundamental significance in conserving water.

Acknowledgements

This chapter is dedicated with affection to my wife, Patricia, for her constant encouragement and support. I thank E. J. Braun and E. Skadhauge for their critical comments on the manuscript. Various portions of my research have been supported by grants from the National Science Foundation (GB-3844, GB-6865, GB-28665); and Moorhead State University research funds. Certain features of kidney structure were elucidated through studies carried out at the Mid-Pacific Marine Laboratory. Support was provided from Atomic Energy Commission Contract AT(29-2)-226, Hawaii Institute of Marine Biology. I am very grateful to the many individuals who have assisted in obtaining specimens over the past decade, with special recognition to J. R. Jehl, R. D. Ohmart and R. L. Zusi.

References

Akester, A. R. (1964). Radiographic studies of the renal portal system in the domestic fowl (*Gallus domesticus*), *J. Anat.* **98**, 365–376.

Akester, A. R. (1967). Renal portal shunts in the kidney of the domestic fowl, *J. Anat.* **101**, 569–594.

Akester, A. R., Anderson, R. S., Hill, K. J. and Osbaldiston, G. W. (1967). A radiographic study of urine flow in the domestic fowl, *Br. Poult. Sci.* **8**, 209–212.

Akester, A. R. and Mann, S. P. (1969). Adrenergic and cholinergic innervation of the renal portal valve in the domestic fowl, *J. Anat.* **104**, 241–252.

Atterbury, R. R. (1923). Development of the metanephric anlage of chick in allantoic grafts, *Am. J. Anat.* **31**, 409–437.

Barkow, H. (1829). Anatomisch-physiologische Untersuchungen vorzüglich über das Schlagadersystem der Vögel, *Arch. Anat. Physiol.* **12**, 305–496.

Bartholomew, G. A. and Cade, T. J. (1956). Water consumption of House Finches, *Condor* **58**, 406–412.

Bartholomew, G. A. and Cade, T. J. (1958). Effects of sodium chloride on the water consumption of House Finches, *Physiol. Zoöl.* **31**, 304–310.

Bartholomew, G. A. and Cade, T. J. (1963). The water economy of land birds, *Auk* 80, 504–539.

Baumel, J. J. (1975). Aves heart and blood vessels. *In* Sisson and Grossman's "The Anatomy of the Domestic Animals" (R. Getty, Ed.), Vol. 2. Saunders, Philadelphia.

Baumel, J. J., King, A. S., Lucas, A. M., Breazile, J. E. and Evans, H. E. (1979). "Nomina Anatomica Avium." Academic Press, London and New York (In press).

Beeuwkes, R. (1971). Efferent vascular patterns and early vascular-tubular relations in the dog kidney, *Am. J. Physiol.* **221**, 1361–1374.

Berger, C. (1966). Mikroskopische und histochemische Untersuchungen an der Niere von *Columba livia aberratio domestica*, *Z. mikrosk.-anat. Forsch.* **74**, 436–456.

Bortolami, R. and Palmieri, G. (1962). Anatomia macroscopica e microscopica degli ureteri in alcuni uccelli, *Riv. Biol.* **55**, 95–146.

Braun, E. (1976). Intrarenal blood flow distribution in the desert quail following salt loading, *Am. J. Physiol.* **231**, 1111–1118.

Braun, E. J. and Dantzler, W. H. (1972). Function of mammalian-type and reptilian-type nephrons in kidney of desert quail, *Am. J. Physiol.* **222**, 617–629.

Braun, E. J. and Dantzler, W. H. (1974). Effects of ADH on single-nephron glomerular filtration rates in the avian kidney, *Am. J. Physiol.* **226**, 1–8.

Cade, T. J. (1965). Survival of the Scaly-feathered Finch *Sporopipes squamifrons* without drinking water, *Ostrich* **36**, 131–132.

Cade, T. J. and Bartholomew, G. A. (1959). Sea-water and salt utilization by Savannah Sparrows, *Physiol. Zoöl.* **32**, 230–238.

Cade, T. J. and Dybas, J. A. (1962). Water economy of the Budgerygah, *Auk* **79**, 345–364.

Cade, T. J., Tobin, C. A. and Gold, A. (1965). Water economy and metabolism of two estrildine finches, *Physiol. Zoöl.* **38**, 9–33.

Calder, W. A. (1964). Gaseous metabolism and water relations of the Zebra Finch, *Taeniopygia castanotis*, *Physiol. Zoöl.* **37**, 400–413.

Calder, W. A. and King, J. R. (1974). Thermal and caloric relations of birds. *In* "Avian Biology" (D. S. Farner and J. R. King, Eds), Vol. 4. Academic Press, New York and London.

Carey, C. and Morton, M. L. (1971). A comparison of salt and water regulation in California Quail (*Lophortyx californicus*) and Gambel's Quail (*Lophortyx gambelii*), *Comp. Biochem. Physiol.* **39A**, 75–101.

Crile, G. and Quiring, D. P. (1940). A record of the body weights and certain organ and gland weights of 3,690 animals, *Ohio J. Sci.* **40**, 219–259.

Dantzler, W. H. (1970). Kidney function in desert vertebrates. *In* "Memoirs of the Society for Endocrinology, No. 18. Hormones and the Environment" (G. K. Benson and J. G. Phillips, Eds), Cambridge University Press, London.

Das, B. K. (1924). On the intra-renal course of the so-called "renal portal" veins in some common Indian birds, *Proc. zool. Soc. Lond.* **50**, 757–773.

Dolezel, S. and Zlabek, K. (1969). Über einen monoaminergen Mechanismus in Nierenpfortadersystem der Vögel, *Z. Zellforsch. mikrosk. Anat.* **100**, 527–535.

Edwards, J. G. (1940). The vascular pole of the glomerulus in the kidney of vertebrates, *Anat. Rec.* **76**, 381–389.

Emery, N., Poulson, T. L. and Kintner, W. B. (1972). Production of concentrated urine by avian kidneys, *Am. J. Physiol.* **223**, 180–187.

Ericsson, J. L. E. and Trump, B. F. (1969). Electron microscopy of the uriniferous tubules. *In* "The Kidney: Morphology, Biochemistry, Physiology" (C. Rouiller and A. F. Muller, Eds), Vol 1. Academic Press, New York and London.

Feinstein, B. (1962). Additional cases of bilobated kidneys in the hornbills, *Auk* **79**, 709–711.

Feldotto, A. (1929). Die Harnkanälchen des Huhnes, *Z. mikrosk.-anat. Forsch.* **17**, 353–370.

Fisher, C. D., Lindgren, E. and Dawson, W. R. (1972). Drinking patterns and behavior of Australian desert birds in relation to their ecology and abundance, *Condor* **74**, 111–136.

Fourman, J. and Moffat, D. B. (1971). "The Blood Vessels of the Kidney." Blackwell, Oxford.

Francis, E. T. B. (1964). Excretory system. *In* "A New Dictionary of Birds" (A. L. Thomson, Ed.). McGraw Hill, New York.

Gilbert, A. B. (1961). The innervation of the renal portal valve of the domestic fowl, *J. Anat.* **95**, 594–598.

Goodchild, W. M. (1956). Biological Aspects of the Urinary System of *Gallus domesticus* with Particular Reference to the Anatomy of the Ureter. M.Sc. thesis, University of Bristol.

Greenwald, L., Stone, W. B. and Cade, T. J. (1967). Physiological adjustments of the Budgerygah (*Melopsittacus undulatus*) to dehydrating conditions, *Comp. Biochem. Physiol.* **22**, 91–100.

Hinman, F., Murphy, W. K. and Weyrauch, H. M. (1939) An experimental study of uretero-intestinal implantation. II. The significance of the normal ureterocloacal arrangement in some reptiles and all Aves, *Surgery Gynec. Obstet.* **69**, 713–716.

Hodges, R. D. (1974). "The Histology of the Fowl." Academic Press, New York and London.

Huber, C. G. (1917). On the morphology of the renal tubules of vertebrates, *Anat. Rec.* **13**, 305–339.

Hüfner, C. G. (1866). Anatomie und Physiologie der Harnkanälchen. Med. Diss., Leipzig.

Hughes, M. R. (1970). Relative kidney size in nonpasserine birds with functional salt glands, *Condor* **72**, 164–168.

Hyrtl, M. (1846). Beiträge zur Physiologie der Harnsekretion, *Z. K. K. Ges. Ärzte, Wien.*

Hyrtl, M. (1863). Über die injektionen der Wirbeltiernieren und deren Ergebnisse, *Sitzungsber. Akad. Wiss. Wien Math. Naturw. Kl.* **47**, 146–204.

Imai, M. and Kokko, J. P. (1974). Sodium chloride, urea, and water transport in the thin ascending limb of Henle, *J. clin. Invest.* **53**, 393–402.

Inoue, H. (1953). An experimental study of the upper urinary tract and cloaca in the fowl, *Acta medica et Biologica* (*Niigata*) **1**, 127–136.

Ivanov, N. M. (1974). Zur Innervation der Harnwege der Wirbeltiere und des Menschen, *Anat. Anz.* **135**, 209–225.

Johnson, O. W. (1968). Some morphological features of avian kidneys, *Auk* **85**, 216–228.

Johnson, O. W. (1974). Relative thickness of the renal medulla in birds, *J. Morph.* **142**, 277–284.

Johnson, O. W. and Mugaas, J. N. (1970a). Some histological features of avian kidneys, *Am. J. Anat.* **127**, 423–435.

Johnson, O. W. and Mugaas, J. N. (1970b). Quantitative and organizational features of the avian renal medulla, *Condor* **72**, 288–292.

Johnson, O. W and Ohmart, R D. (1973a). The renal medulla and water economy in Vesper Sparrows (*Pooecetes gramineus*), *Comp. Biochem. Physiol.* **44A**, 655–661.

Johnson, O. W. and Ohmart, R. D. (1973b). Some features of water economy and kidney microstructure in the Large-billed Savannah Sparrow (*Passerculus sandwichensis rostratus*), *Physiol. Zoöl.* **46**, 276–284.

Johnson, O. W., Phipps, G. L. and Mugaas, J. N. (1972). Injection studies of cortical and medullary organization in the avian kidney, *J. Morph.* **136**, 181–190.

Johnson, O. W. and Skadhauge, E. (1975). Structural-functional correlations in the kidneys and observations of colon and cloacal morphology in certain Australian birds, *J. Anat.* **120**, 495–505.

Kaissling, B., de Rouffignac, C., Barrett, J. M. and Kriz, W. (1975). The structural organization of the kidney of the desert rodent *Psammomys obesus*, *Anat. Embryol.* **148**, 121–143.

King, A. S. (1975). Aves urogenital system. *In* Sisson and Grossman's "The Anatomy of the Domestic Animals" (R. Getty, Ed.), Vol. 2. Saunders, Philadelphia.

King, A. S. and McLelland, J. (1975). "Outlines of Avian Anatomy." Baillière Tindall, London.

King, J. R. and Farner, D. S. (1961). Energy metabolism, thermoregulation and body temperature. *In* "Biology and Comparative Physiology of Birds" (A. J. Marshall, Ed.), Vol. 2. Academic Press, New York and London.

Koike, T. I. and McFarland, L. Z. (1966). Urography in the unanesthetized hydropenic chicken, *Am. J. vet. Res.* **27**, 1130–1133.

Krag, B. and Skadhauge, E. (1972). Renal salt and water excretion in the Budgerygah (*Melopsittacus undulatus*), *Comp. Biochem. Physiol.* **41A**, 667–683.

Kriz, W. and Dieterich, H. J. (1970). The supplying and draining vessels of the renal medulla in mammals, *Proc. 4th Int. Congr. Nephrol.* **1**, 138–144.

Kriz, W. and Koepsell, H. (1974). The structural organization of the mouse kidney, *Z. Anat. EntwGesch.* **144**, 137–163.

Kriz, W., Schnermann, J. and Koepsell, H. (1972). The position of short and long loops of Henle in the rat kidney, *Z. Anat. EntwGesch.* **138**, 301–319.

Kudo, N., Takahata, K. and Sugimura, M. (1971). Das Blutgefässsystem der Nierenkörperchen in den Metanephren, *Jap. J. vet. Res.* **19**, 43–63.

Kurihara, S. and Yasuda, M. (1973). Comparative and topographical anatomy of the fowl LXXIII. Size and distribution of corpuscula renis, *Jap. J. vet. Sci.* **35**, 311–318.

Kurihara, S. and Yasuda, M. (1975a). Morphological study of the kidney in the fowl. I. Arterial system, *Jap. J. vet. Sci.* **37**, 29–47.

Kurihara, S. and Yasuda, M. (1975b). Morphological study of the kidney in the fowl. II. Renal portal and venous systems, *Jap. J. vet. Sci.* **37**, 363–377.

Kuroda, N. (1963). A fragmental observation on avian kidney, *Misc. Rep. Yamashina Inst. Orn. Zool.* **3**, 280–286.

Lasiewski, R. C. and Dawson, W. R. (1967). A re-examination of the relation between standard metabolic rate and body weight in birds, *Condor* **69**, 13–23.

Lee, C. P. (1964). Water Balance in the Zebra Finch (*Taeniopygia castanotis*). Doctoral dissertation, Duke University.

Lindgren, H. (1868). Ueber den Bau der Vogelnieren, *Z. rationelle Med.* **33**, 15–35.

Liu, Hin-Ching. (1962). The comparative structure of the ureter, *Am. J. Anat.* **111**, 1–15.

Longley, J. B., Burtner, H. J. and Monis B. (1963). Mucous substances of excretory organs: a comparative study, *Ann. N.Y. Acad. Sci.* **106**, 493–501.

Lustick, S. (1970). Energetics and water regulation in the Cowbird (*Molothrus ater obscurus*), *Physiol. Zoöl.* **43**, 270–287.

Marshall, E. K. (1934). The comparative physiology of the kidney in relation to theories of renal secretion, *Physiol. Rev.* **14**, 133–159.

Marshall, E. K. and Smith, H. W. (1930). The glomerular development of the vertebrate kidney in relation to habitat, *Biol. Bull., Woods Hole* **59**, 135–153.

McKelvey, R. W. (1963). The presence of a juxtaglomerular apparatus in non-mammalian vertebrates, *Anat. Rec.* **145**, 259–260.

McNabb, F. M. Anne (1969a). A comparative study of water balance in three species of quail—I. Water turnover in the absence of temperature stress, *Comp. Biochem. Physiol.* **28**, 1045–1058.

McNabb, F. M. Anne (1969b). A comparative study of water balance in three species of quail—II. Utilization of saline drinking solutions, *Comp. Biochem. Physiol.* **28**, 1059–1074.

McNabb, F. M. Anne, McNabb, R. A. and Steeves, H. R. (1973). Renal mucoid materials in pigeons fed high and low protein diets, *Auk* **90**, 14–18.

McNabb, R. A. (1974). Urate and cation interactions in the liquid and precipitated fractions of avian urine, and speculations on their physico-chemical state, *Comp. Biochem. Physiol.* **48A**, 45–54.

McNabb, R. A. and McNabb, F. M. Anne (1975). Urate excretion by the avian kidney, *Comp. Biochem. Physiol.* **51A**, 253–258.

Michel, G. and Junge, D. (1972). Zur mikroskopischen Anatomie der Niere bei Huhn und Ente, *Anat. Anz.* **131**, 124–134.

Minock, M. E. (1969). Salinity tolerance and discrimination in House Sparrows (*Passer domesticus*), *Condor* **71**, 79–80.

Moldenhauer, R. R. and Taylor, P. G. (1973). Energy intake by hydropenic Chipping Sparrows (*Spizella passerina passerina*) maintained on different diets, *Condor* **75**, 439–445.

Moldenhauer, R. R. and Wiens, J. A. (1970). The water economy of the Sage Sparrow (*Amphispiza belli nevadensis*), *Condor* **72**, 265–275.

Möllendorf, W. v. (1930). "Handbuch der Mikroskopischen Anatomie des Menschen", Vol. 7. Springer-Verlag, Berlin.

Mouchette, R. and Cuypers, Y. (1959). Etude de la vascularisation du rein de Coq, *Extrait Arch. Biol.* **69**, 577–589.

Nechay, B. R., Boyarsky, S. and Catacutan-Labay, P. (1968). Rapid migration of urine into intestine of chickens, *Comp. Biochem. Physiol.* **26**, 369–370.

Oelofsen, B. W. (1973). Renal function in the penguin (*Spheniscus demersus*) with special reference to the role of the renal portal system and renal portal valves, *Zoologica Africana* **8**, 41–62.

Oelofsen, B. W. (1977). The renal portal valves of the Ostrich (*Struthio camelus*), *S. Afr. J. Sci.* **73**, 57–58.

Ohmart, R. D., McFarland, L. Z. and Morgan, J. P. (1970). Urographic evidence that urine enters the rectum and ceca of the Roadrunner (*Geococcyx californianus*) Aves, *Comp. Biochem. Physiol.* **35**, 487–489.

Ohmart, R. D. and Smith, E. L. (1970). Use of sodium chloride solutions by the Brewer's Sparrow and Tree Sparrow, *Auk* **87**, 329–341.

Ohmart, R. D. and Smith, E. L. (1971). Water deprivation and use of sodium chloride solutions by Vesper Sparrows (*Pooecetes gramineus*), *Condor* **73**, 364–366.

Oksche, A., Farner, D. S., Serventy, D. L., Wolff, F. and Nicholls, C. A. (1963). The hypothalamo-hypophyseal neurosecretory system of the Zebra Finch (*Taeniopygia castanotis*), *Z. Zellforsch. mikrosk. Anat.* **58**, 846–914.

Pak Poy, R. K. F. and Robertson, J. S. (1957). Electron microscopy of the avian renal glomerulus, *J. biophys. biochem. Cytol.* **3**, 183–192.

Plakke, R. K. and Pfeiffer, E. W. (1964). Blood vessels of the mammalian renal medulla, *Science, N.Y.* **146**, 1683–1685.

Policard, A. and Lacassagne, A. (1910). Recherches histophysiologiques sur le rein des oiseaux, *C. r. Ass. Anat.* **12**, 57–65.

Poulson, T. L. (1965). Countercurrent multipliers in avian kidneys, *Science, N.Y.* **148**, 389–391.

Poulson, T. L. (1969). Salt and water balance in Seaside and Sharp-tailed Sparrows, *Auk* **86**, 473–489.

Poulson, T. L. and Bartholomew, G. A. (1962a). Salt balance in the Savannah Sparrow, *Physiol. Zoöl.* **35**, 109–119.

Poulson, T. L. and Bartholomew, G. A. (1962b). Salt utilization in the House Finch, *Condor* **64**, 245–252.

Prong, L. A., Bjoraker, D. G. and Harvey, R. B. (1969). Comparison of the renal medullary vascular systems of dog and chinchilla, *Microvascular Res.* **1**, 275–286.

Quiring, D. P. (1962). Organ weights: birds. *In* "Growth" (P. L. Altman and D. S. Dittmer, Eds), Fed. Amer. Soc. Exp. Biol., Washington, D.C.

Radu, C. (1975). Les fosses renales des oiseaux domestiques (*Gallus domesticus*, *Meleagris gallopavo*, *Anser domesticus* et *Anas platyrhynchos*), *Anat., Histol., Embryol.* **4**, 10–23.

Radu, C., Popesco, M. and Stoica, M. (1973). Quelques aspects microscopiques de l'innervation adrénergique et cholinergique des vaisseaux sanguins intra-rénaux et de l'urétére chez les oiseaux domestiques (*Gallus domesticus*, *Meleagris gallopavo*, *Anser domesticus* et *Anas platyrhynchos*), *Anat., Histol., Embryol.* **2**, 127–135.

Rennick, B. R. and Gandia, H. (1954). Pharmacology of smooth muscle valve in renal portal circulation of birds, *Proc. Soc. exp. Biol. Med.* **85**, 234–236.

Rensch, B. (1948). Organproportionen und Körpergrösse bei Vögeln und Säugetieren, *Zool. Jb.* **61**, 337–412.

Riboisiere, J. (1910). Recherches organometriques en fonction du régime alimentaire sur les oiseaux. *In* "Collection de Morphologie Dynamique" (F. Houssay, Ed.), Vol. 2. Hermann, Paris.

Romanoff, A. L. (1960). "The Avian Embryo." Macmillan, New York.

Rouiller, C. (1969). General anatomy and histology of the kidney. *In* "The Kidney: Morphology, Biochemistry, Physiology" (C. Rouiller and A. F. Muller, Eds), Vol. 1. Academic Press, New York and London.

Schmidt-Nielsen, Bodil and O'Dell, R. (1961). Structure and concentrating mechanism in the mammalian kidney, *Am. J. Physiol.* **200**, 1119–1124.

Shoemaker, V. H. (1972). Osmoregulation and excretion in birds. *In* "Avian Biology" (D. S. Farner and J. R. King, Eds), Vol. 2. Academic Press, New York and London.

Siller, W. G. (1971). Structure of the kidney *In* "Physiology and Biochemistry of the Domestic Fowl" (D. J. Bell and B. M. Freeman, Eds), Vol. 1. Academic Press, New York and London.

Siller, W. G. and Hindle, Ruth M. (1969). The arterial blood supply to the kidney of the fowl, *J. Anat.* **104**, 117–135.

Skadhauge, E. (1972). Salt and water excretion in xerophilic birds. Symp. Zool. Soc. London, No. 31.

Skadhauge, E. (1973). Renal and cloacal salt and water transport in the fowl (*Gallus domesticus*), *Dan. Med. Bull.* **20**, suppl. 1, 1–82.

Skadhauge, E. (1974a). Renal concentrating ability in selected West Australian birds, *J. exp. Biol.* **61**, 269–276.

Skadhauge, E. (1974b). Cloacal resorption of salt and water in the Galah (*Cacatua roseicapilla*), *J. Physiol., Lond.* **240**, 763–773.

Skadhauge, E. (1975). Renal and cloacal transport of salt and water. Symp. Zool. Soc. London, No. 35.

Skadhauge, E. (1976a). Cloacal absorption of urine in birds, *Comp. Biochem. Physiol.* **55A**, 93–98.

Skadhauge, E. (1976b). Water conservation in xerophilic birds, *Israel J. Med. Sci.* **12**, 732–739.

Skadhauge, E. and Bradshaw, S. D. (1974). Saline drinking and cloacal excretion of salt and water in the Zebra Finch, *Am. J. Physiol.* **227**, 1263–1267.

Skadhauge, E. and Schmidt-Nielsen, Bodil. (1967). Renal medullary electrolyte and urea gradient in chickens and turkeys, *Am. J. Physiol.* **212**, 1313–1318.

Smith, H. W. (1951). "The Kidney. Structure and Function in Health and Disease." Oxford University Press, New York.

Smyth, M. and Bartholomew, G. A. (1966). The water economy of the Black-throated Sparrow and the Rock Wren, *Condor* **68**, 447–458.

Sokabe, H. (1974). Phylogeny of the renal effects of angiotensin, *Kidney Internat.* **6**, 263–271.

Sokabe, H. and Ogawa, M. (1974). Comparative studies of the juxtaglomerular apparatus. *In* "International Review of Cytology" (G. H. Bourne, J. F. Danielli and K. W. Jeon, Eds), Vol. 37. Academic Press, New York and London.

Spanner, R. (1925). Der Pfortaderkreislauf in der Vogelniere, *Morph. Jb.* **54**, 560–632.

Sperber, I. (1944). Studies on the mammalian kidney, *Zool. Bidr. Upps.* **22**, 249–432.

Sperber, I. (1949). Investigations on the circulatory system of the avian kidney, *Zool. Bidr. Upps.* **27**, 429–448.

Sperber, I. (1960). Excretion. *In* "Biology and Comparative Physiology of Birds" (A. J. Marshall, Ed.), Vol. 1. Academic Press, New York and London.

Sturkie, P. D. (1976). Kidneys, extrarenal salt excretion, and urine. *In* "Avian Physiology" (P. D. Sturkie, Ed.). Springer-Verlag, New York.

Sutherland, L. E. (1966). Immunological and Functional Aspects of Juxtaglomerular Cells. Doctoral dissertation, Univ. of Toronto, Ontario.

Sykes, A. H. (1971). Formation and composition of urine. *In* "Physiology and Biochemistry of the Domestic Fowl" (D. J. Bell and B. M. Freeman, Eds), Vol. 1. Academic Press, New York and London.

Tchang, Li. Koue (1923). Recherches Histologiques sur la Structure du Rein des Oiseaux. Thesis, Lyon.

Tisher, C. C., Bulger, R. E. and Valtin, H. (1971). Morphology of renal medulla in water diuresis and vasopressin-induced antidiuresis, *Am. J. Physiol.* **220**, 87–94.

Van Tyne, J. and Berger, A. J. (1959). "Fundamentals of Ornithology." Wiley, New York.

Vilter, R. W. (1935). The morphology and development of the metanephric glomerulus in the pigeon, *Anat. Rec.* **63**, 371–385.

Waldeyer, A. (1931). Die Entwicklung der Vogelniere mit besonderer Berücksichtigung des Gefässsystems. Untersuchungen am Hühnchen. I. Teil, *Z. Anat. EntwGesch.* **96**, 723–765.

Ward, J. M., McNabb, R. A. and McNabb, F. M. Anne (1975a). The effects of changes in dietary protein and water availability on urinary nitrogen compounds in the rooster, *Gallus domesticus* —I. Urine flow and the excretion of uric acid and ammonia, *Comp. Biochem. Physiol.* **51A**, 165–169.

Ward, J. M., McNabb, R. A. and McNabb, F. M. Anne (1975b). The effects of changes in dietary protein and water availability on urinary nitrogen compounds in the rooster, *Gallus domesticus* —II. Diurnal patterns in urine flow rates, and urinary uric acid and ammonia concentrations, *Comp. Biochem. Physiol.* **51A**, 171–174.

Willoughby, E. J. (1968). Water economy of the Stark's Lark and Grey-backed Finch-lark from the Namib Desert of South West Africa, *Comp. Biochem. Physiol.* **27**, 723–745.

5
Female genital organs

A. B. GILBERT

Agricultural Research Council's Poultry Research Centre, King's Buildings, West Mains Road, Edinburgh, Scotland

---

CONTENTS

I. Introduction . . . 237

II. The egg . . . 239
  A. Central yolk-mass . . . 241
  B. Albumen . . . 250
  C. Shell . . . 251

III. Reproductive organs . . . 258
  A. Embryological formation of the reproductive organs . . . 261
  B. The left ovary . . . 268
  C. The right gonad . . . 302
  D. Accessory reproductive organs . . . 304

Acknowledgements . . . 337

References . . . 338

I. Introduction

The prime function of the reproductive organs in the sexually mature adult is to produce viable offspring and is accomplished in birds by the formation of a fertile egg.

The need to protect the embryo from the effects of an arid environment was a major problem during the evolutionary migration of vertebrates from water to land. Amphibians have not been entirely successful in this and most return to water to breed, although there are some exceptions, e.g. the Tree Frog (*Rhacophorus*). The other classes of land vertebrates have solved the problem either by producing a cleidoic egg, as in reptiles and birds, or by developing placental viviparity, as in most mammals. The advantage of the cleidoic egg (oviparity) is that the aqueous microenvironment of the egg is almost totally isolated from the exterior. Through evolution there has been a general trend towards a larger and more complex form of egg. This trend has reached its peak in the telolecithal or megalecithal eggs of modern reptiles and birds. The larger and more complex eggs become, however, the fewer that can be produced. Hence it is very important that a greater proportion of the embryos reach maturity, i.e. to the stage when they can survive as independent, free organisms. To this end certain fish, amphibians and reptiles have taken the protection of the embryo a stage further and the egg is retained inside the body until the young are about to hatch (ovoviviparity). In some of these vertebrates the production of a cleidoic egg has been abandoned and replaced by a form of true viviparity: yolk sac placentae and allanto-chorionic placentae are found in some reptiles, the vascular membranes being opposed to the maternal tissue of the oviduct (Matthews and Marshall, 1956). However, the large yolk sac must still be the major source of nutrition for the embryo. No true viviparity occurs in amphibians or fish but in two species of toad the embryos develop attached to highly vascular ridges in pouches of the skin and some bony fish have extensions of either the gills or the vascular system, which become associated with the ovary or ovarian follicles. For further consideration of viviparity see Matthews and Marshall (1956) and Hogarth (1976).

Unlike the other classes of vertebrates, birds have not evolved away from the purely egg-laying type of reproduction, possibly because "pregnancy" might disturb flight, although flying mammals (bats) apparently experience no difficulties (Bellairs, 1960). Instead of viviparity, birds have developed other methods, including incubation, to aid the survival of their embryos (Welty, 1962).

For all terrestrial animals, fertilization has to be internal. Consequently, complex organs have been developed in males for the introduction of semen into the female, and the female oviduct has evolved as an organ capable of sustaining, for a certain length of time, spermatozoa in a fertilizable state. This latter aspect is highly developed in birds, bats and some reptiles (Lake, 1975).

The reproductive organs are concerned therefore in carrying out several different functions, though these are interrelated. As in other vertebrates the ovary of a bird not only produces the female gamete but it also produces sex-steriod hormones which play such an important part in reproduction. The oviduct is directly involved in a number of functions including the survival of spermatozoa and their transport to the female gamete, the transport of the fertile zygote to the outside and the production of much of the shelled-egg.

In any consideration of reproduction in birds the domestic fowl dominates all other species since it is the only bird which has been systematically studied. Other birds have been investigated but these studies are usually, by comparison, relatively superficial. Notwithstanding this, a considerable amount of information on avian reproduction is available and a number of specialist reviews on the subject have appeared over the last half century, many of which are cited in the recent surveys by Jordanov (1969), Hodges (1974), King (1975) and Murton and Westwood (1977), etc. and in the relevant chapters in Bell and Freeman (1971) and Farner and King (1973).

II. The egg

Reconciliation between mammalian and avian terminology is often difficult. In mammals the "egg" is colloquially used as a synonym for both the ovum and the oocyte whereas usually in birds the "egg" is the separate, fully-shelled mass which is a product of both the ovary and oviduct. Unfortunately, there is no suitable scientific term for the egg in this sense and the use of "*ovum*" (the term adopted in the *Nomina Anatomica Avium*), apart from being of dubious validity, is also confusing. Throughout this text "egg" is used in its avian meaning.

The physical characteristics of the shelled-egg are hard to define precisely because they vary from species to species, from bird to bird within a species and even between eggs of one bird. These characteristics are affected by numerous hereditary, physiological and environmental factors (Romanoff and Romanoff, 1949; Romanoff, 1960; Aitken, 1971; Gilbert, 1971a,b,c,d,e; Gilbert and Wood-Gush, 1971; Simkiss and Taylor, 1971). Nevertheless, there have been many attempts to relate the shape, size and colour of eggs to an adaptive function but few have been entirely satisfactory.

Egg colour has been of interest, both popular and scientific, for many years, and an excellent, though unfinished, publication is that by Schönwetter (1960). Colours vary from plain white in many species to almost black in the Chilean Tinamou (*Nothoprocta perdicaria*). Blues, greens and browns predominate but even bright brick-red colours are known as in Cetti's Warbler (*Cettia cetti*). Speckling and streaking overlying the ground colour is common.

The reason why eggs are pigmented or not has aroused much controversy. One view involves the effects of solar radiation on the egg. Thus Montevecchi (1976) suggested that darkly-pigmented eggs suffer more harmful heating effects from sunlight than white eggs and hence white eggs should be more common amongst tropical species. In reality, however, the eggs of tropical birds tend to be darker than those produced by birds of less sunny climes. Possibly, as first suggested by Darwin, the shell pigment actually acts to shield the embryo from the harmful effects of solar radiation (Romanoff and Romanoff, 1949). Further support for this view is the fact that white eggs are produced by birds laying in dark places, e.g. holes, and by birds which cover their eggs. Similar observations were utilized by Welty (1962) to argue convincingly that white is the primitive colour of the eggs of birds (reptilian eggs are invariably white) and that pigmentation subsequently arose for other purposes. According to Lack (1958), the white colour of eggs evolved as an aid to recognition in dimly-lit places.

It must also be accepted that cryptic colouring of eggs occurs, Lack (1958) observing a relationship between egg colour and nesting site. However, Montevecchi (1976) and others found little evidence in some species that cryptic colouring affects the survival value of the eggs. Furthermore, highly camouflaged eggs are often laid in concealed nests, hence obviating the need for protective colouring (Romanoff and Romanoff, 1949). Some eggs have highly-coloured markings, and this led Abercrombie (1931) to suggest that the markings could act as a stimulus which attracts the incubating female to the nest and to sit. Similarly Welty (1962), when considering the experiments of Tschantz (1959) on the Common Murre (*Uria aalge*) which nests in large colonies and which lays a single egg on an exposed rocky ledge, suggested that the highly individualistic markings on the eggs are the means by which the female identifies her own egg or nest-site. The most highly adaptive egg colouration is possibly that of the Cuckoo (*Cuculus canorus*) and specific races of this species lay eggs with colourings and markings remarkably similar to those of their individual host species (Welty, 1962).

The term "egg-shaped" has passed into everyday speech and avian eggs are typically semi-ellipsoids with one pole (*polus obtusus*) slightly flattened and the other pole (*polus acutus*) more pointed (Fig. 5.1). The shape of the egg, however, may differ considerably from this general form. For example, the eggs of the Hooded Merganser (*Mergus cucullatus*) are almost spherical, those of sand-grouse (Pteroclididae) almost cylindrical, and those of grebes (Podicipedidae) extremely pointed at both ends and wide in the middle. Very pointed pear-shaped eggs are laid by plovers (Charadriidae) (Thomson, 1964) and this type of egg is often produced by species which lay four eggs in a nest, possibly to facilitate packing. The suggestion that pear-shaped eggs are also produced by birds nesting on rocky ledges, since this shape prevents the eggs rolling off the ledge (Welty, 1962), is only partly true, some species which nest on rocky ledges laying conventionally shaped eggs. There is some evidence that the shape of the egg is related to the shape of the pelvic bones. Thus species in which females have a dorsoventrally compressed pelvis tend to lay elongated eggs whereas those species in which the pelvis is deep lay eggs which are more spherically shaped (Rensch, 1947).

Within certain limits egg size is usually characteristic of the species (Murton and Westwood, 1977), although even races of the same species may differ considerably. The size is related approximately to the size of the parent and for any given species tends to get larger with increasing age (Romanoff and Romanoff, 1949; Welty, 1962). However, the eggs of precocial species (such as *Gallus*) tend to be relatively larger (10–15% of the maternal body weight) than those of altricial species (5%). The advanced development of precocial species at hatching is related to this. The two extremes in egg size are represented by the Brown Kiwi (*Apteryx australis*) and the Cuckoo (*Cuculus canorus*) which have eggs 25 and 3% respectively of the body weight (Thomson, 1964). The very small egg of *Cuculus* seems to be related to the fact that it parasitizes the nest of birds smaller than itself. The largest known avian egg belonged to the extinct Madagascan Elephant Bird (*Aepyornis*), measured 34×24 cm and had a capacity of over 9 l. According to Romanoff and Romanoff (1949), Welty (1962) and Thomson (1964) the largest egg of a living bird is that of the Ostrich (*Struthio camelus*) which measures $17 \times 13{\cdot}5$ cm, whilst the smallest egg is that of the Vervain Hummingbird (*Mellisuga minima*) which measures $1{\cdot}09 \times 0{\cdot}75$ cm and weighs only 0·5 g.

Despite the great variety in egg shape and size, the basic structure is remarkably similar throughout the class (Romanoff and Romanoff, 1949; Tyler and Simkiss, 1959b; Thomson, 1964; Tyler, 1964, 1965, 1966, 1969) perhaps because, apart from gaseous exchange, the egg must provide the microenvironment and the material needed for the production of a viable chick. Essentially the egg consists of three parts, known colloquially as "yolk", "white" and "shell" (Gilbert, 1971e). However, confusion can arise from the use of these terms because none of the structures they refer to is homogeneous. Furthermore, both "yolk" and "shell" have been used more specifically (Gilbert, 1971e). In the present text "white" is referred to by its correct term albumen, and "shell" is used in its more general sense as the composite structure consisting of the membranes, the calcified portion ("true-shell" or testa) and the cuticle. A more difficult problem is what term to apply to the yolk mass of the laid, fertile egg. To call it by the widely used term "ovum" is incorrect, and the use of "embryo" poses other problems. Furthermore, fertilization is not necessary for the production of eggs in birds; there is a similar mass in an infertile egg, which could lead to pedantic repetition if strict nomenclature is adopted throughout. In the absence of a satisfactory term I propose to use "central yolk-mass" for the colloquial "yolk", i.e. the structure which is derived from the primary oocyte (*ovocytus primarius*), containing the true yolk and the germinal disc, and which ultimately comes to lie at the centre of the fertile or infertile shelled egg (Fig. 5.1). A similar application of the term "yolk sphere" was used by Nickel *et al.* (1977) but this leads to confusion with the "yolk spheres" within the central yolk-mass.

Although the general structure of the egg is similar in all birds, the relative proportions of the various components are considerably dissimilar (Romanoff and Romanoff, 1949). When eggs are grouped according to the proportions of the central yolk-mass and albumen, they fall naturally into two classes. Those in which the central yolk-mass forms between 30–40% of the egg weight belong almost exclusively to the precocial species, whereas those in which it forms less than 20% belong to the altricial species which hatch as naked young (Romanoff and Romanoff, 1949). Similarly the proportion of albumen is greater and that of the shell is less in eggs of altricial and precocial species respectively (Romanoff and Romanoff, 1949).

There are also differences in detailed structure, but unfortunately wild birds have not been subjected to the scrutiny of the domestic species, e.g. *Gallus*, *Meleagris*, *Anser* and *Anas*, nor even of some of the semi-domesticated ones, e.g. *Coturnix* and game-birds. Though outside the scope here, there are many general descriptions of the chemistry and physico-chemistry of the eggs of domestic species (Romanoff and Romanoff, 1949; Bellairs, 1964; Williams, 1967; Baker, 1968; Cooke, 1968; Shenstone, 1968; Feeney and Allison, 1969; Jordanov, 1969; Gilbert, 1971e; Robinson, 1972).

The following description is concerned with the egg only up to the moment of oviposition: subsequent embryonic development is described by Romanoff (1960).

A. Central yolk-mass

Immediately after ovulation the central yolk-mass (Fig. 5.1) consists basically of the secondary oocyte (*ovocytus secondarius*) surrounded by the perivitelline layer (*lamina perivitellina*) (Figs 5.4, 17, 18). However, in *Gallus* this surrounding layer is insufficiently rigid to maintain the shape of the central yolk-mass (Fig. 5.2), although it is able to do so in *Columba*. In the laid egg the central yolk-mass is not exactly spherical, all three axes being slightly different (Romanoff and Romanoff, 1949).

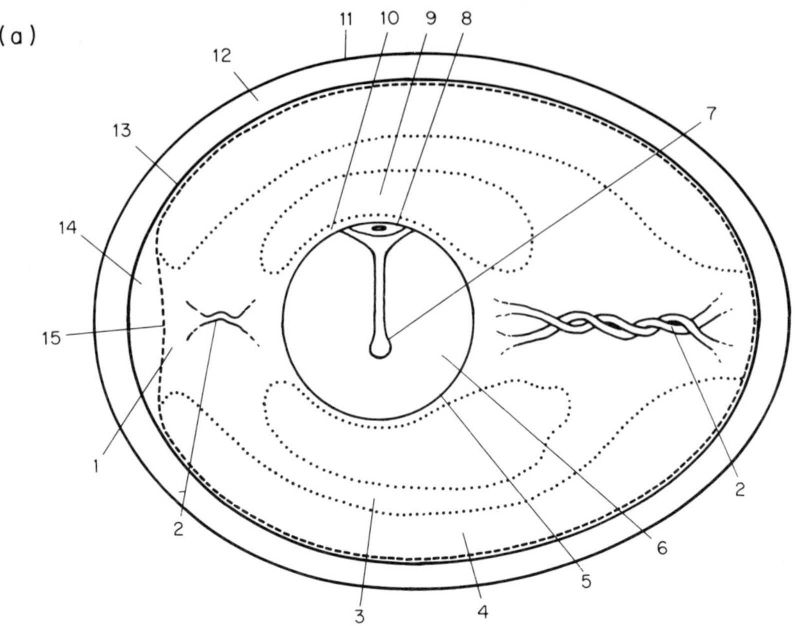

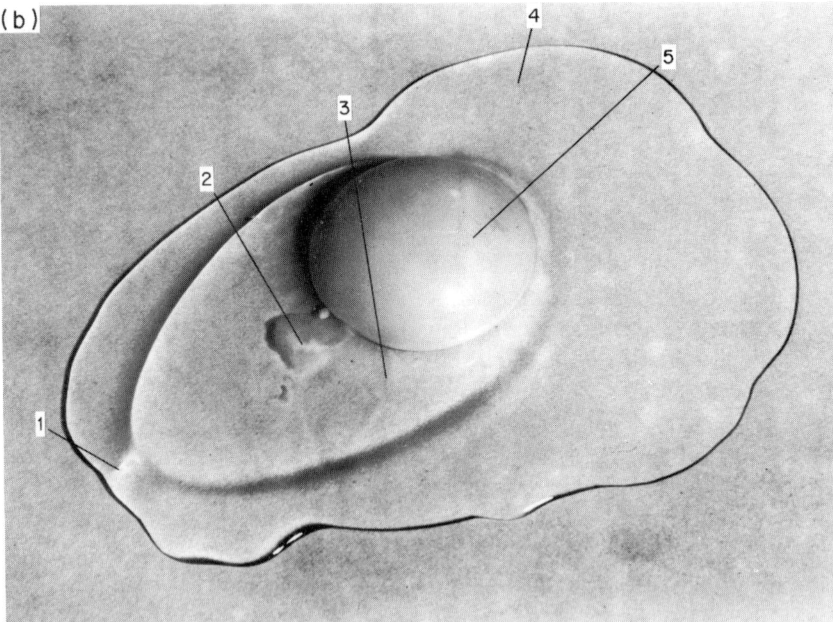

Fig. 5.1. General form of an egg: not all the structures in (a) are to scale. (Redrawn from Gilbert, 1971e). In (b) the two major compartments of the albumen (the outer thin and middle thick) are easily seen. (a): 1, "Ligament"; 2, chalazae; 3, middle thick white (albumen); 4, outer thin white (albumen); 5, yolk membranes; 6, yolk; 7, latebra; 8, germinal disc; 9, inner thin white (albumen); 10, chalaziferous layer; 11, cuticle; 12, calcified portion of shell; 13, outer shell membrane; 14, air space; 15, inner shell membrane. (b): 1, "Ligament"; 2, chalaza; 3, middle thick white (albumen); 4, outer thin white (albumen); 5, central yolk mass.

1. Germinal disc (discus germinalis)

After ovulation the germinal disc (*discus germinalis*) is visible as a small (about 3 mm in diameter in *Gallus*), creamy-white opaque area with dark and light rings (Fig. 5.2). Although it forms only a small part of the secondary oocyte, it identifies the animal pole. Being derived from the cytoplasmic part of the developing primary oocyte, it contains the cytoplasmic inclusions necessary for the maintenance of normal metabolic function and its outer border extends to form an exceptionally thin layer of cytoplasm covering the whole surface of the yolk. At this time the germinal vesicle is no longer clearly visible (Fig. 5.2), although the chromatid material is situated in the lighter central area (Fig. 5.3) (Romanoff and Romanoff, 1949).

The exact site in the body where fertilization occurs is not known for certain. Harper (1904) suggested penetration of the oocyte cytolemma (*cytolemma ovocyti*) by the spermatozoa in *Columba* occurred in the coelomic cavity, though Patterson (1910) favoured the infundibulum of the oviduct. Wherever it takes place, it must occur before the deposition of the albumen by the oviduct (Olsen and Neher, 1948; Bakst and Howarth, 1977b), and hence in *Gallus* (Olsen, 1942, 1952) and *Meleagris* (Olsen and Fraps, 1944) within about 15 min of ovulation. The question of the occurrence of polyspermy is outside the scope of this account (Romanoff, 1960).

The second maturation division occurs after penetration by the sperm (Romanoff, 1960): the second polar body (*polocytus secondarius*) is formed and the chromatid material which remains reconstitutes itself within a definitive nuclear membrane to form the female pronucleus (Olsen, 1942; Olsen and Fraps, 1944; Romanoff, 1960). Hence, as in mammals (Baker, 1972), the ovum (the mature female gamete of Henderson and Henderson, 1957) does not strictly exist as an independent structure. Even if it is accepted that the ovum is the stage after the female pronucleus is formed and before its fusion with that of the male, its life is of short duration. It is therefore unfortunate that the term "ovum" has been used so widely and indiscriminately in texts on avian reproduction, having been applied to the ovarian follicular contents (really the oocyte), to the mass of the fertile or infertile egg, i.e. the central yolk-mass in the present text, and even to the fully-formed, shelled egg.

Fusion of the male and female pronuclei (fertilization of Henderson and Henderson, 1957) occurs to form the zygotic nucleus. A characteristic feature of birds is that the female is the heterogametic sex (Boring and Pearl, 1914) so that the future sex of the offspring is determined before ovulation and not after fertilization as it is in mammals. Discussion of the difference in the sex chromosomes can be found in Romanoff (1960).

Considerable embryological changes occur during the passage of the egg down the oviduct. Division in all avian species is meroblastic (Fig. 5.10) (Romanoff, 1960), the structural details of the early divisions being provided by Gipson (1974) and Bellairs *et al.* (1978). The first cleavage division in *Gallus*, *Meleagris* and *Columba* occurs 3–5h after fertilization (Harper, 1904; Blount, 1909; Patterson, 1910; Olsen, 1942; Olsen and Fraps, 1944) when the egg is situated in the isthmus (Gilbert, 1971d). The second division in *Gallus* occurs rapidly, although in *Columba* it is later. According to Coste (1850) the second division

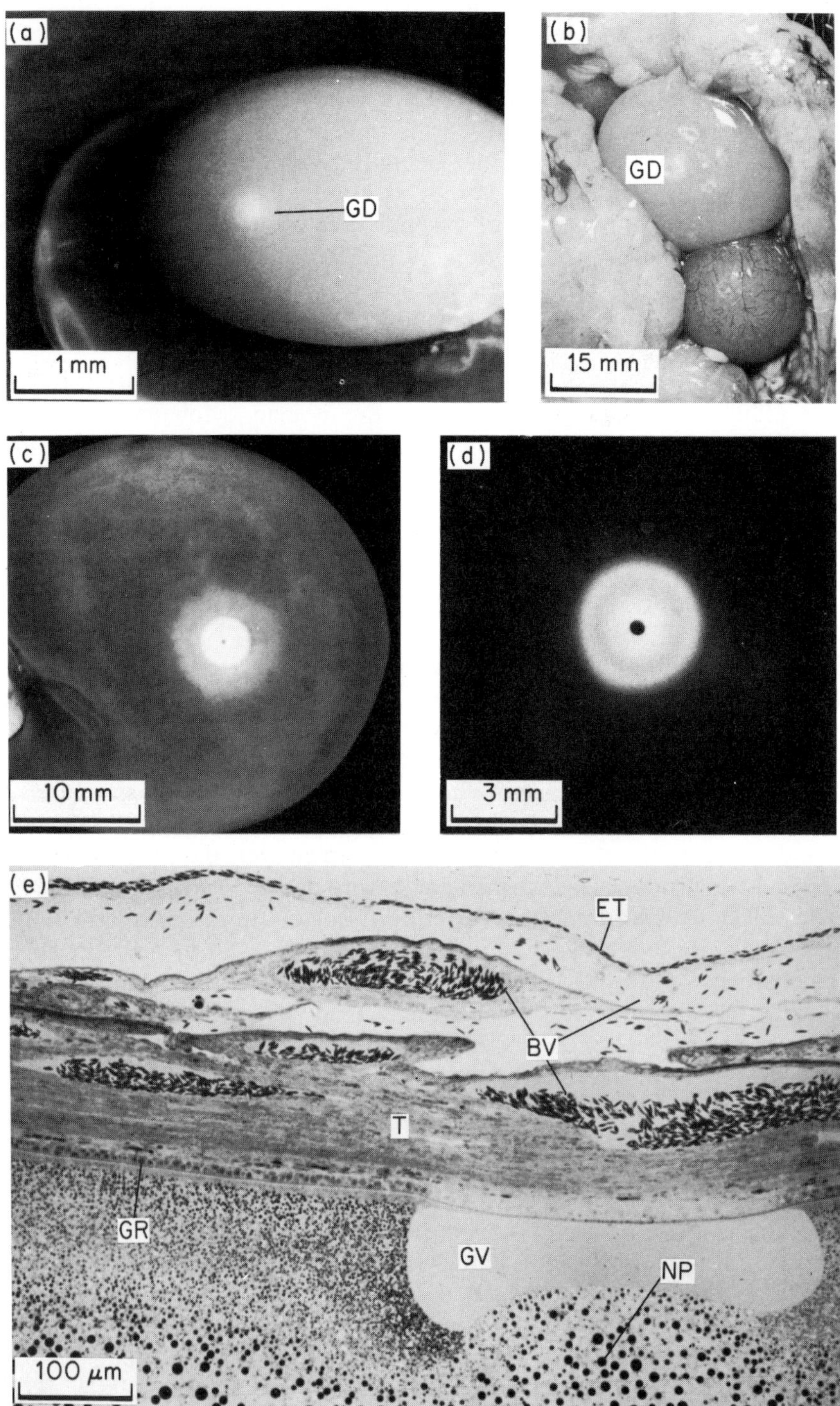

Fig. 5.2. General structure of the germinal disc (GD). (a) Blastodisc of the laid egg (infertile). (b) Germinal disc immediately after ovulation; the germinal vesicle is no longer present (compare with c). (c) The germinal disc from a pre-ovulatory oocyte about 20 h

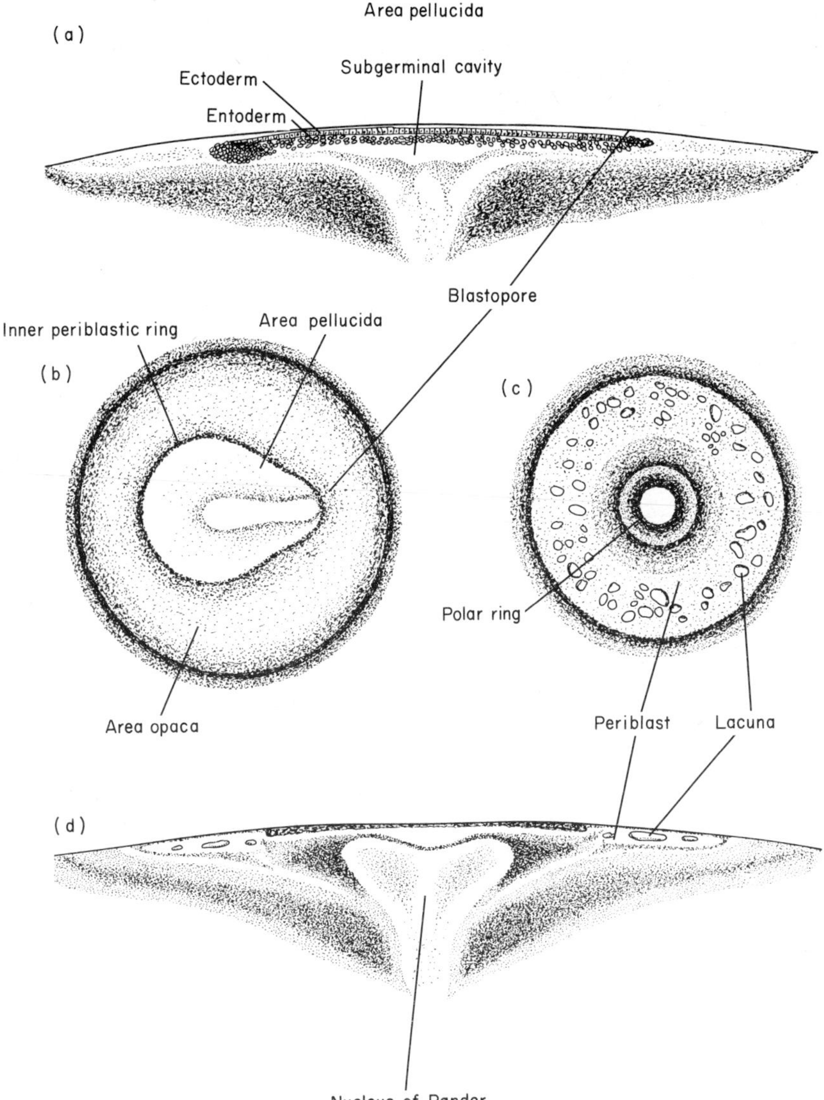

Fig. 5.3. The blastoderm: (A) vertical section, (B) surface view. The blastodisc: (C) surface view, (D) vertical section. After the germinal vesicle breaks down the female chromatid material lies in the central clear area of the blastodisc. (Redrawn from Romanoff and Romanoff, 1949).

---

before ovulation. The yolk has been coloured by feeding the bird a fat-soluble dye and the corona of "white" yolk surrounding the disc region stands out clearly against the darkened "yellow" yolk. The germinal vesicle is in the centre. (d) Details of the germinal disc from a follicle similar to (c). (e) Vertical section through the germinal vesicle. BV, blood vessels; ET, epithelium; GD, germinal disc; GR, granulosa; GV, germinal vesicle; NP, nucleus of Pander; T, theca. Araldite section, toluidine blue.

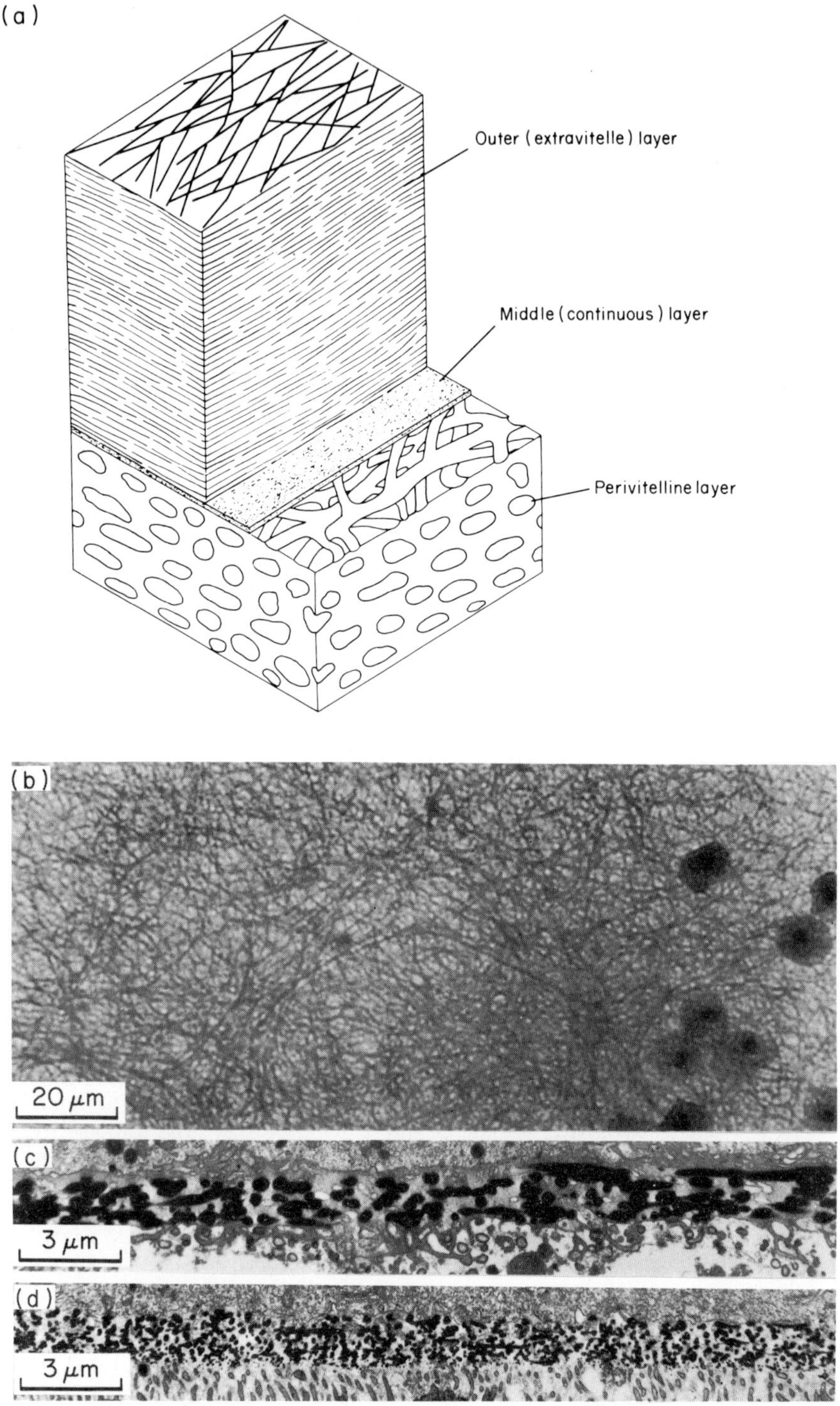

(a)

Outer (extravitelle) layer

Middle (continuous) layer

Perivitelline layer

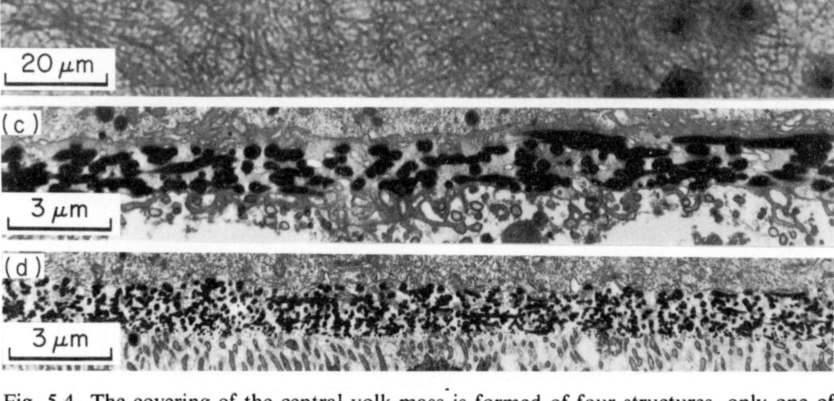

Fig. 5.4. The covering of the central yolk-mass is formed of four structures, only one of which (the oocyte cytolemma) is a true membrane. The other layers are produced by the ovary and oviduct: together they constitute the "yolk membranes". (a) Diagram of the "yolk membranes": the cytolemma is not shown. (Redrawn from Gilbert, 1971e and King

coincides with the first appearance of the shell membranes. The third and fourth divisions take place still within the isthmus so that the egg enters the shell-gland (uterus) at about the 16-cell stage, although great variations are possible (Romanoff, 1960).

Polyspermy is not uncommon. When it occurs the supernummerary male pronuclei also divide, setting up centres of meroblastic division outside the main one. However, these nuclei soon lose their ability to divide and subsequently degenerate (Romanoff, 1960).

Considerable division occurs in the shell gland and the appearance of the blastoderm in the laid egg is very different from the freshly-ovulated disc (Fig. 5.3). Instead of the ringed appearance at ovulation, two areas are distinguishable. These are a central *area pellucida* and a peripheral *area opaca*, which give rise to embryonic and extraembryonic structures respectively (Romanoff and Romanoff, 1949). The size of the blastoderm is about 4·5 mm in *Gallus* (Romanoff, 1960), 2 mm in *Anas* (Chen, 1932), 3·5 mm in the Common Tern (*Sterna hirundo*) and 1·6 mm in the Rook (*Corvus frugilegus*) (Mitrophanow, 1902). The embryo at this stage consists of a double layer of cells overlying the central sub-germinal cavity (Fig. 5.3), although this is probably not a true embryonic cavity (Pasteels, 1940, 1945). The embryo is orientated usually at approximately right-angles to the long axis of the egg with its anterior (cranial) end nearer the pointed end (Romanoff and Romanoff, 1949).

In contrast, little change occurs in the infertile egg, and the germinal disc (now termed blastodisc) (Figs 5.2, 3) still retains its ringed structure, although vacuolation may tend to obliterate this (Kosin, 1944). Since early dead embryos may also show vacuolation, it is often difficult to determine whether an egg at this time is infertile or not.

2. "*Yolk membranes*"

The boundary of the freshly-ovulated secondary oocyte consists of the original oocyte cytolemma (Fig. 5.18) and the covering perivitelline layer (the "inner membrane" of Bellairs *et al.*, 1963) (Figs 5.4, 17, 18) which is probably formed, though this has not been proved, from the granulosa cells of the ovarian follicle. Although earlier work with *Gallus* suggested that the oocyte cytolemma breaks down before ovulation (Bellairs, 1965), this is not so (Bakst and Howarth, 1977a; Perry, Evans and Gilbert, unpublished). The perivitelline layer is about 0·2–0·6 μm thick and is composed of a coarse network of long fibres arranged parallel to the surface of the oocyte with pore spaces of about 2 μm in diameter (Bellairs *et al.*, 1963; Bakst and Howarth, 1977a; Bakst, 1978). These spaces appear to be filled with an amorphous ground substance (Bakst and Howarth, 1977a; Bakst, 1978). In fertile eggs (Lake and Gilbert, 1964a; Fujii, 1976), as well as in oocytes treated with spermatozoa *in vitro* (Bakst and Howarth, 1977b),

1975). (b) Surface view of the perivitelline layer removed from a follicle 1 h before ovulation; a few granulosa cells remain attached. Light green and haematoxylin. (c) The perivitelline layer in section. (Perry, unpublished). (d) A similar section to (c) but in the region of the germinal disc (see also Fig. 5.18). (Perry, unpublished).

numerous spermatozoa can be found embedded within the perivitelline layer. It is interesting that the perivitelline layer in the region of the germinal disc differs from that of the remainder of the follicle (Bakst, 1978; Perry *et al.*, 1978b) (Figs 5.4, 18). Whether or not this difference is related to the penetration of the sperm is conjectural at present but it should be considered together with the observations of Fujii (1976) and Bakst and Howarth (1977b).

Soon after entering the oviduct the two outer layers, i.e. the continuous layer (*lamina continua*) and the extravitelline layer (*lamina extravitellina*) of the composite membrane (erroneously called the "vitelline membrane"), are laid down (Fig. 5.4). Their exact origin in the oviduct is not known, although the extravitelline layer may be derived from the chalaziferous region of the oviduct (Fig. 5.31). This layer (the "outer layer" of Bellairs *et al.*, 1963) in *Gallus* is between 3 and 8 μm thick and is composed of a delicate meshwork of fine fibrils arranged in concentric sub-layers. The exact number of these sub-layers is still not agreed. Whilst Jensen (1969) accepted the suggestion of Bellairs *et al.* (1963) that the number varies, Bain and Hall (1969) considered that there is an outer part which over-lies two fibrous sub-layers. The continuous layer (the "middle layer" of Bellairs *et al.*, 1963) (Fig. 5.4) is a sheet 50–100 nm thick, with little structure except for granules about 7 nm in diameter.

Since it is useful to have a term which can be applied to this composite boundary of the central yolk-mass in the laid egg, even although it is composed of several layers of different origin, King (1975) and Hodges (1974) proposed "yolk membranes". However, only one of these layers is a true cytological membrane. The thickness of the composite structure has been variously determined to be between 4 and 24 μm (Needham, 1931; Moran and Hale, 1936; Bellairs *et al.*, 1963; Bain and Hall, 1969). As pointed out by Hodges (1974), the discrepancies in thickness are probably due to shrinkage at fixation, the larger values being more likely to be correct for the fresh state.

3. Yolk

The detailed structure of the yolk (*vitellus*) has received considerable attention for many years, the earlier work being surveyed by Romanoff and Romanoff (1949) and Romanoff (1960). More recently several critical reviews have appeared on both its structure and chemical composition (Bellairs, 1964, 1967; Jordanov, 1969; Gilbert, 1971e; McIndoe, 1971; Hodges, 1974; Jordanov and Boyadjieva-Michailova, 1974).

Yolk has a definite structural organization, though at first sight its appearance may suggest otherwise (Figs 5.1, 2, 3, 5). The latebra (*centrum latebrae*), a small ball of white yolk (*vitellus albus*), is visible at the centre of the central yolk-mass. It is about 5 mm in diameter and represents the remains of the highly proteinaceous material laid down in the early stages of maturation of the oocyte (Fig. 5.5). Its mass is about 2% of the total yolk. Continuing from the latebra is a long neck (*collum latebrae*) of similar material which extends to the cone shaped disc (*discus latebrae*) (formerly nucleus of Pander) at the surface of the yolk-mass and on which floats the germinal disc (Fig. 5.5). Bellairs (1961, 1964, 1967) described the composition of this yolk material and found a continuous phase

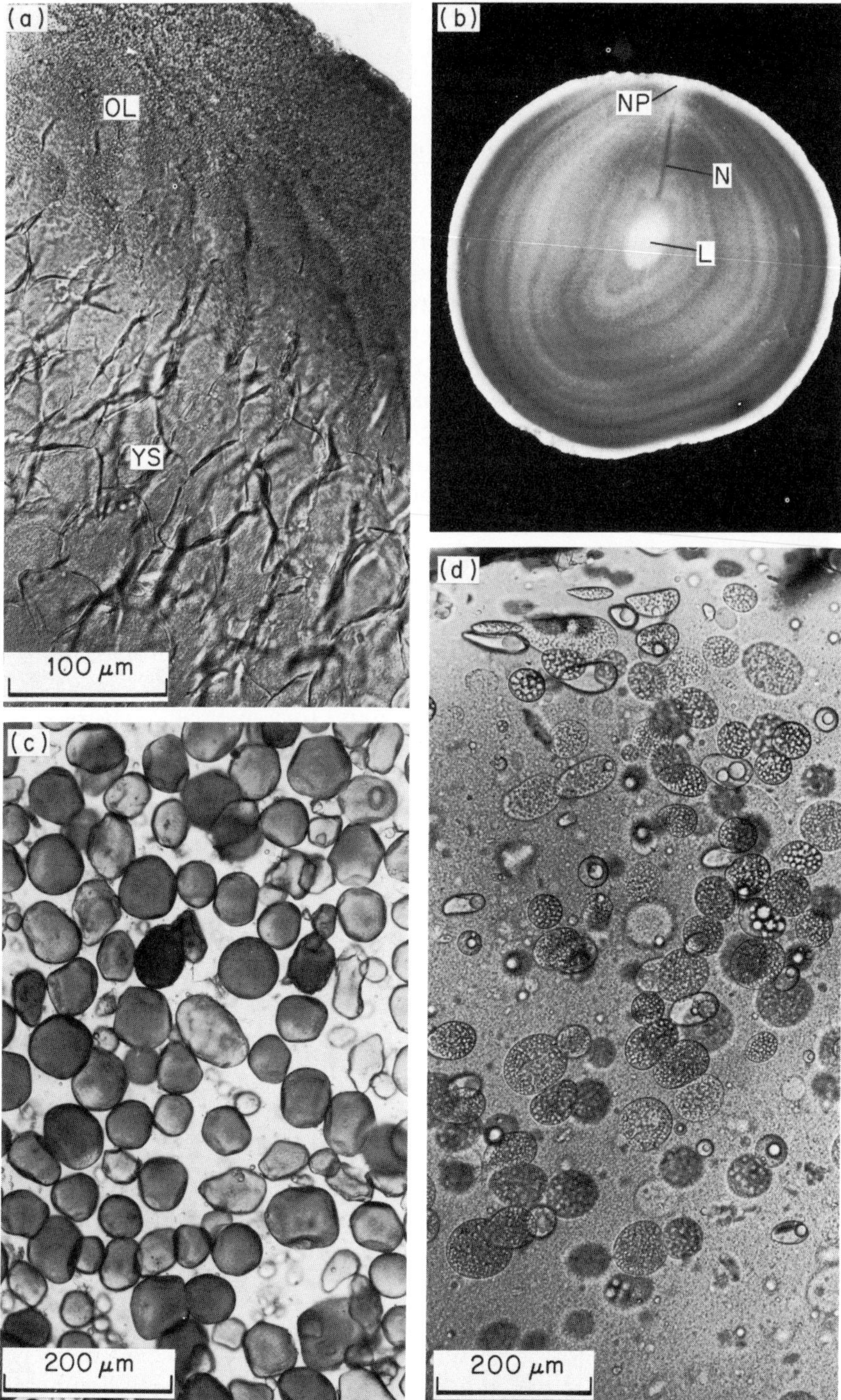

Fig. 5.5. Yolk is not homogeneous but varies considerably from one part to another (see also Fig. 5.2e). (a) Thick section of yellow yolk taken from the central yolk mass illustrated in (b). The outer layer (OL) is devoid of the characteristic yolk spheres (YS). (b) A slice through the central yolk-mass after potassium dichromate fixation to demonstrate the ring structure (Grau, 1976). L, latebra; N, neck; NP, nucleus of Pander. (c) Yolk spheres from dichromate fixed "yellow" yolk separated by agitation. (d) Untreated spheres from yellow yolk. The granular background is formed by disrupted spheres.

in which floated spheres between 4 and 75 μm in diameter, some of the spheres containing sub-droplets. Other contents included the so-called "yolk-spindles", "multivesiculate bodies" and "lining-body vesicles".

The remainder of the yolk consists of the yellow yolk (*vitellus aureus*), a material very different from the white yolk both in chemical composition and structure (Jordanov, 1969; McIndoe, 1971; Gilbert, 1971e). Basically yellow yolk consists of a mass of tightly-packed yolk spheres within an aqueous continuous phase (Fig. 5.5). Only at its outer borders, i.e. the area in direct contact with the vitelline membrane, does this pattern appear different, although no detailed description is available. Certainly within it are several different structures but their precise nature is unknown (Fig. 5.18).

Much has been made in *Gallus* of the concentric colour-stratification of yellow yolk after its first description by Thomson (1859). It is now certain that the dark and pale strata are artefacts of feeding and that they are related to the availability of carotenoid pigments (Gilbert, 1971e). To call them alternating layers of white and yellow yolk is incorrect and grossly misleading. However, yolk is not homogenous throughout. Recent work has shown that, when suitably treated with potassium dichromate, dark brown rings are produced (Fig. 5.5) (Grau, 1976) and similar, though not identical, rings follow treatment with potassium ferricyanide and alizarin red. These rings have been described in *Gallus*, *Coturnix* and the Canada Goose (*Branta canadensis*), and they may be a feature of avian eggs generally. Certainly all wild birds that have been examined so far have them (Grau, personal communication). Bellairs (personal communication) also noticed a layering in the yolk of microscopic droplets.

Yolk is composed of just over 50% solids, 99% of which are proteins. It is also rich in lipoproteins, 30% of the yolk being formed of lipo- and phosphoproteins (Gilbert, 1971e; McIndoe, 1971). Williams (1967) suggested that the composition of yolk may play a dominant role in embryogenesis and certainly it forms the main nutrient source of the developing embryo (Gilbert, 1971f). Since the embryonic yolk sac grows over the surface of the yolk, the nutritional future of the embryo is determined mainly before fertilization. However, despite extensive analytical investigation of yolk, it is not known for certain which of the components form part of the nutritional needs of the embryo and which are there by accident of the mechanisms for forming yolk. Thus, although fat-soluble dyes fed to chickens will be deposited in yolk, these cannot be regarded as normal nutritional requirements of the embryo (Gilbert, 1971e).

B. Albumen

Albumen or egg white forms about 65% of the mass of the egg in precocial species and rather more in the others. It is almost exclusively composed of protein and water (Gilbert, 1971e). Unlike the yolk, albumen has little intrinsic structure though it is divided into compartments (Fig. 5.1) (Romanoff and Romanoff, 1949; Gilbert, 1971e). Immediately surrounding the central yolk-mass is the chalaziferous layer (*stratum chalaziferum*) which is formed of a thick viscous material. During rotation in the oviduct some of the fibrous ovomucin strands of this region become twisted and condensed into the chalazae which lie in the long axis of the egg (Fig. 5.1). Two strands occur at the pointed end of the egg and one strand at the blunt end (Conrad and Phillips, 1938; Scott and Huang, 1941). These strands are a characteristic feature of the avian egg since they are not present in the eggs of reptiles.

The inner layer (*stratum internum*) of the thin (liquid) albumen (*albumen rarum*) extends outside the chalaziferous layer and, in turn, is surrounded by the middle thick dense layer (*albumen densum*) (Fig. 5.1) which is connected at each end of the egg to the shell membranes by the so-called ligament (*albumen polare*). Surrounding these layers, except at the ligaments, is the outer thin layer (*stratum externum*) of thin albumen (Fig. 5.1). The main difference between the various layers is the amount of water and ovomucin which they contain, the chalaziferous layer and the middle dense layer containing appreciably more ovomucin (Shenstone, 1968; Feeney and Allison, 1969). Possibly the inner thin albumen is formed during the twisting of the chalazae, and the outer thin albumen during the plumping (Gilbert, 1971c).

Albumen has at least three important roles. First, it forms a thick aqueous environment for embryonic development which, because of its high protein content, does not dry out easily. Second, many of its proteins have anti-bacterial properties (Board, 1966, 1968), particularly lysozyme and ovotransferrin, and therefore it forms the last defence against bacterial invasion. Finally, in many, if not all birds, it provides some additional nutritional material for the embryo (Sibley, 1960).

C. Shell

There are three parts to what is usually regarded as the shell, viz. the shell membranes, the testa or calcified portion and the cuticle (Fig. 5.6). Earlier references to shell structure have been collected by Romanoff and Romanoff (1949) and more recently Tyler has described the physical aspects of the shells of many species (Tyler and Simkiss, 1959a,b; Tyler, 1964, 1965, 1966, 1969).

1. *Shell membranes*

The shell membranes (*membranae testae*) are two tough, slightly elastic, whitish layers surrounding the albumen (Gilbert, 1971e; Candlish, 1972) and together are about 70 μm thick (Simons and Wiertz, 1965). Though designated membranes, the term is used in its anatomical sense not its cytological one (Fig. 5.6). Together the membranes form about 1 or 2% of the egg mass, although generally their thickness is inversely proportional to the shell thickness.

The inner membrane (*membrana testae interna*) rests on the surface of the albumen and, at the region of the ligaments, is penetrated by mucin fibres from the albumen. In *Gallus* it is about 20 μm thick and is possibly composed of three layers, with the fibres of the middle layer possibly being orientated at right-angles to those of the other two (but see Candlish, 1972). Individual fibres measure about 24 μm by 1 to 5 μm (Fig. 5.6).

The more complex outer membrane (*membrana testae externa*) is 50 μm

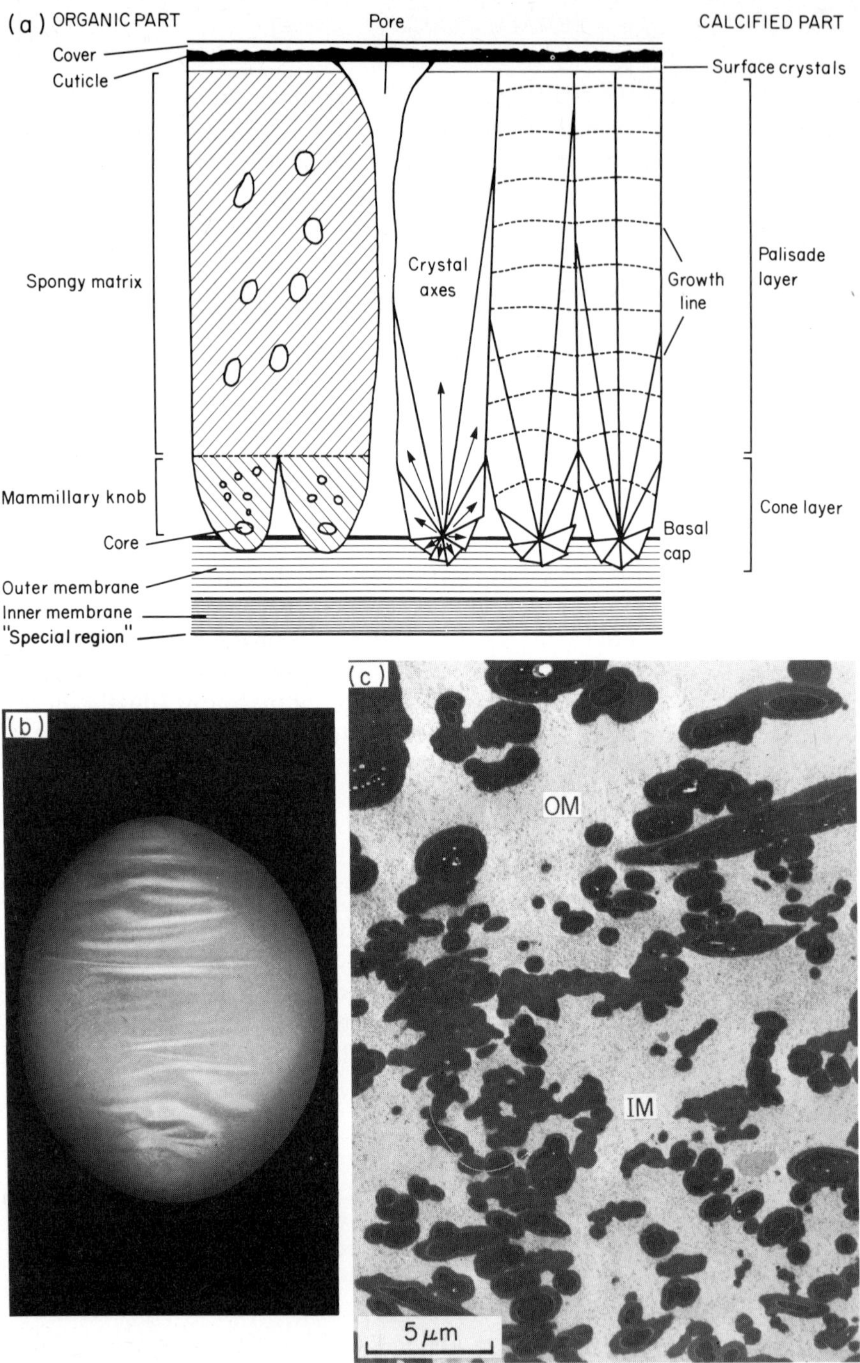

Fig. 5.6. (a) The shell is a complex structure consisting of an organic and an inorganic part, each of which may be subdivided further. However, details differ between species and not all structures are always present. Proportions may also differ from species to species. (Composite drawing from data of Simons and Wiertz, 1970, Simkiss and Taylor, 1971 and Hodges, 1974). (b) When the membranes are formed they are not closely applied to the developing egg but form a loose bag; this allows for the "plumping" process during which the albumen about doubles in volume. (c) Ultrastructural details of the shell membranes: the smaller fibres belong to the inner membrane (IM) and the others to the outer membrane (OM). (Johnston, unpublished).

thick and has shorter (15 μm) and thicker (3 μm) fibres (Fig. 5.6) than those of the inner membrane. Generally, however, fibre thickness is related to the size of the egg and the size of the bird (Becking, 1975). In *Gallus* the fibres may be arranged in six layers, with the fibres in alternate layers possibly orientated at right-angles to each other. The mesh diameter increases towards the outer surface (from 1 to 10 μm). Structurally the fibres appear to be very complex but essentially they are composed of a central proteinaceous core covered with a glycoprotein (Fig. 5.6) (Masshoff and Stolpmann, 1961; Simons and Wiertz, 1963; Bellairs and Boyde, 1969; Balch and Cooke, 1970; Candlish, 1972). However, little is known of the chemical nature of the fibres, although collagen may be present.

Recently Fujii *et al.* (1970) claimed to have identified a third membrane directly adjacent to the albumen and composed of a mucin-like sheet in which were embedded very fine fibres. This may be the "specialized" region (2·7 μm thick) described by Simons and Wiertz (1963) and Bellairs and Boyde (1969), and thought to be part of the inner membrane. Becking (1975) considered this layer to be peculiar to Galliformes.

Immediately after laying, as the egg cools, the membranes at the blunt end of the egg separate to form the air sac (*cella aeria*) (Fig. 5.1), a structure which is absent in reptiles. Later, during embryogenesis, the head comes to lie near this space. The size of the air sac is related to the size of the egg (Romanoff and Romanoff, 1949).

2. *Testa*

The calcified portion of the shell, the *testa*, is perhaps the best-known feature of the avian egg. In all species it forms a rigid support for the fluid contents of the egg (Fig. 5.6) and in some species, e.g. the Crested Francolin (*Francolinus sephaena*) is particularly hard. The testa contributes the most to the thickness of the shell, the extent to which it is developed being characteristic for the species. Relatively thick shells occur in birds which lay large eggs (Tyler and Simkiss, 1959b; Tyler, 1964, 1965, 1966, 1969); *Aepyornis* had a shell of about 4 mm thick whereas that of the eggs of hummingbirds (Trochilidae) is paper thin (60 μm) (Romanoff and Romanoff, 1949). However, shell thickness varies within a species, from egg to egg, and from point to point on the same egg. In *Gallus* it ranges from 270 to 370 μm (Carter, 1968). The testa contains very little water, 98–99 % of it being composed of solid material, mainly calcium carbonate (calcite) (Stewart, 1935; Romanoff and Romanoff, 1949; Terepka, 1963a; Heyn, 1963; Cain and Heyn, 1964). Only 2–3 % of the testa is organic and this is mainly protein (for further details see Gilbert, 1971e), some of the proteins having affinites with reptilian counterparts (von Krampitz *et al.*, 1974).

The following account of the avian testa is a generalized one and is based on the work of Romanoff and Romanoff (1949), Tyler and Simkiss (1959b), Masshoff and Stolpmann (1961), Heyn (1963), Simons and Wiertz (1963, 1966, 1970), Terepka (1963a,b), Tyler (1964, 1965, 1966, 1969), Fujii and Tamura (1969, 1970), Simons (1971), Meyer *et al.* (1973), Fujii (1974), Becking (1975) and Creger *et al.* (1976). The testa is composed basically of two parts, the organic matrix and the calcified deposit (Fig. 5.6). The organic matrix is organized into two separate regions, an inner part adjacent to the shell membranes called the mammillary layer (*stratum mamillarium*), and an outer part, the spongy layer (*stratum spongiosum*), adjacent to the cuticle. The mammillary layer forms about 20–30% of the thickness of the testa except in species with very thin-shelled eggs. It consists of numerous conical shaped knobs, the mammillae (Fig. 5.7). The apices of the mammillae are embedded in the outer shell membrane and, via the apices, fibres from the shell membrane enter the mammillae (Simons, 1971; Bunk and Balloun, 1977). During shell formation the bases of the mammillae fuse together to produce the framework on which the spongy layer is formed (Talbot and Tyler, 1974; Creger *et al.*, 1976). In the centre of each mammilla is a small protein mass (the mammillary core) (Fig. 5.6) which is associated with the fibres from the outer shell membrane (Simons and Wiertz, 1963). The size and shape of the mammillae appear to be characteristic for the species, since in even closely related species the mammillae are markedly different (Romanoff and Romanoff, 1949). The fibres of the mammillae are small and form a very fine meshwork with a mesh size of 100 nm. Embedded within this meshwork are small vesicles (Simons, 1971), larger cavities occurring in the bases of the mammillae (Simons and Wiertz, 1963). The matrix of the spongy layer is composed of fibres of up to 10 μm in length and 10 nm in thickness which are arranged parallel to the surface of the egg. These fibres branch and are associated with small vesicles (Simons and Wiertz, 1963; Simons, 1971). Although the matrix is similar in chemical composition to the organic part of the mammillae, there are slight differences (Simkiss, 1968; Cooke and Balch, 1970a,b). In some water birds and gulls (Laridae) (Romanoff and Romanoff, 1949) the matrix appears to be absent over narrow vertical and horizontal bands, although confirmation of this is required with the electron microscope.

Like the organic part, the crystalline inorganic mass is divisible into several regions (Fig. 5.6). The inner part consists of the basal caps and the cone layer, and corresponds to the mammillary layer. External to this is the palisade layer which forms the main mass of the testa and corresponds almost exactly to the spongy layer except at its outer surface where a thin covering, the surface crystal layer, is formed (Tyler, 1965; Simons and Wiertz, 1970; Simons, 1971; Becking, 1975). Formation of the organic part, it has been claimed, occurs from centres of crystallization in the mammillary cores (Schmidt, 1958a, 1962; Fujii and Tamura, 1969, 1970; Simons, 1971; Wyburn *et al.*, 1973; Fjuii, 1974), the indentations of the cores being associated with the radiating growth in the form of spherulites of the calcite crystals. The initial calcification has recently been shown to occur peripherally round the cores (Bunk and Balloun, 1977). The cones are formed by the outer rapidly-growing crystals which extend laterally until they make contact with adjacent crystals (Fig. 5.7). The palisade layer is essentially a continuation of the crystal formation in the cone layer and hence is formed of irregular "columns" of calcium (Fig. 5.6). The columns are neither single crystals nor homogenous (Terepka, 1963a; Simons, 1971) and they appear to develop first around the organic fibres of the spongy layer (Fujii, 1974). On completion the columns form compact masses of dendritic or spherulitic crystals (Heyn, 1963). In birds generally, vesicular holes appear in

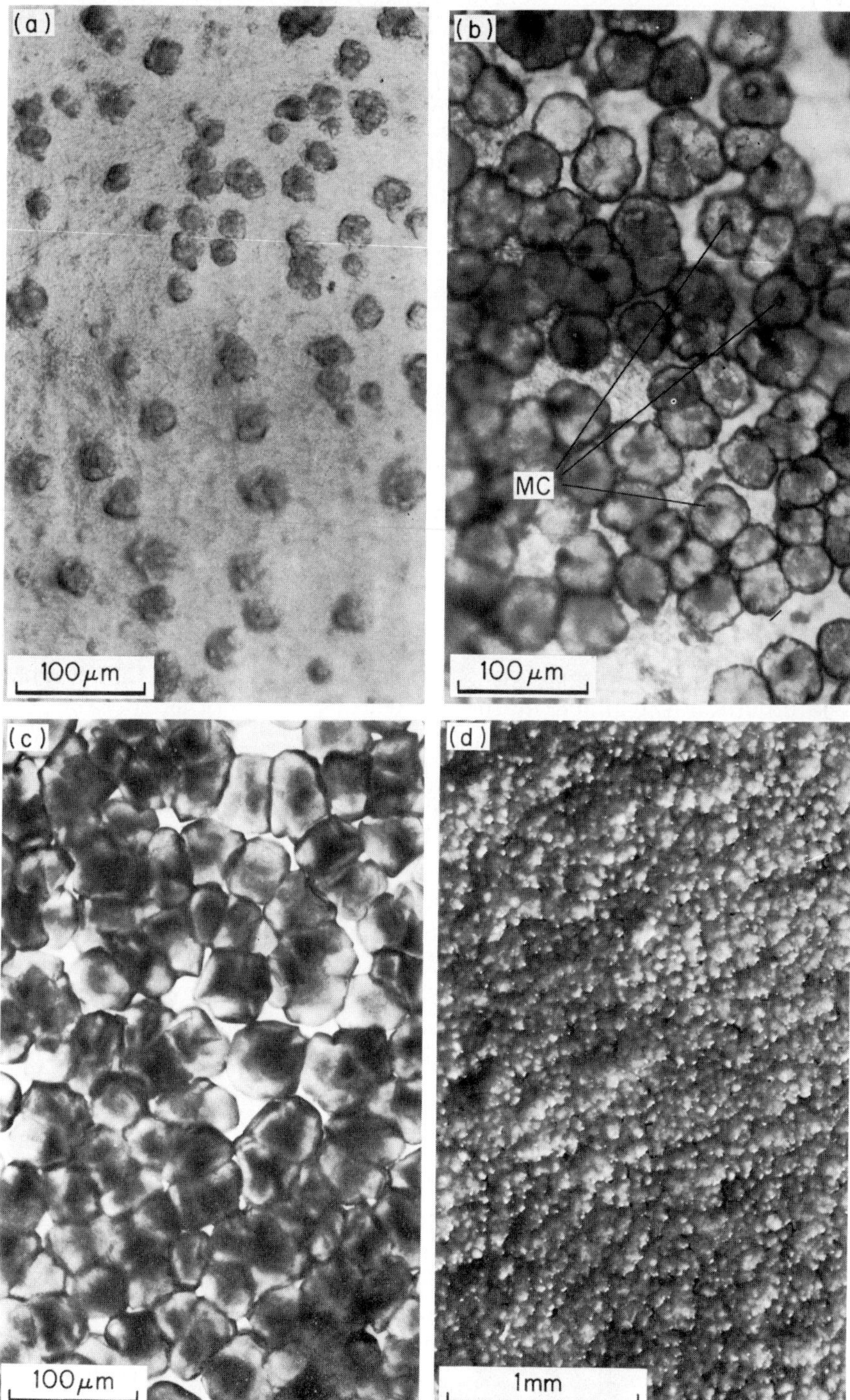

Fig. 5.7. Development of the calcification in the cone layer. (Davidson, unpublished). (a) First calcium carbonate deposit on some of the mammillary cores. Surface view of an egg which had just entered the pouch of the shell gland. (b) Further growth with some crystals just touching; the position of the original mammillary cores (MC) are visible in some. Haematoxylin. (c) Almost complete mosaic of crystals. (d) The completed cone layer at lower magnification. The palisade layer is formed on this.

the palisade layer in the region nearest to the shell membranes. The function of these holes is not known but, since they are largest in tropical species, Becking (1975) speculated that they may be concerned with gaseous exchange. Three types of calcium aggregates were identified in the testa of *Gallus* by Simkiss and Taylor (1971). Immediately outside the centres of spherulite formation, i.e. adjacent to the mammillary cores, radial aggregates, possibly containing traces of aragonite, occur (Erben, 1970). In the cone layer itself there are tubular aggregates. In the palisade layer the aggregates form a fish-bone pattern.

Although the description of the testa provided in this account appears to be applicable to most species which have been studied so far, interspecific differences in detail do occur. In the Emu (*Dromaius novaehollandiae*) and cassowaries (Casuariidae), for example, there appears to be an extra layer situated outside the spongy layer. This extra layer is rich in protein and in most parts is covered by a granular deposit (Thomson, 1964).

In most species small, oval to circular mushroom-shaped pores (*pori testae*) (Fig. 5.6) open on the surface of the testa (Tyler and Simkiss, 1959b; Tyler, 1964, 1965, 1966, 1969). Exceptions include the Emu (*Dromaius novaehollandiae*) and cassowaries (Casuariidae) (Romanoff and Romanoff, 1949; Tyler and Simkiss, 1959b). In *Gallus* there are between 7000 and 17 000 pores, the pores often being grouped more closely at the blunt end of the egg. Unfortunately, little information is available on the number of pores in other species. The size of the pore is generally larger in species with large eggs, auks (Alcidae) being an exception (Romanoff and Romanoff, 1949). In *Gallus* the diameter of the pore at its outer opening is between 15 and 65 μm. The pores are filled with a material similar to that of the cuticle. However, unlike in the cuticle, the pore material contains radial cracks through which gaseous exchange can occur (Tullett *et al.*, 1975). The pores connect with spaces, the canaliculi (*canaliculi testae*), which are between 6 and 20 μm in diameter and which pass through the whole depth of the testa to the membranes (Fujii, 1974). In *Gallus* a single canaliculus is usually associated with each pore, although recent evidence has shown that two or more canaliculi may be involved (Fujii, 1974). In the extinct Moa (*Diornis crassus*) and in the Ostrich (*Struthio camelus*) the canaliculi are branched (Von Nathusius, 1868, 1871; Tyler and Simkiss, 1959b). Branching also occurs in Anatidae, Tyler (1964) finding the branching to be more common in species with thicker shells.

Pigmentation occurs mainly throughout the calcified palisade layer of the shell where it forms the ground-colour of the egg; seldom is it found in the mammillary layer (Romanoff and Romanoff, 1949; Tyler, 1966, 1969). When it occurs on the surface of the cuticle it forms the characteristic mottling of the species. Porphyrins and biliverdin give brown and green-blue hues respectively (Kennedy and Vevers, 1976). In all species examined the pigments are deposited as crystals (Baird *et al.*, 1975) throughout the calcified region, although they are more heavily deposited in the outer parts. Gulls (Laridae) seem to be an exception since pigment occurs only in several bands which lie deep in the palisade layer (Romanoff and Romanoff, 1949). The outer pigment of falcons (Falconidae) and plovers (Charadrii) is so superficial that it can easily be rubbed off, yet in some falcons the deeper-lying pigment may extend even to the shell membranes (Tyler, 1966).

3. Cuticle

The outer covering of the egg, the cuticle (*cuticula*) (Fig. 5.6), imparts to the surface of the egg its characteristic texture and appearance. In general, most avian eggs have a slight sheen to them but this not universally so. Thus the eggs of penguins (Spheniscidae) and gannets (Sulidae) are chalky, in contrast to the highly glossy eggs of woodpeckers (Picidae) and babblers (Timaliinae). The Emu (*Dromaius novaehollandiae*) has ridged eggs whilst those of the Ostrich (*Struthio camelus*) are pitted. In the Anatidae the eggs have a greasy texture. Over the surface of the eggs of cormorants (Phalacrocoracidae) and grebes (Podicipedidae) is a thin chalky film while the eggs of flamingos (Phoenicopteridae) have a powdery surface overlying a layer rich in fat. The adaptive significance of the wide variability of the outer covering of eggs remains generally unknown.

The thickness of the outer covering of the egg varies considerably from species to species. In the Dalmatian Pelican (*Pelecanus crispus*) it is about 100 μm thick while in the Brown Kiwi (*Apteryx australis*) it is about 25 μm (Romanoff and Romanoff, 1949). In *Anas* the covering is only 3 μm thick. The covering is absent in some gulls (Laridae) and small birds (Romanoff and Romanoff, 1949; Becking, 1975). In the Anatidae the outer covering is formed of two or three layers (Tyler, 1964). Whilst Romanoff and Romanoff (1949) imply that three layers occur in all birds, this has been questioned, at least for *Gallus* (Simons and Wiertz, 1963, 1966).

In view of this variability in the outer covering, it is debatable if it is a homologous structure throughout birds generally and hence whether one term should be used for it. Simons (1971) proposed that "cuticle" should be applied only to the organic layer. The extra powdery deposit which occurs in a number of species (Schmidt, 1958b; Tyler, 1964, 1969), including some broiler domestic fowl, he called "cover" (Fig. 5.6). Sometimes the cover may lie beneath the cuticle, e.g. in grebes (Podicipedidae) and pelicans (Pelecanidae) (Tyler, 1969).

The true cuticle has been most studied in *Gallus* in which it is a continuous layer (about 10 μm thick) over the whole shell, including the pores, although varying from egg to egg and from point to point on the egg (Simons and Wiertz, 1963). Part of the cuticle has a foamy structure, the other part being more compact (Simons and Wiertz, 1963, 1966, 1970; Simons, 1971). Chemically it is an organic layer composed of about 90% peptide with galactose, fucose, mannose and hexosamine (Simkiss, 1958; Baker and Balch, 1962; Cooke and Balch, 1970b), and richer in tyrosine, lysine and cystine than the matrix of the testa (Simkiss and Taylor, 1971). In some species it contains lipid material (Romanoff and Romanoff, 1949; Simkiss, 1958; Hasiak *et al.*, 1970).

The cuticle, together with the testa and the membranes, provides not only the mechanical strength to support the fluid internal components but also acts as a deterent to bacterial infection. Only 1·0% of the pores of the egg of the domestic fowl are considered to be penetrable by bacteria (Board, 1966, 1968; Board and Board, 1967). The cuticle is also responsible for conferring on the egg its water-repellant properties (Board and Halls, 1973a,b; Board, 1974) as well as reducing water loss, an important feature for terrestrially breeding animals.

III. Reproductive organs

In all vertebrate classes the reproductive organs arise as bilateral primordia and this symmetry usually persists into the adult. However, fusion of the early primordia to form a single ovary occurs in lampreys and some teleosts (Franchi *et al.*, 1962), though fusion may be incomplete in the latter. In some elasmobranchs, myxinoids, reptiles, monotremes and bats (Okkelberg, 1921; Garde, 1930; Matthews, 1937, 1950; Chieffi, 1955, 1959) the early symmetrical pattern is not maintained and one ovary fails to develop.

Failure of one ovary to develop is also typical of birds, the characteristic pattern being for only the left ovary (*ovarium sinistrum*), and the left oviduct (*oviductus sinister*), to develop fully and to become functional in the adult (Fig. 5.8). On the right side the embryonic gonadal tissue (*gonadum dextrum*) usually fails to develop further, remaining small and non-functional and, significantly, containing little ovarian tissue. The right oviduct (*oviductus dexter*) is usually no more than a minute vestige. However, not all species follow this typical pattern. The normal vertebrate pattern of two ovaries is a common, though not universal, feature found in birds of prey such as eagles, hawks and buzzards (Accipitridae), falcons (Falconidae) and New World vultures (Cathartidae), a fact so well known as to be included in most accounts of avian reproduction (Groebbels, 1937; Van Tyne and Berger, 1958; Marshall, 1961; van Tienhoven, 1961; Welty, 1962; Darling and Darling, 1963; Thomson, 1964; Gilbert, 1967, 1971b; King, 1975). What is not so well-known is that the bilateral condition occurs quite frequently in individuals outside the birds of prey (Fig. 5.9). Thus there is evidence of two ovaries in many species belonging to at least 16 orders in which the norm is usually accepted to be one ovary. In some species the frequency with which two ovaries occur is quite high (Kummerlöwe, 1930a,b, 1931a,b; Groebbels, 1937; Kinsky, 1971) and in the Brown Kiwi (*Apteryx australis*) two functional ovaries are present normally (Kinsky, 1971). Moreover, as Kinsky (1971) pointed out, the occurrence of bilateral symmetry in the ovaries may be more common than is usually accepted because post-mortem examination often tends to concentrate on the left side only, it being generally assumed that the right ovary is absent. Double oviducts are undoubtedly more rare. Again they predominate in Falconiformes, but have been found in Galliformes, Ciconiiformes, Anseriformes, Ralliformes, Columbiformes and Psittaciformes (Kinsky, 1971).

No satisfactory explanation has been given for the tendency towards the evolution of only single ovaries in birds. Welty (1962) and Darling and Darling (1963) suggested it might be a weight-saving adaptation to flight but this cannot unreservedly be accepted in view of the frequent occurrence of two ovaries in birds of prey. It is even more difficult to reconcile the relatively common occurrence of two ovaries with the rarity of two oviducts. Van Tienhoven (1961) suggested that any oocyte from the right ovary would be unlikely to enter the left oviduct and hence could not be formed ultimately into an egg. When such "internal laying" occurs, i.e. the entry of the oocyte into the coelom, it is not detrimental (Sturkie, 1955; Wood-Gush and Gilbert, 1965a), although it does not appear to convey any evolutionary advantage to the animal. However unlikely the crossing of oocytes from right to left seems to be, the evidence of

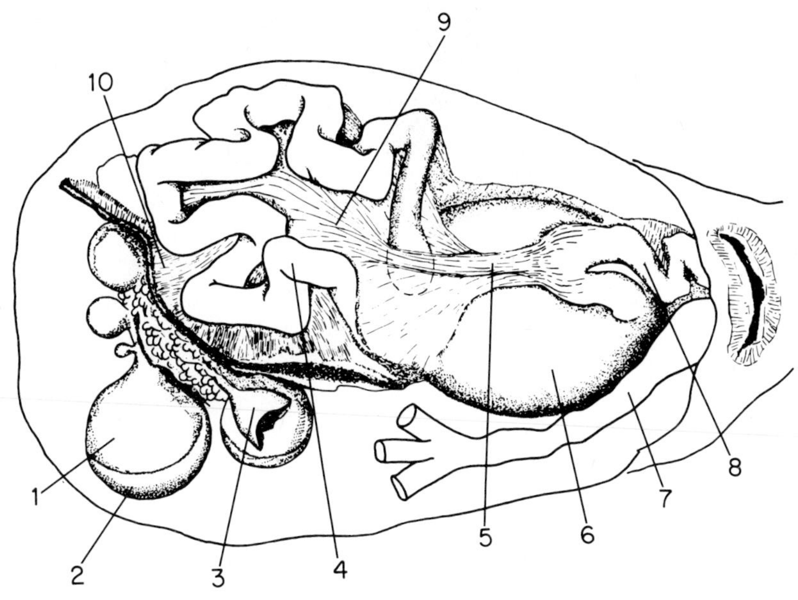

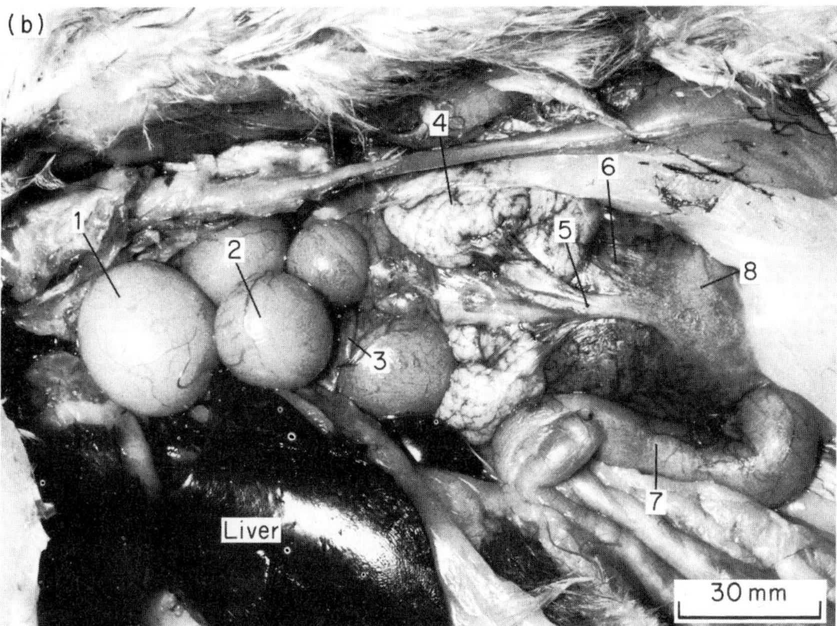

Fig. 5.8. General anatomical relationships of the reproductive organs in the abdominal cavity of the domestic fowl. The diagram is redrawn from King (1975). (a): 1, Follicle; 2, stigma; 3, post-ovulatory follicle; 4, magnum; 5, muscular cord; 6, shell gland; 7, gut; 8, vagina; 9, ventral ligament; 10, infundibulum. (b): 1, Follicle; 2, stigma; 3, post-ovulatory follicle; 4, magnum; 5, muscular cord; 6, shell gland; 7, gut; 8, vagina.

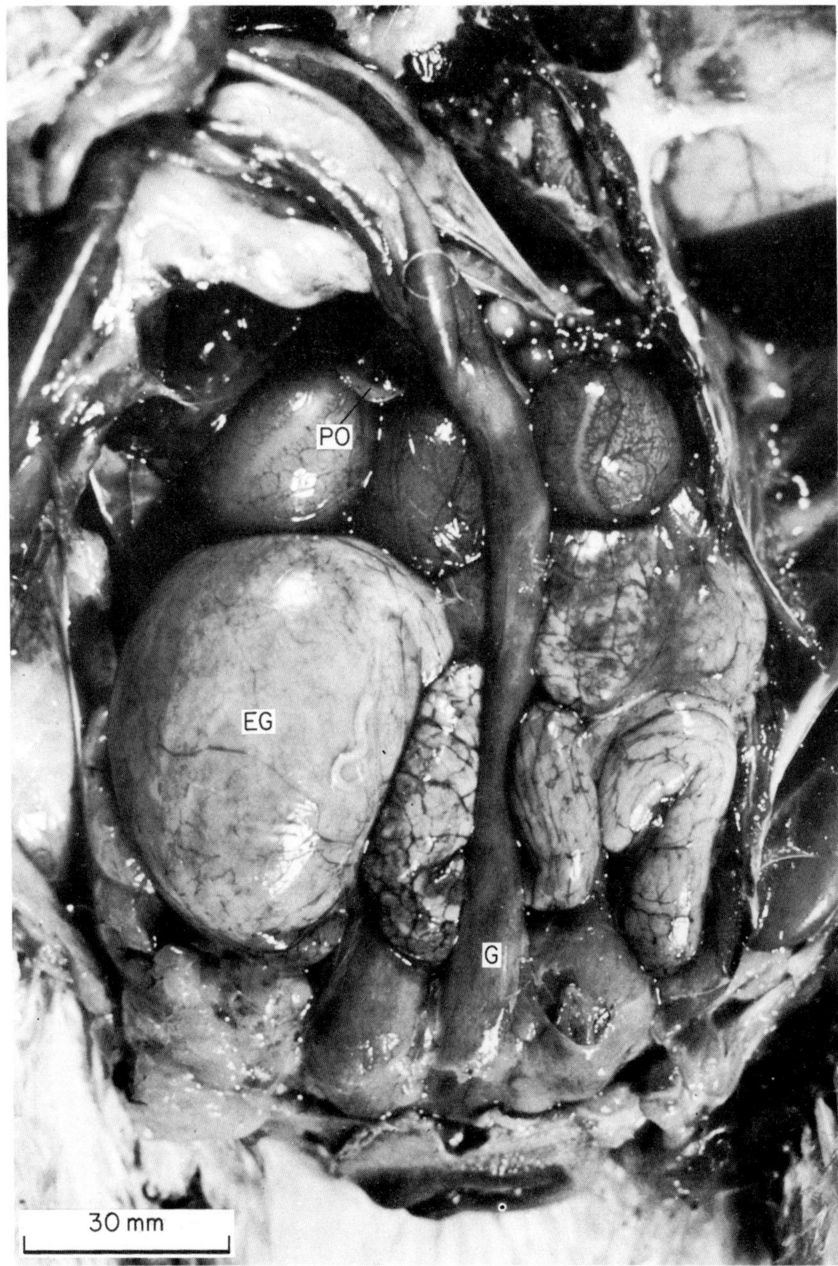

Fig. 5.9. The abdominal cavity of a domestic fowl with two functional ovaries and oviducts. A developing egg (EG), derived from the post-ovulatory follicle (PO) of the right ovary, is in the magnum of the right oviduct. Of interest is the arrangement of the cloaca to accept both oviducts and the centrally inserted gut (G) (compare with Figs 5.8 and 5.40).

Bickford (1965) shows that it does occur. In the Brown Kiwi (*Apteryx australis*), in which two ovaries are normally found, the oviduct is specially placed to engulf oocytes from either side (Kinsky, 1971). Witschi (1935b) concluded that the asymmetry of the oviduct was the primary adaptation for the aviatic life and that ovarian reduction was a subsequent adaptation to the loss of the right oviduct: as such it has not been completely eradicated in the class. Kummerlöwe (1955), on the other hand, took the opposite view that the loss of the ovary came first in evolution.

A. Embryological formation of the reproductive organs

Most accounts of embryogenesis use a chronological system to describe development but this has disadvantages because not all embryos develop at the same rate even under the same incubation conditions. Moreover, conditions may differ from one incubator to another (Hamburger and Hamilton, 1951). A more satisfactory method is to use standard morphological criteria (Hamburger and Hamilton, 1951) but this is not always feasible. The present text has adopted the chronological system, mainly because most of the original work is given in this form, but it is acceptable where only comparative rates of development are being considered.

Embryogenesis in *Gallus* has been studied for many years; unfortunately this is not so for most avian species and these have been grossly understudied. Thus it is not possible to give a completely comparative account of the development of the reproductive organs in birds. The following is essentially that for *Gallus* and is based on the accounts of Swift (1914, 1915), Brode (1928), Domm (1939), Lillie (1952), Brambell (1956), van Limborgh (1957), Romanoff (1960) and Witschi (1961).

Unfortunately, there is still confusion over the terminology used in the formation of the female gamete in vertebrates. In the present account the terminology follows closely that of Baker (1972) for mammals. The primordial germ cells are the extra-embryonic cells which give rise to the oogonia in the definitive ovary. The oogonia are capable of mitotic division and at the end of this phase, when they enter the prophase of the first meiotic division, they are regarded as oocytes. "Oogenesis" is considered here to be the complete process of maturation of the female gamete (Henderson and Henderson, 1957), though others have used oogenesis in a more restricted sense to exclude the oocyte (Franchi *et al.*, 1962).

As in all vertebrates the avian reproductive system is closely associated with the urinary system in early embryogenesis (Fig. 5.10). These systems are located dorsal to the other organs, develop retroperitoneally and later project into the embryonic coelom. The pronephros appears first in the formation of the urogenital system with its associated pronephric duct. But soon after its formation, the pronephros degenerates together with the cranial part of the duct. The caudal part of the duct persists, however, as the mesonephric (Wolffian) duct (Lillie, 1952) and this, in turn, appears to induce the formation of the mesonephros (Romanoff, 1960). Once the mesonephros is formed, the remainder of the urogenital system develops.

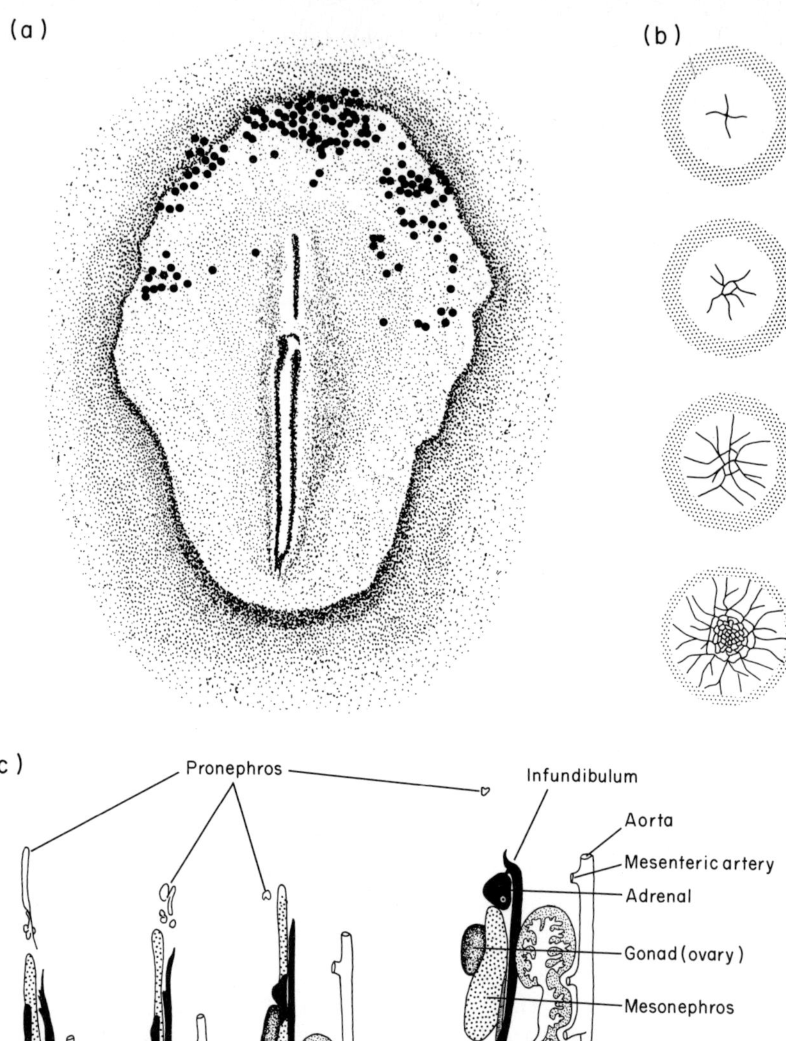

Fig. 5.10. Embryogenesis. (a) Distribution of the primordial germ cells in the blastoderm at 18 h incubation. (Composite diagram from data of Swift, 1914, Willier, 1937, Romanoff, 1960 and Fujimoto *et al.*, 1976). (b) Early cleavage furrows in *Gallus*. (Redrawn from Romanoff and Romanoff, 1949). Details of the ultrastructure of cleavage furrows are given by Gipson (1974). (c) Embryological structure on the left urogenital system of passerines of 4, 5, 6 and 8 days incubation. (Redrawn from Witschi, 1956). See also Fig. 5.40.

The urogenital ridge becomes obvious about the fourth day of incubation when it forms as a peritoneal thickening of the median surface of the mesonephros. Basically it is a thickening derived from the intermediate mesodern and is covered with coelomic epithelium. The female gonads and oviducts are formed from this ridge after it becomes invaded by the primordial germ cells.

1. The ovary

The primordial germ cell in vertebrates is a characteristically large turgid round to oval cell with blunt pseudopodia (Brambell, 1956). In the bird it is 10–12 μm in diameter and, hence, much larger than most somatic cells at this stage of development. The nucleus of the germ cell is strongly demarcated, and in *Gallus* is claimed to have a characteristically fragmented nucleolus (Fujimoto *et al.*, 1975, 1976). The cell contains glycogen (Meyer, 1964) and its yolk material persists after that of most somatic cells has been lost (Brambell, 1956; Romanoff, 1960), although there is a reduction in the yolk content during its life (Swift, 1914). The germ cell in birds and reptiles appears capable of undergoing mitotic division both before and after it reaches the gonadal development site (Swift, 1914; Jordan, 1917; Risley, 1933; Witschi, 1935b; Clawson and Domm, 1963b), although not all accept this view (Firket, 1913; Swift, 1915; Goldsmith, 1928).

Despite some conflicting data, Brambell (1956), Romanoff (1960) and Franchi *et al.* (1962) concluded that the preponderance of the evidence indicates that the primordial germ cell in the bird should be regarded as the reproductive cell which gives rise to all the definitive sex cells of the adult, either male or female.

By the time the urogenital ridge is formed, the primordial germ cell has had a long history, a characteristic common to most, if not all, vertebrates (Brambell, 1956). The cells first appear in embryo domestic fowl at about 18 h ("primitive-streak" stage) lying free in the space between the ectoderm and endoderm, although Matsumoto (1932) allegedly found them earlier, at the anterior and antero–lateral border (the so-called "germinal crescent") between the area pellucida and the area opaca (Fig. 5.10) (Swift, 1914; Reagan, 1916; Benoit, 1931; Dantschakoff *et al.*, 1931; Romanoff, 1960; Fujimoto *et al.*, 1976). Since the germ cells are probably derived from the germ-wall entoderm (Brambell, 1956; Romanoff, 1960), and thus are extra-embryonic in origin, their origin in birds is similar to that in reptiles (Brambell, 1956).

From this position the germ cells migrate to the gonadal development area. In the majority of vertebrates this occurs by an extra-vascular route whereas in birds and lizards migration occurs through the circulatory system (Brambell, 1956). As the mesodermal cells develop and pass between the ectoderm and endoderm they enclose the primordial germ cells in areas in which "blood islands" differentiate. These islands develop into the vascular system which is invaded by the primordial germ cells (Blocker, 1933). At about 33 h incubation the vascular system becomes functional and the primordial germ cells are carried through the embryonic tissues, although in lower vertebrates generally it has been estimated that possibly only about 50% ever reach the area of the gonad (Allen, 1907; Brambell, 1956): in birds many of them are found in such places as the heart and head (Swift, 1914; Meyer, 1964).

In the 21- to 25-somite embryo (50 to 55 h incubation) many of the primordial germ cells collect at the splanchnic mesoderm (Meyer, 1964) either by chemotaxis (Swift, 1914; Firket, 1914; Reagan, 1916; Fujimoto *et al.*, 1976) or by mechanical obstruction to their flow (Dantschakoff, 1931; Blocker, 1933; van Limborgh *et al.*, 1960; Meyer, 1964). From this time they rapidly disappear from the blood vessels, passing through the capillary walls and traversing the splanchnic and somatic mesoderm to appear in the coelomic epithelium (Romanoff, 1960). Up to the third day of incubation in *Gallus* (Matsumoto, 1932; Witschi, 1935b; Vannini, 1945; Abdel-Malek, 1950; Venzke, 1954; Clawson and Domm, 1963a; Meyer, 1964), *Anas* (van Limborgh, 1957, 1958) and *Coturnix* (Didier and Fargeix, 1976) there is a symmetrical accumulation of the primordial germ cells at the gonadal region. On the other hand, in some hawks (Accipitridae) the distribution of the primordial germ cells is asymmetrical from the start of their colonization (Stanley and Witschi, 1940).

After this initial distribution, colonization by germ cells in *Gallus*, *Anas* and *Coturnix* (van Limborgh, 1957, 1958; Clawson and Domm, 1963a; Didier and Fargeix, 1976) becomes unequal, there being between two and five times the number of primordial germ cells in the left ovary than in the right (Firket, 1913, 1914; Swift, 1915; Defretin, 1924; Blocker, 1933; Clawson and Domm, 1963a; Didier and Fargeix, 1976). This unequal distribution is reinforced in *Gallus* (Firket, 1914, 1920; Swift, 1915), the House Sparrow (*Passer domesticus*) and the Red-winged Blackbird (*Agelaius phoeniceus*) (Witschi, 1935b) by a migration of the primordial germ cells from the right ovary to the left. However, no similar migration has been found in *Anas* (van Limborgh, 1957), Cooper's Hawk (*Accipiter cooperii*), the Red-tailed Hawk (*Buteo jamaicensis*) and the Marsh Hawk (*Circus cyaneus*) (Stanley and Witschi, 1940).

Although it is known how the bilateral symmetry arises, the real cause for the asymmetrical distribution of the primordial germ cells has not been positively ascertained (Romanoff, 1960; van Tienhoven, 1961). Various suggestions have been made (Dantschakoff and Guelin-Schedrina, 1933; Stanley and Witschi, 1940; van Limborgh, 1957) but none is entirely satisfactory.

After the primordial germ cells come to rest in the region of the urogenital ridge, they become distributed in two groups. Some cells force their way between the cells of the covering coelomic epithelium while others remain grouped in the mesenchymatous tissue (Meyer, 1964). For this reason the epithelium at this point is often erroneously called the "germinal epithelium" but the epithelial cells do not contribute to the formation or origin of the germ cells (Franchi *et al.*, 1962; Baker, 1972). At this stage the primordial germ cells are not the definitive sex cells for they are of a non-specific nature and can give rise to either male or female sex cells depending on the physiological environment to which they are subjected (Willier, 1937, 1950).

Once the primordial germ cells arrive at their definitive positions in the urogenital ridge true gonadal development occurs, although it is clear that the cells have no direct influence on the development of the gonad (Dulbecco, 1946; Salzgeber, 1950; Marin, 1959; Romanoff, 1960). At about the 5th day of incubation the proliferation of the gonadal tissue begins with the formation of the rete tissue, the origin of which appears to differ in the different vertebrate classes (Brambell, 1956). In birds it is probably derived from the mesenchyme lying between the glomeruli and the gonad rudiment (Firket, 1914; Swift, 1915). The rete cords form connecting strands between the nephric glomeruli and the future sex cords. These latter are formed as budding proliferations of the coelomic ("germinal") epithelium (d'Hollander, 1904; Firket, 1914, 1920; Swift, 1915; Koch, 1926; Dantschakoff, 1933) which grow into the underlying stroma (derived from the mesenchyme) and which carry some of the primordial germ cells with them (Brambell, 1956; Romanoff, 1960). This first multiplication occurs about the 6th day and gives rise to the future medullary tissue. From the 7th day, in *Gallus* the medulla is separated from the epithelium by the connective tissue primary tunica albuginea of mesenchymal origin (Brambell, 1956), the formation of this marking the end of the first proliferation.

In these early stages there is no morphological distinction between the sexes, although even by the 3rd and 4th day sexual differentiation has started (Willier, 1925, 1933; Corinaldesi, 1926). Morphological differentiation begins at about 7 days in *Gallus* (Firket, 1914, 1920; Swift, 1916; Essenberg and Garwacki, 1938), 5 days in the House Sparrow (*Passer domesticus*) (Witschi, 1935a) and 7·5 days in the Mallard (*Anas platyrhynchos*) (Burwell, 1931).

In females the subsequent development lies mainly in the cortical region. In males this part remains undeveloped and it is the medullary region which gives rise to the functional male organ (Franchi *et al.*, 1962). However, the medullary tissue does increase over the next few days in the left ovary of the female and by the 9th day it has differentiated into two layers, a superficial layer and an inner reticular layer with large interstices (Brode, 1928; Witschi, 1956). In *Anas*, and possibly other species, the medulla contains primordial germ cells (Lewis, 1946) which persist for some time before degenerating (Brode, 1928). The epithelium becomes thicker and composed of several layers of columnar cells. The medulla of the right gonad also develops for a while but in *Anas* and *Gallus* there are early signs of degeneration (Romanoff, 1960). The main difference between the left and right gonads at this time is the conspicuous number of primordial germ cells which still remain in the medulla of the right side and which often form solid cords of cells.

It is at this stage, i.e. about the 8th day of incubation, that the primordial germ cells in the left gonad begin their rapid multiplication to form the oogonia (*ovogonia*) and ultimately the oocytes (Swift, 1915; Brode, 1928; Goldsmith, 1928; Willier, 1939). This continues for the next few days (up to the 11th day) until cords of cells are produced which contain some epithelial cells and which penetrate into the underlying medullary tissue. These are the cords of the second proliferation (cortical cords) and give rise to the future cortex. Although this second proliferation is similar in *Anas* (Lewis, 1946), it is not certain that it is common to all birds Blocker (1933), for example, being unable to demonstrate it in the House Sparrow (*Passer domesticus*). After their formation the cortical cords increase in size and new ones are formed. On about the 14th day the cords become separated from the epithelium by the secondary albuginea which is derived, like the primary one, from the mesenchymatous elements of the ovary (Brambell, 1956). After hatching, in *Gallus* this albuginea forms the definitive tunica albuginea (Swift, 1915; Willier, 1939).

Between the 15th day and hatching the epithelium becomes reduced to a single layer of cuboidal or columnar cells. The primary tunica albuginea becomes less conspicuous and at about the time of hatching disappears (Romanoff, 1960; Komarek and Prochazkova, 1970).

Unlike the left ovary, the gonad on the right side shows no such massive proliferation of the future cortical cells. In *Anas* (Lewis, 1946) and *Gallus* (Zawadowsky and Zubina, 1928) the cortical layer remains only one or two cells thick, and this layer progressively breaks up until about day 14 when the cortical areas are no more than small clumps of cells attached to the epithelium. However, the condition of the cortex is variable and primordial germ cells may be present (Brode, 1928). Usually there is no tunica albuginea.

The functional significance of the acid and alkaline phosphatases, and the carboxylases, which have been demonstrated histologically is obscure, particularly since these enzymes are weak or absent in the right gonad (Delforge and Schippers, 1965).

At hatching the interstitial elements of the left medulla consist mainly of the connective tissue cells of the stroma, blood sinusoids (Brode, 1928), undifferentiated large cells (Deol, 1955) and other cells among which are eosinophils (Hodges, 1974). Also present are the interstitial cells (the "luteal" cells of Fell, 1924). These true interstitial cells (*interstitiocyti ovarii*) (Fig. 5.15) are involved in steroidogenesis and are large, round or polygonal cells with a round or oval nucleus. However, not only is their homology with mammalian interstitial cells doubtful (Brambell, 1956) but their origin in avian embryogenesis is still uncertain, as it is in vertebrates generally (Franchi *et al.*, 1962). Benoit (1926) considered that the medullary interstitial cells of birds arise from the embryonic proliferation which produces the primary sex cords (see above) and hence are epithelial. Some of the cortical interstitial cells have also been described as coming from the medullary cells through migration, whereas others in the cortex have been thought to arise from the secondary sex cords (Benoit, 1950). On the other hand, the observations of Narbaitz and de Robertis (1968) and Scheib (1970) support Fell (1924) who suggested that all interstitial cells, whatever their anatomical location within the ovary, are derived from one cell source which is itself derived from the primary sex cords.

There may be other sources of steroidogenic cells in the adult, at least in some species. These cells, if they exist, should be clearly distinguished from the general population of embryonically derived interstitial cells. Marshall and Coombs (1957) described so-called "interstitial cells" in the Rook (*Corvus frugilegus*) which were derived from atretic follicles. Similarly, Guraya (1976) claimed that the stromal cells of the theca interna in the pre-vitellogenic atretic follicles of *Columba* and the Collared Turtle Dove (*Streptopelia decaocto*) hypertrophied to give rise to the "interstitial gland cells". He described these cells as remaining after atresia and being steroidogenic. If they exist, they would form a second population of interstitial cells derived from embryonic mesenchymal tissue and would be distinct from the normal interstitial cells derived from the embryonic epithelium. However, the evidence presented by Guraya (1976) is not conclusive and the hypertrophy that he claimed may be, in fact, hyperplasia of the thecal interstitial cells, despite his belief that the cells are different. It is a pity that all these authors use terms like "interstitial cells" and "interstitial gland cells" for these supposedly different cell types, since this only leads to further confusion. With the uncertainty at present as to the true affinities of interstitial cells generally, the adjectival use of "cortical", "medullary" and "thecal" should more prudently be used to denote their anatomical location only, not their origin.

2. *Accessory reproductive organs*

The accessory reproductive ducts convey the reproductive cells to the outside and consist of the male mesonephric (Wolffian) duct derived from the original pronephric duct, and the paramesonephric (Müllerian) duct. During early embryogenesis both ducts are present in the male and female (Figs 5.10, 40). Another accessory reproductive structure present in the early embryos of both sexes is the genital tubercle which in the male ultimately gives rise to the copulatory organs (Hashimoto, 1930; Wolff, 1950). In the female embryo both the mesonephric duct and the genital tubercle usually either regress or do not develop further. On the other hand, the left paramesonephric duct continues its development until it ultimately becomes the functional oviduct of the adult (Fig. 5.10).

The oviduct is originally derived from the urogenital ridge. In *Gallus* its precursor, the tubal ridge, appears on both sides on the 4th day of incubation as a groove-like invagination of the coelomic epithelium at the cranial end of the mesonephros. By the 5th day this invagination has become closed over caudally although cranially it remains open to the coelomic cavity via the ostium. During its development the tubal ridge remains closely associated with the mesonephric duct and shares a common basal lamina with it (Gruenwald, 1952).

The left paramesonephric duct continues to develop, growing caudally until by the 11th day it reaches the cloaca. However, there is no connection between it and the cloaca until well after hatching. Differentiation of the various regions of the duct becomes apparent between the 12th and 13th day (Stoll, 1944). During development the oviduct is supported in the coelom by a double sheet of peritoneum derived from the urogenital ridge (Kar, 1947a).

In *Gallus* the right duct similarly develops for a time but by the 4th to 7th day its size is less than that on the left side. Growth ceases in *Gallus* and *Anas* at about the 8th day of incubation and regression occurs from about the 11th day (Stoll, 1944; Lewis, 1946; Lutz-Ostertag, 1954), although in the Red-tailed Hawk (*Buteo jamaicensis*) regression occurs later (18th day) (Stanley, 1937; Stanley and Witschi, 1940).

It has been suggested by Gruenwald (1942) that this asymmetry of the oviduct, in *Gallus* at least, is a primary one and outside the influence of the ovary since it appears to exist before the gonads are fully differentiated. Nevertheless, gonadal effects on the growth and development of the oviduct are important but their complexities are little understood at present (Witschi, 1961).

Involution of the mesonephros in the female starts between the 8th and 16th day of incubation in *Gallus* (Romanoff, 1960), between the 14th and 15th day in the Alpine Swift (*Apus melba*), and between the 8th and 9th day in the

Blackbird (*Turdus merula*) (Stampfli, 1950). This degeneration continues until at hatching little remains (Firket, 1920). However, in most birds small strands of apparently non-functional tissue persist on the lateral walls of the postcardinal vein. These strands form the *epoophoron* (from the cranial mesonephros) and the *paroophoron* (from the caudal mesonephros) (Franchi *et al.*, 1962). In the male the cranial portion is functional as the epididymis.

The right mesonephric duct persists with a patent lumen and connects with the cloaca (King, 1975) but the system on the left side is more uncertain. Both Domm (1927) and Brode (1928) believed that in *Gallus* the system on the left side is similar to that on the right but Kar (1947a) concluded that it disappeared entirely before hatching. In other birds it persists into the adult (Fig. 5.40).

B. The left ovary

Birds are well known to be seasonal breeders (Baker, 1938a,b; Baker and Ransom, 1938; Thomson, 1964; Immelmen, 1971; Murton and Westwood, 1977). In northern latitudes the breeding season is usually associated with spring, although this is not necessarily so elsewhere. At any time of the year there are some species to be found in the breeding condition. Reproduction in birds is a complex process which Herrick (1907) divided into the following closely interrelated phases: (1) migration, (2) courtship and mating, (3) nest selection and building, (4) egg laying, (5) incubation, (6) care of the young in the nest ("broodiness") and (7) care and training of the young after they have left the nest. Few of these phases can be considered here and further information should be obtained from the reviews by Welty (1962), Cody (1971), Immelmann (1971), Lofts and Murton (1973), Follett and Davis (1975), Lofts (1975), Murton (1975) and Murton and Westwood (1977). More specialized texts on the physiological mechanisms involved in reproduction in the domestic fowl are to be found in Bell and Freeman (1971).

The factors which regulate the breeding cycle are environmental but the precise details are not always understood (Cody, 1971; Immelmann, 1971; Murton and Westwood, 1977). In the early phases light seems to be a major factor in activating the neuroendocrine system responsible for development of the reproductive organs (Jørgensen and Larson, 1967; Jørgensen, 1968; Kobayashi and Wada, 1973; Follett and Davis, 1975; Murton, 1975; Murton and Westwood, 1977) but in the culminative stages (nidification, ovulation and laying) factors other than light (such as rainfall, temperature, food, availability of nest sites and the availability of nest building materials) may be of more importance (Marshall, 1956; Amoroso and Marshall, 1960; Immelmann, 1971; Murton and Westwood, 1977).

The number of breeding seasons in a year, their length and the length of time between them are all variable from species to species. Most birds have only one season in a year, some have two (Marshall, 1956) while a number of sparrows (*Passer* spp.) and the Mourning Dove (*Zenaida macroura*) may have up to four or five (Romanoff and Romanoff, 1949; Welty, 1962; Immelmann, 1971). In species taking more than a year to raise their young, e.g. albatrosses

(Diomedeidae) and the King Penguin (*Aptenodytes patogonica*), breeding does not occur each year (Thomson, 1964; Immelmann, 1971).

The number of eggs in a clutch also varies greatly with the species (Welty, 1962; Murton and Westwood, 1977). Some birds, e.g. bustards (Otididae) and the Sooty Tern (*Sterna fuscata*), lay only one egg whereas nudifugous species tend to lay larger clutches. Individual variation between birds of the same nudifugous species may be considerable (for details of clutch size in over fifty species see Davis, 1955). The factors regulating clutch size are complex and not entirely understood (Welty, 1962; Murton and Westwood, 1977). However, bigger birds tend to lay fewer eggs. Many species, such as the Budgerigar (*Melopsittacus undulatus*), the Common Magpie (*Pica pica*), the Common Crow (*Corvus brachyrhynchos*), the Barn Swallow (*Hirundo rustica*) and the Gentoo Penguin (*Pygoscelis papua*), are determinate layers and produce fixed clutch sizes. Others (among them domestic species) are indeterminate layers and can produce more eggs should any in the nest be destroyed or removed. What is more, the rate of lay in wild birds under these conditions may equal that of *Gallus*. For example, a common Flicker (*Colaptes auratus*) is reputed to have produced 71 eggs in 73 days and a Wryneck (*Jynx torquilla*) is known to have laid one egg a day for 62 days (Welty, 1962).

Knowledge of the morphological changes in the ovary associated with the seasonal reproductive cycle is inadequate since comparative descriptions are few and often superficial (Bissonnette and Zujko, 1936; Bullough, 1942; Cuthbert, 1945; Moreau *et al.*, 1947; Marshall and Coombs, 1957; Dominic, 1961a; Kern, 1972). Regretably much of the information about ovarian structure and function is restricted to the domestic fowl. If there are fundamental differences between the domestic fowl and other species, as has been suggested (Marshall, 1956; Lofts and Murton, 1973), this could lead to difficulties in the use of the domestic fowl as a typical member of the class. One aspect which has been stressed is that in *Gallus* seasons do not appear to be part of the reproductive pattern. However, domestic poultry are housed in conditions which are designed to extend the "season" indefinitely and to produce the maximum number of eggs in one year, and when wild birds are given such conditions their reproductive performance is similarly enhanced (Clayton, 1972). Moreover, the waves of follicular development and atresia in the ovaries of those wild species which have been examined (the Scrub Jay, *Aphelocoma coerulescens*; the Common Magpie, *Pica pica* and the Rook, *Corvus frugilegus*) (Marshall and Coombs, 1957; Erpino, 1969, 1973) suggest that ovarian competence is present throughout the year in these species also, as it is in *Gallus*. What is more, feral chickens do have recognized breeding seasons (McBride *et al.*, 1969). It may therefore be not too unreasonable to extrapolate in a general way from the domestic fowl to those species which have not been extensively studied.

1. The juvenile ovary

It is a fundamental feature of all vertebrates that they pass through an immature, non-reproductive phase during the first part of their independent life, even although the organs are present which would allow reproduction to take place. The length of time that the juvenile state persists in birds depends on the species and the individual. This will be affected by environmental conditions but the right conditions for breeding must be present at the time when the animal is physiologically capable of entering the reproductive phase. Quail (*Coturnix*) may be made to breed within a few weeks of hatching (Stein and Bacon, 1976) and domestic species within a few months. In most wild species breeding occurs usually in the first breeding season after hatching (Immelmann, 1971) but larger birds may take several years to reach sexual activity, eight or more for the Royal Albatross (*Diomedea epomophora*) and fulmars (Procellariidae).

After the embryological differentiation of the left ovary (*ovarium sinistrum*) at about the 14th day of incubation, little major change occurs until just before the breeding season (Fig. 5.11). Consequently, the juvenile ovary at hatching is a small, ribbon-like organ, roughly triangular in shape, with its apex pointing caudally. In colour it is usually pink, turning in time to brownish red.

The size and shape of the immature ovary depends on the species. In *Gallus* it weighs 0·3–0·5 g (Nalbandov and James, 1949; Amin and Gilbert, 1970), in *Coturnix* 0·06–0·09 g (Tanaka *et al.*, 1965) and in the Common Starling (*Sturnus vulgaris*) about 0·008 g (Witschi, 1956). During the first few weeks the ovary grows very slowly reaching in *Gallus* about 1·5 cm in length, 1–1·5 cm in width and 0·3–0·4 cm in depth (Brode, 1928; Gilbert, 1971b). However, this growth of the ovary in *Columba*, *Gallus*, the Jackdaw (*Corvus monedula*) and presumably other birds leads to changes in its shape, the organ becoming broader and more obviously triangular (Koch, 1927; Van den Broeck, 1931). In addition, various folds develop and the cranial part is flexed laterally (Fig. 5.11). However, the primitive shape of ovary is retained in some species, e.g. the Red-breasted Merganser (*Mergus serrator*) and the aquatic domestic birds (Koch, 1927; Grau, 1943). The detailed earlier growth and appearance of the ovary of the House Sparrow (*Passer domesticus*) and *Columba* have been excellently recorded by Kummerlöwe (1930b).

The ovary lies in the cranial part of the body cavity, ventral to the aorta and the caudal vena cava, and adjacent to the cranial extremity of the left kidney and the caudal part of the left lung (Fig. 5.11). Ventrally it is covered by the left abdominal air sac. In *Gallus*, Biswal (1954) found that the left adrenal is closely applied to the ovary, a common connective tissue capsule enclosing both organs. In many adults the adrenal is partly embedded in the ovary. The ovary is suspended from the dorsal body wall by a peritoneal fold, the *mesovarium*. This fold is strengthened by connective tissue and forms the ill-defined hilus (*hilus ovarii*) (Fig. 5.14) (King, 1975) containing blood vessels, nerves and smooth muscle (Deol, 1955).

The arterial supply of the ovary in *Gallus* is variable (Fig. 5.12) (Oribe, 1977 described seven different types) but usually it arises from the left cranial renal artery via its ovario-oviductal branch (Dang-quan-Dien, 1951; Oribe, 1977). In other species, e.g. *Anas* and *Meleagris* there is no ovario-oviductal branch in this region (Hodges, 1965) and the artery to the ovary arises directly from the renal artery. Sometimes supplementary branches arise directly from the aorta (Mauger, 1941; Westphal, 1961; Oribe, 1977) as does the ovarian artery (*a. ovarica*) in some cases (Nalbandov and James, 1949; Oribe, 1977).

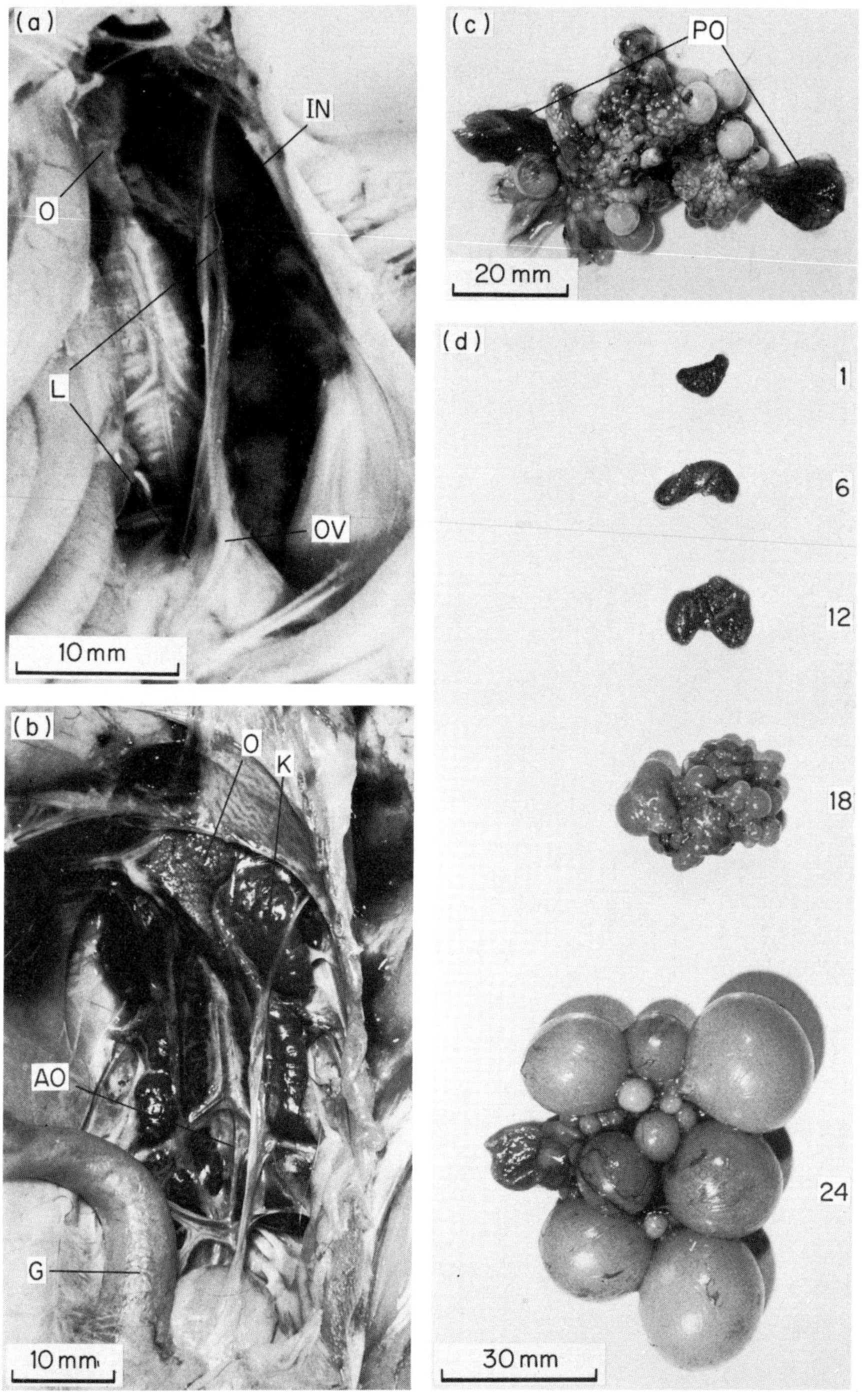

Fig. 5.11. (a) Ovary (O) of a 3-day-old domestic fowl photographed under water to show more clearly the position of the oviduct (OV) and associated ligaments (L). The infundibulum (IN) is well placed to capture the shed oocyte in later life. (b) Six-week-old ovary (O) *in situ* in the abdominal cavity. AO, aorta; G, gut; K, kidney. The ovary has the "speckled" appearance resulting from some follicular growth. (c) Adult ovary with larger follicles removed. PO, post-ovulatory follicle. (d) The difference in appearance of the ovary at 1, 6, 12, 18 and 24 weeks of age.

The ovary is drained via two ovarian veins (*vv. ovaricae*) directly into the caudal vena cava (Fig. 5.11). The cranial ovarian vein joins the left adrenal vein which opens into the left side of the vena cava. The caudal ovarian vein opens into the dorsal surface of the vena cava at about the level of the origin of the common iliac veins (Nalbandov and James, 1949). Blood may also drain via short vessels into the cranial oviductal vein (*v. oviductalis cranialis*). Oribe (1970, 1976) confirmed, in general, the previous accounts and gave details of the vessels within the ovarian medulla and cortex.

The lymphatic tissue of the ovary does not seem to have been described in any

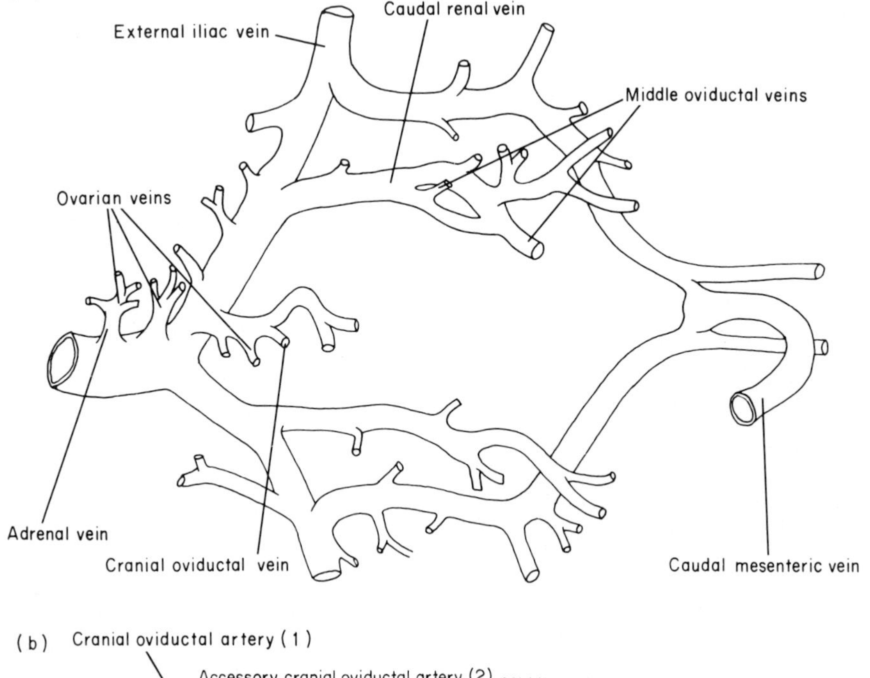

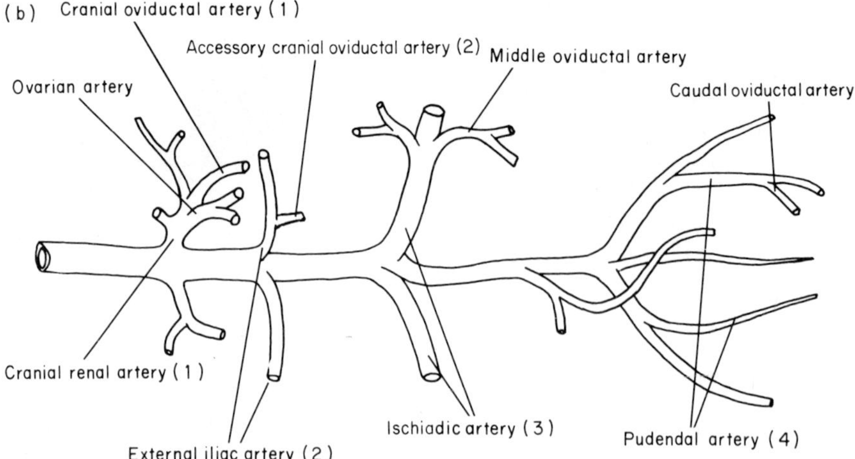

Fig. 5.12. Schematic diagrams of the blood supply of the ovary and oviduct. (a) Veins. (b) Arteries. Numbers refer to the origin of the arterial supply to the ovuduct (see text).

detail but in general it appears to be poorly developed in comparison with that of mammals. Deol (1955) noted the presence of lymph vessels in the medulla of the domestic fowl which ultimately connect with the cranial lymphatic vessels of the mesentery (Dransfield, 1945). Biswal (1954) found lymphocytes scattered throughout the whole ovary in *Gallus* and lymphocytes have been described in the perivascular connective tissue in *Anas* (Das and Biswal, 1968).

The ovary of the domestic fowl becomes increasingly innervated from about the 14th day of embryological development (Oribe *et al.*, 1963). In the adult the innervation is particularly extensive and is associated with the mature follicle (*nervi folliculares*) (Fig. 5.13) (Johnson, 1925; Mauger, 1941; Hsieh, 1951; Biswal, 1954; Bradley and Grahame, 1960; Gilbert, 1965, 1969; Freedman, 1968). Both cholinergic and adrenergic nerves are present (Gilbert, 1965, 1969). The nerves are derived from a complex network lying in and around the adrenals, the hilus of the ovary, and the caudal vena cava (Gilbert 1969) (Fig. 5.14). So complex is this network, that a full description of the component parts has yet to be produced. However, Freedman (1968) traced sympathetic links from the plexus to the sympathetic trunk in the region of the 4th, 5th, 6th and 7th thoracic ganglia and the 1st and 2nd lumbar ganglia. No vagal innervation has been positively identified (Hsieh, 1951), although both Freedman (1968) and Gilbert (1969) suggested that one is present.

During this juvenile period the ovary of the domestic fowl consists of a separate medulla (*medulla ovarii*) and cortex (*cortex ovarii*) (Benoit, 1950; Bradley and Grahame, 1960; Marshall, 1961; Prochazkova and Komarek, 1970), the cortex covering the entire surface of the medulla except at the region of the hilus. In most species the surface is smooth, although it has a granular appearance due to the numerous oocytes present (Fig. 5.11) (Blount, 1945). The medulla is highly vascular, and contains some smooth muscle and nerves. The ovary is covered by the cuboidal superficial epithelium (*epithelium superficiale*) beneath which is the definitive *tunica albuginea* (Bradley and Grahame, 1960).

Between two to five weeks after hatching the smooth appearance of the ovary is broken by the development of a series of grooves which separate the cortical gyri (Fig. 5.15) (King, 1975): with time these grooves increase in number and become deeper. Thereafter the medullary tissue invades the gyri (Prochazkova and Komarek, 1970), a process which continues into sexual maturity.

2. *The mature ovary in the breeding season*

The invasion of the gyri by medullary tissue continues until the distinction between the cortex and medulla is almost completely obliterated (Fig. 5.15), this possibly being helped by an extension of the cortical tissue (Komarek and Prochazkova, 1970). Consequently, instead of there being two separate layers, irregular masses of two sorts are formed. One mass contains mainly oocytes and maturing follicles, and the other medullary type tissue (Hodges, 1974). So different are these masses from the previous immature structures that Prochazkova and Komarek (1970) proposed new terms for them. They called the areas containing the oocytes the parenchymatous zones (*zonae parenchymatosae*) and the areas composed mainly of the medullary tissue the vascular zones (*zonae vasculosae*).

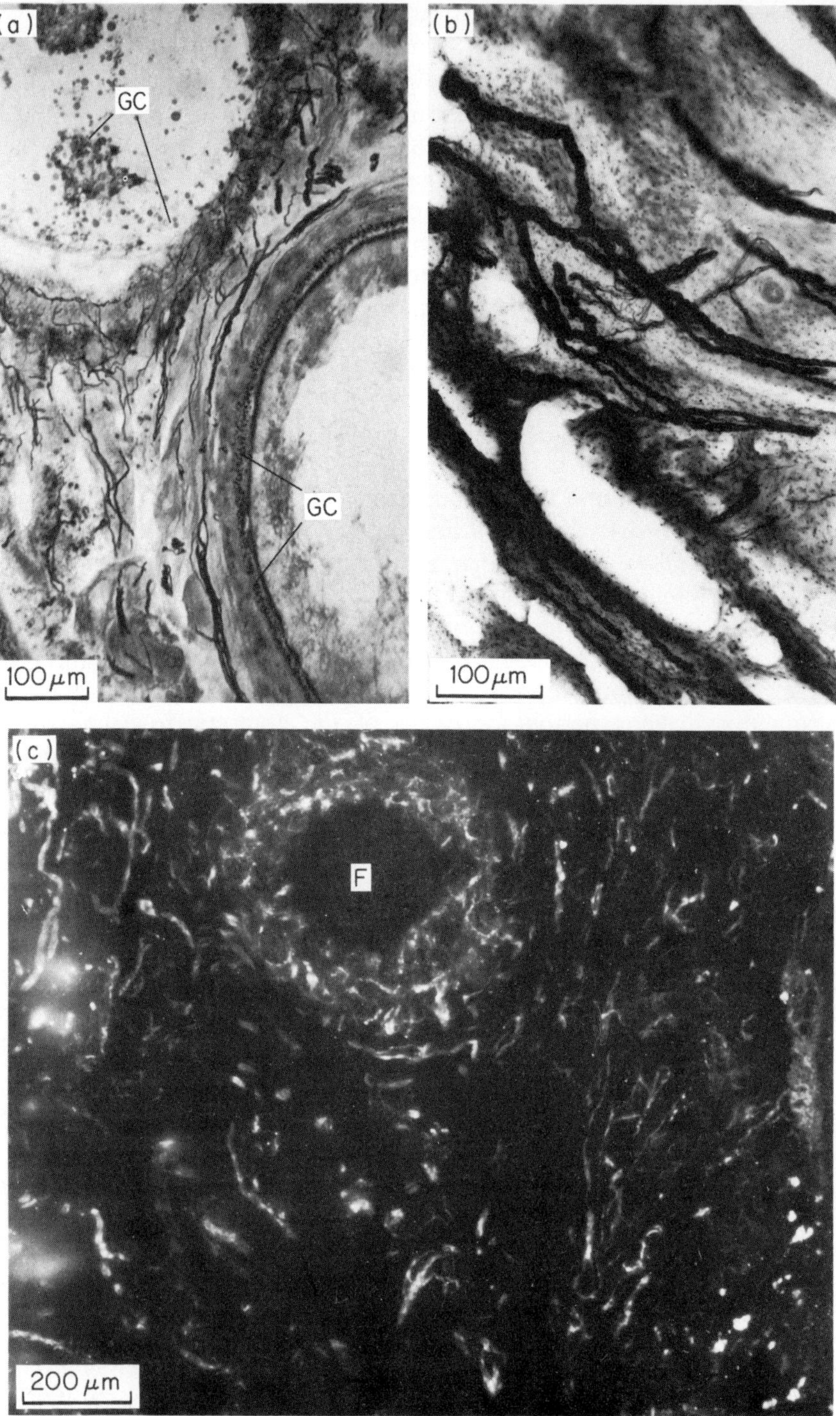

Fig. 5.13. Intrinsic innervation of the ovary. (a) Nerves in the stroma of an adult domestic fowl. Interestingly the upper follicle shows the "invasion" type of atresia with the granulosa cells (GC) mingled with the lipid droplets in the lumen of the follicle. Namba stain. (Gilbert, 1969). (b) Nerves in the pedicle of a follicle lying adjacent to the sinus-like vascular channels. Weddell and Glees silver stain. (Gilbert, 1965). (c) Adrenergic nerves are very prominent, especially in small follicles (F); although not shown, cholinesterase-positive nerves also are present. Fluorescent catecholamine technique. (Gilbert, 1969).

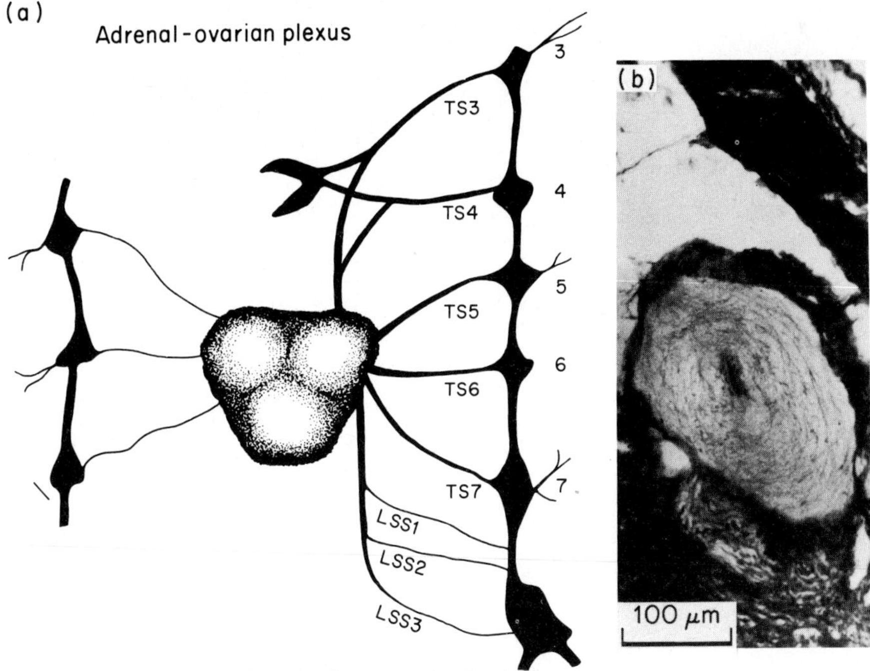

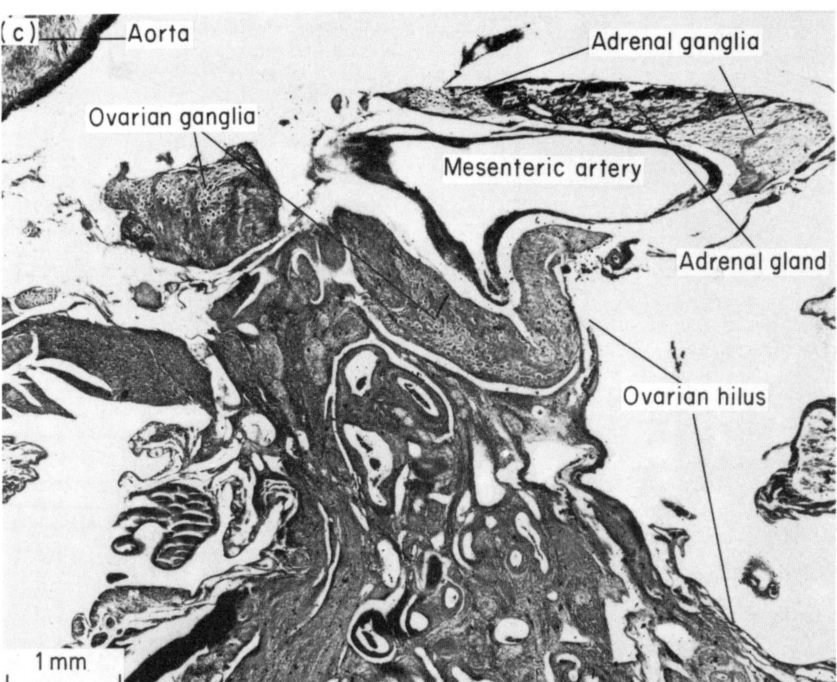

Fig. 5.14. (a) The ovary is particularly well innervated by branches of the adrenal-ovarian plexuses, the left plexus predominating. The thoracic sympathetic ganglia are numbered and the lumbosacral nerves (LSS 1–3) arise from the fused ganglia. (Redrawn from Gilbert, 1971b). (b) A sensory nerve ending, similar to a Herbst corpuscle, adjacent to the ovary of a chicken; its function in this position is unknown. Holmes' silver stain. (Gilbert, 1969). (c) Vertical section through the ovarian hilus. Holmes' silver stain. (Gilbert, 1969).

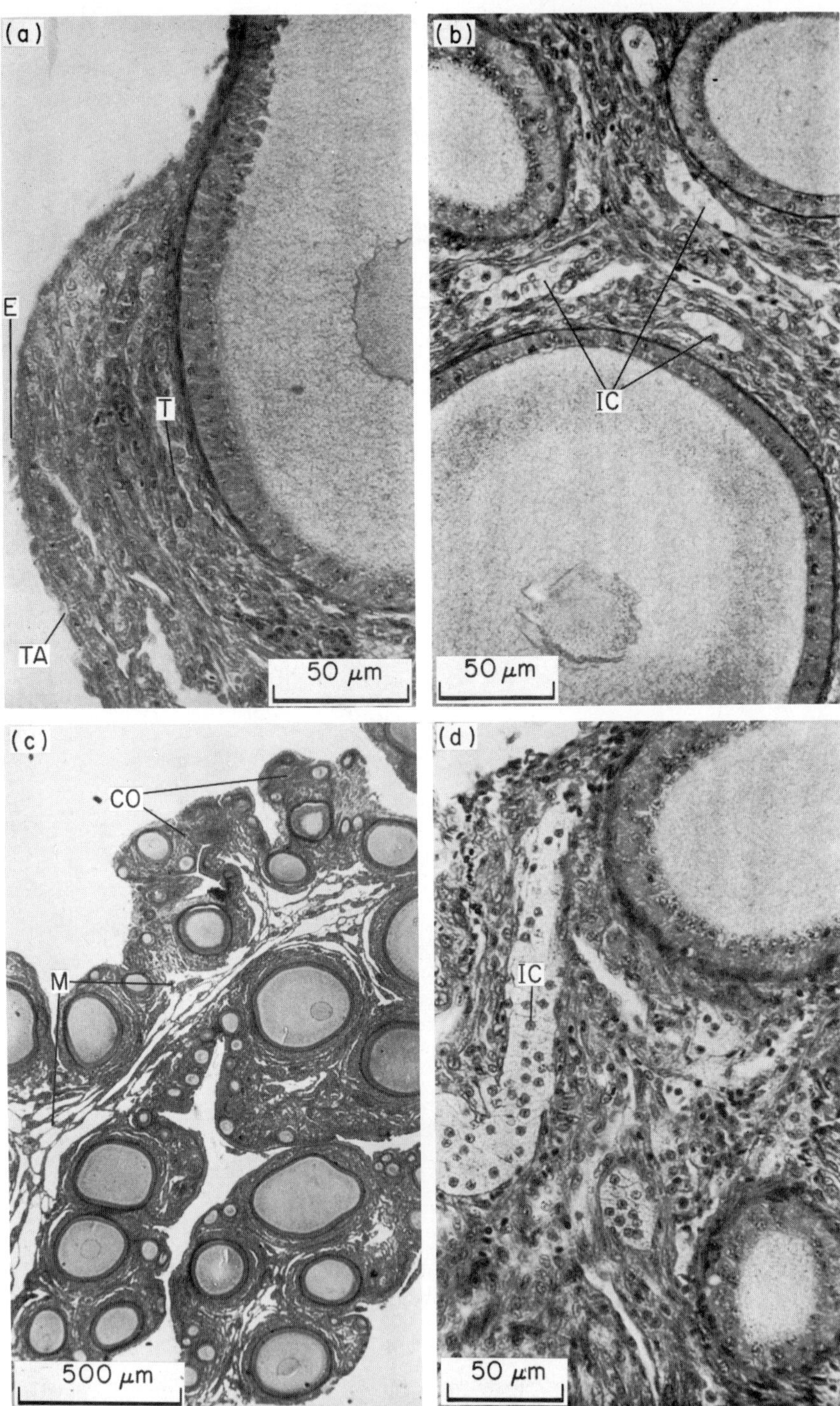

Fig. 5.15. General ovarian and follicular structure. (a) First signs of extrusion from the ovary of a follicle with an undivided theca (T); in many species the stigma forms at the region where the follicle fuses with the ovarian epithelium (E). The thin tunica albuginea (TA) is just visible. (b) Typical appearance of early follicular formation. None has a true

The weight of the ovary increases during this time, from about 0·5 g to 60 g or so in *Gallus* (Bennett, 1947; Romanoff and Romanoff, 1949; Amin and Gilbert, 1970) and from 0·008 g to 1·4 g in the Common Starling (*Sturnus vulgaris*) (Witschi, 1956). However, little increase occurs in the general ovarian mass and the main increase in weight is due almost entirely to the growth of the yolky follicles (Fig. 5.11) (Amin and Gilbert, 1970). As these form, the ovary takes on the "bunch of grapes" appearance which is characteristically associated with birds (Figs 5.8, 11) (Benoit, 1950; Breneman, 1955; Franchi *et al.*, 1962; Lofts and Murton, 1973). However, the final appearance of the ovary depends on the species. In birds which lay few eggs in a clutch there are generally relatively few developing follicles visible. Extreme examples of such an ovary include that of a dove (*Streptopelia* sp.) in which two follicles can be seen (Cuthbert, 1945), and that of the Band-tailed Pigeon (*Columba fasciata*) in which only one follicle is visible (Lofts and Murton, 1973). Although the large yolky, developing oocytes are the most obvious feature of the avian ovary, they form only a small proportion of the total number of oocytes in the ovary (Figs 5.11, 15). Moreover, as for other vertebrates generally, only a minute fraction of the oocytes ever reach this stage, even over the many years of reproductive activity in some long-lived species.

The development of the female gamete can conveniently be divided into several sections including oogenesis, the formation of the follicle and vitellogenesis. So far as it is known there are only two fates for the oocyte once it has started its development; ovulation and the subsequent incorporation into the egg, and atresia. Since the post-ovulatory follicle is functional, it will be considered separately. Atresia, although occurring throughout the life of the animal, is more usually seen in the post-breeding phase of the reproductive cycle and is dealt with there.

(*a*) *Oogenesis.* A major function of the ovary is to produce the female gamete. This involves the oocyte in a maturation process (oogenesis) leading to the cytoplasmic and nuclear changes necessary for the production of the haploid zygote. Remarkably there is almost absolute uniformity in the nuclear stages of the many birds which have been examined (Loyez, 1906). It also involves a growth process in which yolk material is incorporated into the oocyte (vitellogenesis) to provide the nutritional material for the future embryo.

Oogenesis starts in early embryogenesis with the transition of the primordial germ cell into the oogonium. However, the distinction between the two is not clear: Baker (1972) uses the term "oogonium" to apply to the reproductive cells of mammals in the post-differentiated ovary and hence after the primordial germ cells have started dividing. Romanoff (1960) also refers to "oogonia" during this period in birds. Since the oogonia are derived directly from the primordial germ cell, transition forms are likely to occur. Certainly at first the oogonia

---

theca as yet although signs of its formation are visible. Stromal interstitial cells (IC) are present in the region in which the theca will form. (c) General appearance of an immature ovary; the marked distinction between the cortex (CO) and medulla (M) is already obscured to some extent. (d) Typical histological appearance of a large group of stromal interstitial cells (IC).

appear very similar to the primordial germ cells except that they are smaller and have their mitochondria grouped around the "attraction sphere" (Swift, 1915). This complex was termed the "Balbiani body" by d'Hollander (1904), although it has received many names since (Gilbert, 1971b). Its structure has been described by Bellairs (1967) and Guraya (1975).

Subsequent development of the oogonia includes repeated mitotic divisions in which the Balbiani body is involved, and an increase in size. An interesting feature is that the division does not appear to be total, intercellular bridges remaining. This is not peculiar to birds, however, and it is found in vertebrates and invertebrates generally (Skalko *et al.*, 1972).

The distinction between the oogonium and the oocyte is precise because, by definition, once the oogonia reach the prophase of their first meiotic division they become primary oocytes (Franchi *et al.*, 1962; Baker, 1972). The fate of any given oocyte cannot be predicted. Some are destined to have a relatively short life span and to enter the developmental stage during the next breeding season. Others may have a protracted life span and take several years to reach maturity; most never enter the final stages. It is not known whether selection for development occurs in a random way or not and, if the selection is not random, what features convey to the oocyte a preferential choice of selection. Preliminary observations in *Gallus* (Gilbert, 1971b) tend to indicate that, at least in the later stages, preferential selection is taking place and that the complex innervation of the ovary may be involved (Gilbert and Wood-Gush, 1970).

During the developmental period the nucleus completes the several stages of the meiotic prophase (d'Hollander, 1904; Loyez, 1906; Sonnenbrodt, 1908; van Durme, 1914; Goldsmith, 1928; Olsen, 1942; Romanoff, 1960) which appear similar to those of mammals (Franchi *et al.*, 1962). At hatching the nucleus is in the pachytene stage and Goldsmith (1928) considered no further changes occur until just before the final maturation preceding ovulation. On the other hand, Franchi *et al.* (1962) argued convincingly that the nucleus passes through the pachytene stage and that the major portion of the life of the primary oocyte is spent in the diplotene stage, as it is in amphibians and some reptiles. It is during this stage that the oocyte accumulates most of its yolk material; towards the end of it the chromosomes take on the characteristic form of diakinesis.

During the last part of the development of the primary oocyte, changes occur in readiness for the final phases of the first maturation division of the nucleus. About 24 h before ovulation the germinal vesicle (Fig. 5.2), the oocyte nucleus, breaks down (Oslen, 1942). In *Gallus* the first noticeable sign of this is a slight wrinkling of the nuclear membrane (Loyez, 1906) but in *Columba* a clear peri-nuclear zone appears (Harper, 1904).

In *Gallus* (Olsen and Fraps, 1950), *Meleagris* (Olsen and Fraps, 1944), the Barn Swallow (*Hirundo rustica*) (van Durme, 1914) and *Columba* (Harper, 1904; van Durme, 1914), and presumably other species, the completion of the first reduction division occurs a few hours before ovulation, the exact time depending on the species. During this division the first polar body (*polocytus primarius*) is produced and the chromosomes are reduced to the haploid number (see Romanoff, 1960 for diploid counts of many species). The first polar body consists of chromatin material and a small amount of karyoplasm. Theoretically it should be capable of a further division but this has not been seen in birds (Romanoff,

1960). However, according to Harper (1904), it may still retain its kinetic properties since the chromatin material remains massed.

Subsequent to this, the second maturation spindle starts to form but mitosis is arrested in metaphase (Harper, 1904; van Durme, 1914; Olsen, 1942). It is now generally accepted that completion of the division occurs after ovulation and after penetration by the sperm (but see van Durme, 1914), as it is in most vertebrates (Romanoff, 1960; Baker, 1972). Details of this process have been given earlier (The egg).

Waldeyer (1870) first proposed that at hatching the avian ovary contains a finite stock of germ cells which is not increased during the life of the animal. All subsequent evidence has supported this contention (d'Hollander, 1904; Loyez, 1906; Sonnenbrodt, 1908; van Durme, 1914; Brambell, 1926; Fauré-Fremiet and Kaufman, 1928; Marza and Marza, 1935). What is more, if the indirect evidence is accepted, it appears that the ovary at hatching contains all the oocytes it is to have. A full discussion on this point is given by Franchi *et al.* (1962) and Hughes (1963).

A further curious feature of the ovary of the domestic fowl is the apparent considerable reduction in the number of oocytes within the first few weeks after hatching (Brambell, 1926; Fauré-Fremiet and Kaufman, 1928). This reduction is similar to that which occurs in mammals (Franchi *et al.*, 1962) but it is not known whether it is a feature common to other birds and histological details are lacking.

(*b*) *The structure of the follicle.* The ovarian follicle (*folliculus ovaricus*), within which most of the life of the oocyte is spent, has several known functions. First, it supplies the oocyte with yolk material produced by the liver (McIndoe, 1971). It also supports the massive oocyte which, being enclosed in only its own fragile cytolemma, would otherwise soon burst and collapse. Finally, it releases the secondary oocyte (ovulation). The follicle is also almost certainly involved in the production of steroid hormones. Unfortunately, our understanding is often inadequate to precisely relate structure to function.

In the early stages the oocyte has no follicular covering. This period lasts for a few days after hatching (Goldsmith, 1928), up to the so-called "lamp brush" chromosome stage of diplotene (d'Hollander, 1904). Then cells, which are now generally accepted to be derived from the epithelial cells of the embryological ovary carried into the secondary sex cords (Brambell, 1956; Romanoff, 1960; Franchi *et al.*, 1962; Callebaut, 1976), align themselves around the oocyte. By the 4th to 6th day after hatching these have formed a complete ball of cells around the oocyte and constitute the granulosa layer. Other cells, probably derived from mesenchymal elements, later invest this structure and give rise to the theca (Fig. 5.15) (Brambell, 1956). Komarek and Prochazkova (1970) recognized four morphologically distinct stages during this formation. They also described aberrant follicles containing two oocytes and some containing one oocyte with two nuclei. On present evidence it seems likely that the oocyte itself stimulates the follicle to form around it (Brambell, 1956; Franchi *et al.*, 1962).

When first formed the theca consists of a single layer (Fig. 5.15), differentiation into the *theca interna* and *theca externa* (Fig. 5.16) occurring when the follicle reaches a size of about 2 mm (Hodges, 1974). The follicle at this stage is still embedded in a highly vascular parenchymatous tissue. It receives blood

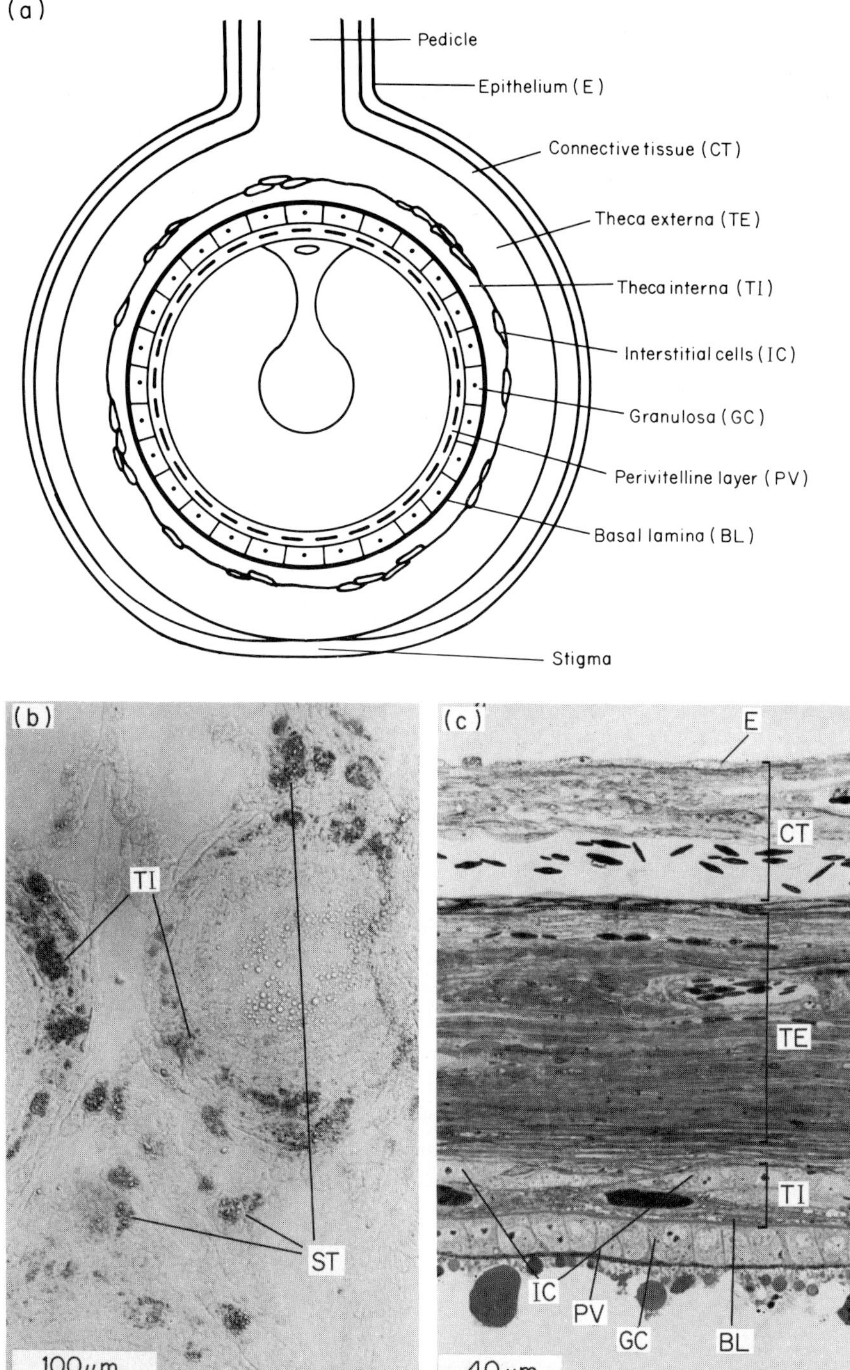

Fig. 5.16. The ovarian follicle of the bird is a complex structure. (a) Follicle and enclosed oocyte; not all layers are to scale. (b) Interstitial cells in both the theca (TI) of the follicle and the ovarian stroma (ST) are positive for steroid dehydrogenase. Tetrazolium technique using dehydroepiandrosterone as substrate. (c) Vertical araldite section (toluidine blue stain) of the follicle to show detail of structures given in (a).

from small arterioles which supply the surrounding capillary network, the latter draining into the extensive venous system of the ovary (Oribe, 1967). As the follicle grows its outer layer comes into contact with the ovarian epithelium. In *Gallus* these fuse to form the stigma (*stigma folliculare*) (Fig. 5.15) (Romanoff, 1960) but this cannot be so in all birds for in some, e.g. the White-crowned Sparrow (*Zonotrichia leucophrys*), the stigma is lacking (Kern, 1972). Further growth causes the follicle to extrude from the surface of the ovary and this leads to the formation of the pedicle (*pedunculus folliculi*) (Figs 5.15, 16). Blood vessels are drawn into the pedicle as it develops (Oribe, 1976).

The basic morphology of the follicle was first established with the light microscope. However, as King (1975) points out, the accounts of these investigations are frequently confusing since in them the components of the follicle are generally named without full appreciation of their anatomical nature. Whilst studies with the electron microscope have tended to correct this situation, difficulties may still arise when comparisons are made with the older literature. Unfortunately these ultrastructural studies (Press, 1964; Bellairs, 1965, 1967; Wyburn *et al.*, 1965a; Greenfield, 1966; Schjeide *et al.*, 1970; Paulson and Rosenberg, 1972) have been mainly concerned with the early stages of growth and have been applied almost exclusively to *Gallus*. The ultrastructure of the mature follicle has recently been described in some detail by Perry *et al.* (1978a,b).

The maturing follicle consists of several distinct regions (Figs 5.16, 17, 18, 19) (Gilbert, 1971b; Hodges, 1974; King, 1975). From inside to outside these are (a) the perivitelline layer (*lamina perivitellina*) (Figs 5.4, 16, 17, 18); (b) the granulosa layer (*stratum granulosum*) (Figs 5.15, 16, 17); (c) the basal lamina (*lamina basalis folliculi*) (Figs 5.16, 17); (d) the theca (*theca folliculi*) (Figs 5.16, 18) consisting of the *theca interna* and the *theca externa*; (e) the connective tissue layer (*tunica superficialis*) (Fig. 5.16); (f) the epithelium (*epithelium superficiale*) (Fig. 5.16). (Note that morphologically the oocyte is distal.)

The junction between the oocyte and the follicle is a complex apposition which changes throughout the life of the oocyte (Fig. 5.18). In small follicles, less than about 2 mm in diameter, the granulosa cells have small interdigitations with the oocyte cytolemma (vitelline membrane) (Fig. 10 of Gilbert, 1971b; Wyburn *et al.*, 1965a). These interdigitations increase in size and complexity, and at a follicular diameter of 7 mm form the characteristic *zona radiata* (Wyburn *et al.*, 1965a; Rothwell and Solomon, 1977) which earlier light-microscopists erroneously designated as a specific structural component of the follicle. There is no equivalent of the corona radiata of mammals. The zona radiata is transient, appearing first in follicles of between 2 and 5 mm diam. (Hodges, 1974; Rothwell and Solomon, 1977) but disappearing by the time they reach 15 mm (Fig. 5.18) (Perry *et al.*, 1978a). Detailed descriptions of this region in smaller follicles have been given by Schjeide and McCandless (1962), Schjeide *et al.* (1963, 1964, 1966), Press (1964), Bellairs (1965), Wyburn *et al.* (1965a), Greenfield (1966), Wyburn and Baillie (1966) and Wyburn *et al.* (1966). Despite this attention to the oocyte–follicle interface, the structural details of which are presumably related to yolk transport, the precise physiological mechanisms whereby yolk material is passed across the cytolemma of the oocyte are still obscure. Moreover, in the sizes of oocytes which have been studied, the

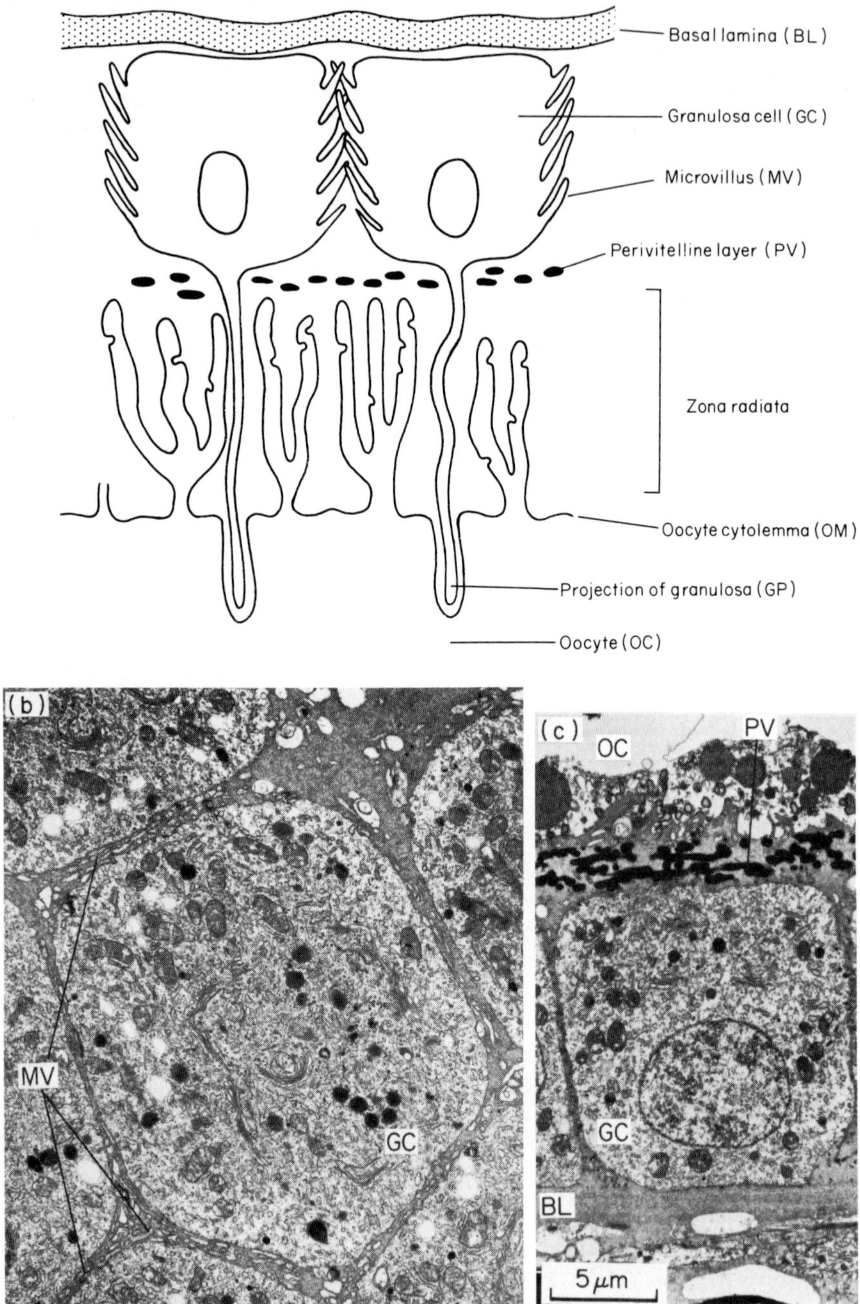

Fig. 5.17. (a) Structure of the granulosa-oocyte region of a follicle of 7 mm diam. (Redrawn from King, 1975). (b) Tangential section through the granulosa cell (GC) to show its its hexagonal shape and microvilli (MV). (Perry, unpublished). (c) Vertical section of a granulosa cell (GC) and part of oocyte (OC). BL, basal lamina; PV, perivitelline layer. (Perry, unpublished).

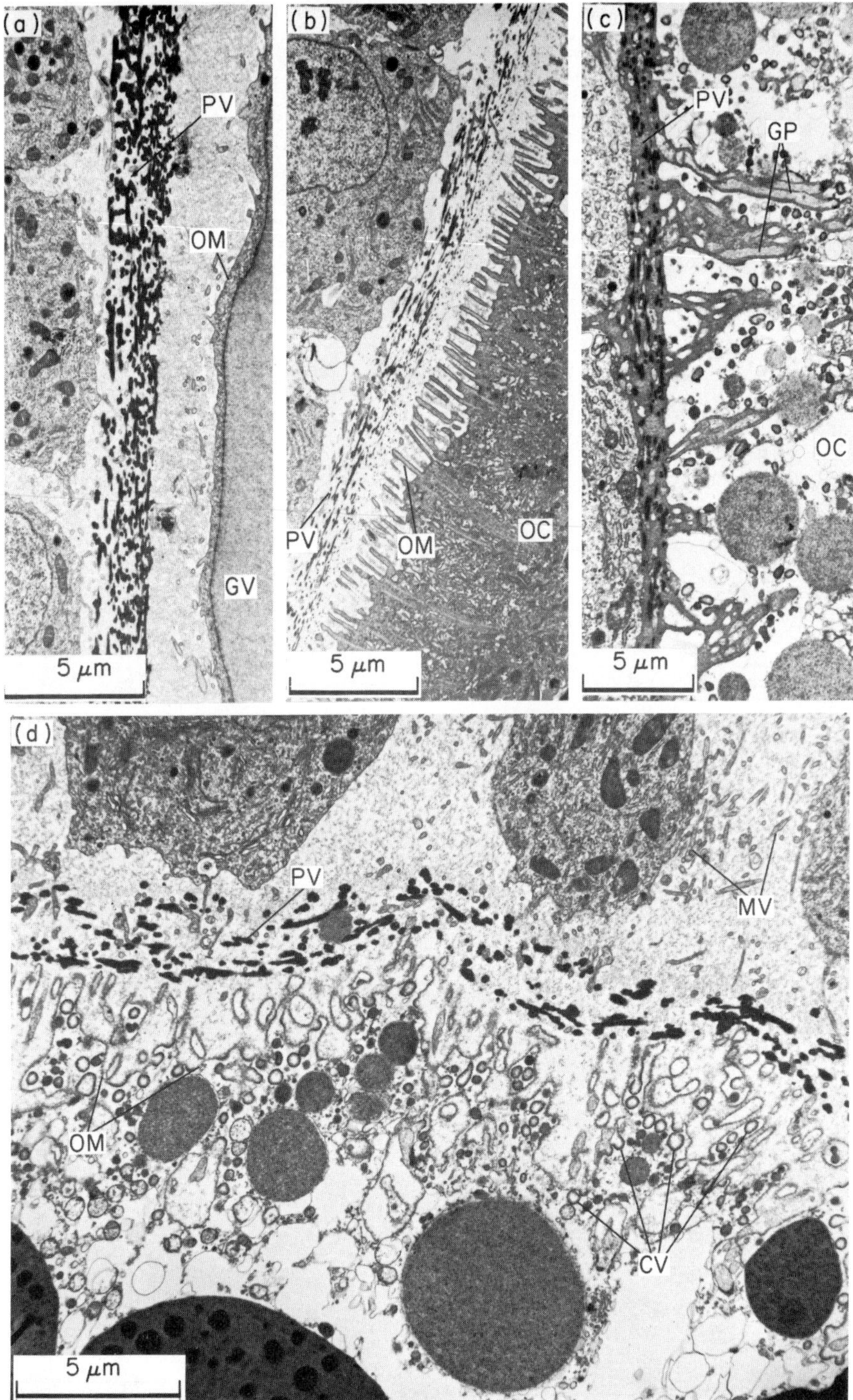

Fig. 5.18. Follicle-oocyte interrelationships. (Perry, unpublished). (a) In the region of the germinal vesicle (GV). (b) In the region of the germinal disc. (c) In the follicular wall. (d) In a special preparation (Gilbert *et al.*, 1977) of the follicular wall to illustrate better the oocyte cytolemma (OM), the microvilli (MV) and the coated vesicles (CV). Note also the complexity of structures in the outer layer of the yolk of the oocyte (OC). GP, projection of granulosa; PV, perivitelline layer.

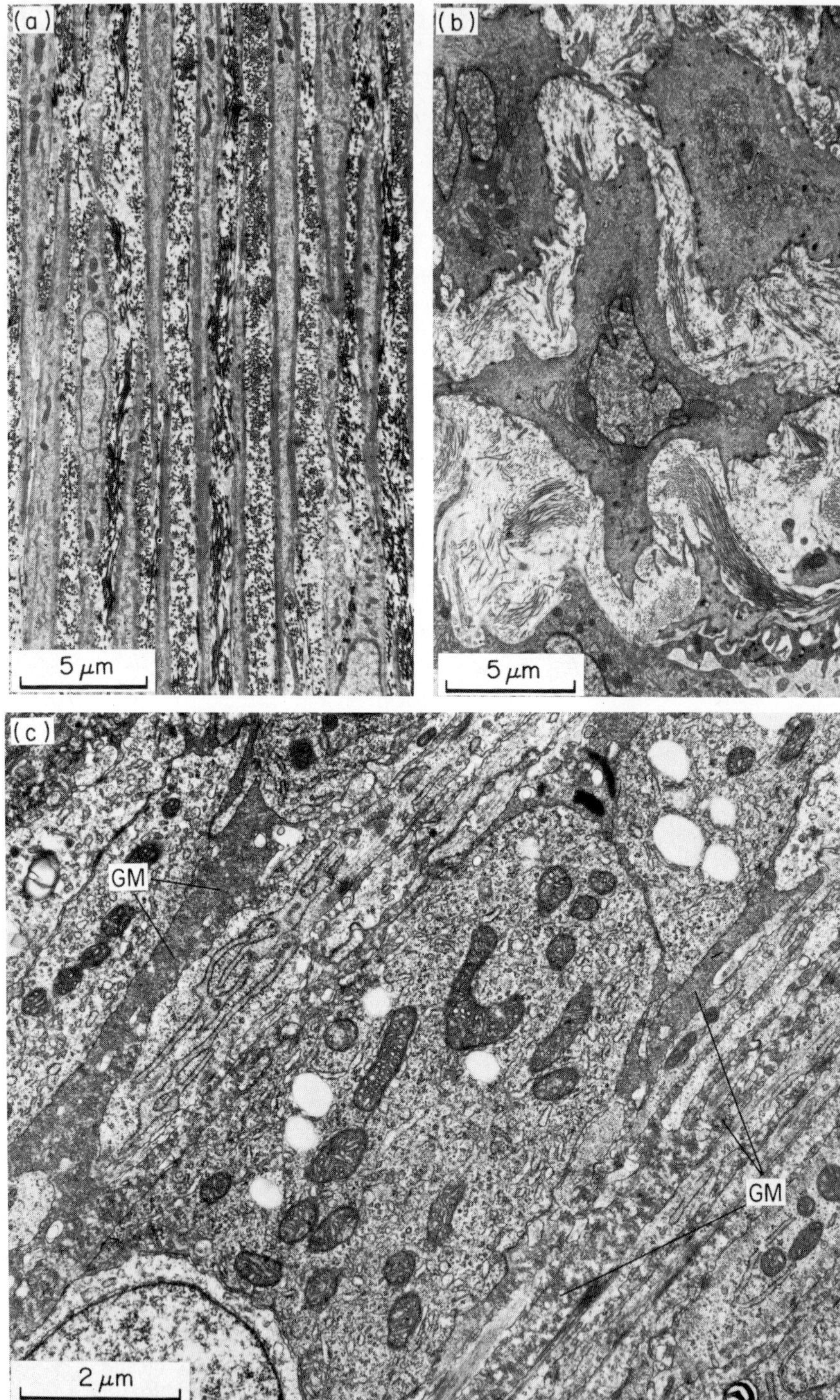

Fig. 5.19. Although seldom considered, the theca is a complex structure. (a) Fibroblasts and interspersed collagen fibres of the theca externa. (Perry, unpublished.) (b) Contracted fibroblast in tangential section. (Perry, unpublished). (c) Interstitial cell in the theca interna of a nearly mature follicle; in comparison with the appearance of these cells after ovulation only a few vacuoles were present (Fig. 5.24). The intercellular material (GM) is a prominent feature of the theca interna.

accumulation of true yolk has barely started. In the larger follicles (greater than 15 mm diam.) the structural elements of the zona appear to be restricted to the larger diverticula on the irregular surface of the oocyte (Fig. 5.18) (Wyburn *et al.*, 1965a; Perry *et al.*, 1978a): whilst numerous coated vesicles (Fig. 5.18) have been found their size was several times larger than those reported for the smaller follicles (Perry *et al.*, 1978a).

Despite the considerable attention to the ultrastructure of the oocyte-follicle border there is still uncertainty about it regarding nomenclature and function. However, the extensive studies by Lin (1968), Schjeide *et al.* (1970), Schjeide and de Vellis (1970), Paulson and Rosenberg (1972) and Jordanov and Boyadjieve-Michailova (1974) have helped to clarify this and should be consulted for a detailed discussion.

The perivitelline layer is acellular (Figs 5.4, 18) (McNally, 1943) and is probably produced by the granulosa cells (Bellairs, 1965; Wyburn *et al.*, 1965a). King (1975) suggested it may be homologous to the zona pellucida of mammals. It first appears when the follicle reaches about 2·5 mm in diameter as electron-dense aggregates between the granulosa cells and the oocyte (Bellairs, 1965). These aggregates extend until they become recognizable as the precursors of the electron-dense rods which finally form the meshwork of the layer (Wyburn *et al.*, 1965a, 1966; Perry *et al.*, 1978a,b). With time the thickness of the layer increases from about 1 to 3 mm (Bellairs *et al.*, 1963; Wyburn *et al.*, 1965a; Perry *et al.*, 1978a). Further details are provided under The egg.

The granulosa layer is sometimes referred to as the "follicular epithelium" (Hodges, 1974; Nickel *et al.*, 1977), a term borrowed from mammalian terminology. However, this is confusing since the avian follicle is also enclosed by the ovarian epithelium. Most of the descriptions of the granulosa layer apply only to relatively small follicles, although Wyburn *et al.* (1965a,b, 1966) did comment on its appearance in larger follicles. At first the granulosa cells (*cellulae strati granulosi*) are columnar and arranged in a manner resembling a pseudostratified epithelium (Romanoff and Romanoff, 1949; Press, 1964; Bellairs, 1965; Wyburn *et al.*, 1965a). As growth proceeds, the cells become less densely packed and appear approximately hexagonal in surface view (Fig. 5.17). Later the cells become progressively more flattened, passing through a cuboidal stage (Fig. 5.17) (Marza and Marza, 1935; Bellairs, 1965; Perry *et al.*, 1978a). At all times the nucleus remains in the part of the cell nearest to the basal lamina with the exception of the cells in the region of the germinal disc (Perry *et al.*, 1978b). Between the cells are spaces which progressively increase in width from 0·1 to 0·5 μm and which in the more mature follicles are filled with a fine granular material. In later stages these spaces are also invaded by fine microvillus-like processes of the granulosa cells which extend for lengths of up to about 3 μm (Figs 5.17, 18). The exceptionally wide intercellular spaces reported by Wyburn *et al.* (1966) and Rothwell and Solomon (1977) were not seen by Perry *et al.* (1978a), and may, in fact, be fixation artefacts since fixation of the follicle is not always easy (Rothwell and Solomon, 1977). The apical border of the granulosa cell is thrown into a series of processes which cross the perivitelline layer and make contact with the oocyte cytolemma (Figs 5.17, 18). The border in contact with the basal lamina is smooth. Adjacent cells may make contact through specific attachment structures (Rothwell and Solomon,

1977; Perry *et al.*, 1978a) suggesting that coordination of their function can occur.

Throughout its life the granulosa cell appears to be highly active, though its precise function in yolk deposition is not known. In the early growth of the follicle these cells conceivably add material to the oocyte through the transosomes ("lining bodies", "unique organelles") (Press, 1964; Bellairs, 1965; Schjeide *et al.*, 1966; Paulson and Rosenberg, 1972) but later they may act to channel the intercellular material towards the oocyte (Wyburn *et al.*, 1965a; Nishimura *et al.*, 1976; Perry *et al.*, 1978a) or even to modify this material as proposed by Grau and Wilson (1964). Their function in steroidogenesis will be considered later.

The basal lamina has been illustrated many times, though little specific attention has been directed towards it (Fig. 5.17). This is surprising since it forms a continuous sheet around the granulosa layer and hence forms the first barrier that yolk material must cross on its way to the oocyte. When first formed its appearance is similar to the basal lamina in other regions and other species but, as the follicle grows, it increases in thickness from 0·2 or 0·3 μm to about 1 μm (Perry *et al.*, 1978a). This is considerably greater than the basal lamina in other sites and suggests that in the avian ovarian follicle it may have an important, and perhaps different, function. Certainly the basal lamina of the follicle allows complex lipoprotein particles to pass through it, although an almost invariable property of most basal laminae is their ability to prevent this passage of lipid material.

The basal lamina of the follicle has signs of a reticular structure and appears to have embedded in it discrete particles (Perry *et al.*, 1978a), which may be lipoprotein (Evans, Perry and Gilbert, unpublished).

The theca interna forms about 25% of the thickness of the entire theca and is composed of various types of cell (Fig. 5.19c). Its division into three regions as proposed by Wyburn *et al.* (1965a) is seldom entirely clear. It is characterized by large fibroblast-like cells arranged in a discontinuous layer over the basal lamina (Perry *et al.*, 1978a). Features of these cells include the endoplasmic reticulum which has large dilated cisternae, and the fine microfilaments, the latter occurring in small numbers. The spaces between the cells contain collagen fibres and may be filled with a fine granular material (Fig. 5.19c). This material is particularly abundant in the region of the basal lamina (Perry *et al.*, 1978a), and Gilbert (1971b) has speculated that it might consist of yolk on its way to the oocyte. The characteristic interstitial cells (*interstitiocyti ovarii*) (Fig. 5.19) occur at the border between the theca externa and theca interna, although some may be entirely within the theca externa (Deol, 1955; Dahl, 1970a,b, 1971a,b; Peel and Bellairs, 1972). In relatively small follicles they appear to be grouped into specific bodies, the so-called "thecal glands" enclosed by a cellular capsule and basal lamina (Dahl, 1970a,b, 1971b). However, Perry *et al.* (1978a) were unable to demonstrate a similar structure in older follicles, the cells in these follicles being elongated and arranged in irregular, ill-defined patches. In micrographs of all follicles the cells contain vacuoles which become fewer with age; the vacuoles may have contained lipid material.

The theca interna is highly vascular (Nalbandov and James, 1949; Oribe,

1968). The presence of both diaphragm fenestrations and larger gaps in the endothelial wall is a curious feature of the capillaries in this region (Perry *et al.*, 1978a), although they are similar to those in the sinusoids of the rat liver (Wisse, 1970) and in the capillaries of the intestinal villi (Clementi and Palade, 1969). The larger gaps are sufficient to allow blood cells to pass through and erythrocytes are often found outside the blood vessels (Rothwell and Solomon, 1977; Perry *et al.*, 1978a). It must be conjecture at present whether or not this escape of blood bathing the outer surface of the basal lamina is part of the mechanism involved in the transport of macromolecules in yolk formation, although this seems likely to be so.

The theca externa forms the main mass of the follicle wall, yet it has not been carefully studied. It is composed of elongated fibroblast-like cells which in radial section of the follicle are arranged in a stratified manner (Fig. 5.19). In tangential sections the cells appear oval. These cells contain a broad zone of microfilaments, 7–9 nm in diameter, aligned parallel to the cell membrane (Perry *et al.*, 1978a). The theca externa also contains numerous collagen fibres and, according to Hodges (1974), some elastic fibres. There is still controversy about whether smooth muscle fibres are present or not (Hodges, 1974). Whilst most investigators described their presence, how they are distributed is not so certain, Guzsal (1966) observing them only at the pedicular end of the follicle, whereas other workers (Phillips and Warren, 1937; Kraus, 1947; Romanoff and Romanoff, 1949) found them throughout the follicle. If smooth muscle cells are present, it is surprising that they are not seen in published electron micrographs. Whilst Hodges (1974) identified them in the electron micrographs of Dahl (1970b) he may have mistaken them for the fibroblast-like cells which in section are similar to smooth muscle cells.

The two outer layers of the follicle consist of the ovarian epithelium and the connective tissue covering.

The blood supply to the follicle is extremely complex (Figs 5.20, 21) (Nalbandov and James, 1949; Oribe, 1968). Usually two, three or four arteries (*aa. pedunculares*) enter the pedicle and on reaching the follicular wall branch repeatedly to form the so-called "middle arterial layer" (*aa. intramurales*) (Oribe, 1968). The location of the middle arterial layer is still doubtful, Nalbandov and James (1949) and King (1975) placing it in the connective tissue layer (Fig. 5.16), and Hodges (1974), on the basis of Oribe's (1968) observations, placing it in the theca externa. From this arterial layer smaller arteries and arterioles pass into the theca interna and break up close to the basal lamina into a capillary network (*rete capillare terminale*).

The venous drainage is even more complex (Fig. 5.20). The sinus-like capillaries (Nishida *et al.*, 1977) drain into larger venules (*vv. intramurales internae*) close to the basal lamina. These internal intramural veins pass through the theca interna and open into a complex system of venules and sinuses forming the so-called "middle venous layer" (*vv. intramurales mediae*). The vessels of this layer drain into a few large veins (*vv. intramurales externae*) in the connective tissue layer. The external intramural veins open into approximately three venous trunks (*vv. pedunculares*) which pass out of the follicle. As with the arteries various locations have been ascribed to the veins: Hodges (1974) places

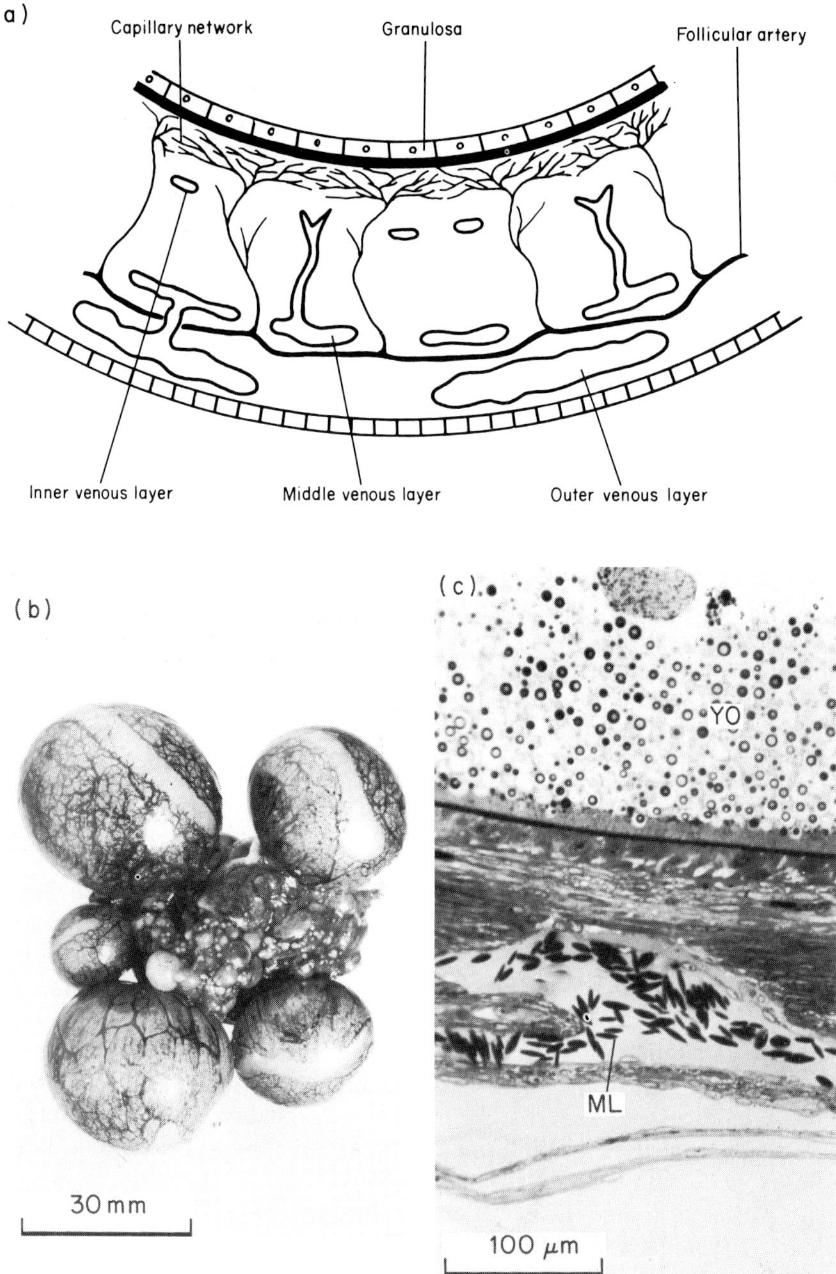

Fig. 5.20. (a) The vascular system of the follicle is complex; although it must be related to the transport to the oocyte of the large masses of yolk material, its precise function is not known. (Redrawn from Gilbert, 1971b and King, 1975). (b) Indian ink injected vascular system of the ovarian follicles; at this magnification only the outer venous layer is visible. (c) Vertical section through the follicular wall showing the middle venous layer (ML) and the large funnel-shaped vessel passing to the inner layer (compare also with Fig. 5.2). The yolk (YO) is that of the surface layer illustrated in Fig. 5.5. Araldite section, toluidine blue stain.

them mainly in the theca interna and theca externa, whereas Nalbandov and James (1949) and King (1975) place both the middle and outer venous layers in the connective tissue layer (Fig. 5.16).

The innervation of the follicle is more extensive than in mammals (Fig. 5.13) (Jacobowitz and Wallach, 1967) and consists of cholinergic and adrenergic fibres, nerve cell bodies and structures which resemble mammalian sensory nerve endings (Gilbert, 1965). The fibres enter through the pedicle (*nervi pedunculares*) and ramify (*nervi intramurales*) throughout all the layers external to the basal lamina. Most fibres appear to innervate blood vessels (Gilbert, 1965), although some are found in association with the interstitial cells (Dahl, 1970b; Perry *et al.*, 1978a). The function of the nerves is not known, although Gilbert and Wood-Gush (1970) suggested that, in part, they may regulate the distribution of blood-borne gonadotrophins throughout the ovary. Ferrando and Nalbandov (1969) obtained evidence that the nerves may be involved in ovulation and such a function has been proposed for them in mammals (Jacobowitz and Wallach, 1967). If sensory structures are present, nothing is known of the stimuli to which they respond.

Ovulation is brought about by a series of events that culminate in the rupture of the follicle along the stigma (Figs 5.21, 22). Whilst these events have been reviewed by Gilbert (1971h), the advances made in radio-immunoassay techniques since that time have altered considerably our concept of the vascular state of the gonadotrophins during the process (Wilson and Sharp, 1976; Etches and Cunningham, 1976).

The stigma is a specialized scar-like region at the apical pole of the follicle in many, though not all, species (Fig. 5.21). Compared with the rest of the follicle it is paler and relatively avascular (Nalbandov and James, 1949). Whilst the stigma is usually elongated and slightly curved, branched and trifoliate forms do occur (Guzsal, 1966). The stigma lacks the connective tissue coat of the other regions of the follicle and the remainder of the wall is thinner than elsewhere (Fig. 5.16) (Guzsal, 1966). While luteinizing hormone is the stimulus for ovulation, it is not known how ovulation is brought about, and there is no information available regarding the structural changes which must occur in the stigma before and at the time of ovulation. Nalbandov (1958, 1961) thought that necrosis followed the local ischemia which occurs a few hours before ovulation but Kraus (1947) was unable to detect any necrotic changes. Unlike in the mammal, there is seldom any bleeding at ovulation, loss of blood being prevented by contraction beforehand of the small arterioles (Lofts and Murton, 1973). Ovulation, by its very nature, terminates yolk deposition, though this may have stopped some time before (Gilbert, 1972). It does not, however, end oogenesis which continues until fertilization (see above).

(*c*) *Vitellogenesis.* As well as undergoing the cytoplasmic and nuclear changes in oogenesis, the oocyte has to accumulate the yolk material (vitellogenesis) necessary for the future growth of the embryo. This yolk is obtained from the follicular blood vessels because the yolk material is formed almost entirely in the liver (Gilbert, 1967; McIndoe, 1971; Lofts and Murton, 1973).

Vitellogenesis is a protracted process occupying a considerable length of time and it is convenient to arbitrarily divide it into various phases (van Durme, 1914; Marza and Marza, 1935; Bellairs, 1964; MacKenzie and Martin, 1967; Gilbert,

1971b), although these are not distinctly separable. Consequently, whilst in *Gallus* Marza and Marza (1935) described three phases, Gilbert (1971b) cited up to nine phases which had been described by other workers. In the Common Starling (*Sturnus vulgaris*) six phases have been observed (Bissonnette and Zujko, 1936).

The early-growth phase (Phase 1 of Marza and Marza, 1935) can last several years and during this time growth may be discontinuous (d'Hollander, 1904; Sonnenbrodt, 1908; Brambell, 1926). The oocyte increases in size slowly by an accumulation of cytoplasm, although a thin layer of lipid material forms near the periphery of the cell (Romanoff, 1960). During this time changes occur in the cytoplasm, including the dispersal of the Balbiani body; these changes have been considered to be related to the preparation for yolk deposition rather than to gametogenesis (Bellairs, 1964, 1967). At the end of this stage the oocyte of the domestic fowl has a diameter of about 1 mm.

Phase 2 (Marza and Marza, 1935) lasts approximately two months. Small membrane-bounded vacuoles appear (Bellairs, 1967) and gradually increase in number to fill the oocyte except for the nucleus and surrounding cytoplasmic inclusions. The nucleus is forced into an eccentric position and hence the bilateral organization of the egg, and the future orientation of the embryo, are determined at this stage, if not earlier (Bartelmez, 1912). Romanoff (1960) argues

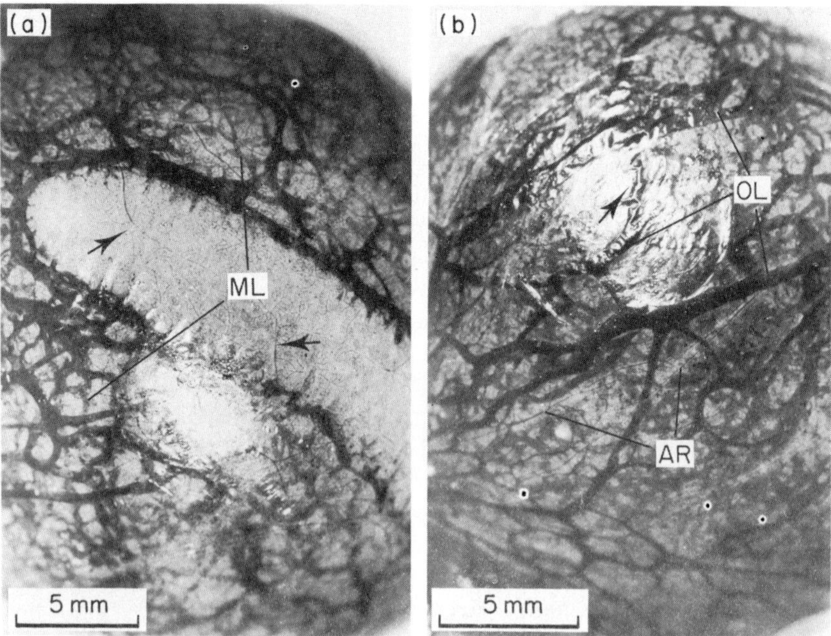

Fig. 5.21. The detailed appearance of the blood vessels in the wall of the follicle, the photograph (a) being taken in the region of the stigma. The stigma is not completely avascular and small vessels cross it (arrows). There is a considerable difference between the distended veins of the outer layer (OL) and the superficial arteries (AR). The vessels of the outer layer also stand out from the surface of the follicle (arrowed in b). Note that the middle layer (ML) is also very extensive. Indian ink injection.

that once the nucleus has become placed in an eccentric position in the oocyte, the events which follow merely serve to reinforce this earlier orientation. Gradually more white-yolk is deposited and in *Gallus* the oocyte reaches a diameter of about 4 mm.

The final phase, the "rapid-growth" stage, is the shortest of all and is probably the best known, for it is during this time that the ovary takes on its characteristic avian appearance. This phase lasts in *Gallus* between 6 and 14 days (Bellairs, 1964; Gilbert, 1971b), in *Meleagris* between 11 and 15 days (Bacon and Cherms, 1968) and in *Coturnix* between 5 and 7 days (Homma *et al.*, 1965; Bacon and Koontz, 1971). Of the little information available in other species 5–7 days has been observed in the Ring-necked Dove (*Streptopelia capicola*) (Cuthbert, 1945), 4 days in the Rook (*Corvus frugilegus*) (Welty, 1962), 8 days in *Columba* (Bartelmez, 1912) and 12 days in the Canada Goose (*Branta canadensis*) (Grau, 1976). From these data it appears that there may be a tendency for larger species to have oocytes which remain longer in the rapid-growth phase; this would be consistent with the generally larger size of the oocyte at ovulation. An extension of the work of Grau (1976) could give the information required to show whether the rate of yolk deposition varies between species.

The structure of yolk has been described earlier (The egg). Similarly the structural aspects of the follicle and oocyte cytolemma in relation to yolk deposition were discussed in The Structure of the Follicle. As yet, the precise

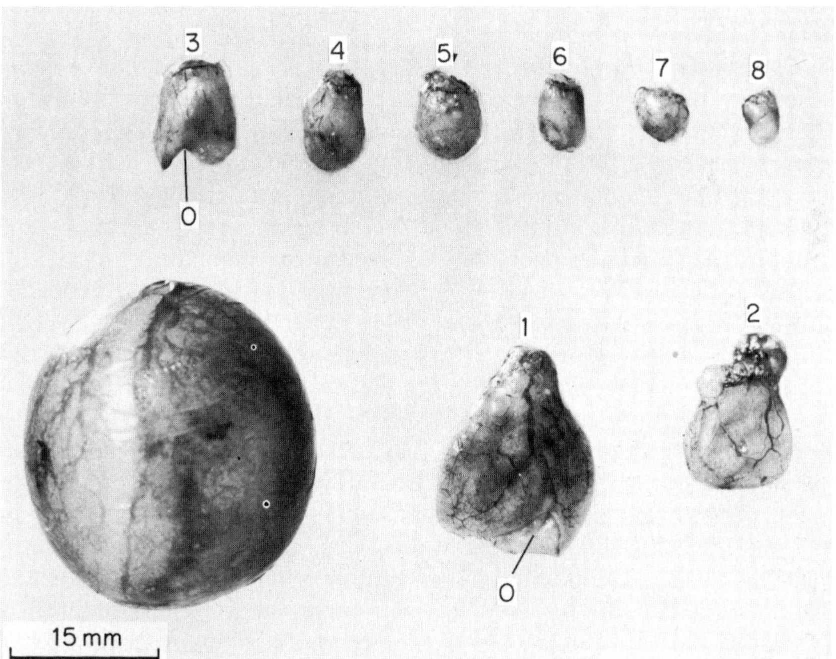

Fig. 5.22. Regression of the post-ovulatory follicle in comparison with a mature pre-ovulatory follicle; time is given in days. The opening (O) is always towards the lower part of the photograph. For the first few days after ovulation the blood vessels are still visible macroscopically.

mechanisms involved in yolk deposition are not understood. Ultimately pinocytosis by the oocyte is likely to be involved to some degree but which substances are transported into the oocyte in this way, and in what quantity, has yet to be determined. However, the work of Cutting and Roth (1973), Roth *et al.* (1976), and Yusko and Roth (1976) has provided the basis for a better understanding of these processes, although the size of their coated vesicles (Woods *et al.*, 1978) is less than those found *in vivo* (Perry *et al.*, 1979a).

Although yolk deposition occurs over most of the surface of the oocyte, there appears to be little or none in the germinal region of the cell. For this reason the disc of the latebra (nucleus of Pander) is formed (Figs 5.2, 3, 5) (Marza and Marza, 1935) and, as yellow yolk continues to be deposited, this floats upwards drawing out the latebra into the characteristic neck (Fig. 5.5) (Romanoff and Romanoff, 1949; Romanoff, 1960). Since the germinal area is paler than the rest of the yolk, it may frequently be distinguished through the wall of the follicle, although it is often obscured by the pedicle (Fig. 5.16).

At the end of the period, i.e. near the time of ovulation, the size of the oocyte is that of the central-yolk mass found in the laid egg. Its size is related to the species and some avian oocytes are among the largest of the animal cells known, that of the extinct Madagascan Elephant Bird reaching 175 mm in diameter (Romanoff and Romanoff, 1949). During this phase the oocyte of the domestic fowl grows from about 5 to 40 mm (Fig. 5.11) (Gilbert, 1971b), that of *Columba* from 5 to 20 mm (Bartelmez, 1912) and that of the White-crowned Sparrow (*Zonotrichia leucophrys*) from 0·5 to 6·6 mm (Kern, 1972). Perhaps the smallest increase in size is found in hummingbirds (Trochilidae) in which the final oocyte diameter is only a few millimetres.

(*d*) *The post-ovulatory follicle.* The structure remaining in the ovary after ovulation is now more usually called the post-ovulatory follicle (*folliculus post ovulationem*) (Figs 5.8, 9, 11, 22). The follicle undergoes a process of regression, sometimes termed "atresia", although Ingram (1962) proposed that this term is better restricted to the non-ovulatory regression described below.

Although a post-ovulatory follicle has been observed in many species, descriptions of it and the processes involved in its regression are lacking for most birds (Stieve, 1919; Hett, 1923; Yocom, 1924; Dominic, 1960a). By far the most intensely studied species has been *Gallus* (Sonnenbrodt, 1908; Pearl and Boring, 1918; Stieve, 1918; Novak and Duschak, 1923; Fell, 1925; Davis, 1942b; Romanoff, 1943; Deol, 1955; Floquet and Grignon, 1964; Aitken, 1966; Guzsal, 1966; Wyburn *et al.*, 1966). Most authors agree that the avian post-ovulatory follicle, although homologous to the mammalian structure, should not be called a corpus luteum (for comparative descriptions of vertebrates see Brambell, 1956, Franchi, 1962 and Harrison, 1962). Thus, although the lumen of the follicle becomes filled with granulosa cells, there is no cellular multiplication (Davis, 1942b; Brambell, 1956; Dominic, 1960a; Aitken, 1966) and the thecal interstitial cells are not involved. Moreover, the endodrine function of the avian post-ovulatory follicle may be considerably different from that of the mammal (Wells and Gilbert, unpublished). However, for a contrary view see Floquet and Grignon (1964) and Guraya and Chalana (1975). According to Aitken (1966) some of the invading fibroblast-like cells may be derived from the theca.

Immediately after ovulation the follicle of the domestic fowl contracts to a cup-shaped structure about half the length of the mature follicle (Fig. 5.22). This reduction in size may be brought about by the contraction of the smooth muscle and the elasticity of the stretched theca (Fig. 5.19). Perhaps also the fine microfilaments of the fibroblast-like cells of the theca (Perry *et al.*, 1978a) confer on them some contractability. This reduction in size occurs rapidly, and during it the basal lamina becomes thrown into folds and may in places become detached from the theca interna (Fig. 5.23) (Aitken, 1966). The granulosa cells are disrupted from their orderly arrangement along the basal lamina and become grouped into clumps or even thrown off into the lumen.

After the first contraction, the fate of the post-ovulatory follicle is one of gradual regression and absorption (Fig. 5.22), as it is in most other non-mammalian vertebrates (Brambell, 1956). This regression must obviously require a functional vascular system, the latter in *Gallus* remaining extensive (Fig. 5.21) (Nalbandov and James, 1949), whilst in the White-crowned Sparrow (*Zonotrichia leucophrys*) hyperaemia has been observed (Kern, 1972). During the later stages the luminal mass becomes invaded by capillaries (Brambell, 1956) and nerve fibres (Gilbert, 1965). Regression in *Gallus* takes 7–10 days and this period seems common to most of the species which have been examined (Fig. 5.22) (Stieve, 1919; Hett, 1923; Davis, 1942b; Romanoff, 1943). However, Kern (1972) calculated that the post-ovulatory follicle he found in the White-crowned Sparrow (*Zonotrichia leucophrys*) could have been between 26 and 50 days old, a length of time closer to the several months which has been cited for the Ring-necked Pheasant (*Phasianus colchicus*) (Kabat *et al.*, 1948) and the Mallard (*Anas platyrhynchos*) (Lofts and Murton, 1973).

During regression the most affected part of the follicle is the theca. As reabsorption proceeds the walls of the "cup" rapidly become shorter. The basal lamina initially appears to swell, then to fragment and finally, by about the 2nd or 3rd day, to disappear (Aitken, 1966; Wyburn *et al.*, 1966). The interstitial cells persist and may still be functional (Peel and Bellairs, 1972). The granulosa cells remain throughout the resorptive phase as a central mass in the lumen, the mass gradually filling the lumen as a result of hypertrophy of the cells and resorption of the wall of the follicle: mitosis does not occur (Brambell, 1956). The granulosa cells, however, do show changes during regression (Fig. 5.23). At first the cells are non-vacuolated (Aitken, 1966) but by 24 h after ovulation they contain appreciable amounts of lipid (Wyburn *et al.*, 1966). This lipogenesis continues causing the hypertrophy of the cells resulting in a distinct yellowing of the central mass of the follicle. After several days degenerative stages are visible and the final phase is one of invasion by phagocytes.

Although this general description seems to apply to most of the species examined, there are some exceptions. Thus Kern (1972) described a rapid disppearance of the granulosa cells in the White-crowned Sparrow (*Zonotrichia leucophrys*). According to Lofts and Murton (1973) in some species hypertrophy of the theca occurs whilst Guraya and Chalana (1975) reported a luminal invasion by the thecal luteal cells in the House Sparrow (*Passer domesticus*).

It must be admitted that our knowledge of the function of the post-ovulatory follicle is restricted almost entirely to one species, the domestic fowl. It is obviously dangerous to extrapolate from one species to all birds but, until the

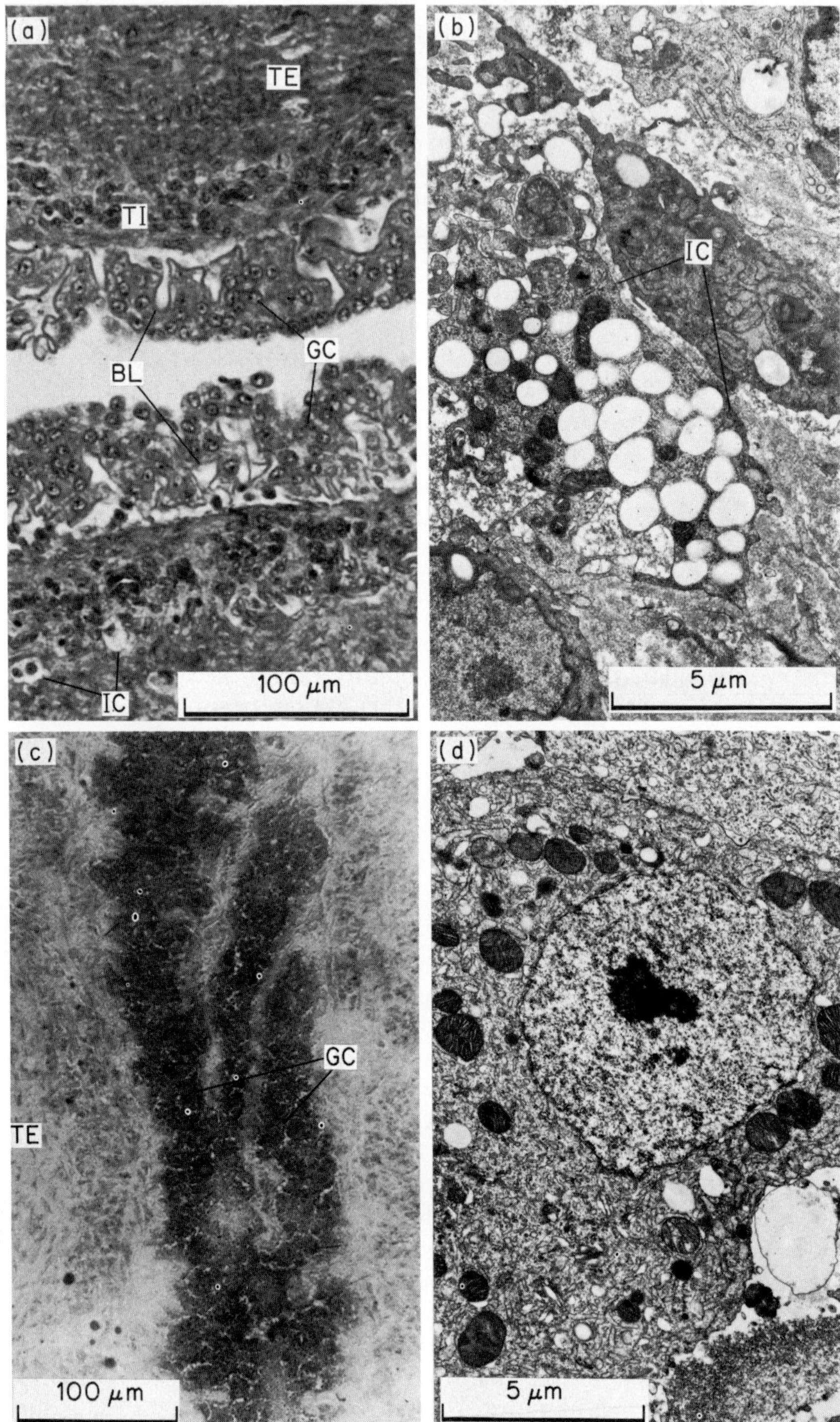

Fig. 5.23. The post-ovulatory follicle is a common feature of the ovary in the post-breeding period. (a) Section through the lumen of a 1 h post-ovulatory follicle. BL, basal lamina; GC, granulosa cells; IC, interstitial cells; TE, theca externa; TI, theca interna. Picro-Mallory. (b) Electron micrograph of an interstitial cell (IC) of a 24 h post-ovulatory

contrary is proved, it seems reasonable to assume a function for the post-ovulatory follicle in them also.

Like the corpus luteum of the mammal, the post-ovulatory follicle of *Gallus* has an effect on the function of the oviduct, directly or indirectly. Excision of the post-ovulatory follicle results in delayed oviposition of the egg derived from that follicle (Rothchild and Fraps, 1944b) and abnormal laying times have been recorded following ligation of the pedicle (Wood-Gush and Gilbert, 1975). The post-ovulatory follicle is also important in the nesting behaviour associated with the laying of the egg derived from that follicle (Wood-Gush and Gilbert, 1975). Moreover, when two ovulations occur within about 4 h of each other, two nestings follow (Scott, 1972). Both oestrogen and progesterone appear to be involved (Wood-Gush and Gilbert, 1975), the inference being that progesterone is derived from the post-ovulatory follicle. However, recently Soliman and Walker (1975), Tanaka (1976) and Tanaka and Goto (1976) have found that the post-ovulatory follicle produces a non-steroidal hormone which affects oviposition. The prostaglandins of the ovary may also be important in oviposition (Day and Nalbandov, 1977).

In contrast with the mammal, it must be concluded that in *Gallus* the post-ovulatory follicle is an active structure during only the first 24 h of its life and that no known function has been found for the older stages. Whether or not the same is true for other birds has yet to be determined.

3. *The ovary in the post-laying period*

Egg-laying normally ceases once the bird completes the clutch size characteristic of her species. She then enters the next phase of her reproductive activity, incubation, which is subsequently followed by broodiness, the care of the young chick. The intermediate post-laying periods lead into another period when the animal is in a non-breeding state, a period which in most cases occupies the majority of the lifetime of the animal.

(*a*) *The period of follicular atresia.* The precise causal relationship between the end of egg laying and the onset of incubation behaviour has not been completely determined. But the change from egg-laying to incubation is associated with considerable changes in the appearance of the ovary, since the normal mechanisms responsible for the growth and maturation of the oocyte (see Gilbert, 1971b,h for a general consideration of these) no longer operate in a way which will lead to ovulation.

A process then occurs for the removal and reabsorption of the developing oocyte and its follicular covering. This process (atresia) is common to most classes of vertebrates (Brambell, 1956; Ingram, 1962; Chieffi and Botte, 1970)

---

follicle. Compare its vacuolation with the cell in Fig. 5.19. (c) The granulosa cells (GC) are intensely positive with the tetrazolium reaction for steroid dehydrogenases but this does not necessarily indicate steroid production. The theca (TE) is negative. (d) Granulosa cell 1 h after ovulation; the first signs of the progressive vacuolation (lipid accumulation) are present even at this stage.

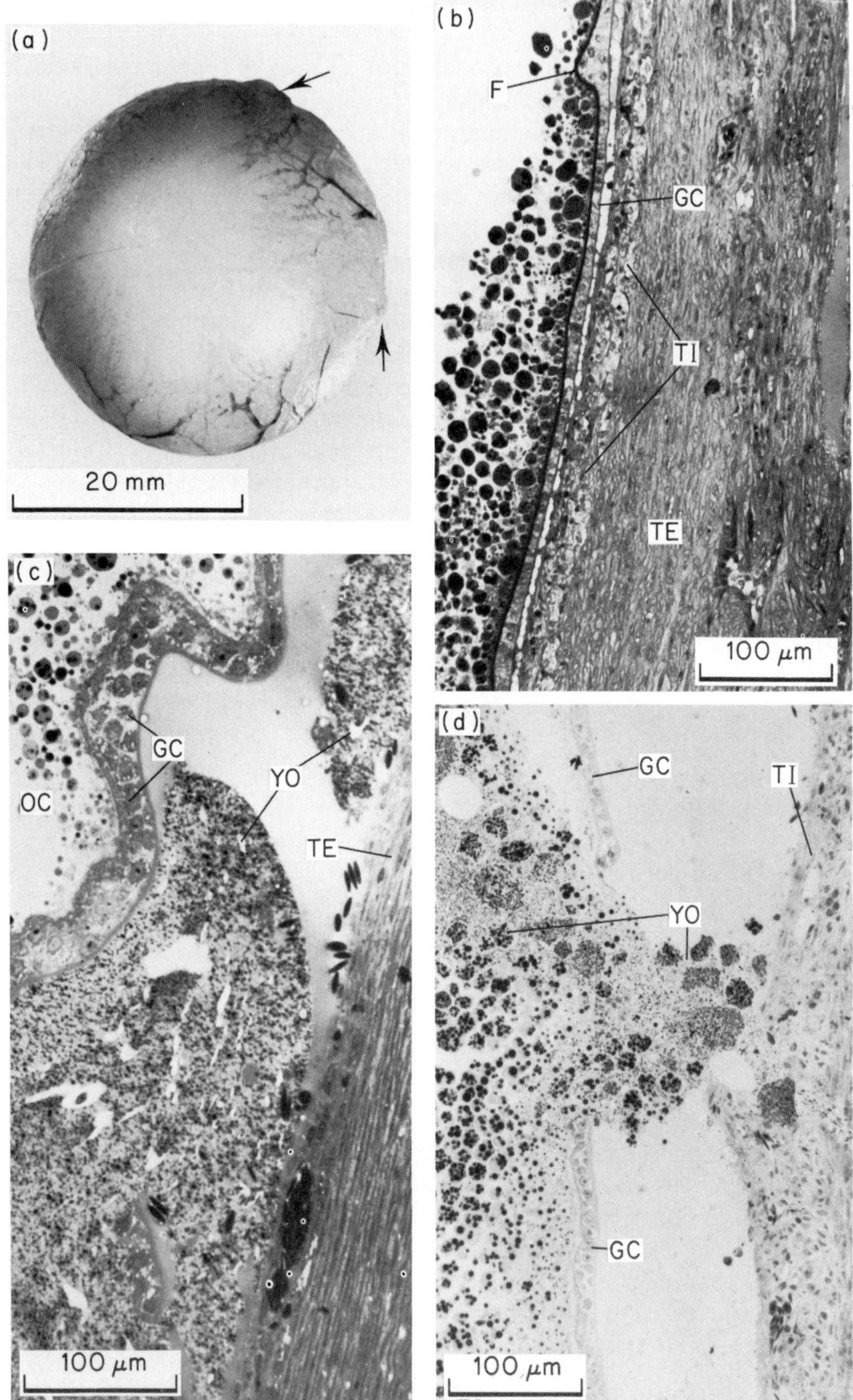

Fig. 5.24. Atresia is a normal feature of bird ovaries: "bursting atresia" (illustrated in Figs 5.24 and 5.25) is more common in the post-laying period whereas "invasion atresia" (illustrated in Figs 5.13 and 5.14) may occur at any time. (a) The first macroscopic signs of bursting atresia are the pale raised spots (arrowed) on the surface of the follicle (compare with normal follicles in Figs 5.11, 5.20 and 5.21). (b) Before the stage of (a) is reached there is considerable internal change; the first noticeable feature is the detachment of the granu-

and atretic follicles (*folliculi atretici*) have been found in all avian species which have been examined for them. Unfortunately, the available information in birds varies considerably from one account to the next, between two and five different types of atretic follicle having been described in different species (Fell, 1925; Davis, 1942a; Deol, 1955; Marshall and Coombs, 1957; Dominic, 1961b; Kern, 1972; Erpino, 1973). Since no comparative investigations have been made it is difficult to know whether these discrepancies in the literature are real or whether they simply reflect differences in technique and interpretation. Following Kern (1972) and Erpino (1973), an attempt has been made to collate the available data in order to provide a simple classification of atresia, though this is not entirely satisfactory.

The most widely found form of atresia has been descriptively called "bursting atresia" (Figs 5.24, 25). Bursting atresia is characterized by the rupture of the thecal wall and the escape of the yolk contents. The liberated yolk may be contained within the ovary or, particularly with larger follicles, it may escape into the body cavity (Dunham and Riddle, 1942; Dominic, 1961b). This type of atresia occurs in follicles of all sizes which contain appreciable amounts of yolk. For this reason, Davis (1942a) considered it to be a specialized type of atresia, differing from the mammalian type, and associated with the removal and reabsorption of large masses of yolk. This indeed may be so, because nearly identical forms have been recognized in monotremes (*Ornithorhynchus*) (Hill and Gatenby, 1926), reptiles (Loyez, 1906; Bragdon, 1952) and the fish *Mystus seenghala* (Sathyanesan, 1960). Despite its widespread occurrence, bursting atresia is the least common form of atresia in birds. Amongst the species in which it has been observed are the Jackdaw (*Corvus monedula*) (Hett, 1923), the Slate-coloured Junco (*Junco hyemalis*) (Rowan, 1930), *Gallus* (Fell, 1925; Bates *et al.*, 1935), the Northern Fulmar (*Fulmarus glacialis*) (Wyne-Edwards, 1939), *Columba* (Dunham and Riddle, 1942; Dominic, 1961b), the White-crowned Sparrow (*Zonotrichia leucophrys*) (Kern, 1972), the Rook (*Corvus frugilegus*) (Marshall and Coombs, 1957), the Scrub Jay (*Aphelocoma coerulescens*) (Erpino, 1973), the Common Starling (*Sturnus vulgaris*) (Bullough, 1942), the Shiny Cowbird (*Molothrus bonariensis*), the Smooth-billed Ani (*Crotophago ani*), the Herring Gull (*Larus argentatus*), the Indigo Bunting (*Passerina cyanea*), the Ruby-throated Hummingbird (*Archilochus colubris*), the Swallow-wing (*Chelidoptera tenebrosa*) and the Barred Woodcreeper (*Dendrocolaptes certhia*) (Davis, 1942a). It is restricted more usually to the post-laying period when more follicles containing yolk are present in the ovary. However, it does occur, at least in domestic species, in response to a variety of stimuli which interrupt normal ovarian function (Stieve, 1918; Dunham and Riddle, 1942; Fraps and Dury, 1942; Fraps and Neher, 1945; Rothchild and Fraps, 1944a, 1945).

---

losa (GC) from the theca interna (TI) and theca externa (TE) and its tendency to fold (F). Araldite section, toluidine blue. (Grau and Gilbert, unpublished). (c) After rupture of the granulosa (see d) the yolk (YO) escapes from the oocyte (OC) which contracts from the theca externa (TE) with the granulosa (GC) attached to it. Araldite section, toluidine blue. (d) Site of rupture of the granulosa (GC) with the first escape of yolk (YO). The granulosa has separated from the theca interna (TI). Araldite section, toluidine blue. (Grau and Gilbert, unpublished).

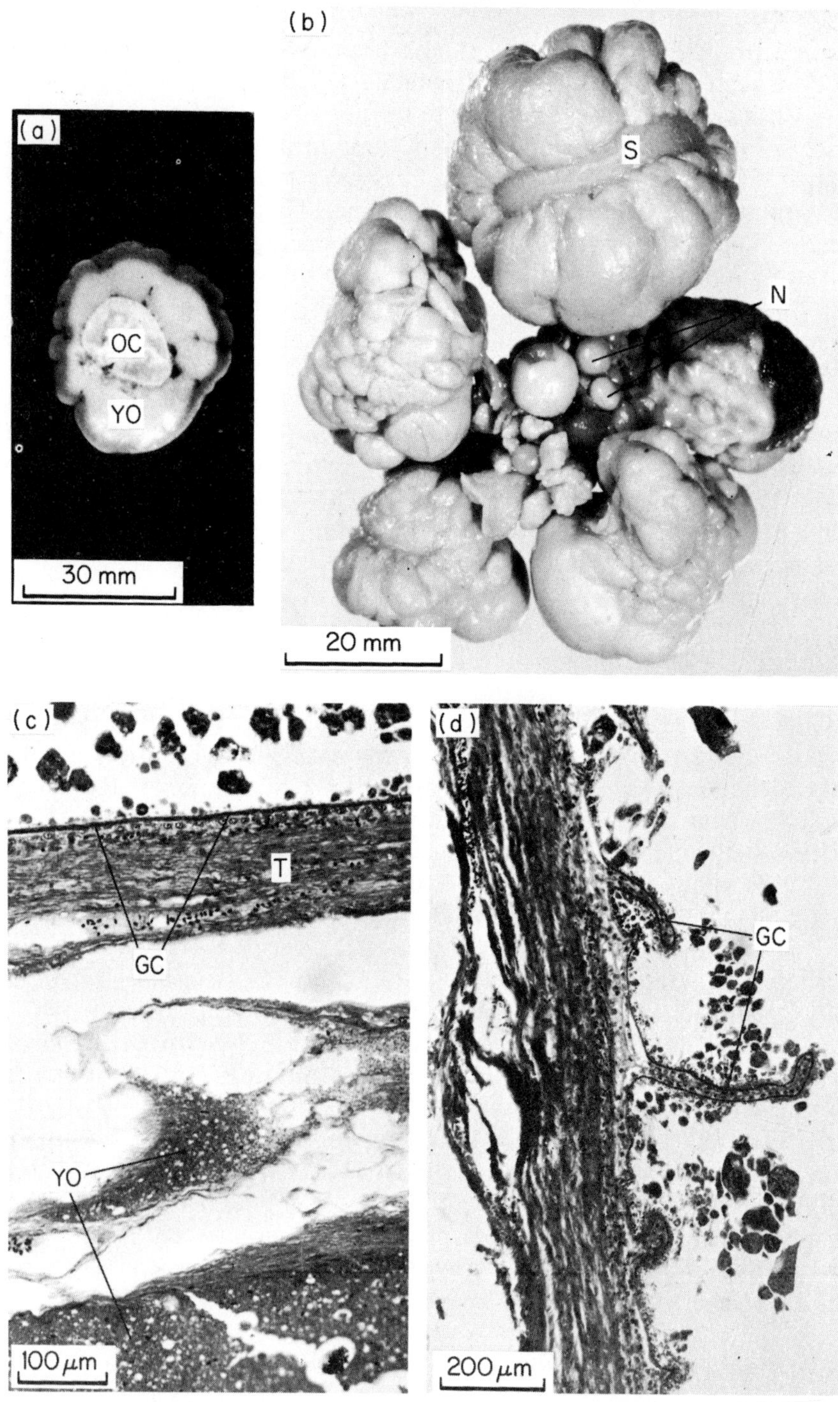

Fig. 5.25. (a) Towards the end of bursting atresia nearly all of the yolk (YO) lies within the outer layers of the follicle external to the theca externa: the theca and the oocyte (OC) remain as a central core. Interior of formalin fixed follicle similar to those in (b). (Grau and Gilbert, unpublished). (b) A typical atretic ovary; the stigma (S) becomes very prominent because in this region the epithelium alone remains attached to the theca since there is no underlying connective tissue for the yolk to flow into. Many of the smaller follicles (N)

The first macroscopic sign of bursting atresia in the larger follicles is the development on the surface of the follicle of minute pale areas (Fig. 5.24) (Fell, 1925; Davis, 1942a). These areas are formed by freed yolk collecting beneath the follicular epithelium. With time the areas become more extensive. Finally, the whole surface of the follicle is thrown into a series of bulges, the follicle as a whole, however, losing its turgidity and becoming flaccid to the touch (Fig. 5.25). The blood vessels of the follicle, apart from a few large vessels, are hidden beneath the outer layer of yolk. Descriptions of the histological changes in this type of atresia have been given many times (Dubuisson, 1906; Pearl and Boring, 1918; Fell, 1925; Dunham and Riddle, 1942; Davis, 1942a; Deol, 1955; Marshall and Coombs, 1957; Dominic, 1961b; Kern, 1972; Erpino, 1973). In *Gallus* (Grau and Gilbert, unpublished) and *Columba* (Dominic, 1961b) the first microscopic signs are a separation of the oocyte cytolemma from the granulosa layer and the formation of spaces between the theca interna and the basal lamina (Fig. 5.24). The granulosa layer ruptures allowing yolk to escape from the oocyte (Fig. 5.24). Following this, shrinkage of the oocyte takes place. Finally the theca ruptures and allows the freed yolk to collect beneath the follicular epithelium, thus producing the macroscopic pale spots on the surface of the follicle (Figs 5.24, 25). After the rupture of the theca, the theca and oocyte contract further and thus sink deeper beneath the surface of the yolk. Up to this stage the process is rapid, less than 24 h (Grau and Gilbert, unpublished). Thereafter invasion of the mass is carried out by phagocytes possibly derived from the hyperplastic granulosa layer and is accompanied by an increase in the connective tissue and an involution of the blood vessels (Fell, 1925; Davis, 1942a; Dominic, 1961b; Kern, 1972). Davis (1942a) and Dominic (1961b) divided this period into five and six merging phases, respectively.

The final fate of the follicle is not entirely certain. In some cases a hyaline degeneration may occur, which leaves a small scar-like area within the ovarian tissue similar to the mammalian corpus albicans (Fell, 1925; Dominic, 1961b). However, both Fell (1925) and Davis (1942b) commented on the similarity which is usually seen between the final stages of regression in the post-ovulatory follicle and those in bursting atresia. Fell (1925) and Aitken (1966) considered that either could give rise to the yellow bodies (*corpora aurea*) described throughout the ovary by Pearl and Boring (1918). On the other hand, complete removal without any scar seems possible, since Davis (1942a) asserted that resting ovaries do not show any remnants of atretic follicles. Liquifaction atresia, described by Kern (1972) in the White-crowned Sparrow (*Zonotrichia leucophrys*), seems likely to be a variant of bursting atresia. According to Kern (1972) it is similar to bursting atresia, except that the yolk is more liquid and is green in colour. It is also similar to the yolky atresia described by Erpino (1969) in the Common Magpie (*Pica pica*).

The other main type of atresia is invasion atresia which, in contrast to bursting

---

have a normal appearance. (c) Despite the efflux of yolk (YO) some areas of the granulosa (GC) and theca (T) appear to retain their normal relationships for a time. Picro-Mallory. (d) An early stage of atresia: yolk has escaped from the oocyte, and adjacent regions of the granulosa (GC) are drawn inwards as a series of folds, a continuation of the process illustrated in Fig. 5.24. Picro-Mallory.

atresia, is characterized by an invasion of cells, either from the granulosa layer or theca, and a reabsorption of yolk *in situ*. The processes are therefore more similar to follicular atresia in mammals than to bursting atresia (Davis, 1942a). Invasion atresia is almost invariably associated with the smaller, less yolky, follicles. Precise details differ between the species, and variants occur within a species, but the general type has been found in many different birds (Loyez, 1906; Stieve, 1919; Fell, 1925; Davis, 1942a; Marshall and Coombs, 1957; Dominic, 1961b; Erpino, 1969, 1973; Kern, 1972; Eroschenko and Wilson, 1974; Guraya, 1976). It can be separated into two types according to whether the invasion is by granulosa cells or thecal cells. Invasion by granulosa cells occurs in two forms, a lipoglandular form (Kern, 1972; Erpino, 1973; Guraya, 1976) and a glandular form (Marshall and Coombs, 1957; Kern, 1972; Erpino, 1973). Histologically both of these forms are similar (Fig. 5.13); the oocyte disrupts and the granulosa cells invade the degenerating mass. In the lipoglandular form the granulosa cells become highly lipoidal and phagocytic, in contrast to the non-lipoidal cells in the glandular form. Subsequent invasion by connective tissue obliterates the lumen and the follicular wall becomes fibrous. Later the follicular wall also becomes obliterated, so leaving no scar. The second type of invasion, i.e. by thecal cells, has been described in the Rook (*Corvus frugilegus*) by Marshall and Coombs (1957) and in the Tree Sparrow (*Passer montanus*) by Lofts and Murton (1973), and it may be similar to an invasion described in *Gallus* by Deol (1955). In this type, the thecal cells become highly lipoidal. Hence Marshall and Coombs (1957) proposed the term "lipoidal atresia" for this type.

Various other structures have been included in atresia by a number of authors but it is doubtful if they have always been correct in doing so. Among these structures are the tracts of lymphocytes and fibroblasts which have been found to occur suddenly in the ovaries of several species including the Rook (*Corvus frugilegus*), the Puffin (*Fratercula arctica*) and the Dovekie (*Alle alle*), and which obliterate all follicles in their path (Marshall and Coombs, 1957). Various cyst-like structures have also been mentioned (Fell, 1925; Dominic, 1961b). Whilst the origin and fate of these structures is unknown, invasion by connective tissue may occur at times. The rare formation of a hyaline structure similar in appearance to the corpus albicans of mammals has been described in the Rook (*Corvus frugilegus*) by Marshall and Coombs (1957). Finally, Kern (1972) warned that his "lipoidal atresia" of small oocytes may not be a correct description.

(*b*) *The inter-breeding period.* The resting ovary of those species which have been examined is similar macroscopically to that of the juvenile ovary and can only be distinguished from it with difficulty, if at all.

Histologically the presence of scars of the post-ovulatory follicle would indicate its adult age, but these are not easy to distinguish from the scars of atretic follicles (Fell, 1925; Davis, 1942a; Marshall and Coombs, 1957; Kern, 1972). Furthermore, atresia is not a positive indication since it is now well recognized that atresia can occur in waves throughout the life of the bird from hatching, although it is more prominent at about the breeding or post-breeding period (Brambell, 1926; Fauré-Fremiet and Kaufman, 1928; Moreau *et al*., 1947; Marshall and Coombs, 1957; Erpino, 1969, 1973; Kern, 1972; Lofts and Murton, 1973). This being so there must also, conversely, be a continual process of oogenesis and vitellogenesis during non-breeding periods. This is completely unlike the male bird in which gametogenic activity is arrested during non-breeding periods (Marshall and Coombs, 1957).

Interestingly the interstitial cells in the resting ovary of adult domestic fowl appear to be atrophic and non-secretory (Peel and Bellairs, 1972) which should distinguish them from those of the juvenile ovary, although this has not been studied in great depth. If this difference exists, it could be a means whereby the adult non-breeding ovary could be distinguished from the juvenile ovary.

4. *Steroidogenesis*

It is unequivocal that steroidogenesis is an important function of the avian ovary (Baillie *et al.*, 1966; Gilbert, 1971g; Lofts and Murton, 1973; Murton and Westwood, 1977), as it is in mammals. It must, however, be accepted that there is no definitive proof that any particular cell produces a specific hormone.

If the evidence of Woods and Weeks (1969) is accepted [they found $\Delta^5-3\beta$-hydroxysteroid dehydrogenase ($-$HSD) in embryos of 2 days incubation], steroidogenesis is possible in the embryo before the morphological differentiation of the ovary. Moreover, further support for this view was provided by recent culture experiments which showed that the undifferentiated gonad in *Anas* and *Gallus* can synthesize steroids from precursor material (Haffen, 1975). Positive evidence for early steroid production was obtained when both oestrogens and androgens were isolated from the gonadal tissue of embryo domestic fowl of 6 and 10 days incubation (Gallien and Le Foulgoc, 1957, 1958; Boucek *et al.*, 1966; Cedard and Haffen, 1966; Noumura, 1966; Wolff *et al.*, 1966; Haffen and Cedard, 1968; Weniger, 1968) and *Coturnix* embryos (Kannankeril and Domm, 1968; Sayler *et al.*, 1970). These findings have since been confirmed by the organ-culture experiments of Guichard *et al.* (1977). This production of steroids in early development coincides with the first proliferation of the germinal cords which, it has been suggested, give rise to the interstitial cells. Moreover, it has been shown that the embryonic interstitial cells contain large amounts of cholesterol and lipid which are possible precursors of steroid metabolism (De Simone-Santoro, 1967; Scheib and Haffen, 1968; Narbaitz and de Robertis, 1968). The interstitial cells also have the appearance at the ultrastructural level of steroid-producing cells (Peel and Bellairs, 1972).

This link between the interstitial cells and steroidogenesis in the embryo is continued in the adult. In the Rook (*Corvus frugilegus*) the development of the stromal interstitial cells closely parallels the development of sexual displays known to be steroid dependant (Marshall and Coombs, 1957). In *Gallus* these cells show greater hydroxysteroid dehydrogenase activity (Fig. 5.16) when the animal enters the reproductive phase, and the ovary does produce oestrogens, progesterones and androgens (Boucek and Savard, 1970). The work of Dahl (1970c, 1971a,c) and Peel and Bellairs (1972) is again consistent with the fact that the interstitial cells produce steroids.

Despite this, linking particular hormones with either the stromal or thecal cells is difficult. Woods and Domm (1966) claimed to have shown androgen production is associated with the stromal interstitial cells by using a fluorescent antibody technique, and a key enzyme in androgenesis, $\Delta^5-3\beta-$HSD, has also been found in these cells (Botte, 1963; Chieffi and Botte, 1965; Narbaitz and de Robertis, 1968). However, both thecal and stromal cells are positive for $17\beta-$HSD when androgen output is high, as occurs in moulting hens.

The source of oestrogen is less clear, although again there are strong indications that the thecal interstitial cells, apart probably from those in the large yolky follicles, are involved (Senior and Furr, 1975). The evidence for this is discussed by Lofts and Murton (1973) and Murton and Westwood (1977) and again depends on the interpretation of the presence and activity of the $\Delta^5-3\beta-$ and $17\beta-$HSD enzymes in relation to known steroid output.

Unlike the other hormones, progesterone production has not been generally attributed to the interstitial cells. The most likely source of this hormone is the granulosa layer (Gilbert, 1971g; Lofts and Murton, 1973), although Peel and Bellairs (1972) concluded that the cells of the granulosa layer do not have the structural inclusions of steroid producing cells. However, care must be exercised in using only morphological grounds to ascribe function to a particular structure especially since the experimental work of Dahl (1971d,e) does indicate that the granulosa cells are involved in steroidogenesis. Progesterone is apparently produced by the pre-ovulatory follicle (Furr, 1969). It may be produced also by the post-ovulatory follicle (Layne, 1957; Furr, 1969) since this follicle contains HSD enzymes (Fig. 5.23). However, recent evidence suggests that this is unlikely (Dick *et al.*, 1978).

Much discussion has centred around the role of the atretic follicles in steroidogenesis, and both Fraps (1955) and Marshall and Coombs (1957) believed that atretic follicles produce progesterone or a similar substance that could be involved in ovulation. However, the evidence is less than conclusive. While cholesterol, and both $\Delta^5-3\beta-$ and $17\beta-$HSD are present (Marshall and Coombs, 1957; Chieffi and Botte, 1970), Sayler *et al.* (1970) found a reduction in the potential for steriod production in *Coturnix* as atresia progressed. According to Marshall and Coombs (1957) and Erpino (1973) the cells remaining at the end of atresia in the Rook (*Corvus frugilegus*) and the Scrub Jay (*Aphelocoma coerulescens*) may be concerned in steroidogenesis. Guraya (1976) considered that in columbid ovaries the cells of the theca interna of atretic follicles hypertrophied into steroid producing cells. However, neither Kern (1972) nor Erpino (1969) could find any evidence for this in other species.

In conclusion it must be agreed with Lofts and Murton (1973) that there is still a need for more information before the location of the sites for the production of each steroid can be precisely defined.

C. *The right gonad*

1. *The undeveloped gonad*

In most individuals of the majority of species the right gonad (*gonadum dextrum*) does not develop after the early period of incubation. Thus it is composed of mainly medullary tissue, although in *Gallus* and perhaps other species there are small areas of cortex. Consequently the right gonad resembles a testis more than an ovary. Typically the right gonad is small and difficult to find, and in *Gallus* is present as an ill-defined strip of tissue no more than 5 or 6 mm long by 1 mm wide on the right, ventral side of the caudal vena cava. Although the position of the gonad is not often mentioned in descriptions of other species, it is probably similar to that of *Gallus*.

Despite the usually quiescent existence of the right gonad it can undergo a rapid development process in some species should the left ovary be damaged or destroyed either accidentally or experimentally. Exceptions to this include *Meleagris* (Domm, 1939). When development of the right gonad occurs it is not straightforward and the result depends to a large extent on the age at which the damage to the left gonad takes place. If damage in *Gallus* takes place before 30 days after hatching, the gonad develops into a testis or ovotestis, either of which can produce spermatozoa (Domm, 1939; Kornfeld and Nalbandov, 1954; Kornfeld, 1958; Nalbandov, 1958). Clearly up to this time the male-like medullary tissue and enclosed primordial germ cells must retain their potential for development. After 30 days the incidence of ovotestes or ovarian-like structures increases, and some of these structures it has been claimed, have been able to produce follicles and to ovulate (Nalbandov, 1959).

Histologically both types of gonad differ considerably from normal ovaries. In most cases testicular and ovarian tissues are intermingled, and are present in amounts which can vary widely.

2. *The functional right ovary*

Although the more usual feature of the avian reproductive system is the presence of only one fully developed ovary (and oviduct), sometimes the right gonad may develop into an ovary. When a right ovary is present, it appears morphologically similar to the left ovary (von Faber, 1958) and could be functional (Domm, 1939). However, positive anatomical proof of function has been obtained in only a few birds. According to Kinsky (1971) one specimen of the Goshawk (*Accipiter gentilis*) was reported by Stieve (1924) to have a ruptured follicle in the right ovary and ruptured follicles were found in the Peregrine Falcon (*Falco peregrinus*) by White (1969). Kinsky (1971) himself reported a functional right ovary as the norm in the Brown Kiwi (*Apteryx australis*). A similar ovary was also claimed to have been found in *Anas* by Chappelier (1913). In the author's laboratory one specimen of domestic fowl was obtained with two ovaries both of which were apparently functional, a post-ovulatory follicle being visible in the right ovary (Fig. 5.9). From a study of the laying record of this bird and an examination of the post-ovulatory follicles present in each side it seemed that ovulation occurred mainly alternately from each ovary.

When two ovaries are present, the left is characteristically the larger. However, the right ovary has been reported to be the larger one in the Marsh Hawk (*Circus cyaneus*) (Brodkorb, 1928), the Brown Kiwi (*Apteryx australis*) (Kinsky, 1971), the Sparrow-hawk (*Accipiter nisus*) and the Common Kestrel (*Falco tinnunculus*) (Gunn, 1912). Of the Falconiformes, the hawks (Accipitridae) tend to have the largest right ovaries. However, in 50% of the Sparrow-hawks (*Accipiter nisus*) examined by Stieve (1924) the right ovary was equal in size to the left. In falcons (Falconidae) the right ovary is intermediate in size. The smallest right ovary occurs in vultures and buzzards (Stanley and Witschi, 1940). Strangely, owls (Strigidae) have the typical avian pattern of a single ovary which is situated on the left side. In falcons, when two ovaries are present, both are arranged symmetrically and parallel, though in other groups the arrangement may not be so regular (Gunn, 1912).

D. *Accessory reproductive organs*

The accessory reproductive organs in birds consist, as in other vertebrates, of the left and right paramesonephric (Müllerian) ducts and the left and right mesonephric (Wolffian) ducts. In the normal way the paramesonephric ducts arise in the embryo as paired structures but only the left develops and differentiates fully into the functional adult organ associated with the left ovary.

The male ducts (mesonephric ducts) are also formed in the female embryo but they do not develop thereafter.

Unlike the ovary which produces little, if any, of the material transported to the oocyte, the oviduct actively synthesizes the organic proteinaceous material added to the egg on its passage down the oviduct. In the case of *Gallus* this means the production and secretion of about 3 g of protein with each egg. The amount of protein secreted by the oviduct in other species is proportional to the size of the egg. The internal and external structure of the oviduct clearly reflects this activity.

1. The left oviduct

After the oocyte and its perivitelline covering are shed at ovulation, they are transformed into an egg by being engulfed by the infundibulum and by having deposited on them various layers before finally being oviposited. All of these stages require the active participation of the oviduct. Fertilization is not a necessary prerequisite for egg formation in birds and they are different in this respect from mammals in which successful fertilization must occur before subsequent development takes place.

Because no avian group has become viviparous, the left oviduct (*oviductus sinister*) resembles more closely the primitive form than does that of the placental mammal. However, the formation of the cleidoic egg has involved the specialization and differentiation of successive regions of the oviduct in birds and reptiles (Aitken and Solomon, 1976), a process which has been paralleled very closely in the egg-laying monotremes (*Echidna* and *Ornithorhynchus*) (Hill and Hill, 1933). The generalized morphology and overall function of the avian oviduct have been known for many years, and this knowledge has been summarized, at least for *Gallus*, in a series of recent reviews (Aitken, 1971; Gilbert, 1971c,d; Gilbert and Wood-Gush, 1971; Simkiss and Taylor, 1971; Hodges, 1974; King, 1975). Information on other species is rare and, when available, is mainly for the more common laboratory or domestic birds, such as *Columba*, *Meleagris*, *Anas* and *Coturnix* (Chakravorti and Sadhu, 1959; Verma and Cherms, 1964;

Das *et al.*, 1965; Tamura and Fujii, 1966a,b; Dalrymple *et al.*, 1968; Fitzgerald, 1969; King, 1975), although occasional reports are available for other species, e.g. the Canary (*Serinus canaria*) and the Black Kite (*Milvus migrans*) (Chakravorti and Sadhu, 1961a; Hutchison *et al.*, 1968).

Although far from exhaustive, the available information led Marshall (1961) to comment on the remarkable uniformity of structure throughout the class. Differences, where they exist, are often of a minor nature and their functional significance is largely unknown.

(*a*) *The oviduct in the non-breeding condition.* In general terms, the adult oviduct in the non-breeding season is similar to that of the juvenile bird (Ljungkvist, 1967; Yu *et al.*, 1972; Yu and Marquardt, 1973a; Boogaard and Finnegan, 1976), although, as noted by Kinsky (1971) in the Brown Kiwi (*Apteryx australis*), small differences in appearance may exist in some species between the oviduct of the adult post-breeding condition and that of the juvenile. Nevertheless, they are sufficiently similar to be considered together.

The oviduct is usually visible as a small, pale pink, semi-translucent structure lying on the ventral surface of the left kidney (Fig. 5.27). In this form it consists of a narrow tube stretching from the region of the ovary to the cloaca. Its actual length is consequently dependent on the size of the animal and the species. It is composed of the following seven layers (Fig. 5.26) (Hodges, 1974; King, 1975): epithelium; lamina propria (Fig. 5.32); inner connective tissue layer; circular muscle layer; outer connective tissue layer; longitudinal muscle layer; serosa.

Even in the non-breeding condition, the oviduct is divided into the five regions listed below in craniocaudal order although these can be distinguished only with some difficulty (Aitken, 1971; Hodges, 1974; King, 1975) (Fig. 5.27).

(1) *Infundibulum.*
(2) *Magnum.*
(3) *Isthmus.*
(4) Shell gland (*uterus*).
(5) *Vagina.*

Most of the above terms for the regions are acceptable, although there is no entirely satisfactory term for the shell gland (*uterus*). "Uterus" has become well entrenched in the literature but, because it implies an unjustified homology with the mammalian structure (Richardson, 1935; Marshall, 1961), it would be better if it were not used. "Shell gland", although descriptive, tends towards cumbersome terminology when applied to the nerves and blood vessels of the region. However, it is used here in preference to "uterus" since it clearly distinguishes the region of the oviduct from the mammalian organ. Similarly, the use of mammalian terms, such as "pre-ampulla" and "ampulla" (Rhea *et al.*, 1974), should be avoided.

The oviduct is suspended in the coelom by a double-layered sheet of peritoneum which divides at the oviduct into a dorsal ligament (*lig. dorsale oviductus*) and a ventral ligament (*lig. oviductus ventrale*) (Figs 5.8, 11, 29) (Curtis, 1910; Kar, 1947a). The dorsal ligament is attached to the dorsal wall of the coelomic cavity and runs in *Gallus* diagonally from the 4th thoracic rib (the 6th rib in *Coturnix*, Fitzgerald, 1969; the last rib in the Budgerigar, *Melopsittacus undulatus*, Evans, 1969) to the cranial division of the left kidney,

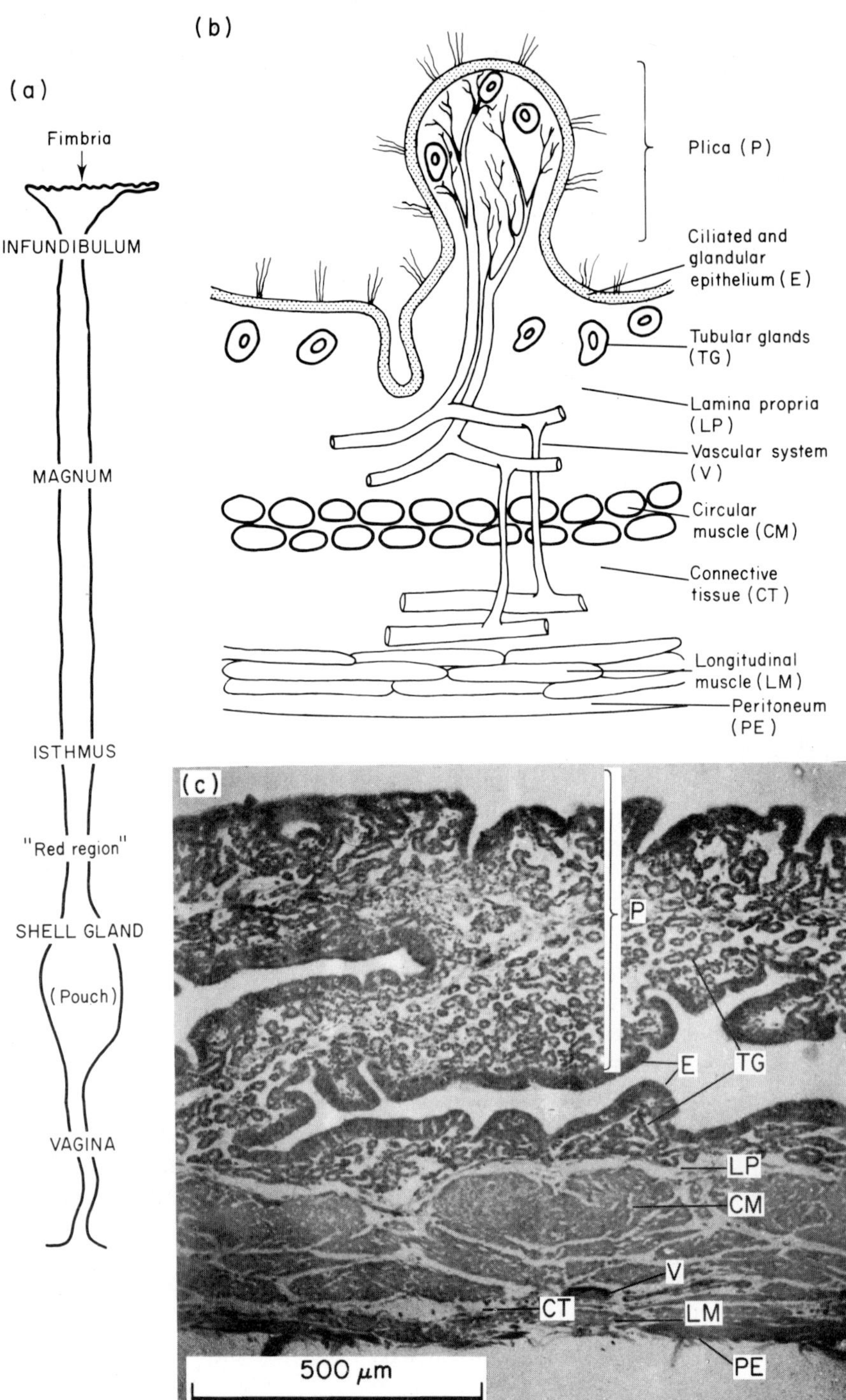

Fig. 5.26. (a) Scale diagram of the oviduct of the domestic fowl: the proportions of the region vary only slightly from species to species. (b) Schematic diagram of the wall of the oviduct to show its basic structure; this, however, may vary from region to region (see text). (c) Transverse section through the wall of the shell gland to illustrate the features drawn in (b). Haematoxylin and eosin.

passing close to the lateral surface of the ovary. For the rest of its length it runs parallel to the longitudinal axis of the kidney to reach the cloaca, being attached to the whole length of the oviduct including the vagina. At the cranial end of the ligament a secondary fusion occurs with the left abdominal air sac along with a narrow strip beginning dorsally and cranially near the infundibulum. Further caudally the medial surface of the ligament fuses with the rectum. The ventral ligament is the continuation of the dorsal ligament beyond the oviduct and therefore its ventral border is free. It is not so extensive as the dorsal ligament, stretching only from the caudal tip of the infundibular funnel to the cranial portion of the vagina. Both dorsal and ventral ligaments contain smooth muscle which blends with the outer longitudinal muscle of the oviduct. These oviductal muscles are poorly developed in the immature bird but are capable of activity (Chen and Hawes, 1970).

The blood supply to the juvenile oviduct has not been studied in detail but generally is similar to that of the oviduct in the adult, though less extensive (Fig. 5.12) (Boogaard and Finnegan, 1976).

In the juvenile bird there is no communication between the lumen of the oviduct and the cloaca, unlike in the non-breeding adult. The occluding plate between the oviduct and cloaca only breaks down at the onset of the first breeding season (Greenwood, 1935; Kar, 1947b), the breakdown possibly being controlled by oestrogens (Kar, 1947b). Histologically the epithelium in the juvenile remains undifferentiated and secretory cells cannot be demonstrated (Boogaard and Finnegan, 1976). Tubular glands are absent and the surface bordering the lumen is smooth with only minor depressions. However, the glandular grooves of the infundibulum are present.

Growth after hatching is slow, but progressive, and is brought about mainly by hyperplasia, although some hypertrophy does occur (Fig. 5.27) (Yu and Marquardt, 1973a). In the older breeds of domestic fowl, even up to 20 weeks of age, i.e. just before sexual maturity and laying, the oviduct weighs only about 5 g (Romanoff and Romanoff, 1949) and has a length of about 15 cm (Giersberg, 1922). In the earlier maturing modern breeds, however, considerable growth occurs before this (Fig. 5.27). During this growth there is a gradual morphological development of the tubular glands (Yu and Marquardt, 1973a), although their secretory contents remain sparse until about the onset of laying.

A complete description of the resorption process which follows laying does not appear to exist. However, the available evidence suggests that regression is essentially the reverse process to growth (Yu and Marquardt, 1973a, 1974; Eroschenko and Wilson, 1974). The earliest structures affected are the glandular elements which lose their secretory granules. This is followed by a gradual reduction in the total mass of the oviduct, the evidence from *Coturnix* and *Gallus* indicating that there is a decrease in the number of cells during this process (Fitzgerald, 1969; Yu and Marquardt, 1974). Ultimately the tubular glands disappear and the juvenile condition is reached. Undoubtedly these changes are brought about by a withdrawal of oestrogen but lack of progesterone may also play a part (Yu *et al.*, 1972; Palmitter, 1973; Eroschenko and Wilson, 1974; Boogaard and Finnegan, 1976).

(*b*) *The adult oviduct in the breeding season.* The change in the environmental conditions to those suitable for reproductive activity affects the output of gonadotrophins. This brings about changes in steroid production by the ovary (van Tienhoven, 1961; Gilbert, 1967; Lofts and Murton, 1973) which, in turn, stimulates the rapid growth and differentiation of the oviduct. Undoubtedly the oestrogens are the most important of the steroid hormones in the seasonal hypertrophy of the oviduct and they are known to cause growth and differentiation of the oviduct in many species including *Gallus* (Juhn and Gustavson, 1930, 1932; Common *et al.*, 1947, 1948; Kar, 1947b; Brant, 1953; Brant and Nalbandov, 1956; Oades and Brown, 1965; Ljungkvist, 1967), *Columba* (Riddle and Tange, 1928; Riddle, 1942; Chakravorti and Sadhu, 1963), the House Sparrow (*Passer domesticus*) (Keck, 1934), the Black-crowned Night Heron (*Nycticorax nycticorax*) (Noble and Wurm, 1940), *Coturnix* (Sandoz *et al.*, 1975) and the Common Starling (*Sturnus vulgaris*) (Witschi and Fugo, 1940). Differentiation of the glandular components is also under the control of oestrogens (Kohler *et al.*, 1969). However, for the full maturation process and secretion, hormones other than oestrogens, such as progesterone and androgens, are required as has been demonstrated in *Gallus* (Common *et al.*, 1947; Hertz *et al.*, 1947; Hertz, 1950; Mason, 1952; Brant and Nalbandov, 1952, 1956; Brant, 1953; Lorenz, 1954; Nalbandov, 1959; Oades and Brown, 1965; Ljungkvist, 1967; O'Malley *et al.*, 1969; Cox and Sauerwein, 1971; Palmitter and Wrenn, 1971; Yu and Marquardt, 1973b), a dove (Columbidae) (Lehrman and Brody, 1957), *Coturnix* (Sandoz *et al.*, 1975), the Canary (*Serinus canaria*) (Hutchison *et al.*, 1968), the Common Starling (*Sturnus vulgaris*) (Benoit, 1950) and *Columba* (Chakravorti and Sadhu, 1963).

Despite the overwhelming evidence for the importance of the steroid hormones generally, the precise role of each in protein formation and secretion has not been fully determined. Hence so far progesterone has been positively found to be important in the formation of only one specific protein (avidin) (O'Malley *et al.*, 1969), although Palmitter (1971) claimed that both conalbumin and ovalbumin were affected by it. On the other hand, oestrogen is important in the formation of several specific proteins. For a fuller discussion on these points see Gilbert (1971c), Schimke *et al.* (1975, 1977) and Sutherland *et al.* (1977).

(i) General morphology. The increase in size of the oviduct from the juvenile or non-breeding state to that of the breeding adult is considerable (Fig. 5.27). In *Gallus*, for example, it increases in length from 15 cm to up to about 86 cm (Giersberg, 1922), although with modern breeds 60 cm appears to be a more usual length (Fig. 5.40). Along with the increase in length is a corresponding increase in diameter (Romanoff and Romanoff, 1949). So enlarged does the oviduct become that it fills the left half of the abdominal cavity and may protrude into the right (Fig. 5.8). Full topographical descriptions of the oviduct of the domestic fowl are given by Kern (1963) and King (1975). Similar overall increases in size have been noted in *Meleagris* (Verma and Cherms, 1964; Dalrymple *et al.*, 1968), *Coturnix* (Fitzgerald, 1969), the House Sparrow (*Passer domesticus*) (Witschi, 1961), *Anas* (Das *et al.*, 1965) and the Brown Kiwi (*Apteryx australis*) (Kinsky, 1971), and so may be common to most birds. However, the illustrations by Lofts and Murton (1973) of the Green-winged Teal (*Anas crecca*) and the Tree Sparrow (*Passer montanus*) suggest that the hypertrophy which occurs in these species may be relatively small. There is also

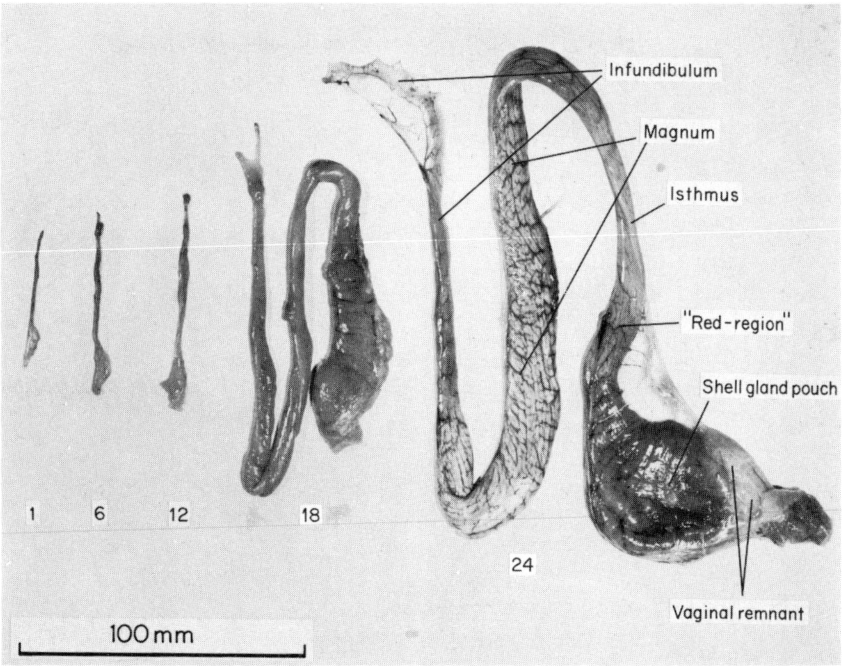

Fig. 5.27. The appearance of the oviducts (less vagina) of the birds the ovaries of which are shown in Fig. 5.11. Age in weeks.

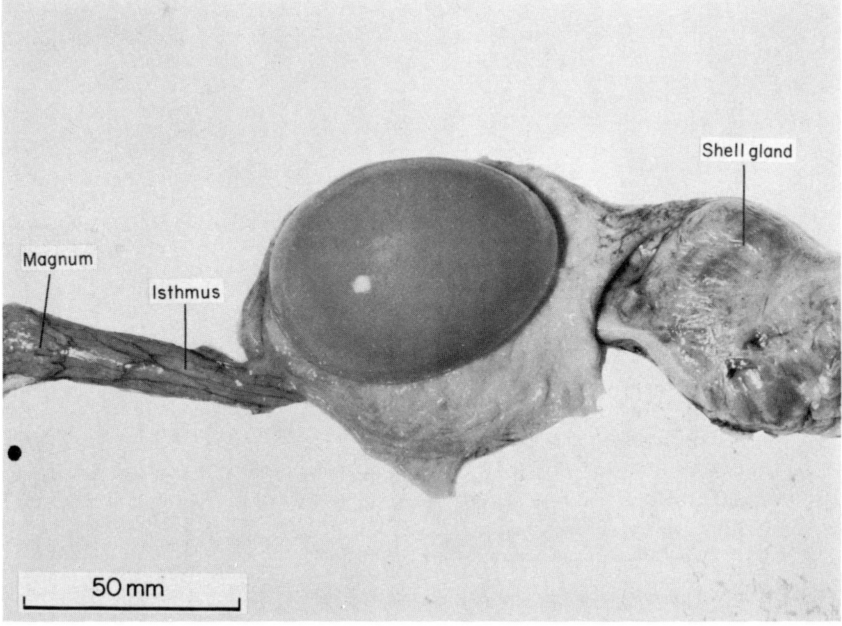

Fig. 5.28. Membraneous egg *in situ* in the isthmus/"red region". The egg-shape has already been produced even though no calcified shell is present. (Gilbert, 1974).

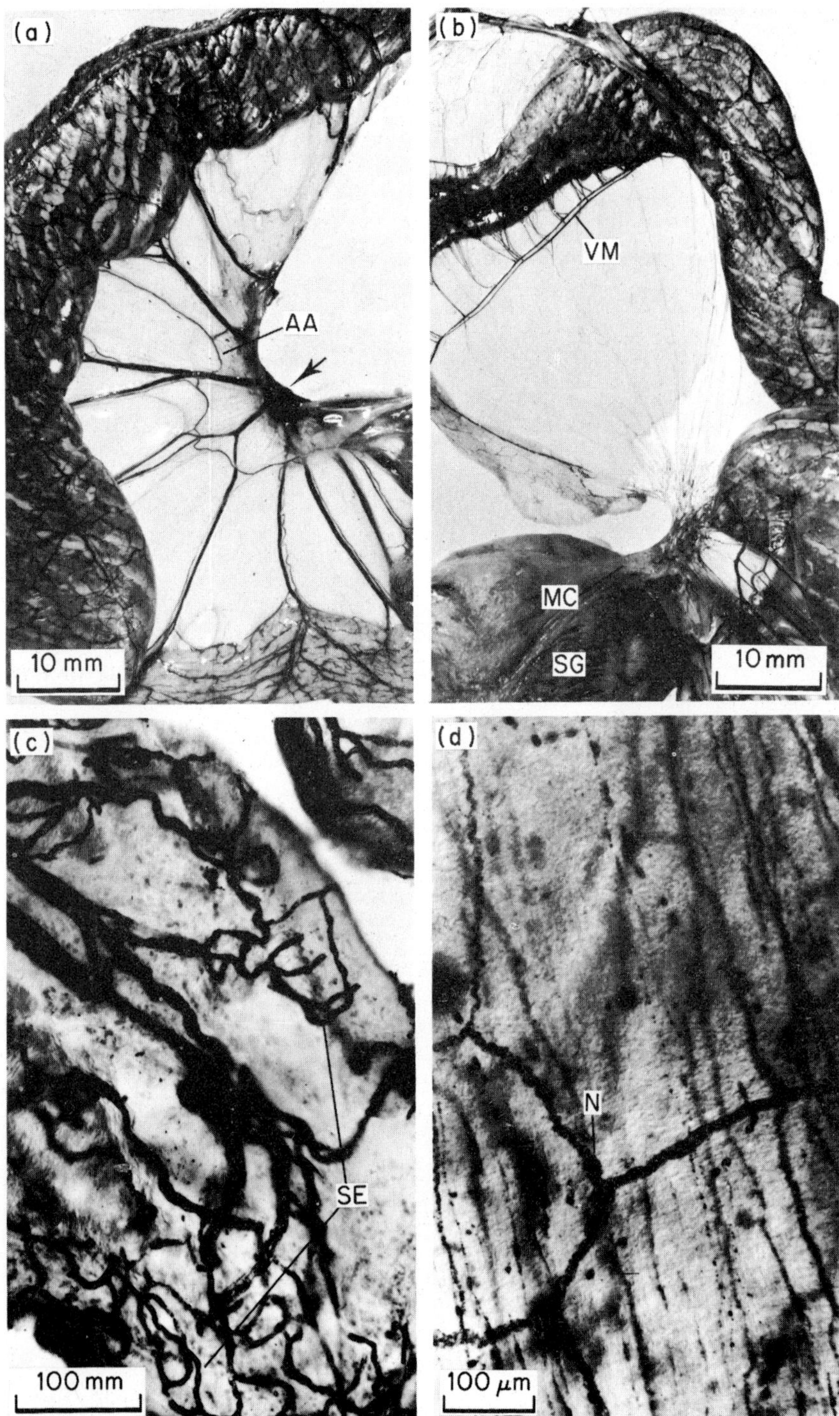

Fig. 5.29. (a) The extensive vascular system of the oviduct drains via several major veins in the dorsal ligament into the caudal renal vein (arrowed). There is no comparable vein to the anastomosing artery (AA). Indian ink injected. (b) The ventral ligament and muscular cord (MC). The ventral marginal artery and vein (VM) run along the length of the oviduct,

some evidence that the relative proportions of the different regions vary from species to species (Verma and Cherms, 1964; Fitzgerald, 1969).

Although growth occurs in the whole oviduct, it is not uniform throughout. Proportionately the magnum increases most (Yu *et al.*, 1972) whereas, surprisingly, the shell gland and vagina are relatively smaller in the breeding adult than in the juvenile (Fig. 5.27). The enlargement of the oviduct is brought about both by a massive hyperplasia (600-fold) and a smaller degree of cellular hypertrophy (5-fold) (Yu and Marquardt, 1974).

With the increase in its length, the oviduct is thrown into folds since it is firmly fixed to the roof of the abdominal cavity by the dorsal ligament. As a consequence of the growth of the oviduct, the dorsal ligament becomes fan-shaped (Fig. 5.29) and, in the region of the vagina, is obscured by connective tissue. The ventral ligament extends with the oviduct and is reinforced by the development of smooth muscle which becomes progressively thicker caudally. This smooth muscle culminates in the solid muscular cord (*funiculus musculosus*) which in *Gallus* is about 5 mm in diameter, and is fused to the ventral surface of the shell gland and the cranial portion of the vagina (Figs 5.8, 29) (Curtis, 1910; Kar, 1947a).

The position of the serosal ligaments in the adult is important in respect of the entry of the oocyte into the oviduct (Fig. 5.11). While "internal laying", the loss of the oocyte into the peritoneal cavity, is not an uncommon feature of *Gallus* (and conceivably may occur in other species) (Cole and Hutt, 1953; Hutt *et al.*, 1956; Wood-Gush and Gilbert, 1965a,b, 1970; Gilbert and Wood-Gush, 1971), it is reduced by the anatomical relationships around the ovary. In *Gallus* these relationships have been described as forming the so-called "ovarian pocket" (Curtis, 1910; Surface, 1912; Bradley and Grahame, 1960), a similar arrangement being found in *Coturnix* (Fitzgerald, 1969). The ligaments close the coelomic opening of the ovarian pocket and bring the ostium of the infundibulum (*ostium infundibulare*) into a suitable position to accept the shed oocyte. In the juvenile, the small intestine and caeca often invade the pocket (King, 1975) but in the breeding adult they are pushed out by the growth of the oviduct. An unusual feature of the Brown Kiwi (*Apteryx australis*) is the arrangement of the infundibulum which spreads across the whole width of the body cavity in the region of the ovary (Kinsky, 1971). This is probably a specialized adaptation to the two functional ovaries which are almost invariably present in this species.

With the growth of the oviduct, its internal structure undergoes considerable change, although the basic arrangement of seven layers is not affected. The greatest alteration, and the greatest difference between the regions, is found in the glandular elements of the epithelium and the tubular glands.

---

although only parts are visible in this preparation. Indian ink injected. (c) The intrinsic vascular system of the lamellae of the shell gland consists of a central artery and vein connected to a sub-epithelial plexus (SE). (See also Figs 5.34b, 5.36a and 5.37a.) Indian ink injected. (d) Intrinsic innervation of the shell gland; neurons (N) are present in the chicken but these may not be a universal feature of birds. Methylene blue, whole mount. (Reproduced with permission from Gilbert and Lake, 1963).

The epithelium is composed of two main types of cell (*epitheliocytus*), the cilated cell and the secretory "goblet"-like cell (Figs 5.31, 32, 34, 36, 37, 38). However, despite this anatomical distinction, the ciliated cells must not be regarded as totally non-secretory. In histological sections the two types of cell appear to be arranged approximately alternately, although in *Gallus* Guzsal (1968) described hexagonal clusters of cells in which six secretory cells surround a central ciliated cell. However, this supposed hexagonal arrangement has not been supported by recent studies with the scanning electron microscope (Fujii, 1975; Bakst and Howarth, 1975) in which the ciliated cells were found to be the predominant cell type and in which the proportion of ciliated cells to secretory cells varied between the different regions of the oviduct.

The cilia of the cells vary from region to region, being short and straight in the infundibulum and utero–vaginal junction, long in the magnum and shell-gland (Fujii, 1975), and short and wavy in the vagina. The size of the cilia in the isthmus is in some doubt since, although Fujii (1975) described them as being short, Blom (1973) noted that they were longer here than in any other region. In all regions of the oviduct the cilia tend to beat towards the cloaca (Mimura, 1937). In the magnum and shell gland, however, the direction of beating is relatively erratic, and in a specialized tract in the magnum which runs along the attachment of the dorsal ligament, the cilia beat exclusively in the direction of the infundibulum (King, 1975). According to Parker (1930) the cilia in *Columba* may beat in either direction. The function of the cilia is not known, although Thomson (1964) implied that they were solely responsible for the transport of the egg. As Draper *et al.* (1968) pointed out, however, this seems hardly likely, although they could possibly help the passage of the egg along the oviduct. According to Fujii (1975) the cilia, at least in the region of the utero-vaginal glands and the infundibulum, will act to prevent the ascent of spermatozoa and, hence, will cause the spermatozoa to enter the so-called "sperm-host glands". However, it seems also unlikely that the cilia have an effect on the movement of spermatozoa up the oviduct, although both in *Gallus* and *Meleagris* dead spermatozoa do reach the upper levels of the oviduct by some means (Howarth, 1971).

The glandular elements of the oviduct are remarkably constant throughout its length, and consist of the secretory epithelial cells already described and the multicellular tubular glands (Figs 5.32, 33, 35, 36). Whilst regional differences occur in the histological staining properties of the glandular cells and their granular contents, attempts to relate these differences to function do not always lead to meaningful results. Recently the glandular cells have been the subject of studies at the ultrastructural level but unfortunately not all accounts agree, discrepancies occurring in both the observations of different authors and the way the observations are interpreted.

The unicellular secretory glands occur in considerable numbers throughout the whole oviduct (Aitken and Johnston, 1963), although Das and Biswal (1968) noted that in *Anas* there are few in the infundibulum. The glands are most numerous and largest in the magnum (Fujii *et al.*, 1965). In general terms they produce mucin-like material or other similar mucopolysaccharides (Guzsal, 1968).

The multicellular glands are of two kinds, glandular-grooves and tubular glands. Both kinds of gland are located between the secondary folds of the mucosa. The glandular-grooves are found only in the infundibulum (Fig. 5.31) (Surface, 1912; Richardson, 1935; Aitken and Johnston, 1963) and appear as shallow, concave plates of glandular-like cells. However, there is still some doubt that these cells are truly secretory. The tubular glands are present throughout the oviduct except in the cranial part of the infundibulum, the translucent part of the isthmus (Fig. 5.32) and the caudal vagina (Fig. 5.38). They reach their greatest development in the magnum where their characteristic granular material varies with the position of the egg, the granules presumably being the precursors of the egg albumen (Figs 5.32, 33). The tubular glands open into the grooves between the secondary folds and the first portion of the duct is lined with cells which are a continuation of those in the surface epithelium (Schwartz, 1969; Wyburn *et al.*, 1970; Makita and Sandborn, 1970, 1971). An intermediate zone of transitional cells lies between the "mucin" (epithelial) cell type and the serous (gland) cell type (Ulrich and Sandoz, 1972). Secretory granules are prominent and the secretion is generally merocrine (Schwartz, 1969; Fertuck and Newstead, 1970; Wyburn *et al.*, 1970; Sandoz *et al.*, 1971). However, Aitken (1971) suggested that an apocrine type of secretion may occur in the tubular glands of the shell gland. On the other hand, the swollen tips of the microvilli which he described could possibly be fixation artefacts.

One consistent feature of the secretory cells is the presence of desmosomes or other junctional complexes (Fig. 5.35) (Hodges, 1974) which indicates that some degree of coordination from cell to cell is possible.

Specialized tubular infoldings of the epithelium occur in the infundibulum and vagina. These have been called, descriptively, "sperm-host glands" (Figs 5.38, 39). That they are involved in the survival of spermatozoa in the oviduct is beyond doubt (Lake, 1975) but it is highly questionable that they are glands in the strictest sense.

The mucosa of the oviduct is thrown into a series of longitudinal primary folds (*plicae primariae*) which are clearly visible macroscopically (Fig. 5.30). A precise description of the folds has been given in *Gallus* by Surface (1912) and Bradley and Grahame (1960) and in *Coturnix* by Fitzgerald (1969), and few differences appear to be present. In the infundibulum, isthmus, vagina and the cranial part of the shell gland these primary folds carry many secondary folds (*plicae secundariae*) which involve the epithelium only (Surface, 1912; Blom, 1973). Scanning electron microscopy of the folds in *Gallus* by Blom (1973) and Bakst and Howarth (1975) has shown them to vary in size and shape and the way they are directed along the oviduct. It also demonstrated differences in the appearance and distribution of the microvilli and gland openings. The muscle layers vary in thickness from region to region and in the infundibulum consist only of a relatively small number of bundles (Romanoff and Romanoff, 1949). In other regions the muscle layers act in passing the developing egg along the oviduct and spontaneous *in vivo* motility has been recorded (Crossley *et al.*, 1975). Aid in this transport of the egg may be given by the muscles of the dorsal and ventral ligaments (Curtis, 1910). The muscle layers may also be involved in the transport of spermatozoa up the oviduct (Lorenz, 1964) and anti-peristaltic waves have been observed (Crossley *et al.*, 1975) but there is still controversy on this point. The lamina propria (*lamina propria mucosae*) (Fig. 5.32) lies between

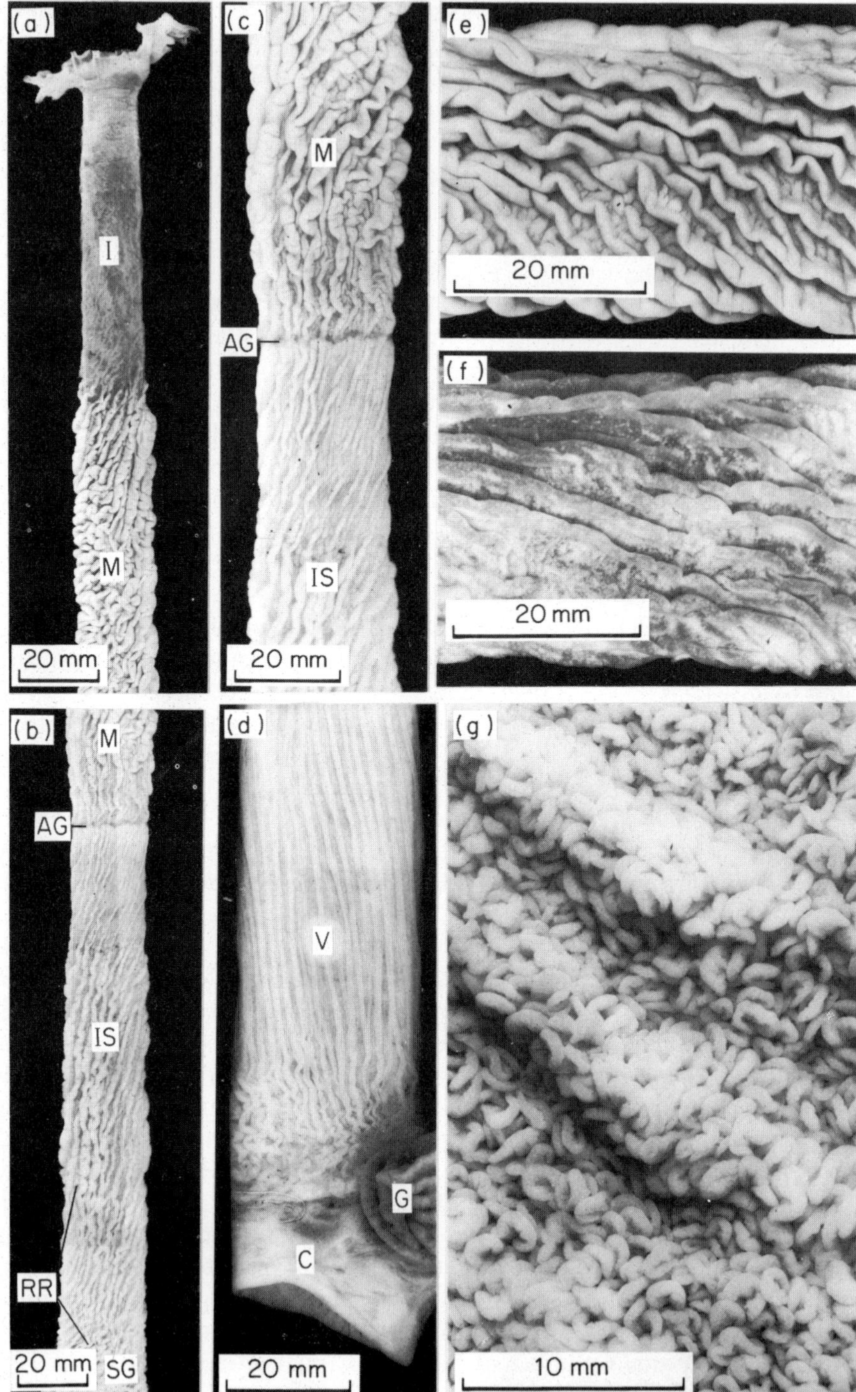

Fig. 5.30. Surface plicae of the everted oviduct. (a) Infundibulum (I) and magnum (M). (b) Caudal magnum (M), isthmus (IS) showing aglandular part (AG), "red-region" (RR) and the pouch of the shell gland (SG). (c) Junction between the magnum (M) and isthmus (IS) with the aglandular region (AG). (d) The relationships between the vagina (V), gut (G) and cloaca (C) (the cloaca is not complete). (e) Magnum before an egg has passed. (f) Similar region to (e) after an egg has passed. The obvious discharge of material is noticeable when compared with (e). (g) Shell gland, "pouch".

the musocal epithelium and the inner muscle layer and in most regions of the oviduct is obscured by the tubular glands which lie in it.

It is not surprising that such an active organ as the oviduct has an extensive blood supply (Fig. 5.29). Descriptions in various detail are available for *Gallus* (Westphal, 1961; Freedman and Sturkie, 1963a; Hodges, 1965; Nickel *et al.*, 1977), *Anas* (Hodges, 1965; Ma, 1972), *Meleagris* (Hodges, 1965) and *Coturnix* (Fitzgerald, 1969). King (1975) reviewed the literature on the first three species and rationalized the nomenclature which had been used. The basic arrangement of the blood vessels seems constant between the species examined (with some exceptions given below), there being some degree of individual variation within a species. How far this general description applies to birds as a whole is not known. The arterial supply to the oviduct is derived from the general arterial system on the left side and there appears to be four possible positions from which this occurs (Fig. 5.12): *position* (*1*), the ovario-oviductal branch of the left cranial renal artery; *position* (*2*), the left external iliac artery; *position* (*3*), the left ischiadic artery; *position* (**4**), the pudendal artery. The oviductal arteries (*aa oviductales*) are four in number in *Meleagris* (Hodges, 1965), *Anas* (Hodges, 1965; Ma, 1972) and *Coturnix* (Fitzgerald, 1969), and five in *Anser* (Gertner, 1969). In *Gallus* there are usually four arteries, although a fifth is sometimes present (Hodges, 1965). In *Meleagris*, *Gallus*, *Anas* and *Coturnix* not all the available positions are occupied since two oviductal arteries arise from the pudendal artery. King (1975) described the following oviductal arteries in *Gallus* as arising from the positions given above: *position* (*1*), the cranial oviductal artery (*a. oviductalis cranialis*); *position* (*2*), in some birds an extra cranial artery (Hodges, 1965) arises from here but usually there is no artery; *position* (*3*), the middle oviductal artery (*a. oviductalis media*); *position* (*4*), the caudal oviductal artery (*a. oviductalis caudalis*) and the vaginal artery (*a. vaginalis*). In *Anas* (Hodges, 1965; Ma, 1972), *Meleagris* (Hodges, 1965) and *Coturnix* (Fitzgerald, 1969) there is no artery in position (1), and the artery in position (2) was called by Fitzgerald (1969) and King (1975) the "cranial artery". However, this artery corresponds in position to the "extra" middle artery (Hodges, 1965) which is sometimes present in the domestic fowl, but not to the position of the cranial artery in the fowl described by King (1975) and Nickel *et al.* (1977). The difficulty is further aggravated by the presence as a normal feature in *Anser* of arteries in both positions (1) and (2) (Gertner, 1969), that in position (2) being called by Gertner the "cranial artery". If the arteries in positions (1) and (2) are called the cranial artery and accessory cranial artery (*a. oviductalis cranialis accessorius*) respectively, then this would help to bring into conformity the nomenclature for *Meleagris*, *Anas* (King, 1975), *Coturnix* (Fitzgerald, 1969) and *Anser* (Gertner, 1969): the "extra middle artery" occasionally found in *Gallus* would then be called the accessory cranial artery which would avoid confusion with the middle oviductal artery arising from the ischiadic artery, i.e. position (3). This suggestion is adopted in the following description which is based mainly on *Gallus* (Fig. 5.12): details of individual variation, and the slightly different distribution of the smaller vessels between the species, are to be found in the original papers.

The cranial artery enters the dorsal ligament, its distribution differing in *Anser* and *Gallus*. In *Anser* it supplies almost exclusively the infundibulum whereas in

Gallus its distribution is more similar to the accessory cranial artery of *Anser*, *Meleagris* and *Anas*. In *Gallus* a branch of the cranial artery traverses the dorsal ligament as the anastomosing artery (Fig. 5.29) (*a. anastomotica*) which gives off further branches to the artery running along the oviduct (*a. oviductalis marginalis dorsalis*). At the shell gland the anastomosing artery continues into the uterine arteries.

The ventral marginal oviductal artery (*a. oviductalis marginalis ventralis*) (Fig. 5.29) extends along the ventral ligament, and receives branches from the middle oviductal artery and from the arteries supplying the shell gland. Both the ventral and dorsal marginal oviductal arteries supply mainly the magnum.

In all four domestic species, the middle oviductal artery usually arises from the left ischiadic artery, though it may branch from the left middle renal artery: it travels in the dorsal ligament, here in *Gallus* receiving the anastomosing artery, and ends at the cranial part of the shell gland where it divides into the right and left uterine arteries (*a. uterina medialis* and *lateralis*). In *Columba* the cranial and accessory cranial arteries may be missing; then the cranial regions of the oviduct are supplied by branches of the middle oviductal artery (Baumel, personal communication).

The caudal oviductal artery arises from the left pudendal artery and divides into two vessels supplying the shell gland. The vaginal artery is short and may arise from either the pudendal artery or from a small branch passing to the shell gland. The distribution of its branches is not entirely clear; Hodges (1965) claimed they supply only the vagina but according to Freedman and Sturkie (1963a) they run to the shell gland. In *Anser* they may supply both regions of the oviduct (Gertner, 1969).

The veins of the oviduct are essentially counterparts of the arteries (Fig. 5.12). However, there are no veins comparable to the anastomosing artery (Fig. 5.29), and three to five middle oviductal veins (*vv. oviductales*) may be present (Fig. 5.29) (Freedman and Sturkie, 1963a; Hodges, 1965). The principle cranial veins ultimately empty into the caudal vena cava, while the caudal veins have access to either the hepatic portal or renal portal systems.

Details of the intrinsic blood vessels of the oviduct have been provided in *Gallus* only for the shell gland (Hodges, 1965, 1966) and the vagina in the region of the sperm-host glands (Gilbert *et al.*, 1968). Both descriptions are similar and this similarity may indicate that the arrangement occurs in other regions of the oviduct. The main arteries form a plexus between the two muscle layers and this sends branches to a secondary arterial plexus in the lamina propria (Fig. 5.26). From here arterioles pass through the centre of one of the mucosal folds and enter a capillary network around the tubular glands and beneath the epithelium (Figs 5.29, 34, 36, 37). However, it is possible that not all birds follow this arrangement. Thus the arrangement of the blood vessels in the Black Kite (*Milvus migrans*) appears to be more complex than in *Gallus* (Chakravorti and Sadhu, 1961a), although a full description for *Milvus* is not available.

The extrinsic nerve supply of the oviduct has generally received little attention, the earlier descriptions of Johnson (1925), Mauger (1941), Hsieh (1951) and Bradley and Grahame (1960) being very superficial. A more detailed investigation by Freedman and Sturkie (1963b) dealt only with the more caudal regions of the oviduct but from this study it seems certain that both sympathetic and parasympathetic nerves are involved. On the basis of the observations of Mauger (1941), Hodges (1974) suggested that the infundibulum receives branches from the ovarian nerve plexus, and that the magnum is innervated by the renal and aortic plexuses. He proposed that the innervation of the shell gland and vagina arises from the pelvic plexus, although according to Freedman and Sturkie (1963b) it comes from the aortic plexus via the hypogastric nerve. The parasympathetic nerve supply is derived from the eighth to eleventh lumbosacral spinal nerves (King, 1975).

The intrinsic innervation of the oviduct appears to be more extensive in the shell gland and its junction with the vagina than elsewhere (Fig. 5.29) (Gilbert and Lake, 1963). It consists of typical plexuses of fibres distributed at two levels, one plexus beneath the serosa and the other in the muscle tunic. The plexuses appear to be ganglionated in *Gallus* (Biswal, 1954; Gilbert and Lake, 1963) but not in *Anas* (Das and Biswal, 1968).

The function of the motor nerves has not been completely elucidated. Where they are associated with muscles, their function seems obvious and the action of autonomic mimetic drugs supports this view (Gilbert, 1971d; Verma and Walker, 1974). Consequently it is likely that they have a role in the transport of the egg down the oviduct and in oviposition (Gilbert, 1971d). On the other hand, secretion of the oviduct does not appear to be under neural control (Sturkie and Weiss, 1950; Sturkie *et al.*, 1954). No specific sensory endings were found by Gilbert and Lake (1963), although some apparently free endings were observed. That sensory stimuli may be important at times is indicated by the disruption of normal function when a foreign object is placed in the caudal regions of the oviduct (van Tienhoven, 1953; Sykes, 1953a; Lake and Gilbert, 1961, 1964b; Gilbert *et al.*, 1966).

There is little evidence for the presence of lymphoid tissue in the oviduct. In *Gallus* lymphatic nodules may occur in the lamina propria (Biswal, 1954), and scattered lymphocytes and plasma cells are almost invariably present. Aitken (1971) regarded these as indications of a previous infection rather than a normal anatomical feature and their absence from the oviduct of *Anas* (Das and Biswal, 1968) is consistent with this view. For an account of the lymph vessels in *Gallus* see Dransfield (1945).

From the nature of the secretory activity of the oviduct it would be expected to contain many enzyme systems. Diculesco (1961) reported on the histochemistry of 16 enzymes throughout the oviduct and Gilbert (1971c) discussed the possible roles of many in oviductal function. However, although there is an expanding literature on oviductal enzymes, few positive conclusions about their functions can be drawn from the studies so far. Certainly the functions of some are clear. For example, the steroid-dependent RNA- and DNA-methylases and polymerases of the magnum must be involved in the production of the albumen proteins (McGuire and O'Malley, 1968; Hacker, 1969; Müller *et al.*, 1974), and the increase in adenyl cyclase activity concomitant with the production of avidin is also suggestive of such a function (Kissel *et al.*, 1970; Rosenfield and O'Malley, 1970). In contrast the role of other enzymes is uncertain, although many theories have been put forward. The epithelial secretory cells in the infundibulum and cranial magnum contain ATP-ase (Rhea *et al.*, 1974) and NTP-ase (Anderson *et al.*, 1974), and it has been claimed (Anderson and Rosenberg, 1976) that these are transferred to the egg as it passes through the oviduct and that they are important for early embryonic development. Similarly the high concentration of sulphur-activating enzymes in the isthmus has been linked with the production of the membranes (Suzuki and Strominger, 1959, 1960a,b,c). However, there is no positive proof of the function of either group of enzymes in the way claimed.

In the past much has been made of the presence of alkaline phosphatase, acid phosphatase and carbonic anhydrase in the shell gland in relation to calcification (Common, 1941; Lauwers *et al.*, 1970; Snapir and Perek, 1970). However, according to Solomon (1975a) the alkaline phosphatase in adult domestic fowl is present only as an accident of the circulatory system, being restricted to the endothelial lining of the blood vessels. This limited distribution was recently confirmed by Aire and Steinbach (1976). In *Columba* and *Gallus*, however, Chakravorti and Sadhu (1959, 1961b) reported intense alkaline phosphatase activity in the secretory cells in all regions of the oviduct. It is difficult to reconcile these two opposed views. The uncertainty of the role of carbonic anhydrase still exists (Simkiss and Taylor, 1971; Simkiss, 1975), although its location in the soluble fraction (Bernstein *et al.*, 1968) of the secretory cells of the tubular glands (Diamanstein and Schluens, 1964) has now been established. The involvement in *Coturnix* of a further enzyme (a calcium-magnesium activated ATP-ase) in calcium transport in the shell gland was recently proposed by Pike and Alvarado (1975).

The roles of many other enzymes isolated from the oviduct are still speculative, although much is known of their biochemistry. Amongst these enzymes are the poly (A) polymerase reported by Müller *et al.* (1975), the glycopeptide hydrolysing enzymes studied by Tarentino and Maley (1969) and the enzymes involved in carbohydrate metabolism reported by Campbell *et al.* (1971). Undoubtedly many more enzymes have yet to be isolated and inevitably their properties will be known before their function. Yet studies of this kind, particularly if they can be related to specific cell populations, will ultimately lead to a better understanding of the functions of the various secretory cells of the oviduct.

(ii) Regional morphology of the oviduct. In view of the detailed reviews recently published on the histology and ultrastructure of the oviduct (Aitken, 1971; Hodges, 1974), a full anatomical description is not given. Instead, only the functional aspects of the morphology will be considered.

(*a*) *Infundibulum*. The *infundibulum* (Fig. 5.31) has three known functions and a possible fourth; (a) it engulfs the shed oocyte; (b) it produces the first layer of albumen; (c) it transports the developing egg to the next region; (d) since fertilization occurs in the infundibulum, it may be involved in this also.

The oocyte does not enter the oviduct of its own accord but has to be actively engulfed by the infundibulum (Fig. 5.2). To this end the cranial portion of the infundibulum consists of a wide-mouthed, thin walled, flexible funnel with flared lips (*fimbriae infundibulares*) (Figs 5.27, 30) and flattened in a dorsoventral direction. It has no direct connection to the ovary (King, 1975).

There is no doubt as to the importance of the activity of the infundibulum (Warren and Scott, 1935). Inanimate objects can be transferred from the coelomic cavity to the oviduct (Sturkie, 1965) and entire follicles may be torn

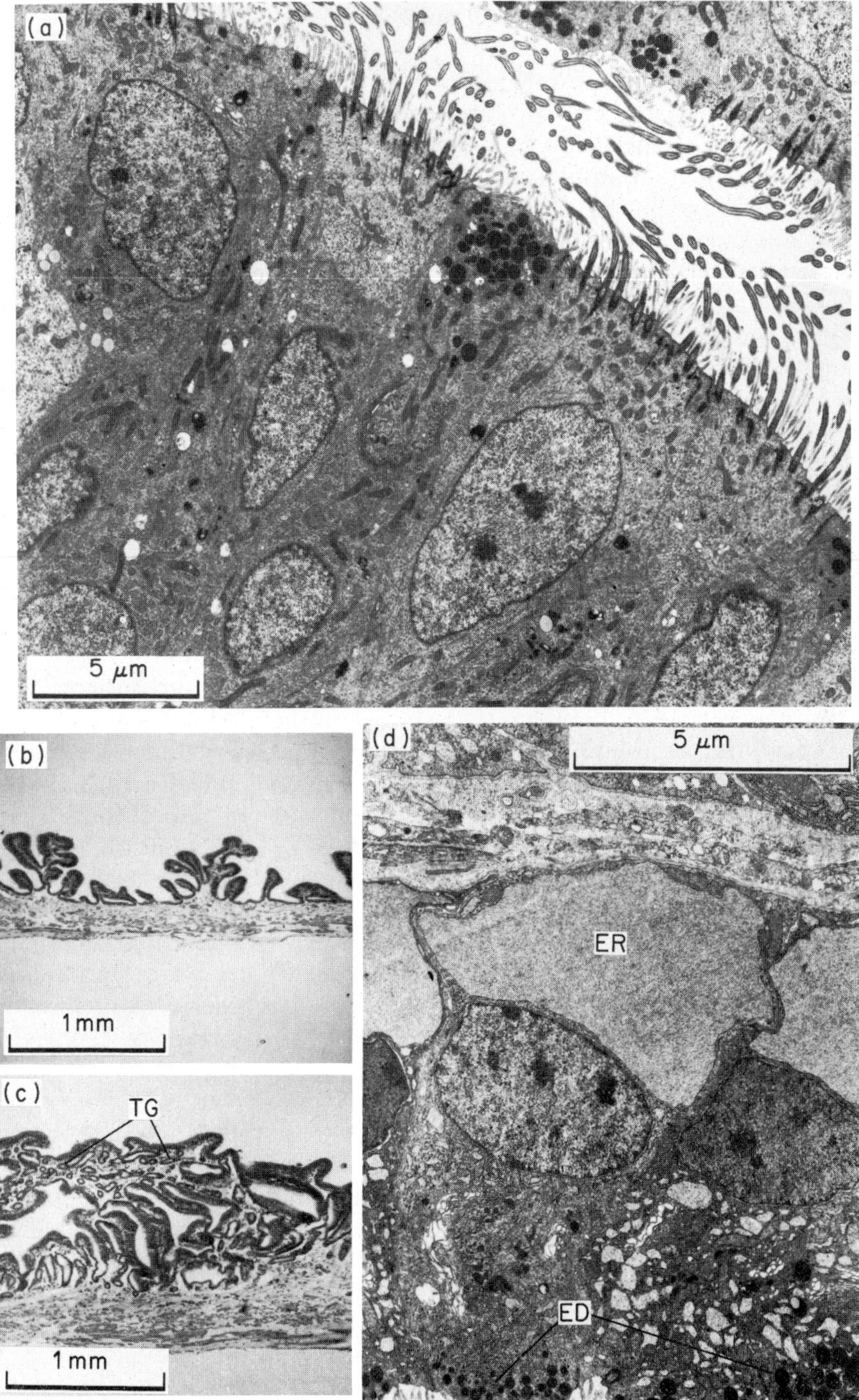

Fig. 5.31. The infundibulum. (a) Epithelium of the neck region. (Johnston, unpublished). (b) Cranial portion of the infundibulum. Haematoxylin and eosin. (Davidson, unpublished). (c) Caudal region with the development of the simple tubular glands (TG). Haematoxylin and eosin. (Davidson, unpublished). (d) Typical cells of the glandular grooves with the infranuclear dilated endoplasmic reticulum (ER) filled with a fine granular material. The electron dense granules (ED) of the cells are adjacent to the lumen of the infundibulum. (Johnston, unpublished).

from the ovary and transformed into "eggs" (Hutt, 1939, 1946; Cole, 1946). Gilbert (1968) described an "egg" in the shell gland which contained a complete follicle attached to the ovary by a much extended pedicle.

From these observations it is clear that the infundibular activity is independent of ovulation *per se* but the precise control mechanism has not been discovered. Neural mechanisms may be important but attention has been drawn to the possible role of histamine in the function of the infundibulum just after ovulation (Meyer and Sturkie, 1974). Certainly a prominent feature of the infundibulum is the presence of mast cells (Wight, 1970) which are absent from most other regions of the oviduct. The infundibular muscles must be involved and perhaps their unusual arrangement in bundles (Romanoff and Romanoff, 1949) may give a greater degree of flexibility than the more typical oviductal arrangement of longitudinal and circular layers. As pointed out by Aitken (1971), the engorgement of the blood vessels and the activity of the ligaments may also play a part.

Fertilization must occur before the oocyte is covered by albumen and it may be significant that the typical oviductal glandular cells are mainly absent from the funnel of the infundibulum (Richardson, 1935; Aitken and Johnston, 1963; Guzsal, 1968; Hodges, 1974), only ciliated cells being found in any number (Fig. 5.31). The importance in *Gallus* of the infundibular sperm-host glands (Van Drimmelen, 1946, 1949) has yet to be determined. Whilst there is no doubt that sperm may be found in them their significance is obscure (Lake, 1975), particularly since they are absent in *Meleagris* (Verma and Cherms, 1965).

The developing egg spends only a few minutes in the funnel before passing on to the tubular portion (*tubus infundibularis*) (Fig. 5.31). Here in *Gallus* the first layer of albumen is deposited (Scott and Huang, 1941), although this may not be so in *Coturnix* (Tamura and Fujii, 1966a). The first layer of albumen contributes to the perivitelline layer and may also form the chalazae (but see Aitken, 1971 and Gilbert, 1971c). It is in the tubular portion that the infundibular glands first appear in quantity. Here also the ridges become pronounced with secondary folds, and the glandular grooves (*fossae glandulares infundibuli*) become more complex (Surface, 1912; Aitken and Johnston, 1963). The tubular glands (*glandulae tubi infundibularis*) (Fig. 5.31) appear near to the junction with the magnum and may contain two (Aitken and Johnston, 1963; Makita and Kiwaki, 1968) or three (Aitken, 1971) types of secretory cell. In *Gallus* (Richardson, 1935) and *Columba* (Dominic, 1960b) the gland cells of the grooves have some similarities with those of the tubular glands and the epithelium, although Aitken (1971) considered them to be a distinct type. Hodges (1974), however, doubted the secretory role of the gland cells of the grooves. The typical secretory cell of the epithelium is present in large numbers in the infundibulum and often obscures the ciliated cells (Fig. 5.31) (Hodges, 1974) which may, however, also be secretory (Makita and Kiwaki, 1968; Aitken, 1971).

The ridges are arranged in a slight, but distinct, spiral fashion (Fig. 5.30) and so may start the egg on its spiral course which may lead to the formation of the chalazae. The arrangement of the muscle layers in the tubular part is typical.

(*b*) *Magnum*. The *magnum* (Figs 5.27, 29, 30) is the longest region of the oviduct and apart from the shell gland, contains the egg for the longest time.

It is highly distensible and its wall is generally markedly thicker than that of the infundibulum because of the well-developed nature of the folds (in *Gallus* they reach up to 4·5 mm in height). The diameter of the magnum is reduced for the most caudal few centimetres of its length (Fig. 5.40). The large quantity of proteinaceous material in the secretory glands gives the magnum a milky-white colour.

The magnum has two functions. First, it transports the egg to the isthmus. For this purpose the muscle layers are well-developed and the outer longitudinal layer is arranged in a slight spiral (Hodges, 1974). Its second and major function is to produce the remainder of the albumen (Gilbert, 1971c). Suggestions have been made that the albumen is deposited in consecutive layers (Asmundson, 1931; Asmundson and Jervis, 1933; Asmundson and Burmester, 1936; Scott and Burmester, 1939), the inner thin albumen being derived from the secretions of the cranial magnum and the middle thick layer being formed in the caudal magnum (Asmundson and Burmester, 1938; Scott and Burmester, 1939). Certainly there are slight, but distinct, anatomical differences between the two parts: compared with the caudal part, the cranial magnum has a lower epithelium and more tubular glands, and the higher ridges have a smaller number of secondary folds (Aitken, 1971; Hodges, 1974).

As would be expected, the magnum is richly endowed with secretory cells in both the epithelium and tubular glands (*glandulae magni*) (Figs 5.32, 33). These cells contain large numbers of secretory granules and there seems no doubt that they are involved in albumen formation since they discharge their granules with the passage of an egg and replenish them in the interval between eggs (Figs 5.30, 33). Further support for this view comes from Wyburn *et al.* (1970) and Wyburn (1974) who found these cells to be abnormal in domestic fowl laying "watery-white" eggs in which the protein content is low.

However, of the many proteins in albumen (Gilbert, 1971c), few have been positively linked to a specific cell type. Moreover, Wyburn *et al.* (1970) pointed out that it is not known whether all of the different kinds of secretory cell have, in fact, been identified. Within the tubular glands, three types of secretory cell have been described (types A, B and C), although types A and C are probably the same cell in a different secretory state (Fig. 5.33) (Wyburn *et al.*, 1970). Type A(C) is thought to secrete ovalbumin, the most common of the proteins of albumen. Type B, it has been suggested, is involved in lysozyme production (Wyburn *et al.*, 1970). Both these functions are supported by the experimental evidence of Kohler *et al.* (1968) and Oka and Schimke (1969). According to Schimke *et al.* (1977) the tubular glands produce ovotransferrin (conalbumin) and ovomucoid. Avidin is formed in the epithelial cells (Kohler *et al.*, 1968; O'Malley *et al.*, 1969; Tuohimaa, 1975), contrary to the suggestions of Wyburn *et al.* (1970) and Sandoz *et al.* (1971) that the epithelial cells appear to secrete only a mucin-like material.

(*c*) *Isthmus*. The boundary between the *isthmus* and the magnum is delineated by a narrow translucent zone (*pars translucens isthmi*) (Figs 5.30, 32) in which tubular glands are absent (Aitken, 1971; Hodges, 1974).

In the isthmus proper the ridges increase in height after their reduction in the caudal magnum but never reach the size of the ridges in the main part of the

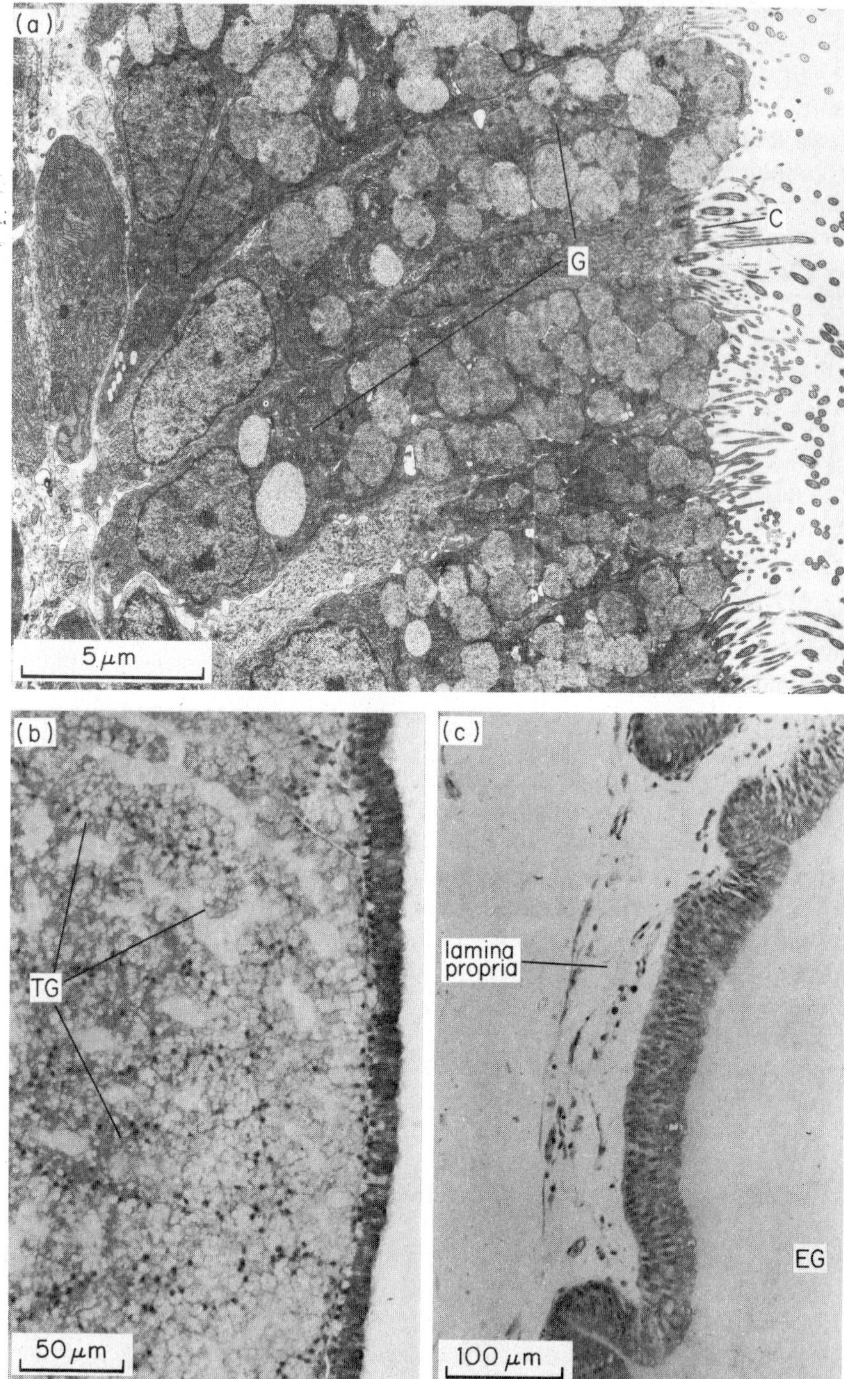

Fig. 5.32. Magnum. (a) Epithelial ciliated (C) and glandular (G) cells. (Johnston, unpublished). (b) The glandular cells of the epithelium stain positively with alcian blue (and PAS). The tubular glands (TG) are fully charged with secretory granules. Alcian blue, haematoxylin and eosin. (Davidson, unpublished). (c) The magnum–isthmus region with an egg (EG); the fine fibrous-like albumen is just discernible. There are no tubular glands here (compare with b) and the structure of the lamina propria is more clearly seen. Haematoxylin and eosin. (Davidson, unpublished).

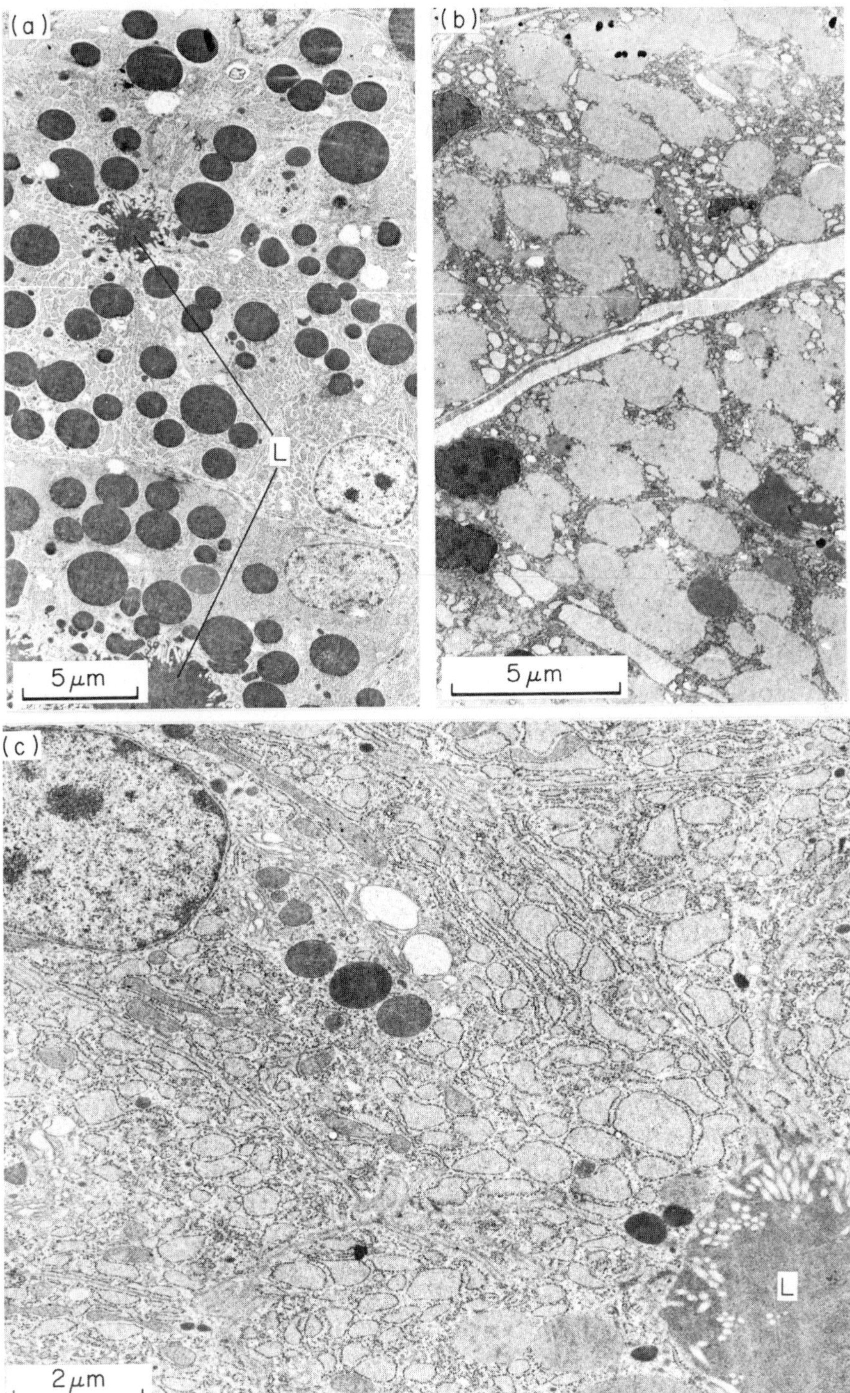

Fig. 5.33. Tubular glands of magnum. (a) Type "A" cells full of electron-dense material surrounding the lumen of the glands. (Johnston, unpublished). (b) Cells similar to type "B" cells of Wyburn *et al.* (1970). (Johnston, unpublished). However, if Sandoz *et al.* (1971) are correct, they may be type "A" cells in the process of reforming their secretory granules. (c) Type "C" cells: it has been suggested that these cells are the discharged "A" cells at an early stage in their process of protein resynthesis. The extensive endoplasmic reticulum is a characteristic feature of the cells. The lumen (L) of the gland is filled with material of less electron density than in (a). (Johnston, unpublished).

magnum (Fig. 5.30). Compared to the magnum, the ridges are less obviously arranged in a spiral manner (King, 1975) and the crypts between the ridges are more open (Richardson, 1935).

The glandular elements in the isthmus are distinctive. The tubular glands (*glandulae isthmi*) (Figs 5.34, 35) are more numerous than elsewhere but the secretory cells do not appear to pass through such clear secretory phases as they do in the magnum (Hodges, 1974). Whilst it has been suggested that the secretory cells of the epithelium may consist of two distinct types (Draper *et al.*, 1972) they never fully develop into the "goblet" cell type typical of elsewhere. The tubular glands are regarded as forming the membrane fibre-cores (Khairallah, 1966; Hoffer, 1971; Simkiss and Taylor, 1971; Draper *et al.*, 1972) and the epithelial cells the fibre-mantle (Candlish, 1972). It has also been proposed that the mammillary cores of the testa are a product of the isthmus (Richardson, 1935; Hertelendy and Taylor, 1964; Simkiss and Taylor, 1971; Creger *et al.*, 1976) but the earlier suggestion (Johnston *et al.*, 1963; Simkiss, 1968) of the cells involved in this is not supported by the work of Solomon (1975b). Wyburn *et al.* (1973) and Stemberger *et al.* (1977) place the formation of the mammillary cores in the epithelial cells of the so-called "red-region" (Figs 5.30, 36), a 4 cm-long stretch of oviduct lying between the generally accepted isthmus and the pouch of the shell gland (Draper *et al.*, 1972). The red-region has not only caused confusion regarding where the mammillary cores are formed but also where initiation of calcification takes place, since there is still controversy whether the red-region is part of the isthmus (Richardson, 1935; Davidson *et al.*, 1968; Draper *et al.*, 1972) or part of the shell gland (Surface, 1912; Aitken, 1971; King, 1975; Stemberger *et al.*, 1977). Recent evidence suggests that the initial deposits in calcification are laid down in the true isthmus, although they are probably not calcium carbonate (Stemberger *et al.*, 1977). The first deposit of calcium carbonate occurs in the red-region (Stemberger *et al.*, 1977), the granular cells of the epithelium probably being involved (Davidson, 1973). The continued deposition of calcium carbonate leads to the formation of the cone layer (Figs 5.6, 7), subsequent calcification taking place in the shell gland.

The movement of the egg in the isthmus appears to be more complex than in other regions. Although the isthmus is the shortest region, the egg still spends in it between one and two hours. Moreover, movement within the isthmus is not continuous, since after the egg enters the region it remains stationary for a while (Richardson, 1935). The muscle layers are involved in this movement and are well developed, particularly the longitudinal layer. According to Giersberg (1922), there is a thickened sphincter-like reinforcement of the circular muscle at the junction of the isthmus with the shell gland.

At the end of its stay in the isthmus the egg has, for the first time since its entry into the oviduct, a definite boundary. What is not often realized is that the typical egg-shape is a product of the membranes and not the calcified shell (Fig. 5.28).

(*d*) *Shell gland*. Of the total time the developing egg spends in the oviduct, about 80% of it is passed in the shell gland (*uterus*). When the egg reaches the shell gland it consists of the central yolk-mass, the solid (mainly protein) portion of the albumen and the shell membranes. Since it has not received the watery

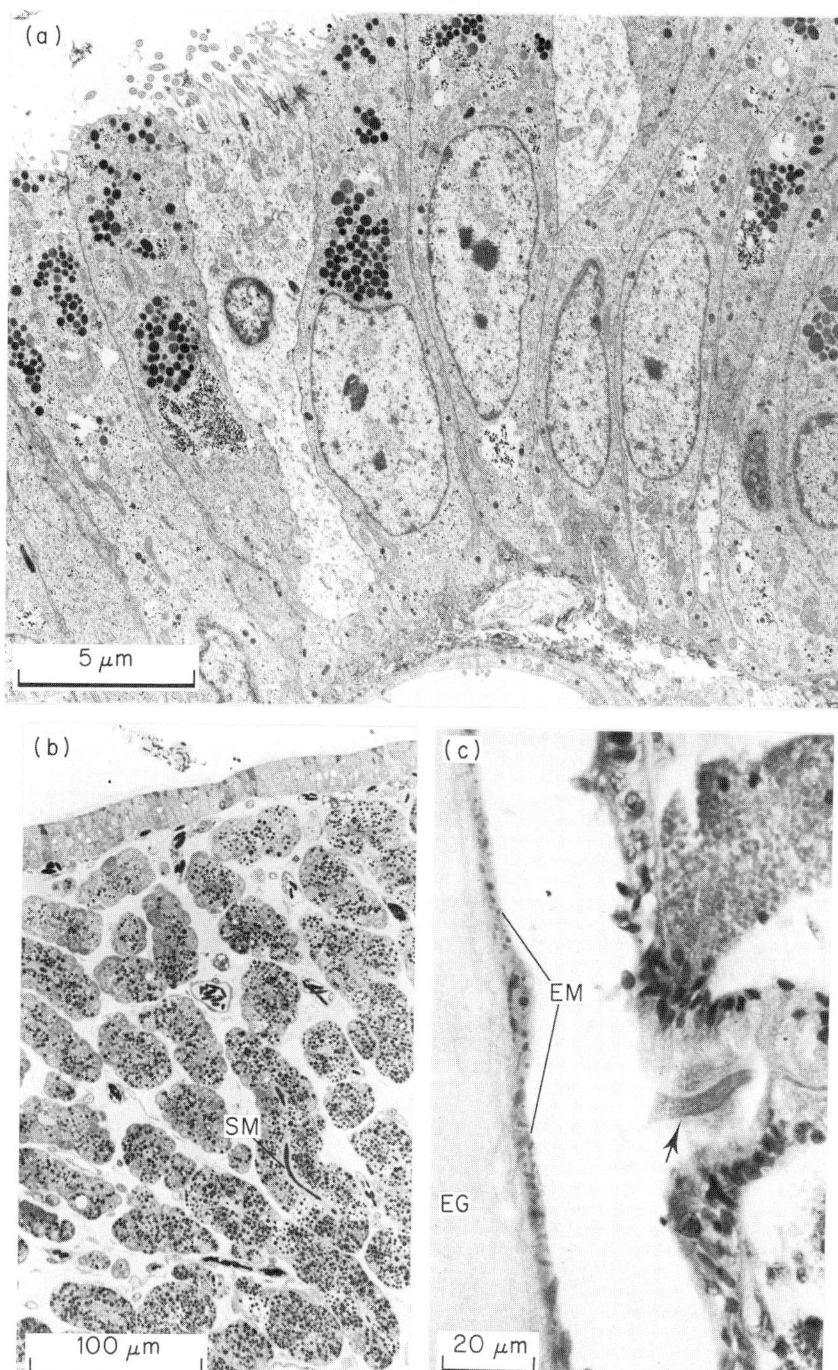

Fig. 5.34. Isthmus. (a) Epithelium. (Johnston, unpublished). (b) Generalized area showing the tubular glands and their secretory product which is in the form of a fibre (SM). Araldite section, azure blue II. (Johnston, unpublished). (c) Extrusion of the tubular gland secretion (arrowed) into the lumen of the cranial isthmus. An egg (EG) is present and the first deposition of the membrane (EM) has occurred. Haematoxylin and phloxin. (Reproduced with permission from Candlish, 1972).

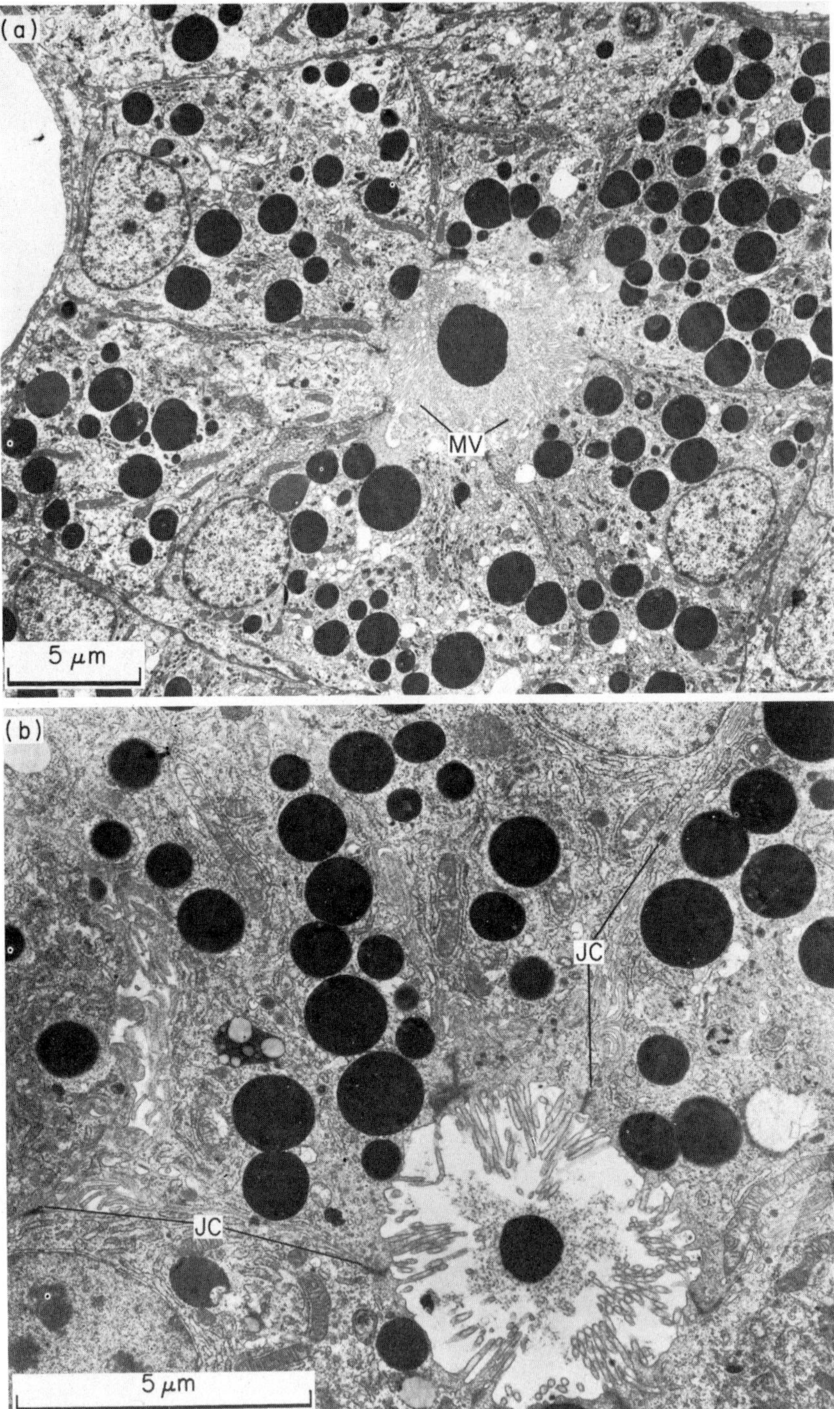

Fig. 5.35. Isthmus. (a) Transverse section of a tubular gland during the secretory phase. The microvilli (MV) are extensive, almost filling the lumen except for the fibre strand (Fig. 5.34). (b) A similar section to (a) but during the phase after an egg has passed. Both the granules and the microvilli appear to be fewer. Junctional complexes (JC) suggest that some coordination of activity may be possible between the cells. (Johnston, unpublished).

component of the albumen, a process colloquially called "plumping", it has the appearance of being contained in a loose and wrinkled covering (Fig. 5.6).

Subsequent development involves the addition of the watery fluid to the albumen and the formation of the various other components of the shell, i.e. the testa, the cuticle and the pigment. These complex secretory functions make it difficult to positively relate the structure of the shell gland to function. Moreover, the fact that the shell gland has been studied in such detail at the ultrastructural level is itself a further complication because there is a wealth of detailed information which has ultimately to be related to function. At present this cannot be done with confidence, nor is it possible to make general conclusions about birds as a class from information obtained from essentially two species, *Coturnix* and *Gallus*.

Macroscopically the shell gland is divisible into two parts if the so-called "red-region", the cranial tube-shaped portion (*pars cranialis uteri*) is included (Figs 5.27, 30). Through this cranial portion the egg passes rapidly (Johnston *et al.*, 1963). The second and main part is the caudal pouch-like section (*pars major uteri*) (Figs 5.8, 27, 29, 30). Internally Fujii (1963) distinguished a third, funnel-shaped part (*recessus uterinus*). The ridges of the shell gland are not so obviously continuous as they are in the other regions of the oviduct for they are broken-up by transverse grooves, thus forming flat leaf-like projections (*lamellae uterinae*) which protrude into the lumen (Fig. 5.30).

As in the other regions of the oviduct, the epithelium contains two cell populations, here termed apical and basal (Fig. 5.37). The two populations are distinguished from each other by the position of their nuclei, the restricted luminal border of the basal cells and the cilia of the apical cells. The apical cells contain secretory granules which appear coincidentally with the onset of the formation of the shell matrix and which disappear during matrix formation (Richardson, 1935; Johnston *et al.*, 1963; Breen and de Bruyn, 1969). They also all basal cells undergo secretory phases associated with the end of calcification when they become particularly long and slender (Johnston *et al.*, 1963).

The cells of the basal type may vary from species to species because in *Coturnix* they are uniform throughout the shell gland (Tamara and Fujii, 1966b) whereas in *Gallus* those of the cranial portion appear to contain a different mucopolysaccharide to those in the caudal part (Johnston *et al.*, 1963). However, all basal cells undergo secretory phases associated with the end of calcification, and their microvilli also change in a similar way to those of the apical cells.

The tubular glands (*glandulae uterinae*) (Fig. 5.37) are complex, branched structures and open to the surface by short closely packed ducts. The ducts are lined by rows of polygonal cells (Richardson, 1935). The secretory cells are pale-staining and at the end of calcification contain many vacuoles. Ultrastructurally the cells possess few granules, and at their apical border is a complex arrangement of microvilli which become much swollen and distended during the shell-deposition process. Between the cells are complex interdigitations. In the cranial region of the shell gland the tubular glands have features which are a mixture of those of the isthmus and those of the shell gland. However, a specialized type of gland, the so-called "junctional zone" gland, containing particularly complex granules is also present (Johnston *et al.*, 1963).

There has been much speculation about the roles of the various glandular

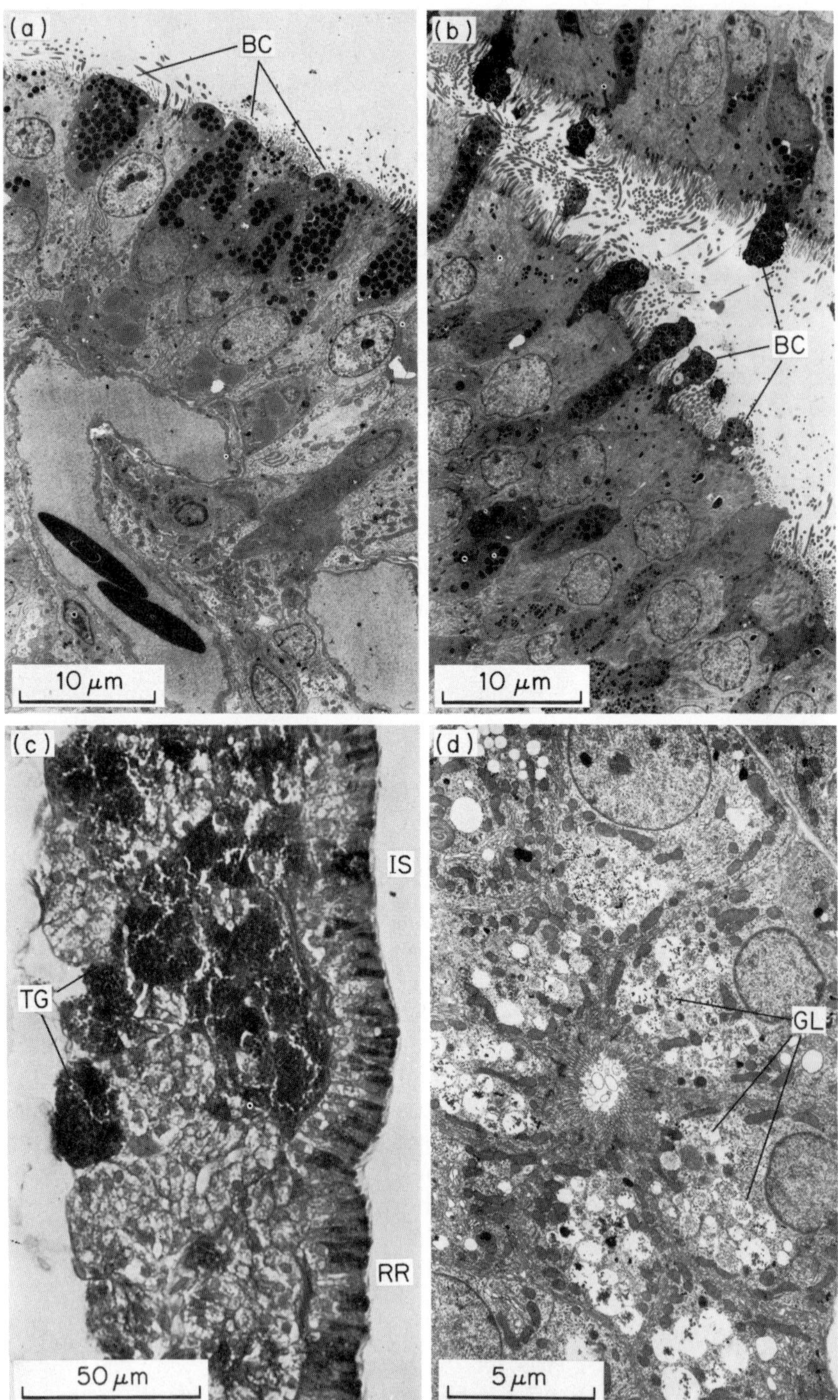

Fig. 5.36. "Red-region" (tubal region of shell gland). (a) Epithelium with the curious glandular "bleb" cells (BC). (Johnston, unpublished). (b) Similar to (a) but the protrusion of the "bleb" cells (BC) into the lumen of the oviduct is more obvious. (Johnston, unpublished). The functional distinction between stage (a) and stage (b) is not known. (c) Mucosa at the junction of the isthmus (IS) and the "red-region" (RR). The granular cells

elements in the completion of the shelled egg but much of the evidence is contradictory (see Aitken, 1971, Wyburn *et al.*, 1973 and Hodges, 1974 for a development of the arguments). The addition of the watery fluid to the egg (plumping) occurs during the first six or eight hours of its stay in the shell gland (Gilbert, 1971c). This fluid is thought to be secreted by the tubular glands of the red-region (Wyburn *et al.*, 1973) but it must flow into the pouch region for it to be incorporated into the egg. The significance of the glycogen in the tubular glands of this region (Fig. 5.36) is uncertain (Draper *et al.*, 1972).

The formation of the main part of the shell is a complex process because both a mineral and a proteinaceous component have to be produced, and laid down approximately simultaneously. The cells responsible for the transport of the calcium may lie in the tubular glands (Richardson, 1935; Johnston *et al.*, 1963; Breen and de Bruyn, 1969; Wyburn *et al.*, 1973) and it has been suggested that the swollen appearance of the microvilli of the gland cells is connected with this transport. On the other hand, the experimental evidence of Gay and Schraer (1971) places calcium transport in the ciliated epithelial cells and there is other, but less definite, supporting evidence for this (Nevalainen, 1969; Aitken, 1971).

The protein component of the matrix in *Gallus*, it has been suggested, is secreted by the apical cells (Fig. 5.37) of the epithelium (Hodges, 1974) and in support of this the existence of cyclical secretory phases in the cells have been described. Tamura and Fujii (1966b), however, could not see such phases in *Coturnix* and Wyburn *et al.* (1973) even queried their existence in *Gallus*. The mucopolysaccharide component which, together with the protein, forms the final matrix may be produced in the basal cells (Breen and de Bruyn, 1969).

The source of the cuticle is not known. One suggestion is that it is derived from the microvilli of the apical cells (Johnston *et al.*, 1963; Tamura and Fujii, 1966b; Breen and de Bruyn, 1969; Hodges, 1974), although others (Tyler and Simkiss, 1959a; Baker and Balch, 1962) have attributed it to the basal cells (Fig. 5.37). Fujii (1963) even thought it was produced in the tubular glands of the third funnel-shaped part.

Pigment is deposited throughout the shell but again the cells responsible have not been positively identified. Circumstantial evidence in *Gallus* (Aitken, 1971) and *Coturnix* (Tamura *et al.*, 1965) at present favours the apical cells but Baird *et al.* (1975) argued against the specificity of the apical cells in this connection and suggested instead that the whole epithelium may be involved.

The only conclusion to be drawn from these studies of the shell gland is the difficulty in attempting to ascribe function to structure from morphological examinations unsupported by experimental evidence.

The major portion of the shell gland forms a pouch in which the egg remains during the greater part of its stay in the oviduct. Rotation of the egg may occur here and could be the cause of the streaking of the pigment of the eggs of many species (Welty, 1962). Usually eggs are found lying in this portion of the oviduct

---

of the epithelium are PAS-positive. The tubular glands (TG) of the isthmus contain acidophilic granules. PAS, haematoxylin and light green. (Davidson, unpublished). (d) Unlike those of any other region, the tubular glands of the "red-region" contain quantities of glycogen (GL). (Johnston, unpublished).

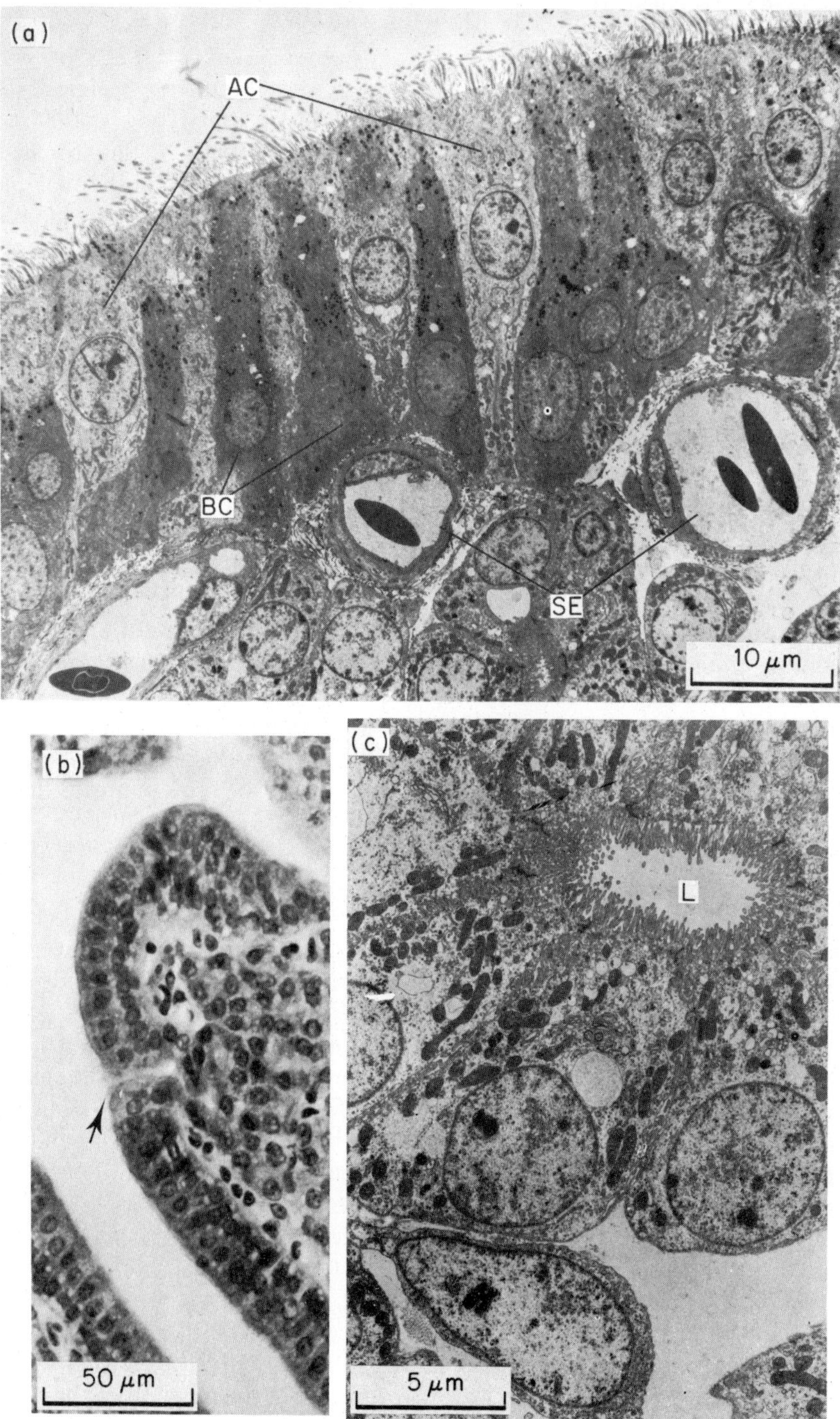

Fig. 5.37. Shell gland (pouch). (a) Epithelium with apical ciliated cells (AC) and basal cells (BC). Parts of the sub-epithelial blood plexus are visible (SE). (Johnston, unpublished). (b) The tubular glands open through the epithelium via short ducts (arrowed). Haematoxylin and eosin. (c) Tubular gland with prominent microvilli. There are none of the large granules found in the magnum or isthmus and no glycogen. The lumen (L) contains no visible secretory product. (Johnston, unpublished).

with their pointed end directed caudally and like this are laid. In some species, e.g. gulls (Laridae) and ducks (Anatidae), a rotation of the egg occurs just before oviposition through 180° in the horizontal plane and about the short axis (Thomson, 1964). Eggs in these birds therefore are laid blunt end first. In *Gallus* the situation is not clear. Whilst Bradfield (1951) claimed that rotation occurred, this cannot always be so because many domestic fowl lay eggs pointed end first (Olsen and Byerly, 1932; Wood-Gush and Gilbert, 1969). In view of the variability in rotation between and within species, its biological significance is obscure. Movement within the shell gland must be brought about by its muscle layers but whether or not the muscular cord of the ventral ligament is involved has yet to be determined.

(*e*) *Vagina.* The vagina is a relatively short, narrow, muscular duct which cranially leads from the shell gland and caudally joins the cloaca at a narrow opening (*ostium cloacale oviductus sinistri*). This region of the oviduct is tightly folded upon itself, the folds being bound together by connective tissue (Fujii, 1963). At its cranial end the vagina is delineated by the utero-vaginal sphincter (*m. sphincter vaginae*), although it is questionable that a true sphincter exists here (Aitken, 1971). The circular muscle of the vagina is well formed but the longitudinal layer is reduced to no more than scattered bundles (Surface, 1912) which may be related to the fact that the vagina has to be highly distensible to allow the passage of the rigid shelled egg.

The mucosa of the vagina is thrown into numerous long and narrow longitudinal ridges which contain conspicuous secondary folds (Fig. 5.38). Unlike in the other regions there are no secretory tubular glands, though secretory cells are present in the epithelium.

In the cranial portion of the vagina, adjacent to the shell gland, are the sperm-host glands (*fossulae spermaticae*) (Figs 5.38, 39) which are important to the survival of spermatozoa in the oviduct (Lake, 1975). In those species which have been examined, e.g. *Gallus*, *Anser*, *Meleagris*, *Anas*, the Mallard (*Anas platyrhynchos*), *Coturnix*, the Ring-necked Pheasant (*Phasianus colchicus*), a dove (Columbidae) and the Helmet Guineafowl (*Numida meleagris*), fertile eggs are produced up to several weeks after a single insemination (Lake, 1935) and it is likely that this will be found to be a common feature of all birds. Despite considerable attention, little is known of the way in which spermatozoa enter the glands, how they survive in them, and how and at what times they are released. For a full discussion of these aspects see Lake (1975). The glands are branched tubular structures in the lamina propria of the mucosal folds (Fujii and Tamura, 1963; Fujii, 1963; Bekhtina and Diaguileva, 1964; Bobr *et al.*, 1964) and are similar in appearance to those of the shell gland, although they have a larger diameter. They are lined by tall, columnar cells which have basal nuclei and apical microvilli. The blood supply of the glands is well-developed (Gilbert *et al.*, 1968). No contractile elements or nerve fibres appear to be associated with the glands (Van Krey *et al.*, 1967; Gilbert *et al.*, 1968; Tingari and Lake, 1973). Within the glands, the spermatozoa appear to be freely separated from each other (Figs 5.38, 39) and do not make contact with the microvilli of the gland cells (Schindler *et al.*, 1967; Burke *et al.*, 1972; Tingari and Lake, 1973). There cannot, therefore, be any direct transfer of "nutrients" or other protective substances from the gland to the sperm. However, the glands

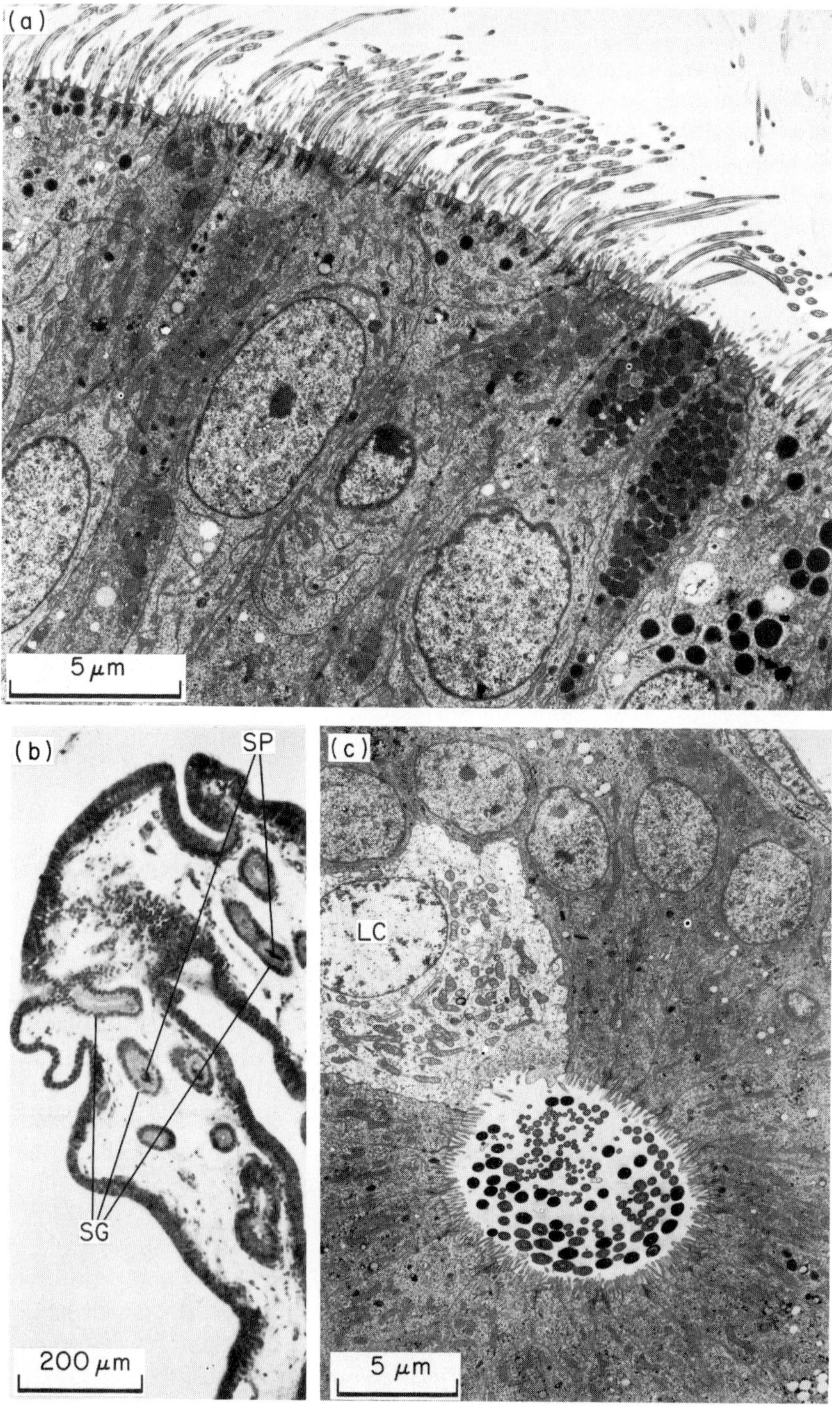

Fig. 5.38. Vagina. (a) The surface epithelium of the vagina has fewer granular secretory cells than other regions of the oviduct. (Johnston, unpublished). (b) Although there are no true secretory tubular glands, simple tubular infoldings (SG) (sperm-host glands) are present in the cranial section and these contain spermatozoa (SP). Haematoxylin and eosin. (c) Transverse section of a sperm-host gland containing spermatozoa. The significance of the obviously different large cell (LC) is unknown. (Johnston, unpublished).

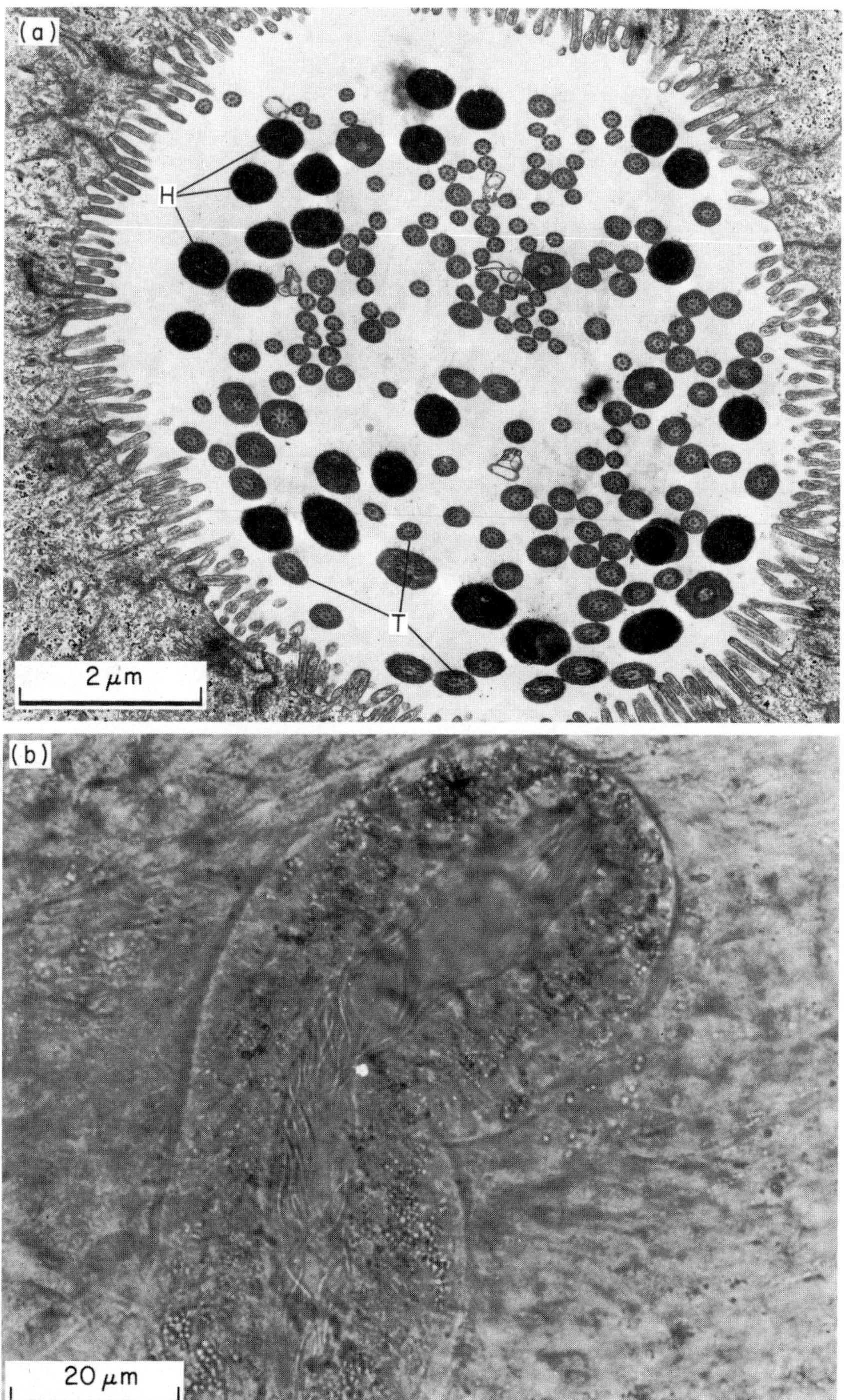

Fig. 5.39. Vaginal sperm-host glands. (a) Details of the spermatozoa in the gland; both tails (T) and heads (H) are visible. (Johnston, unpublished). (b) Living preparation of a complete sperm-host gland; the wave-like swimming motion of the spermatozoa indicates that the lumen is fluid-filled.

must be fluid-filled if the sperm are to survive in them and, if transfer occurs, it could be through this fluid medium. Enzymes, particularly those concerned with oxidative pathways, are found in the glands of *Coturnix* which indicates some energy transport mechanisms may be operating (Von Sinowatz *et al.*, 1976). Moreover, as in other vertebrates, the sperm-storage sites in *Coturnix* are positive for steroid dehydrogenases (Von Sinowatz *et al.*, 1976) which might suggest an important role for these enzymes in the function of the glands.

Ultimately the egg has to be voided from the oviduct, and oviposition results from the coordinated activity of both the shell gland and the vagina. Although the precise mechanisms are not understood (Gilbert, 1971d), it is likely that the contraction of the musculature of the shell gland forces the egg into the vagina through the relaxed utero-vaginal sphincter. It seems reasonable to suppose that the complex innervation of these regions is involved in some way. Supporting evidence for this perhaps is that prostaglandins cause contraction of the muscle of the shell gland and relaxation of the sphincter (Verma *et al.*, 1976), and recent observations suggest that they do play an important role in the transport of the egg through the oviduct and in oviposition (Talo and Kekäläinen, 1976; Wechsung and Houvenaghel, 1976; Day and Nalbandov, 1977). It is interesting that the avian vaginal sphincter appears to act in a manner analogous to the cervix of mammals since experimental relaxation of the sphincter in *Gallus* leads to premature oviposition (Lake and Gilbert, 1964b) just as an incompetent cervix in mammals leads to abortion. The vagina by itself appears to be incapable of expelling the egg (Sykes, 1953b, 1955a). Instead, distension of the vagina by the egg brings about the "bearing down" reflex (Sykes, 1954, 1955b) which results in the egg being forced through the cloaca. The processes involved in oviposition vary in duration from species to species. In some birds, e.g. the Brown-headed Cowbird (*Molothrus ater*) and the Cuckoo (*Cuculus canorus*) they take only a few seconds (Welty, 1962), whereas in *Meleagris* and *Anser* they last several hours.

The time of oviposition depends on the species but little is known of the physiological mechanisms, or the associated structures, which cause different birds to lay at different times of the day. Many song birds, including finches (Fringillidae), wrens (Troglodytidae) and some warblers (Parulidae), lay around sunrise (Thomson, 1964) and laying early in the day seems common to passerines generally (Davis, 1955). However, egg laying can be found at other times. Domestic fowl, although usually laying in the morning, may extend their egg-laying throughout the afternoon. Pigeons (Columbidae) and pheasants (Phasianidae) usually lay in the early afternoon and evening respectively. The American Coot (*Fulica americana*) lays around midnight and the Ringed Plover (*Charadrius hiaticula*) will oviposit at any time of day or night (Welty, 1962).

Both Welty (1962) and Thomson (1964) suggested that the time interval between eggs results from the time required for the egg to pass through the oviduct. This hardly seems likely in those species in which several days intervene between eggs and it would seem to be more realistic to assume that the interval between eggs reflects the time interval between ovulations, as it does in *Gallus* and other domestic species (Gilbert, 1971d). If so, the time interval between eggs gives an indication of general ovarian competence, although as pointed out by Gilbert and Wood-Gush (1971), the oviposition rate in *Gallus* does not always accurately reflect the ovulation rate. Whilst the interval between successive eggs varies between the species, most passerines have approximately a 24 h cycle (Welty, 1962). A similar interval occurs in *Gallus* except that its exact length depends on the rate of lay and the position of the egg in the sequence (Gilbert and Wood-Gush, 1972). In other species longer intervals are not uncommon (Welty, 1962; Thomson, 1964; Murton and Westwood, 1977), 44–46 h being cited for *Columba*, 48 h for the Raven (*Corvus corax*), the Ostrich (*Struthio camelus*), the Cuckoo (*Cuculus canorus*) and the Greater Rhea (*Rhea americana*), 3 days for cassowaries (Casuariidae), 5 days for the Andean Condor (*Vultur gryphus*) and 6–7 days for the Blue-faced Booby (*Sula dactylatra*). The longest interval between successive eggs of a clutch seems to have been recorded for the Brown Kiwi (*Apteryx australis*); in the various studies that have been made on this species the average laying interval was found to be between 21 and 44 days, whilst the range of actual laying intervals was between 11 and 57 days (Kinsky, 1971). Kinsky suggested that this long interval is related to the way in which ovulation occurs alternately from each of the two ovaries but it seems reasonable to ask whether or not these eggs should be regarded as forming part of the same clutch.

2. *The right oviduct*

The right oviduct (*oviductus dexter*) develops from the paramesonephric duct in the same way as that on the left side until the 4th to 7th day of incubation, although some asymmetry is visible even at this early stage (Gruenwald, 1942). About the 7th day of incubation degeneration occurs cranially in the right oviduct and progresses in a caudal direction. By the 18th day only small vestiges of the oviduct remain attached to the cloaca (Fig. 5.40). In adult domestic fowl these vestiges persist throughout life (Webster, 1948; Williamson, 1965) and may contain small areas of oviductal tissue, even at times small recognizable tubes (Winter, 1958). Under normal hormonal stimulation these oviductal rudiments may become cystic and can reach a considerable size (Fig. 5.40) (Winter, 1958, Williamson, 1964, 1965), although the cystic fluid is of plasma, not oviductal, origin (Nemirovsky and Narbaitz, 1970). Histologically both the cystic and tubal oviducts appear to contain tissue which is mainly of the magnum and only rarely does infundibular tissue occur. Tissue of the shell gland or vagina is never found in the right oviduct (Winter, 1958).

The incidence of two morphologically adult oviducts is rare in *Gallus* (Crew, 1931; Webster, 1948; Winter, 1958; Sell, 1959) and the evidence for a functional right oviduct is almost non-existent (King, 1975). However, for full details of a bird with a functional right oviduct see Bickford (1965). In the author's laboratory one bird has been examined which had functional right and left oviducts (Fig. 5.9) as well as two functional ovaries, the right oviduct at death actually containing a developing egg in the magnum. McBride (1962) suggested that the development of the right oviduct is controlled by a single pair of genes, the heterozygote leading to the cystic form and the dominant homozygote resulting in a fully formed oviduct.

The description for *Gallus* is probably pertinent to birds generally though

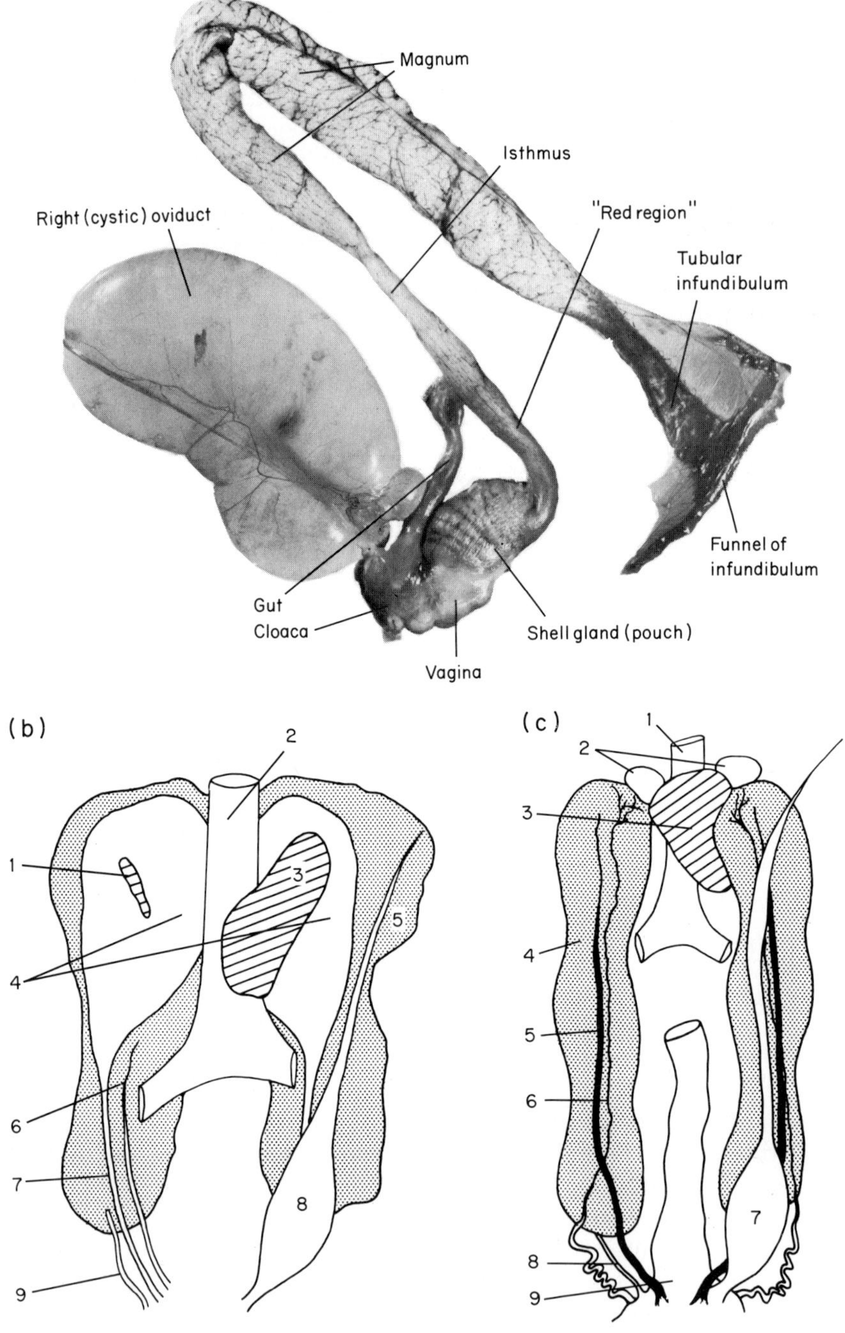

Fig. 5.40. (a) Cystic right oviduct of a laying domestic fowl showing its relationship with the gut and the normal left oviduct (compare with Figs 5.8 and 5.9). (b) Diagram of the female reproductive organs and associated male ducts just before hatching. (Redrawn from Romanoff, 1960). (c) Stylized diagram of the avian urogenital organs in the juvenile stage. (Redrawn from Witschi, 1961). In many species the male ducts become prominent just before the breeding season. (b): 1, Right ovary; 2, vena cava; 3, left ovary; 4, mesonephros; 5, metanephros; 6, ureter; 7, Wolffian duct; 8, oviduct; 9, right Müllerian duct. (c): 1, Vena cava; 2, adrenal; 3, left ovary; 4, metanephros; 5, ureter; 6, Wolffian duct; 7, oviduct; 8, right Müllerian duct; 9, gut.

specific details are lacking for most species. However, according to Kinsky (1971) seldom are even vestiges of a right oviduct found in wild birds, and in the Brown Kiwi (*Apteryx australis*) he examined, only 6% contained vestiges. Hence, where double oviducts are mentioned in the literature, it is not certain what level of development is meant and probably most right oviducts were non-functional and vestigial (Kinsky, 1971). Unfortunately, few authors reported the actual state of the right oviduct, so that no positive conclusions can be drawn. However, whatever this state is, elements of right oviducts have been reported in over 30 species belonging to many orders, among them Falconiformes, Ciconiiformes, Anseriformes, Ralliformes, Columbiformes and Psittaciformes (Kinsky, 1971). Reports of functional right oviducts in species other than *Gallus* are equally rare, Guhn (1912) describing such an oviduct in the Sparrow-hawk (*Accipiter nisus*) and Common Kestrel (*Falco tinnunculus*), whilst Chappelier (1913) claimed to have found one in *Anas*. Enlarged and possibly functional right oviducts have also been reported in the Pekin Duck (Kamar and Yamani, 1962) and the Ring-necked Pheasant (*Phasianus colchicus*) (Purohit *et al.*, 1977).

3. *Mesonephric ducts*

In female birds generally, the mesonephric (Wolffian) ducts arise during embryonic life (Fig. 5.10), although their subsequent development is arrested. However, vestiges of the ducts remain as a permanent feature (Fig. 5.40) (Marshall, 1961). *Gallus* in this respect may be unusual, for it is uncertain whether or not a mesonephric duct is present on the left side (compare Domm, 1927, Brode, 1928 and Kar, 1947a).

In the non-breeding seasons the mesonephric ducts occur as small thread-like strands of tissue on the ventral surface of the kidney close to the ureter. Generally the ducts have a patent lumen (King, 1975). The ducts have no known function, although they are able to respond to the androgenic hormones produced by the female towards the beginning of the breeding season. At this time they become quite prominent in many species, e.g. the House Sparrow (*Passer domesticus*), finches (Fringillidae), the Common Starling (*Sturnus vulgaris*) and the Black-crowned Night-heron (*Nycticorax nycticorax*), and may even contain some secretory products (Callow and Parkes, 1935; Noble and Wurm, 1940; Witschi and Fugo, 1940; Bailey, 1953; Witschi, 1961).

The *tubuli epoophori* (homologues of the male *ductuli aberrantes*) are found in *Gallus* associated with the capsule of the adrenal gland. They form a group of secretory cells which Budras (1972) claimed produced steroid hormones at the onset of sexual maturity.

Acknowledgements

I am indebted to the many friends and colleagues who have helped me in the preparation of this chapter. In particular I wish to thank the staff of the Poultry Research Centre who contributed so much: Maida F. Davidson who provided the histological material, R. K. Field and E. Armstrong of the photographic section who produced the illustrations, and J. Coltherd (the Librarian) who collected the many publications I requested. I also owe a great debt of gratitude to Margaret M. Perry (PRC) and H. S. Johnston (Department of Anatomy, Glasgow University) for granting me permission to use their unpublished electron micrographs of the ovary and oviduct respectively. My thanks are also due to the staff of the Sub-department of Ornithology, British Museum (Natural History) for answering my many queries.

References

Abdel-Malek, E. T. (1950). Early development of the urino genital system in the chick, *J. Morph.* **86**, 599–626.

Abercrombie, R. G. (1931). Colour of bird's eggs, *Naturalist, Hull* 105–108.

Aire, T. A. and Steinbach, J. (1976). Histochemical studies of the development of alkaline and acid phosphatase activity in the ovary and oviduct of the fowl (*Gallus domesticus*), *Acta anat.* **95**, 207–217.

Aitken, R. N. C. (1966). Post-ovulatory development of ovarian follicles in the domestic fowl, *Res. vet. Sci.* **7**, 138–142.

Aitken, R. N. C. (1971). The oviduct. *In* "The Physiology and Biochemistry of the Domestic Fowl" (D. J. Bell and B. M. Freeman, Eds), Vol. 3, 1237–1289. Academic Press, London and New York.

Aitken, R. N. C. and Johnston, H. S. (1963). Observations on the fine structure of the infundibulum of the avian oviduct, *J. Anat.* **97**, 87–99.

Aitken, R. N. C. and Solomon, Sally E. (1976). Observations on the ultrastructure of the oviduct of the Costa Rican Green Turtle (*Chelonia Mydas* L.), *J. exp. mar. Biol. Ecol.* **21**, 75–90.

Allen, B. M. (1907). A statistical study of the sex cells of *Chrysemis marginata*, *Anat. Rec.* **1**, 64–65.

Amin, S. O. and Gilbert, A. B. (1970). Cellular changes in the anterior pituitary of the domestic fowl during growth, sexual maturity and laying, *Br. Poult. Sci.* **11**, 451–458.

Amoroso, E. C. and Marshall, F. H. A. (1960). External factors in sexual periodicity. *In* "Marshall's Physiology of Reproduction" (A. S. Parkes, Ed.), Vol. 1. Longmans Green and Co., London.

Anderson, B. and Rosenberg, M. D. (1976). Variations of oviductal lumen ATPase with the ovulatory cycle of the hen, *Biology Reprod.* **14**, 253–255.

Anderson, Beverley, Kim, Norma B., Rhea, R. P. and Rosenberg, M. D. (1974). Nucleoside triphosphate hydrolase activities of avian vitelline membrane and oviductal lumen, *Devl Biol.* **37**, 306–316.

Asmundson, V. S. (1931). The formation of the hen's egg. I. Function of the parts of the oviduct, *Scient. Agric.* **11**, 590–602.

Asmundson, V. S. and Burmester, B. R. (1936). The secretory activity of the parts of the hen's oviduct, *J. exp. Zool.* **72**, 225–246.

Asmundson, V. S. and Burmester, B. R. (1938). The effect of resecting a part of the uterus on the formation of the hen's egg, *Poult. Sci.* **17**, 126–130.

Asmundson, V. S. and Jervis, J. G. (1933). The effect of resection of different parts of the oviduct on the formation of the hen's egg, *J. exp. Zool.* **65**, 395–420.

Bacon, W. L. and Cherms, F. L. (1968). Ovarian follicular growth and maturation in the domestic turkey, *Poult. Sci.* **47**, 1303–1314.

Bacon, W. L. and Koontz, M. (1971). Ovarian follicular growth and maturation in coturnix quail, *Poult. Sci.* **50**, 233–236.

Bailey, R. E. (1953). Accessory reproduction organs of male fringillis birds. Seasonal variations and responses to various sex hormones, *Anat. Rec.* **115**, 1–20.

Baillie, A. H., Ferguson, M. M. and Hart, D. M. (1966). "Developments in Histochemistry." Academic Press, London and New York.

Bain, Joan M. and Hall, Janice M. (1969). Observations on the development and structure of the vitelline membrane of the hen's egg: an electron microscope study, *Aust. J. biol. Sci.* **22**, 653–665.

Baird, T., Solomon, Sarah E. and Tedstone, D. R. (1975). Localization and characterization of egg shell porphyrins in several avian species, *Br. Poult. Sci.* **16**, 201–208.

Baker, C. M. A. (1968). The proteins of egg white. *In* "Egg Quality: a Study of the Hen's Egg" (T. C. Carter, Ed.), 67–108. Oliver and Boyd, Edinburgh.

Baker, J. R. (1938a). Latitude and egg seasons in Old World Birds, *Tabul biol.* **15**, 333–370.

Baker, J. R. (1938b). The evolution of breeding seasons. *In* "Evolution" (G. R. de Beer, Ed.), 161–177. Clarendon Press, Oxford.

Baker, J. R. and Balch, D. A. (1962). A study of the organic material of the hen's egg shell, *Biochem. J.* **82**, 352–361.

Baker, J. R. and Ranson, R. M. (1938). The breeding seasons of Southern Hemisphere birds in the Northern Hemisphere, *Proc. zool. Soc. Lond.* **108**, 101–141.

Baker, T. G. (1972). Oogenesis and ovarian development. *In* "Reproductive Biology" (H. Balin and S. Glasser, Eds), 398–437. Excerpta Medica, Amsterdam.

Bakst, M. R. (1978). Scanning electronmicroscopy of the vitelline membrane of the hen ovum, *J. Reprod. Fert.* **52**, 361–364.

Bakst, M. and Howarth, B. (1975). SEM preparation and observation of the hen's oviduct, *Anat. Rec.* **181**, 211–226.

Bakst, M. R. and Howarth, B. (1977a). The fine structure of the hen's ovum at ovulation. *Biology Reprod.* **17**, 361–369.

Bakst, M. R. and Howarth, B. (1977b). Hydrolysis of the hen's perivitelline layer by cock sperm *in vitro*, *Biology Reprod.* **17**, 370–379.

Balch, D. A. and Cooke, R. A. (1970). A study of the composition of hen's egg shell membranes, *Annls Biol. anim. Biochim. Biophys.* **10**, 13–25.

Bartelmez, G. W. (1912). The bilaterality of the pigeon's egg. A study in egg organisation from the first growth of the oocyte to the beginning of cleavage, *J. Morph.* **23**, 269–328.

Bates, R. W., Lahr, E. L. and Riddle, O. (1935). The gross action of prolactin and follicle stimulating hormone on the mature ovary and sex accessories of the fowl, *Am. J. Physiol.* **111**, 361–368.

Becking, J. H. (1975). The ultrastructure of the avian egg shell, *Ibis* **117**, 143–151.

Bekhtina, V. G. and Diaguileva, G. E. (1964). Fate of spermatozoids in genital tract of hens, *Archs Anat. Histol. Embryol.* **46**, 25–32.

Bell, D. J. and Freeman, B. M. (1971). "Physiology and Biochemistry of the Domestic Fowl", Vol. 3. Academic Press, London and New York.

Bellairs, Ruth (1960). Development of Birds. *In* "Biology and Comparative Physiology of Birds" (A. J. Marshall, Ed.), Vol. 1, 127–188. Academic Press, London and New York.

Bellairs, Ruth (1961). The structure of the yolk of the hen's egg as studied by electron microscopy. I. The yolk of the unincubated egg, *J. biophys. biochem. Cytol.* **11**, 207–225.

Bellairs, Ruth (1964). Biological aspects of the yolk of hen's eggs. *In* "Advances in Morphogenesis" (M. Abercrombie and J. Brachet, Eds), Vol. 4, 217–272. Academic Press, London and New York.

Bellairs, Ruth (1965). The relationship between the oocyte and follicle in the hen's ovary as shown by electron microscopy, *J. Embryol. exp. Morph.* **13**, 215–233.

Bellairs, Ruth (1967). Aspects of the development of yolk spheres in the hen's oocyte, studied by electron microscopy, *J. Embryol. exp. Morph.* **17**, 267–281.

Bellairs, Ruth and Boyde, A. (1969). Scanning electron microscopy of the shell membranes of the hen's egg, *Z. Zellforsch. mikrosk. Anat.* **96**, 237–249.

Bellairs, Ruth, Harkness, M. L. and Harkness, R. D. (1963). The vitelline membrane of the hen's egg: a chemical and ultrastructural study, *J. Ultrastruct. Res.* **8**, 339–359.

Bellairs, Ruth, Lorenz, F. W. and Dunlop, Tania (1978). Cleavage in the chick embryo, *J. Embryol. exp. Morph.* **43**, 55–69.

Bennett, C. H. (1947). Relation between size and age of the gonads in the fowl from hatching date to sexual maturity, *Poult. Sci.* **26**, 99–104.

Benoit, J. (1926). Sur l'origine des cellules interstitielles de l'ovaire de la poule, *C.r. Séanc. Soc. Biol.* **94**, 873–874.

Benoit, J. (1931). Destruction des gonocytes primaires dans le blastoderme du poulet par les rayons ultra-violets, aux premiers stades du développement embryonaire, *Proc. Int. Congr. Sex. Res.* **2**, 162–170.

Benoit, J. (1950). Organes uro-génitaux. *In* "Traité de Zoologie" (P. P. Grasse, Ed.), Vol. 15, 341–377, Masson, Paris.

Bernstein, R. S., Nevalainen, T., Schraer, R. and Schraer, H. (1968). Intracellular distribution and role of carbonic anhydrase in the avian (*Gallus domesticus*) shell gland mucosa, *Biochim. biophys. Acta* **159**, 367–376.

Bickford, A. A. (1965). A fully formed and functional right oviduct in a Single Comb White Leghorn pullet, *Avian Dis.* **9**, 464–470.

Bissonette, T. H. and Zujko, A. J. (1936). Normal progressive changes in the ovary of the starling (*Sturnus vulgaris*) from December to April, *Auk* **53**, 31–50.

Biswal, G. (1954). Additional histological findings in the chicken reproductive tract, *Poult. Sci.* **33**, 843–851.

Blocker, H. W. (1933). Embryonic history of the germ-cells in *Passer domesticus* (L.), *Acta zool. Stockh.* **14**, 111–152.

Blom, L. (1973). Ridge pattern and surface ultrastructure of the oviducal mucosa of the hen (*Gallus domesticus*), *K. danske Vidensk. Selsk. Skr.* **20**, 1–15.

Blount, Mary (1909). The early development of the pigeon's egg with special reference to polyspermy and the origin of the periblast nuclei, *J. Morph.* **20**, 1–64.

Blount, W. P. (1945). "Sexing Day-old Chicks", 2nd edn. Poultry World, London.

Board, Patricia A. and Board, R. G. (1967). A method of studying bacterial penetration of the shell of the hen's egg, *Lab. Pract.* **16**, 471–482.

Board, R. G. (1966). Review article: the course of microbial infection of the hen's egg, *J. appl. Bact.* **29**, 319–341.

Board, R. G. (1968). Microbiology of the egg. *In* "Egg Quality: a Study of the Hen's Egg" (T. C. Carter, Ed.) Oliver and Boyd, Edinburgh.

Board, R. G. (1974). Microstructure, water resistance and water repellancy of the pigeon egg shell, *Br. Poult. Sci.* **15**, 415–419.

Board, R. G. and Halls, N. A. (1973a). The cuticle: a barrier to liquid and particle penetration of the shell of the hen's egg, *Br. Poult. Sci.* **14**, 69–97.

Board, R. G. and Halls, N. A. (1973b). Water uptake by eggs of Mallard and guineafowl, *Br. Poult. Sci.* **14**, 311–314.

Bobr, L. Wanda, Lorenz, F. W. and Ogasawara, F. X. (1964). Distribution of spermatozoa in the oviduct and fertility in domestic birds. I. Residence sites of spermatozoa in fowl oviduct, *J. Reprod. Fert.* **8**, 39–47.

Boogaard, C. L. and Finnegan, C. V. (1976). The effects of estradiol and progesterone on the growth and differentiation of the quail oviduct, *Can. J. Zool.* **54**, 324–325.

Boring, A. M. and Pearl, R. (1914). The odd chromosome in the spermatogenesis of the domestic chicken, *J. exp. Zool.* **16**, 53–83.

Botte, V. (1963). La localizzozione della steroide-3β-olo-deidrongenasi nell'ovario di pollo, *R.c. Inst. Sci., Univ. Camerino* **4**, 205–209.

Boucek, R. J. and Savard, K. (1970). Steroid formation by the avian ovary *in vitro* (*Gallus domesticus*), *Gen. comp. Endocr.* **15**, 6–11.

Boucek, R. J., Györi, E. and Alvarez, R. (1966). Steroid dehydrogenase reactions in the developing chick adrenal and gonadal tissues, *Gen. comp. Endocr.* **7**, 292–303.

Bradfield, J. R. G. (1951). Radiographic studies on the formation of the hen's eggshell, *J. exp. Biol.* **28**, 125–140.

Bradley, O. C. and Grahame, T. (1960). "The Structure of the Fowl", 4th edn. Oliver and Boyd, Edinburgh.

Bragdon, P. E. (1952). Corpus luteum formation and atresia in the common garter snake (*Thannophis siptalis*), *J. Morph.* **91**, 413–437.

Brambell, F. W. R. (1926). The oogenesis of the fowl (*Gallus bankiva*), *Phil. Trans. R. Soc. Ser. B* **214**, 113–151.

Brambell, F. W. R. (1956). Ovarian Changes. *In* "Marshall's Physiology of Reproduction" (A. S. Parkes, Ed.), Vol. 1, 397–542. Longmans Green and Co., London.

Brant, J. W. A. (1953). Morphology and albumen secretion of the avian oviduct as controlled by sex hormones, *Biol. Abstr.* (1954) **28**, 618–619.

Brant, J. W. A. and Nalbandov, A. V. (1952). Role of sex hormones in the secretory activity of the oviducts of hens, *Poult. Sci.* **31**, 908–909.

Brant, J. W. A. and Nalbandov, A. V. (1956). Role of sex hormones in albumen secretion by the oviduct of chickens, *Poult. Sci.* **35**, 692–700.

Breen, P. C. and de Bruyn, P. P. H. (1969). The fine structure of the secretory cells of the uterus (shell gland) of the chicken, *J. Morph.* **128**, 35–66.

Breneman, W. R. (1955). Reproduction in birds: the female, *Mem. Soc. Endocr.* **4**, 94–113.

Brode, M. D. (1928). The significance of the symmetry of the ovaries of the fowl, *J. Morph.* **46**, 1–56.

Brodkorb, P. (1928). Paired ovaries in the marsh hawk, *Auk* **45**, 211.

Van den Broek, A. J. P. (1931). Gonaden und Ausführungsgänge. *In* "Handbuch der vergleichenden Anatomie der Wirbeltiere" (L. Bolk, E. Göppert, E. Kallius and W. Lubosch, Eds), Bd. VI. Urban und Schwartzenberg, Berlin.

Budras, K. D. (1972). Das Epoophoron der Henne und die Transformation Seiner Epithelzellen in Interrenal-und Interstitialzellen, *Ergebn. Anat. EntwGesch.* **46**, 1–70.

Bullough, W. S. (1942). The reproductive cycles of the British and Continental races of the starling, *Phil. Trans. R. Soc. Ser. B* **231**, 165–246.

Bunk, M. J. and Balloun, S. L. (1977). Structure and relationship of the mammillary core to membrane fibres and initial calcification of the avian eggshell, *Br. Poult. Sci.* **18**, 617–621.

Burwell, L. I. (1931). Early differentiation of the duck gonad. Inaugural Dissertation, Chicago University.

Burke, W. H., Ogasawara, F. X. and Fuqua, C. L. (1972). A study of the ultrastructure of the utero-vaginal sperm-storage glands of the hen, *Gallus domesticus*, in relation to a mechanism for the release of spermatozoa, *J. Reprod. Fert.* **29**, 29–36.

Cain, C. J. and Heyn, A. N. J. (1964). X-ray diffraction studies of the crystalline structure of the avian egg shell, *Biophys. J.* **4**, 23–39.

Callebaut, M. (1976). Origin of ovarian follicle cells in birds, *Experientia* **32**, 1337–1339.

Callow, R. K. and Parkes, A. S. (1935). Growth and maintenance of the fowl's comb by androsterone, *Biochem. J.* **29**, 1414–1423.

Campbell, L. D., Yu, J. Y-L., Stothers, S. C. and Marquardt, R. R. (1971). Sex hormone control mechanisms. II. Influence of estrogen and progesterone on the activities of key enzymes involved in the carbohydrate metabolism of chicken (*Gallus domesticus*) oviducts, *Can. J. Biochem.* **49**, 201–206.

Candlish, J. K. (1972). The role of the shell membranes in the functional integrity of the egg. *In* "Egg Formation and Production" (B. M. Freeman and P. E. Lake, Eds). British Poultry Science, Edinburgh.

Carter, T. C. (1968). The hen's egg: estimation of egg mean and flock mean shell thickness, *Br. Poult. Sci.* **9**, 343–357.

Cedard, L. and Haffen, K. (1966). Transformations de la dehydroepiandrosterone par les gonades embryonaires de poulet cultivée *in vitro*, *C.r. hebd. Séanc. Acad. Sci., Paris* **263**, 430–433.

Chakravorti, K. P. and Sadhu, D. P. (1959). Some aspects of the histochemical studies on the oviduct of the pigeon, *Columba livia* Gmelin, *Proc. zool. Soc.* **12**, 9–14.

Chakravorti, K. P. and Sadhu, D. P. (1961a). Some aspects of the histological and histochemical studies on the oviduct of the kite, *Milvus migrans govinda* Sykes, *Anat. Anz.* **110**, 160–164.

Chakravorti, K. P. and Sadhu, D. P. (1961b). Some aspects of the histological and histochemical studies on the oviduct of the hen, *Proc. zool. Soc.* **14**, 27–32.

Chakravorti, K. P. and Sadhu, D. P. (1963). Histological and histochemical studies of the hormone stimulated oviduct of the pigeon, *Columba livia* Gmelin, *Acta Histochem.* **16**, 343–356.

Chappelier, A. (1913). Persistence et dévelopment des organes genitaux droits chez les femelles adultes des oiseaux, *Bull. Scient. Fr. Belg.* **47**, 361–376.

Chen, B. K. (1932). The early development of the duck's egg with special reference to the origin of the primitive streak, *J. Morph.* **53**, 133–187.

Chen, T. W. and Hawes, R. O. (1970). Genital tract motility in the domestic hen, *Poult. Sci.* **49**, 640–649.

Chieffi, G. (1955). Sull'origine dell'assimmetria dell'ovario negli embrioni di *Scylliorhinus canicula*, *Monitore zool. ital.* **63**, 31–41.

Chieffi, G. (1959). Sex differentiation and experimental sex reversal in elasmobranchs, *Archs Anat. micr. Morph. exp.* **48**, 21–36.

Chieffi, G. and Botte, V. (1965). The distribution of some enzymes involved in the steroidogenesis of hen's ovary, *Experientia* **21**, 16–20.

Chieffi, G. and Botte, V. (1970). The problem of "luteogenesis" in non-mammalian vertebrates, *Boll. Zool. agr. Bachic.* **37**, 85–102.

Clawson, R. C. and Domm, L. V. (1963a). Developmental changes in glycogen content of primordial germ cells in chicken embryo, *Proc. Soc. exp. Biol. Med.* **112**, 533–537.

Clawson, R. C. and Domm, L. V. (1963b). The glycogen content of primordial germ cells in the White Leghorn chick embryo, *Anat. Rec.* **145**, 218–219.

Clayton, G. A. (1972). Effects of selection on reproduction in avian species, *J. Reprod. Fert., Suppl.* **15**, 1–21.

Clementi, R. and Palade, G. E. (1969). Intestinal capillaries: I. Permeability to peroxidase and ferritin, *J. Cell Biol.* **41**, 33–58.

Cody, M. L. (1971). Ecological aspects of Reproduction. *In* "Avian Biology" (D. S. Farner and J. R. King, Eds), Vol. I, 462–512. Academic Press, New York and London.

Cole, R. K. (1946). Stimulation of the avian oviduct by an ovarian fragment, *Poult. Sci.* **25**, 473–475.

Cole, R. K. and Hutt, F. B. (1953). Normal ovulation in non-laying hens, *Poult. Sci.* **32**, 481–492.

Cook, W. H. (1968). Macromolecular components of egg yolk. *In* "Egg Quality: A Study of the Hen's Egg" (T. C. Carter, Ed.). Oliver and Boyd, Edinburgh.

Cooke, A. S. and Balch, D. A. (1970a). Studies on membrane, mammillary cores and cuticle of the hen's egg, *Br. Poult. Sci.* **11**, 345–352.

Cooke, A. S. and Balch, D. A. (1970b). The distribution and carbohydrate composition of the organic matrix of hen egg shell, *Br. Poult. Sci.* **11**, 353–365.

Common, R. H. (1941). The carbonic anhydrase activity of the hen's oviduct, *J. agric. Sci., Camb.* **31**, 412–414.

Common, R. H., Rutledge, W. A. and Bolton, W. (1947). The influence of gonadal hormones on serum riboflavin and certain other properties of blood and tissue in the domestic fowl, *J. Endocr.* **5**, 121–130.

Common, R. H., Rutledge, W. A. and Hale, W. (1948). Observations on the mineral metabolism of pullets. VIII. The influence of gonadal hormones on retention of calcium and phosphorus, *J. agric. Sci., Camb.* **38**, 64–80.

Conrad, R. M. and Phillips, R. E. (1938). The formation of the chalazae and inner white of the hen's egg, *Poult. Sci.* **17**, 143–146.

Corinaldesi, F. (1926). Lo sviluppo autonomo del corpo genitale del pollo, *Monitore zool. ital.* **37**, 207–212.

Coste, M. (1850). Recherches sur la segmentation de la cicatricule chez les oiseaux, les reptiles ecailleux et les poissons cartilagineux, *C.r. hebd. Séanc. Acad. Sci., Paris* **30**, 638–642.

Cox, R. F. and Sauerwein, H. (1970). Studies on the mode of action of progesterone on chicken oviduct epithelium. I. Morphological changes associated with early differentiation of the tissue, *Exp. cell. Res.* **61**, 79–90.

Creger, C. R., Phillips, H. and Scott, J. T. (1976). Formation of an egg shell, *Poult. Sci.* **55**, 1717–1723.

Crew, F. A. E. (1931). Paired oviducts in the fowl, *J. Anat.* **66**, 100–103.

Crossley, J., Ayuy, A. and Ferrando, G. (1975). Some physiological characteristics of motility in the *in vivo* avian oviduct, *Biology Reprod.* **13**, 495–498.

Curtis, M. R. (1910). The ligaments of the oviduct of the domestic fowl, *Bull. Maine agric. Exp. Stn* **176**, 1–20.

Cuthbert, N. L. (1945). The ovarian cycle in the ring dove (*Streptopelia risoria*), *J. Morph.* **77**, 351–372.

Cutting, J. A. and Roth, T. F. (1973). Changes in specific sequestration of protein during transport into the developing oocyte of the chicken, *Biochim. Biophys. Acta* **298**, 951–955.

Dahl, E. (1970a). Studies on the fine structure of ovarian interstitial tissue. 2. The ultrastructure of the thecal gland of the domestic fowl, *Z. Zellforsch. mikrosk. Anat.* **109**, 195–211.

Dahl, E. (1970b). Studies on the fine structure of ovarian interstitial tissue. 3. The innervation of the thecal gland of the domestic fowl, *Z. Zellforsch. mikrosk. Anat.* **109**, 212–226.

Dahl, E. (1970c). Studies on the fine structure of ovarian interstitial tissue: effects of clomiphene on the thecal gland of the domestic fowl, *Z. Zellforsch. mikrosk. Anat.* **109**, 227–244.

Dahl, E. (1971a). Studies of the fine structure of the ovarian interstitial tissue. 5. Effects of gonadotropins on the thecal gland of the domestic fowl, *Z. Zellforsch. mikrosk. Anat.* **113**, 133–156.

Dahl, E. (1971b). Studies of the fine structure of ovarian interstitial tissue. 1. A comparative study of the fine structure of the ovarian interstitial tissue in the rat and the domestic fowl, *J. Anat.* **108**, 275–290.

Dahl, E. (1971c). Studies on the fine structure of ovarian interstitial tissue. 4. Effects of steroids on the thecal gland of the domestic fowl, *Z. Zellforsch. mikrosk. Anat.* **113**, 111–132.

Dahl, E. (1971d). Effect of steroids on the granulosa cells in the domestic fowl, *Z. Zellforsch. mikrosk Anat.* **119**, 179–187.

Dahl, E. (1971e). The effects of clomiphene on the granulosa cells of the domestic fowl, *Z. Zellforsch. mikrosk. Anat.* **119**, 188–194.

Dalrymple, J. R., Macpherson, J. W. and Friars, G. W. (1968). The reproductive tract of the turkey hen (a biometrical study), *Can. J. comp. Med.* **32**, 435.

Dang-quan-Dien (1951). Contribution a l'anatomie des artères de la poule domestique. Thesis, Faculty of Medicine and Pharmacology, University of Lyon.

Dantschakoff, W. (1931). Keimzelle und Gonade. I.A. Von der entodermalen Wanderzelle bis zur Urkeimzelle in der Gonade, *Z. Zellforsch. mikrosk. Anat.* **13**, 448–510.

Dantschakoff, W. (1933). Keimzelle und Gonade. 5. Sterilisierung der Gonaden im Embryo mittels Röntgenstrahlen. *Z. Zellforsch. mikrosk. Anat.* **18**, 56–109.

Dantschakoff, W. and Guelin-Schedrina, A. (1933). Keimzelle und Gonade. VI. Asymmetrie der Gonaden beim Huhn—A. Primäre quantitative Asymmetrie der Gonadenanlagen, *Z. Zellforsch. mikrosk. Anat.* **19**, 50–78.

Dantschakoff, W., Dantschakoff, W. Jnr. and Bereskina, L. (1931). Keimzelle und Gonade. 1A. Identität der Urkeimzellen und der entodermalen Wanderzellen, *Z. Zellforsch. mikrosk. Anat.* **14**, 323–375.

Darling, L. and Darling, L. (1963). "Birds." John Dickinson, Northampton.

Das, L. N. and Biswal, G. (1968). Microanatomy of reproductive tract of domestic duck (*Anas boscas*), *Indian vet. J.* **45**, 1003–1009.

Das, L. N., Mishra, D. B. and Biswal, G. (1965). Comparative anatomy of the domestic duck (*Anas boscas*), *Indian vet. J.* **42**, 320–326.

Davidson, M. F. (1973). Staining properties of the luminal epithelium of the isthmus and shell gland of the oviduct of the hen *Gallus domesticus* during the passage of an egg, *Br. Poult. Sci.* **14**, 631–633.

Davidson, M. F., Draper, M. H. and Leonard, E. M. (1968). Structure and function of the oviduct of the laying hen, *J. Physiol., Lond.* **196**, 9P–10P.

Davis, D. E. (1942a). Bursting of avian follicles at the beginning of atresia, *Anat. Rec.* **82**, 153–161.

Davis, D. E. (1942b). Regression of the avian post-ovulatory follicle, *Anat. Rec.* **82**, 297–307.

Davis, D. E. (1955). Breeding Biology of Birds. *In* "Recent Studies in Avian Biology" (A. Wolfson, Ed.). University of Illinois Press, Urbana.

Day, S. L. and Nalbandov, A. V. (1977). Presence of prostaglandin F(PGF) in hen follicles and its physiological role in ovulation and oviposition, *Biology Reprod.* **16**, 486–494.

Defretin, R. (1924). Origine et migration des gonocytes chez le poulet, *C.r. Séanc. Soc. Biol.* **91**, 1082–1084.

Delforge, J. P. and Schippers, Maria (1965). Activités enzymatiques dans les ébauches des deux gonades femelles de *Gallus domesticus* pendent l'ontogénèse et après l'éclosion, *Acta anat.* **61**, 355–378.

Deol, G. S. (1955). Studies on the structure and function of the ovary of the domestic fowl. Ph.D. Thesis, University of Edinburgh.

Diamanstein, T. and Schluens, J. (1964). Lokalisation und Bedeutung der Karboanhydrase im Uterus von Legenhennen, *Acta histochem.* **19**, 292–302.

Dick, H. R., Culbert, J., Wells, J. W., Gilbert, A. B. and Davidson, Maida F. (1978). Steroid hormones in the postovulatory follicle of the domestic fowl *Gallus domesticus*, *J. Reprod. Fert.* **53**, 103–107.

Diculesco, I. (1961). Recherches histoenzymologiques sur l'oviducte de la poule, *Annls Histochim.* **6**, 179–246.

Didier, E. and Fargeix, N. (1976). Aspects quantitatifs du peuplement des gonades par les cellules germinales chez l'embryon de Caille (*Coturnix coturnix japonica*), *J. Embryol. exp. Morph.* **35**, 637–648.

Dominic, C. J. (1960a). A study of the post-ovulatory follicle in the ovary of the domestic pigeon *Columba livia*, *J. zool. Soc. India* **12**, 27–33.

Dominic, C. J. (1960b). On the secretory activity of the funnel of the avian oviduct, *Curr. Sci.* **29**, 274–275.

Dominic, C. J. (1961a). The annual reproductive cycles of the domestic pigeon, *Columba livia. J. scient. Res., Banaras Hindu Univ.* **11**, 272–306.

Dominic, C. J. (1961b). A study of the atretic follicles in the ovary of the domestic pigeon, *Proc. natn. Acad. Sci., India* **31B**, 273–286.

Domm, L. V. (1927). New experiments on ovariotomy and the problem of sex inversion in the fowl, *J. exp. Zool.* **48**, 31–173.

Domm, L. V. (1939). Modifications in sex and secondary sexual characters in birds. *In* "Sex and Internal Secretions" (E. Allen, C. H. Danforth and E. A. Doisey, Eds), 2nd edn. Williams and Wilkins Co., Baltimore.

Dransfield, J. W. (1945). The lymphatic system of the domestic fowl, *Br. vet. J.* **101**, 171–179.

Draper, M. H., Johnston, H. S. and Wyburn, G. M. (1968). The fine structure of the oviduct of the laying hen, *J. Physiol.* **196**, 7P–8P.

Draper, M. H., Davidson, M. F., Wyburn, G. M. and Johnston, H. S. (1972). The fine structure of the fibrous membrane forming region of the isthmus of the oviduct of *Gallus domesticus*, *Q. Jl exp. Physiol.* **57**, 297–309.

Van Drimmelen, G. C. (1946). "Sperm nests" in the oviduct of the domestic hen, *Jl S. Afr. vet. med. Ass.* **17**, 42–52.

Van Drimmelen, G. C. (1949). Structure of the sperm-nests in the oviduct of the domestic hen, *Nature, Lond.* **163**, 950–951.

Dubuisson, H. (1906). Contribution à l'étude du vitellus, *Archs Zool. exp. gén.* **5**, 153–402.

Dulbecco, R. (1946). Suiluppo di gonadi in asseza di cellule sessuali nell'embrioni di pollo. Sterilizozione complete mediante esposizione a raggi allo studio di linea primitiva, *Rc. Accad. naz. Lincei, Ser. VIII.* **1**, 1211–1213.

Dunham, H. H. and Riddle, O. (1942). Effects of a series of steroids on ovulation and reproduction in pigeons, *Physiol. Zool.* **15**, 383–394.

Van Durme, M. (1914). Nouvelles recherches sur la vitellogenese des oeufs d'oiseaux aux stades d'accroissement de maturation, de fécondation et du début de la segmentation, *Archs. Biol., Paris*, **29**, 71–200.

Erben, H. K. (1970). Ultrastrukturen und Mineralisation rezenter und fossilier Eischalen bei Vögeln und Reptilien, *Biomineralization* **1**, 1–66.

Eroschenko, V. P. and Wilson, W. O. (1974). Histological changes in the regressing reproductive organs of sexually mature male and female Japanese quail, *Biology Reprod.* **11**, 168–179.

Erpino, M. J. (1969). Seasonal cycle of reproduction physiology in the Black-billed Magpie, *Condor* **71**, 267–279.

Erpino, M. J. (1973). Histogenesis of atretic ovarian follicles in a seasonally breeding bird, *J. Morph.* **139**, 239–250.

Essenberg, J. M. and Garwacki, J. K. (1938). Origin and development of definitive germ cells in the domestic fowl, *West. J. Surg. Obstet. Gynec.* **46**, 145–152.

Etches, R. J. and Cunningham, F. S. (1976). The inter-relationship between progesterone and luteinizing hormone during the ovulation cycle of the hen (*Gallus domesticus*), *J. Endocr.* **71**, 51–58.

Evans, H. E. (1969). The anatomy of the budgerigar. *In* "Diseases of Cage and Aviary Birds" (Margaret L. Petrak, Ed.). Lea and Febiger, Philadelphia.

von Faber, H. (1958). Geschlechtsunterschiede in Wachstum und in den innersekretorischen Organen des Sperbers, *Accipiter nisus nisus* L., *Acta Endocr., Copenh.* **28**, 49–416.

Farner, D. S. and King, J. R. (1973). "Avian Biology", Vols 1–5. Academic Press, New York and London.

Fauré-Frèmiet, E. and Kaufman, L. (1928). La loi de décroissance progessive du taux de la porte chez la poule, *Annls Physiol. Physiochim. biol.* **4**, 64–122.

Feeney, R. E. and Allison, R. G. (1969). "Evolutionary Biochemistry of Proteins." John Wiley and Sons, New York.

Fell, Honor B. (1924). Histological studies on the gonads of the fowl. II. The histogenesis of the so-called luteal cells in the ovary, *J. exp. Biol.* **1**, 293–312.

Fell, Honor B. (1925). Histological studies on the gonads of the fowl. III. The relation of the "luteal" cells of the ovary in the fowl to the tissue occupying the atretic and discharged follicles, *Q. Jl microsc. Sci.* **69**, 591–609.

Ferrando, G. and Nalbandov, A. V. (1969). Direct effect on the ovary of the adrenergic blocking drug dibenzyline, *Endocrinology* **85**, 38–42.

Fertuck, H. C. and Newstead, J. D. (1970). Fine structural observations on magnum mucosa in quail and hen oviducts, *Z. Zellforsch. mikrosk. Anat.* **103**, 447–459.

Firket, J. (1913). Recherches sur les gonocytes primaires (Urgeschlechtzellen) pendant la période d'indifférence sexuelle et le développement de l'ovaire chez le poulet, *Anat. Anz.* **44**, 166–175.

Firket, J. (1914). Recherches sur l'organogénèse des glands sexuelles chez les oiseaux, *Archs Biol., Paris* **29**, 201–351.

Firket, J. (1920). Recherches sur l'organogénèse des glandes sexuelles chez les oiseaux, *Archs Biol., Paris* **30**, 393–516.

Fitzgerald, T. C. (1969). "The Coturnix Quail: Anatomy and Histology." Iowa State University Press, Ames.

Floquet, A. and Grignon, G. (1964). Étude histologique du follicle post-ovulatoire chez la poule, *C.r. Séanc. Soc. Biol.* **158**, 132–135.

Follett, B. K. and Davis, D. T. (1975). Photoperiodicity and neuroendocrine control of reproduction in birds, *Symp. zool. Soc. Lond.* **35**, 199–224.

Franchi, L. L. (1962). The structure of the ovary–vertebrates. *In* "The Ovary" (S. Zuckerman, Anita M. Mandl and P. Eckstein, Eds), Vol. 1, 121–142. Academic Press, London and New York.

Franchi, L. L., Mandl, Anita M. and Zuckerman, S. (1962). The development of the ovary and the process of oogenesis. *In* "The Ovary" (S. Zuckerman, Anita M. Mandl and P. Eckstein, Eds), Vol. 1, 1–88. Academic Press, London and New York.

Fraps, R. M. (1955). The varying effect of sex hormones in birds, *Mem. Soc. Endocr.* **4**, 205–219.

Fraps, R. M. and Dury, A. (1942). The occurrence of regression in ovarian follicles of the hen following administration of a luteinizing extract, *Anat. Rec.* **84**, 521–522.

Fraps, R. M. and Neher, B. (1945). Interruption of ovulation in the hen by subcutaneously administered non-specific substances, *Endocrinology* **37**, 407–417.

Freedman, S. L. (1968). The innervation of the suprarenal gland of the fowl (*Gallus domesticus*), *Acta anat.* **69**, 18–25.

Freedman, S. L. and Sturkie, P. D. (1963a). Blood vessels of the chicken uterus (shell gland), *Am. J. Anat.* **113**, 1–7.

Freedman, S. L. and Sturkie, P. D. (1963b). Extrinsic nerves of the chicken uterus (shell gland), *Anat. Rec.* **147**, 431–437.

Fujii, S. (1963). Histological and histochemical studies on the oviduct of the domestic fowl with special reference to the region of the uterovaginal junction, *Archum histol. jap.* **23**, 447–459.

Fujii, S. (1974). Further morphological studies on the formation and structure of hen's eggshell by scanning electron microscopy, *J. Fac. Fish. Anim. Husb. Hiroshima Univ.* **13**, 29–56.

Fujii, S. (1975). Scanning electron microscopical observation on the mucosal epithelium of hen's oviduct with special reference to the transport mechanism of spermatozoa through the oviduct, *J. Fac. Fish. Anim. Husb. Hiroshima Univ.* **14**, 1–13.

Fujii, S. (1976). Scanning electron microscopical observations on the penetration mechanisms of fowl spermatozoa into the ovum in the process of fertilization, *J. Fac. Fish. Anim. Husb. Hiroshima Univ.* **15**, 85–92.

Fujii, S. and Tamura, T. (1963). Location of sperms in the oviduct of domestic fowl with special reference to storage of sperms in the vaginal gland. *J. Fac. Fish. Anim. Husb. Hiroshima Univ.* **5**, 145–163.

Fujii, S. and Tamura, T. (1969). Scanning electron microscopy of the hen's egg shell, *J. Fac. Fish. Anim. Husb. Hiroshima Univ.* **8**, 85–98.

Fujii, S. and Tamura, T. (1970). Scanning electron microscopy of shell formation in hen's eggs, *J. Fac. Fish. Anim. Husb. Hiroshima Univ.* **9**, 65–81.

Fujii, S., Tamura, T. and Kunisaki, H. (1965). Histochemical study of mucopolysaccharides in goblet cells of the chicken oviduct, *J. Fac. Fish. Anim. Husb. Hiroshima Univ.* **6**, 25–35.

Fujii, S., Tamura, T. and Okamoto, T. (1970). Scanning electron microscopy of shell-membrane formation in hen's eggs, *J. Fac. Fish. Anim. Husb. Hiroshima Univ.* **9**, 139–150.

Fujimoto, T., Ukeshima, A. and Kiyofuji, R. (1975). Light and electron-microscope studies on the origin and migration of the primordial germ cells in the chick, *Acta. anat. nippon* **50**, 22–40.

Fujimoto, T., Ukeshima, A. and Kiyofuji, R. (1976). The origin, migration and morphology of the primordial germ cells in the chick embryo, *Anat. Rec.* **185**, 139–154.

Furr, B. J. A. (1969). A study of gonadotrophins and progestins in the domestic fowl. Ph.D. Thesis, University of Reading.

Gallien, L. and Le Foulgoc, M. T. (1957). Détection par fluorimétrie et colorimétrie de stéroides sexuels dans les gonades embryonnaires de poulet, *C.r. Séanc. Soc. Biol.* **151**, 1088–1089.

Garde, M. L. (1930). The ovary of ornithorhynchus with special reference to follicular atresia, *J. Anat.* **64**, 422–453.

Gay, Carol V. and Schraer, H. (1971). Autoradiographic localisation of calcium in the mucosal cells of the avian oviduct, *Calc. Tiss. Res.* **7**, 201–211.

Gertner, M. (1969). A tojólúd reprodukciós szervének artériarendszere. (The arterial system of the reproductive organ in the goose), *Agrártud. Agyet. mezögtud. Kar. Közl., Gödöllö*, 73–82.

Giersberg, H. (1922). Untersuchungen über Physiologie und Histologie des Eierleiters der Reptilian und Vogel: nebst einem Beitrag zur Fasergenese, *Z. wiss. Zool.* **120**, 1–97.

Gilbert, A. B. (1965). Innervation of the ovarian follicle of the domestic hen, *Q. Jl exp. Physiol.* **50**, 437–445.

Gilbert, A. B. (1967). The formation of the egg in the domestic chicken. *In* "Advances in Reproductive Physiology" (Anne McLaren, Ed.), Vol. 1, 111–180. Logos Press, London.

Gilbert, A. B. (1968). An observation bearing on the ovarian-oviduct relationship in the domestic hen, *Br. Poult. Sci.* **9**, 301–302.

Gilbert, A. B. (1969). Innervation of the ovary of the domestic hen, *Q. Jl exp. Physiol.* **54**, 409–411.

Gilbert, A. B. (1971a). The female reproductive effort. *In* "Physiology and Biochemistry of the Domestic Fowl" (D. J. Bell and B. M. Freeman, Eds), Vol. 3, 1153–1162. Academic Press, London and New York.

Gilbert, A. B. (1971b). The Ovary. *In* "Physiology and Biochemistry of the Domestic Fowl" (D. J. Bell and B. M. Freeman, Eds), Vol. 3, 1163–1208. Academic Press, London and New York.

Gilbert, A. B. (1971c). Egg albumen and its formation. *In* "Physiology and Biochemistry of the Domestic Fowl" (D. J. Bell and B. M. Freeman, Eds), Vol. 3, 1291–1329. Academic Press, London and New York.

Gilbert, A. B. (1971d). Transport of the egg through the oviduct and oviposition. *In* "Physiology and Biochemistry of the Domestic Fowl" (D. J. Bell and B. M. Freeman, Eds), Vol. 3, 1345–1352. Academic Press, London and New York.

Gilbert, A. B. (1971e). The egg: its physical and chemical aspects. *In* "The Physiology and Biochemistry of the Domestic Fowl" (D. J. Bell and B. M. Freeman, Eds), Vol. 3, 1379–1399. Academic Press, London and New York.

Gilbert, A. B. (1971f). The egg in reproduction. *In* "Physiology and Biochemistry of the Domestic Fowl" (D. J. Bell and B. M. Freeman, Eds), Vol. 3, 1401–1409. Academic Press, London and New York.

Gilbert, A. B. (1971g). The endocrine ovary in reproduction. *In* "Physiology and Biochemistry of the Domestic Fowl" (D. J. Bell and B. M. Freeman, Eds), Vol. 3, 1449–1468. Academic Press, London and New York.

Gilbert, A. B. (1971h). Control of ovulation. *In* "The Physiology and Biochemistry of the Domestic Fowl" (D. J. Bell and B. M. Freeman, Eds.), Vol. 3, 1225–1235. Academic Press, London and New York.

Gilbert, A. B. (1972). The activity of the ovary in relation to egg production. *In* "Egg Formation and Production" (B. M. Freeman and P. E. Lake, Eds). T. and A. Constable, Edinburgh.

Gilbert, A. B. (1974). Poultry. *In* "Reproduction in Farm Animals" (E. S. E. Hafez, Ed.), 3rd edn. Lea and Febiger, Philadelphia.

Gilbert, A. B. and Lake, P. E. (1963). Terminal innervation of the uterus and vagina of the domestic hen, *J. Reprod. Fert.* **5**, 41–48.

Gilbert, A. B. and Wood-Gush, D. G. M. (1971). Ovulatory and ovipository cycles. *In* "The Physiology and Biochemistry of the Domestic Fowl" (D. J. Bell and B. M. Freeman, Eds), Vol. 3, 1353–1378. Academic Press, London and New York.

Gilbert, A. B. and Wood-Gush, D. G. M. (1970). Observations on ovarian transplants in the domestic fowl and their bearing on normal ovarian function, *Res. vet. Sci.* **11**, 156–160.

Gilbert, A. B., Evans, A. J., Perry, M. M. and Davidson, M. F. (1977). A method for separating the granulosa cells, the basal lamina and the theca of the preovulatory ovarian follicle of the domestic fowl (*Gallus domesticus*), *J. Reprod. Fert.* **50**, 179–181.

Gilbert, A. B., Lake, P. E. and Wood-Gush, D. G. M. (1966). Aspects of the physiology of the transport of the ovum through the oviduct of the domestic hen. *In* "Physiology of the Domestic Fowl" (C. Horton-Smith and E. C. Amoroso, Eds). Oliver and Boyd, Edinburgh.

Gilbert, A. B., Reynolds, M. F. and Lorenz, F. W. (1968). Distribution of spermatozoa in the oviduct and fertility in domestic birds. VII. Innervation and vascular supply of the uterovaginal spermhost glands of the domestic hen, *J. Reprod. Fert.* **17**, 305–310.

Gipson, Ilene (1974). Electron microscopy of early cleavage furrows in the chick blastodisc, *J. Ultrastruct. Res.* **49**, 331–347.

Goldsmith, J. B. (1928). The history of the germ cells in the domestic fowl, *J. Morph.* **46**, 275–315.

Grau, C. R. (1976). Ring structure of avian egg yolk, *Poult. Sci.* **55**, 1418–1422.

Grau, C. R. and Wilson, B. W. (1964). Avian oogenesis and yolk deposition, *Experientia* **20**, 26.

Grau, H. (1943). Anatomie der Hausvögel. *In* "Ellenberger und Baum's vergleichenden Anatomie der Haustiere" (O. F. Zeitzschmann and H. Grau, Eds), 18th edn. Springer-Verlag, Berlin.

Greenfield, M. L. (1966). The oocyte of the domestic chicken shortly after hatching studied by electron microscopy, *J. Embryol. exp. Morph.* **15**, 297–316.

Greenwood, A. W. (1935). Perforation of the oviduct in the domestic fowl, *Trudy Din. Razv.* **10**, 81–90.

Groebbels, F. (1937). "Der Vogel." Gebrüder Borntraeger, Berlin.

Gruenwald, P. (1942). Primary asymmetry of the growing Müllerian ducts in the chick embryo, *J. Morph.* **71**, 299–305.

Gruenwald, P. (1952). Development of the excretory system, *Ann. N.Y. Acad. Sci.* **55**, 142–146.

Guichard, A., Scheib, D., Haffen, K. and Cedard, L. (1977). Radioimmunoassay of steroid hormones produced by embryonic chick gonads during organ culture, *J. Steroid Biochem.* **8**, 599–602.

Gunn, T. E. (1912). On the ovaries of certain British birds, *Proc. zool. Soc. Lond.* **1**, 63–79.

Guraya, S. S. (1975). Balbiani's vitelline body in the oocytes of vitellogenic and nonvitellogenic females of the domestic fowl: a correlative cytological and histochemical study, *Acta morph. hung.* **23**, 251–261.

Guraya, S. S. (1976). Morphological and histological observations on follicular atresia and interstitial gland tissue in columbid ovary, *Gen. comp. Endocr.* **30**, 534–538.

Guraya, S. S. and Chalana, R. K. (1975). Histochemical observations on the corpus luteum of the House Sparrow (*Passer domesticus*), ovary, *Gen. comp. Endocr.* **27**, 271–275.

Guzsal, E. (1966). Histological studies on the mature and post-ovulation ovarian follicle of fowl, *Acta vet. hung.* **16**, 37–44.

Guzsal, E. (1968). Histochemical study of goblet cells of the hen's oviduct, *Acta vet. hung.* **18**, 251–256.

Hacker, B. (1969). Estrogen-induced transfer RNA methylase activity in chick oviduct, *Biochim. biophys. Acta* **186**, 214–216.

Haffen, Katy (1975). Sex differentiation of avian gonads *in vitro*, *Am. Zool.* **15**, 257–272.

Haffen, Katy and Cedard, Lise (1968). Étude, en culture organotypique *in vitro*, du métabolisme de la dehydroepiandrosterone et de la testosterone radioactives, par les gonades normales et intersexuées de l'embryon de poulet, *Gen. comp. Endocr.* **11**, 220–234.

Hamburger, V. and Hamilton, H. L. (1951). A series of normal stages in the development of the chick ovary, *J. Morph.* **88**, 49–92.

Harper, E. H. (1904). The fertilization and early development of the pigeon's egg, *Am. J. Anat.* **3**, 349–386.

Harrison, R. J. (1962). *In* "The Ovary" (S. Zuckerman, Anita M. Mandle and P. Eckstein, Eds), Vol. 1, 143–187. Academic Press, London and New York.

Hasiak, R. J., Vadehra, D. V. and Baker, R. C. (1970). Lipid composition of the egg exteriors of the chicken (*Gallus domesticus*), *Comp. Biochem. Physiol.* **37**, 429–435.

Hashimoto, D. (1930). The development of the phallus in the fowl, *Jap. J. Zool.* **3**, 101.

Henderson, I. F. and Henderson, W. D. (1957). *In* "Dictionary of Scientific Terms" (J. H. Kenneth, Ed.), 6th edn. Oliver and Boyd, Edinburgh.

Herrick, H. F. (1907). Analysis of the cyclical instincts of birds, *Science* **25**, 725–726.

Hertelendy, F. and Taylor, T. G. (1964). The citric acid content of blood plasma and tissues of the domestic fowl, *Comp. Biochem. Physiol.* **11**, 173–182.

Hertz, R. (1950). Endocrine and vitamin factors in hormone-induced tissue growth, *Tex. Rep. Biol. Med.* **8**, 154–158.

Hertz, R., Larsen, C. D. and Tullner, W. W. (1947). Inhibition of estrogen-induced tissue growth with progesterone, *J. natn. Cancer Inst.* **8**, 123–126.

Hett, J. (1923). Das Corpus luteum der Dohle, *Arch. mikrosk. Anat. EntwMech.* **97**, 718–833.

Heyn, A. N. J. (1963). The crystalline structure of calcium carbonate in the avian egg shell, *J. Ultrastruct. Res.* **8**, 176–188.

Hill, J. P. and Gatenby, J. B. (1926). The corpus luteum of monotremata, *Proc. zool. Soc. Lond.* **2**, 715–761.

Hill, C. J. and Hill, J. P. (1933). The development of the Monotremata. 1. The histology of the oviduct during gestation, *Trans. zool. Soc. Lond.* **21**, 413–476.

Hodges, R. D. (1965). The blood supply to the avian oviduct with special reference to the shell gland, *J. Anat.* **99**, 485–506.

Hodges, R. D. (1966). The functional anatomy of the avian shell gland. *In* "Physiology of the Domestic Fowl" (C. Horton-Smith and A. C. Amoroso, Eds). Oliver and Boyd, Edinburgh.

Hodges, R. D. (1974). *In* "The Histology of the Fowl". Academic Press, London and New York.

Hoffer, Anita P. (1971). The ultrastructure and cytochemistry of the shell membrane-secreting region of the Japanese Quail oviduct, *Am. J. Anat.* **131**, 253–288.

Hogarth, P. J. (1976). "Viviparity." Edward Arnold, London.

d'Hollander, F. (1904). Recherches sur l'oogénèse et sur la structure et la signification du noyau, vitellin de Balbiani chez les oiseaux, *Archs Anat. microsc.* **7**, 117–180.

Homma, K., Wilson, W. O. and McFarland, L. Z. (1965). Yolk dye deposition as an index of ovum maturation in *Coturnix*, *Am. Zool.* **5**, 194.

Howarth, B. Jnr. (1971). Transport of spermatozoa in the reproductive tract of turkey hens, *Poult. Sci.* **50**, 84–89.

Hsieh, T. M. (1951). The sympathetic and parasympathetic nervous system of the fowl. Ph.D. Thesis, Edinburgh University.

Hughes, G. C. (1963). The population of germ cells in the developing female chick, *J. Embryol. exp. Morph.* **11**, 513–536.

Hutchison, Rosemary E., Hinde, R. A. and Bendon, B. (1968). Oviduct development and its relation to other aspects of reproduction in domestic canaries, *J. Zool., Lond.* **155**, 87–102.

Hutt, F. B. (1939). An intrafollicular ovum laid by a fowl, *Poult. Sci.* **18**, 276–278.

Hutt, F. B. (1946). A pedunculate double-yolked hen's egg containing an intrafollicular ovum, *Auk* **63**, 171–174.

Hutt, F. B., Goodwin, K. and Urban, W. D. (1956). Investigation of non-laying hens, *Cornell Vet.* **46**, 257–273.

Immelman, K. (1971). Ecological aspects of periodic reproduction. *In* "Avian Biology" (D. S. Farner and J. R. King, Eds), Vol. 1, 342–389. Academic Press, New York and London.

Ingram, D. L. (1962). Atresia. *In* "The Ovary" (S. Zuckerman, Anita M. Mandl and P. Eckstein, Eds), Vol. 2, 247–273. Academic Press, London and New York.

Jacobowitz, D. and Wallach, E. E. (1967). Histochemical and chemical studies of the autonomic innervation of the ovary, *Endocrinology* **81**, 1132–1139.

Jensen, Cynthia (1969). Ultrastructural changes in the ovarian vitelline membrane during embryonic development, *J. Embryol. exp. Morph.* **21**, 467–484.

Johnson, J. S. (1925). The innervation of the female genitalia in the common fowl, *Anat. Rec.* **29**, 387.

Johnston, H. S., Aitken, R. N. C. and Wyburn, G. M. (1963). The fine structure of the uterus of the domestic fowl, *J. Anat.* **97**, 333–334.

Jordan, H. E. (1917). Embryonic history of the germ cells of the loggerhead turtle (*Caretta caretta*), *Pap. Tortugas Lab.* **11**, 313–344.

Jordanov, J. S. (1969). "On the Cytobiology of the Hen's Egg." Publishing House of the Bulgarian Academy of Science, Sofia.

Jordanov, J. and Boyadjieva-Michailova, A. (1974). Ultrastructural aspects of lipoprotein passage through oocyte envelopes and storage in ooplasm during avian vitellogenesis, *Acta anat.* **89**, 616–632.

Jørgensen, C. B. (1968). Central nervous control of adenohypophyseal functions. *In* "Perspectives in Endocrinology—Hormones in the Lives of Lower Vertebrates" (E. J. W. Barrington and C. B. Jørgensen, Eds). Academic Press, New York and London.

Jørgensen, C. B. and Larson, L. O. (1967). Neuroendocrine mechanisms in lower vertebrates. *In* "Neuroendocrinology" (L. Martini and W. F. Ganong, Eds), Vol. 2, 485–528. Academic Press, New York and London.

Juhn, M. and Gustavson, R. G. (1930). The production of female genital subsidiary characters and plumage sex characters by injection of human placental hormones in fowls, *J. exp. Zool.* **56**, 31–36.

Juhn, M. and Gustavson, R. G. (1932). The response of a vestigial Müllerian duct to the female hormone and the persistence of such rudiments in the male fowl, *Anat. Rec.* **52**, 299–308.

Kabat, C., Buss, I. O. and Meyer, R. K. (1948). The use of ovulated follicles in determining eggs laid by the ring-necked pheasant, *J. Wildl. Mgmt* **12**, 399–416.

Kamar, G. A. R. and Yamani, K. A. (1962). The physical causes of paired oviducts in ducks, *J. Reprod. Fert.* **4**, 99–101.

Kannankeril, J. V. and Domm, L. V. (1968). Development of the gonads in the female Japanese Quail, *Am. J. Anat.* **123**, 131–146.

Kar, A. B. (1947a). Studies on the ligaments of the oviduct in the domestic fowl, *Anat. Rec.* **97**, 175–192.

Kar, A. B. (1947b). Responses of the oviduct of the female fowl to injection of diethylstilboestrol and the mechanism of perforation of the oviduct in the domestic fowl, *Poult. Sci.* **26**, 352–363.

Keck, W. N. (1934). The control of the secondary sex characters in the English sparrow, *Passer domesticus*, *J. exp. Zool.* **67**, 315–347.

Kennedy, G. Y. and Vevers, H. G. (1976). A survey of avian egg shell pigments, *Comp. Biochem. Physiol.* **55B**, 117–123.

Kern, D. (1963). Die Topographie der Eingeweide der Körpenhöhle des Haushühnes (*Gallus domesticus*) unter besonderer Berücksichtigung der Serosa—and Gekröseverhältnisse. Inaug. Dissertation, Universität Giessen.

Kern, M. D. (1972). Seasonal changes in the reproductive system of the female White-crowned Sparrow (*Zonotrichia leucophrys gambelli*) in captivity and in the field, *Z. Zellforsch. mikrosk. Anat.* **126**, 297–319.

Khairallah, L. (1966). The fine structure of the tubular glands in the isthmus of the oviduct of the hen. Ph.D. Thesis, Boston University Graduate School.

King, A. S. (1975). Aves urogenital system. *In* Sisson and Grossman's "The Anatomy of Domestic Animals" (R. Getty, Ed.), Vol. 2, 5th edn. Saunders, Philadelphia.

Kinsky, F. C. (1971). The consistent presence of paired ovaries in the kiwi (*Apteryx*) with some discussion of this condition in other birds, *J. Orn., Lpz.* **112**, 334–357.

Kissel, J. H., Rosenfeld, M. G., Chase, L. R. and O'Malley, B. W. (1970). Response of chick oviduct adenylcyclase to steroid hormones, *Endocrinology* **86**, 1019–1023.

Kobayashi, H. and Wada, M. (1973). Neuroendocrinology in birds. *In* "Avian Biology" (D. S. Farner and J. R. King, Eds), Vol. 3, 287–347. Academic Press, New York and London.

Koch, W. (1926). Untersuchungen über die Entwicklung des Eierstockes der Vögel. 1. Die post-embryonale Entwicklung der Form und des Aufbaues des Eierstockes beim Haushühn. *Z. mikrosk.-anat. Forsch.* **7**, 1–52.

Koch, W. (1927). Untersuchungen über die Entwicklung des Eierstockes der Vögel. II. Die phylogenetische, Bedeutung der Form des Eierstockes bei den Vögeln, *Zool. Anz.* **71**, 299–303.

Kohler, P. O., Grimley, P. M. and O'Malley, B. W. (1968). Protein synthesis: differential stimulation of cell specific proteins in epithelial cells of the chick oviduct, *Science, N.Y.* **160**, 86–87.

Kohler, P. O., Grimley, P. M. and O'Malley, B. W. (1969). Estrogen induced cyto-differentiation of the ovalbumin secreting glands of the chick oviduct, *J. Cell. Biol.* **40**, 8–27.

Komarek, V. and Prochazkova, E. (1970). Growth and differentiation of the ovarian follicles in the post natal development of the chicken, *Acta vet., Brno.* **39** 11–16.

Kornfeld, W. (1958). Endocrine influence upon the growth of the rudimentary gonad of the fowl, *Anat. Rec.* **130**, 619–638.

Kornfeld, W. and Nalbandov, A. V. (1954). Endocrine influences on the development of the rudimentary gonad of the fowl, *Endocrinology* **55**, 751–761.

Kosin, I. L. (1944). Macro—and microscopic methods of detecting fertility in incubated hen's eggs, *Poult. Sci.* **23**, 266–269.

von Krampitz, G., Engels, J., Heindl, I., Heinrich, A., Hamm, M. and Faust, R. (1974). Biochemische Untersuchungen an Eischalen, *Arch. Geflügelk.* **6**, 197–205.

Kraus, S. D. (1947). Observations on the mechanism of ovulation in the frog, hen and rabbit, *West. J. Surg. Obstet. Gynec.* **55**, 424–437.

van Krey, H. P., Ogasawara, F. X. and Pangborn, J. (1967). Light and electron microscope studies of possible sperm gland emptying mechanisms, *Poult. Sci.* **46**, 69–78.

Kummerlöwe, H. (1930a). Vergleichende Untersuchungen über das Gonadensystem weiblicher Vögel. I. *Columba livia domestica*, *Z. mikrosk.-anat. Forsch.* **21**, 1–156.

Kummerlöwe, H. (1930b). Vergleichende Untersuchungen über das Gonadensystem weiblicher Vögel. 2. *Passer domesticus* (L.), *Z. mikrosk.-anat. Forsch.* **22**, 259–413.

Kummerlöwe, H. (1931a). Vergleichende Untersuchungen über das Gonadensystem weiblicher Vögel. 3. Ausgewählte Beispiele aus Verschiedenen Vogelordnungen, *Z. mikrosk.-anat. Forsch.* **24**, 455–631.

Kummerlöwe, H. (1931b). Vergleichende Untersuchungen über das Gonadensystem weiblicher Vögel. 4. Uber zwei singende Kanerienvögelweibchen und über ein Amselweibchen mit ungewöhnlich intensiver Schnabelfärbung, *Z. mikrosk.-anat. Forsch.* **25**, 311–319.

Kummerlöwe, H. (1955). Das Urogenital System (Nieren und Geschlechtsapparat) der Vögel, *Veröff naturw. Ver Osnabr.* **27**, 86–101.

Lack, D. (1958). The significance of colour in Turdine eggs, *Ibis* **100**, 145–166.

Lake, P. E. (1975). Gamete production and the fertile period with particular reference to domesticated birds, *Symp. zool. Soc. Lond.* **35**, 225–244.

Lake, P. E. and Gilbert, A. B. (1961). Factors affecting uterine motility in the domestic hen, *J.Reprod. Fert.* **4**, 211.

Lake, P. E. and Gilbert, A. B. (1964a). The effect of a foreign object in the shell gland on the fertility of the domestic hen, *J. Reprod. Fert.* **8**, 272–273.

Lake, P. E. and Gilbert, A. B. (1964b). The effect on egg production of a foreign object in the lower oviduct regions of the domestic hen, *Res. vet. Sci.* **5**, 39–45.

Lauwers, H., Boedts, F. and Geerinckx, P. (1970). Localization, histochimique de l'anhydrase carbonique dans l'oviducte de la poule, *Histochemie* **24**, 307–314.

Layne, D. S. (1958). On the nature of the gonadal hormones of the domestic fowl. Ph.D. Thesis, McGill University.

Lehrman, D. S. and Brody, P. N. (1957). Oviduct response to estrogen and progesterone in the Ring Dove (*Streptopelia risoria*), *Proc. Soc. exp. Biol. Med.* **19**, 373–375.

Lewis, L. B. (1946). A study of the effects of some sex hormones on the embryonic reproductive system of the White Pekin duck, *Physiol. Zool.* **19**, 282–329.

Lillie, F. R. (1952). "Lillie's Development of the Chick" (Revised by H. L. Hamilton), 3rd edn. Holt, Rinehart and Winston, New York.

van Limborgh, J. (1957). De ontwikkeling van de assymmetrie der gonaden by het embryo van de eend. Ph.D. Thesis, University of Utrecht.

van Limborgh, J. (1958). Number and distribution of the primary germ cells in duck embryos in the 28- to 36-somite stages, *Acta morph. neerl.-scand.* **2**, 119–133.

van Limborgh, J. (1968). Le prémier indice de la différentiation sexuelle des gonades chez l'embryon de poulet, *Archs Anat. microsc. Morph. exp.* **57**, 79–90.

van Limborgh, J., van Deth, J. H. M. G. and Tacoma, J. (1960). The early gonadal capillary systems of the duck embryo; structure and velocity of the blood, *Acta morph. neerl.-scand.* **3**, 35–47.

Lin, R. I-San (1968). Studies on coated vesicles. Ph.D. Thesis, University of California.

Ljungkvist, H. I. (1967). Light and electron microscopical study of the effect of oestrogen on the chicken oviduct, *Acta endocr., Copnh.* **56**, 391–402.

Lofts, B. (1975). Environmental control of reproduction, *Symp. zool. Soc. Lond.* **35**, 177–197.

Lofts, B. and Murton, R. K. (1973). Reproduction in birds. *In* "Avian Biology" (D. S. Farner and J. R. King, Eds), Vol. 3 1–107. Academic Press, New York and London.

Lorenz, F. W. (1954). Effects of estrogen on domestic fowl and applications in the poultry industry, *Vitams. Horm.* **12**, 235–275.

Lorenz, F. W. (1964). Recent research on fertility and artificial insemination of domestic birds. 5th Congr. Int. Reprod. Anim. Fecond. Art. Vol. 4, pp. 7–32.

Loyez, Marie (1906). Recherches sur le développement ovarien des oeufs meroblastiques à vitellus nutritif abondent, *Archs Anat. micr. Morph. exp.* **8**, 69–397.

Lutz-Ostertag, Y. (1954). Contribution à l'étude du developpement et de la regression des canaux de Müller chez l'embryon d'Oiseau, *Bull. biol. Fr. Belg.* **88**, 333–412.

Ma, R. C. S. (1972). Vascular supply of the oviduct of domestic chicks, *Mem. Coll. Agric., natn. Taiwan Univ.* **13**, 182–187.

McBride, G. (1962). The inheritance of right oviduct development in the domestic hen. 12th Wld's Poult. Congr., pp. 77–79.

McBride, G., Parera, I. P. and Foenander, F. (1969). The social organisation and behaviour of the feral domestic fowl, *Anim. Behav. Monogr.* **2**, 127–181.

McGuire, W. L. and O'Malley, B. W. (1968). Ribonucleic acid polymerase activity of the chick oviduct during steriod-induced synthesis of a specific protein, *Biochim. biophys. Acta* **157**, 187–194.

Mclndoe, W. M. (1971). Yolk synthesis. *In* "Physiology and Biochemistry of the Domestic Fowl" (D. J. Bell and B. M. Freeman, Eds), Vol. 3, 1209–1223. Academic Press, London and New York.

Mackenzie, S. L. and Martin, W. G. (1967). The macromolecular composition of the hen's egg yolk at successive ages of maturation, *Can. J. Biochem.* **45**, 591–601.

McNally, E. H. (1943). The origin and structure of the vitelline membrane of the domestic fowl's egg, *Poult. Sci.* **22**, 40–43.

Makita, T. and Kiwaki, S. (1968). The fine structure of the infundibulum of the quail oviduct, *Jap. J. zootech. Sci.* **39**, 246–254.

Makita, T. and Sandborn, E. B. (1970). Scanning electron microscopy of secretory granules in albumen gland cells of the laying hen oviduct, *Exptl Cell Res.* **60**, 477–480.

Makita, T. and Sandborn, E. B. (1971). Identification of ultra-cellular components by scanning electron microscopy, *Exptl Cell Res.* **62**, 211–214.

Marin, G. (1959). Gonadogenesi ni assenza di gonociti e ozione radiozione ionizzanti sol differenziamento sessuale dell'embrione di pollo, *Arch. ital. Anat. Embriol.* **64**, 211–235.

Marshall, A. J. (1961). Reproduction. *In* "Biology and Comparative Physiology of Birds" (A. J. Marshall, Ed.), Vol. 2, 169–213. Academic Press, New York and London.

Marshall, A. J. and Coombs, C. J. F. (1957). The interaction of environmental, internal and behavioural factors in the Rook (*Corvus f. frugilegus* Linnaeus), *Proc. zool. Soc. Lond.* **128**, 545–589.

Marshall, F. H. A. (1956). The breeding season. *In* "Marshall's Physiology of Reproduction" (A. S. Parkes, Ed.), Vol. 1, 1–42. Longmans Green and Co., London.

Marza, V. D. and Marza, R. V. (1935). The formation of the hen's egg I–IV, *Q. Jl microsc. Sci.* **78**, 134–189.

Mason, R. C. (1952). Synergistic and antagonistic effects of progesterone in combination with estrogens on oviduct weight, *Endocrinology* **51**, 570–572.

Masshoff, W. and Stolpmann, H. J. (1961). Licht und Elektronenmikroskopische Untersuchungen an der Schalenhaut und Kalkschale des Hühnereies, *Z. Zellforsch. mikrosk. Anat.* **55**, 818–832.

Matsumoto, T. (1932). On the early localization and history of the so-called primordial germ cells in the chick embryo, *Sci. Rep. Tohoku Univ.* **7**, 89–127.

Matthews, L. H. (1937). The female sexual cycle in the British Horseshoe bats, *Rhinolophus ferrum-equinum insulanus* Barret-Hamilton and *R. hipposideros minutus* Montagu, *Trans. zool. Soc. Lond.* **23**, 224–255.

Matthews, L. H. (1950). Reproduction in the basking shark, *Celorhinus maximus* (Gunner), *Phil. Trans. R. Soc. Ser. B* **234**, 247–316.

Matthews, L. H. and Marshall, F. H. A. (1956). Cyclical changes in the reproductive organs of the lower vertebrates. *In* "Marshall's Physiology of Reproduction" (A. S. Parkes, Ed.), Vol. 1, 156–225. Longmans Green and Co., London.

Mauger, H. M. (1941). The autonomic innervation of the female genitalia in the domestic fowl and its correlation with the aortic branchings, *Am. J. vet. Res.* **2**, 447–452.

Meyer, D. B. (1964). The migration of primordial germ cells in the chick embryo, *Devl Biol.* **10**, 154–190.

Meyer, M. J. and Sturkie, P. D. (1974). Diurnal rhythm of histamine in blood and oviduct of the domestic fowl, *Comp. gen. Pharmac.* **5**, 225–228.

Meyer, R., Baker, R. C. and Scott, M. L. (1973). Effects of hen eggshell and other calcium sources upon egg shell strength and ultrastructure, *Poult. Sci.* **52**, 949–955.

Mimura, H. (1937). Studies on the ciliary movement of the oviduct in the domestic fowl. 1. The direction of the ciliary movement, *Folia anat. jap.* **15**, 287–295.

Mitrophanow, P. (1902). Beiträge zur Entwicklung der Wesservögel, *Z. wiss. Zool.* **71**, 189–210.

Montevecchi, W. A. (1976). Field experiments on the adaptive significance of avian eggshell pigmentation, *Behaviour* **58**, 26–39.

Moran, T. and Hale, H. P. (1936). Physics of the hen's egg. I. Membranes in the egg, *J. exp. Biol.* **13**, 35–40.

Moreau, R. E., Wilk, A. L. and Rowan, W. (1947). Moult and gonad cycles of three species of birds at 5° south of the equator, *Proc. zool. Soc. Lond.* **117**, 345–364.

Müller, W. E. G., Totsuka, A., Nusser, I., Obermeier, J., Rhode, H. J. and Zahn, R. K. (1974). Poly (adenosine diphosphate-ribose) polymerase in quail oviduct. Changes during estrogen and progesterone induction, *Nucleic Acids Res.* **1**, 1317–1327.

Müller, W. E. G., Totsuka, A., Kroll, M., Nusser, Ingrid and Zahn R. K. (1975). Poly (A) polymerase in quail oviduct: changes during estrogen induction, *Biochim. biophys. Acta* **383**, 147–159.

Murton, R. K. (1975). Ecological adaptation in avian reproductive physiology, *Symp. zool. Soc. Lond.* **35**, 149–175.

Murton, R. K. and Westwood, N. J. (1977). "Avian Breeding Cycles." Clarendon Press, Oxford.

Nalbandov, A. V. (1958). "Reproductive Physiology." W. H. Freeman and Co., San Francisco.

Nalbandov, A. V. (1959). Role of sex hormones in the secretory function of the avian oviduct. *In* "Comparative Endocrinology" (A. Gorbman, Ed.). John Wiley and Sons, New York.

Nalbandov, A. V. (1961). Mechanisms controlling ovulation of avian and mammalian follicles. *In* "Control of Ovulation" (C. A. Villee, Ed.). Pergamon Press, London.

Nalbandov, A. V. and James, M. F. (1949). The blood vascular system of the chicken ovary, *Am. J. Anat.* **85**, 347–378.

Narbaitz, R. and de Robertis, E. M. (1968). Postnatal evolution of steroidogenic cells in the chick ovary, *Histochemie* **15**, 187–193.

Von. Nathusius, W. (1868). Über die Hüllen, welche den Dotter des Vögeleies umgeben, *Z. wiss. Zool.* **18**, 225–270.

Von. Nathusius, W. (1871). Über die Eischalen von Aepiornis, Dinornis, Apteryx und einigen Crypturiden, *Z. wiss. Zool.* **21**, 330–355.

Needham, J. (1931). The relations between yolk and white in hen's egg. V. The osmotic properties of the isolated vitelline membrane, *J. exp. Biol.* **8**, 330–344.

Nemirovsky, N. S. and Narbaitz, R. (1970). Immuno-electrophoretic study of secretion in cystic oviducts of female and intersexual chicks, *Revue can. Biol.* **29**, 227–231.

Nevalainen, T. J. (1969). Electron microscope observations on the shell gland mucosa of calcium deficient hens (*Gallus domesticus*), *Anat. Rec.* **164**, 127–140.

Nickel, R., Schummer, A. and Seiferle, E. (1977). "Anatomy of the Domestic Birds" (Translated by W. G. Siller and P. A. L. Wight). Paul Parey, Berlin.

Nishida, T., Seki, M., Mochizuki, K. and Seta, S. (1977). Scanning electron microscopic observations on the microvascular architecture of the ovarian follicles in the domestic fowl, *Jap. J. vet Sci.* **39**, 347–352.

Nishimura, M., Urakawa, N. and Iwata, M. (1976). An electron microscopical study on ^{203}Hg. transport in the ovarian tissue of laying Japanese Quail, *Jap. J. vet. Sci.* **38**, 83–92.

Noble, G. K. and Wurm, M. (1940). The effect of testosterone proprionate in the Black-crowned Night Heron, *Endocrinology* **26**, 837–850.

Noumura, T. (1966). Steroid biosynthesis by the gonads of chick embryos, *Am. Zool.* **6**, 598.

Novak, J. and Duschak, F. (1923). Die Veränderungen der Follikelhüllen beim Haushuhn nach dem Follikelsprung, *Z. Anat. EntwMech.* **69**, 483–492.

Oades, J. M. and Brown, W. O. (1965). The study of the water-soluble oviduct proteins of the laying hen and female chick treated with gonadal hormones, *Comp. Biochem. Physiol.* **14**, 475–489.

Oka, T. and Schimke, R. T. (1969). Progesterone antagonism of estrogen-induced cytodifferentiation in chick oviduct, *Science, N.Y.* **163**, 83–85.

Okkelberg, P. (1921). The early history of the germ cells in the brook lamprey *Entosphenos wilderi* (Gage) up to and including the period of sexual differentiation, *J. Morph.* **35**, 1–151.

Olsen, M. W. (1942). Maturation, fertilisation and early cleavage of the hen's egg, *J. Morph.* **70**, 513–533.

Olsen, M. W. (1952). Intra-ovarian insemination in the domestic fowl, *J. exp. Zool.* **119**, 461–481.

Olsen, M. W. and Byerly, T. C. (1932). Orientation of the hen's egg in the uterus and during laying, *Poult. Sci.* **11**, 266–271.

Olsen, M. W. and Fraps, R. M. (1944). Maturation, fertilisation and early cleavage of the egg of the domestic turkey, *J. Morph.* **74**, 297–309.

Olsen, M. W. and Fraps, R. M. (1950). Maturation changes in the hen's ovum, *J. exp. Zool.* **114**, 475–489.

Olsen, M. W. and Neher, B. H. (1948). The site of fertilisation in the domestic fowl, *J. exp. Zool.* **109**, 355–366.

O'Malley, B. W., McGuire, W. L., Kohler, P. O. and Korenmann, S. G. (1969). Studies on the mechanism of steroid hormone regulation of synthesis of specific proteins, *Recent Prog. Horm. Res.* **25**, 105–160.

Oribe, T. (1967). Studies on distribution of blood vessels of domestic fowls. I. Process of the forming of blood vessels around the follicles, *Jap. J. zootech. Sci.* **38**, 46–53.

Oribe, T. (1968). Studies on distribution of blood vessels of domestic fowls. II. On the distribution of blood vessels of the mature follicles, *Jap. J. zootech. Sci.* **39**, 228–234.

Oribe, T. (1970). Studies on distribution of blood vessels of domestic fowls. III. On the distribution of blood vessels of the ovarian stroma, *Jap. J. zootech. Sci.* **41**, 329–335.

Oribe, T. (1976). Studies on distribution of blood vessels of ovary of domestic fowl. IV. On the arterial distribution of ovarian stroma of laying hens by vascular casts, *Bull. Hiroshima Agr. coll.* **5**, 317–327.

Oribe, T. (1977). Studies on distribution of ovarian blood vessels of domestic fowl. V. On the various types of ovarian artery in laying hens, *Bull. Hiroshima Agr. coll.* **5**, 439–446.

Oribe, T., Kawata, S. and Ogawa, S. (1963). Study on the innervation of the female gonads in the domestic fowl, *J. Tokyo Soc. vet. zootech. Sci.* **13**, 5–9.

Palmitter, R. D. (1971). Interaction of estrogen, progesterone and testosterone in the regulation of protein synthesis in chick oviduct, *Biochemistry* **10**, 399–404.

Palmitter, R. D. (1973). Role of ovalbumin messenger ribonucleic acid synthesis in the oviduct of estrogen primed chicks, *J. biol. Chem.* **248**, 8260–8270.

Palmitter, R. D. and Wrenn, J. T. (1971). Interaction of estrogen and progesterone in chick oviduct development. III. Tubular gland cell cytodifferentiation, *J. Cell. Biol.* **50**, 598–615.

Parker, G. H. (1930). Ciliary systems in the oviduct of the pigeon, *Proc. Soc. exp. Biol. Med.* **27**, 704–706.

Pasteels, J. (1940). Aperçu comparatif de la gastrulation chez les chordes, *Biol. Rev.* **15**, 59–106.

Pasteels, J. (1945). On the formation of the primary entoderm of the duck (*Anas domestica*) and on the significance of the bilaminer embryo in birds, *Anat. Rec.* **93**, 5–21.

Patterson, J. T. (1910). Studies on the early development of the hen's egg. I. History of the early cleavage and of the accessory cleavage, *J. Morph.* **21**, 101–134.

Paulson, J. and Rosenberg, M. D. (1972). The function and transport of lining bodies in developing ovarian oocytes, *J. Ultrastruct. Res.* **40**, 25–43.

Pearl, R. M. and Boring, A. M. (1918). Sex studies. X. The corpus luteum in the ovary of the domestic fowl, *Am. J. Anat.* **23**, 1–35.

Peel, E. T. and Bellairs, Ruth (1972). Structure and development of the secretory cells of the hen's ovary, *Z. Anat. EntwGesch.* **137**, 170–187.

Perry, Margaret M., Gilbert, A. B. and Evans, A. J. (1978a). Electron microscope observations on the ovarian follicle of the domestic fowl during the rapid growth phase, *J. Anat.* **125**, 481–497.

Perry, Margaret M., Gilbert, A. B. and Evans, A. J. (1978b). The structure of the germinal disc region of the hen's ovarian follicle during the rapid growth phase, *J. Anat.* **127**, 379–392.

Phillips, R. E., and Warren, D. C. (1937). Observations concerning the mechanisms of ovulation in the fowl, *J. exp. Zool.* **76**, 117–135.

Pike, J. W. and Alvarado, R. H. (1975). Ca^{2+}-Mg^{2+}-activated ATPase in the shell gland of Japanese Quail (*Coturnix coturnix japonica*), *Comp. Biochem. Physiol.* **51**, 119–125.

Press, N. (1964). An unusual organelle in avian ovaries, *J. Ultrastruct. Res.* **10**, 528–546.

Prochazkova, E. and Komarek, V. (1970). Growth of the zona vasculosa and zona parenchymatosa in postnatal development of the ovary in the chicken, *Acta vet., Brno* **39**, 3–10.

Purohit, V. D., Basrur, Parvathi K. and Reinhart, B. S. (1977). Persistent right oviduct of the Ring-necked pheasant, *Br. Poult. Sci.* **18**, 177–178.

Reagan, F. P. (1916). The results and possibilities of early embryonic castration, *Anat. Rec.* **11**, 251–267.

Rensch, B. (1947). "Neuere Probleme der Abstammungslehre." Ferdinand Enke Verlag, Stuttgart.

Rhea, R. P., Anderson, Beverley, Kim, Norma B. and Rosenberg, M. D. (1974). Biochemical, electron microscopical and cytochemical studies of ATPase localisation in avian, murine and human oviducts, *Fert. Steril.* **25**, 788–808.

Richardson, K. C. (1935). The secretory phenomenon in the oviduct of the fowl, including the process of shell formation examined by microincineration technique, *Phil. Trans. R. Soc. Ser. B.* **225**, 149–195.

Riddle, O. (1942). Cyclic changes in blood calcium, phosphorus and fat in relation to egg laying and estrogen production, *Endocrinology* **31**, 498–506.

Riddle, O. and Tange, M. (1928). Studies on the physiology of reproduction in birds. XXV. The action of the ovarian and placental hormone in the pigeon, *Am. J. Physiol.* **87**, 97–109.

Risley, P. L. (1933). Contributions on the development of the reproductive system in *Sternotherus odoratus* (Latreille). I. The embryonic origin and migration of the primordial germ cells, *Z. Zellforsch. mikrosk. Anat.* **18**, 459–492.

Robinson, D. S. (1972). Egg white glycoproteins and the physical properties of egg white. *In* "Egg formation and production" (B. M. Freeman and P. E. Lake, Eds). British Poultry Science, Edinburgh.

Romanoff, A. L. (1943). Growth of the avian ovum, *Anat. Rec.* **85**, 261–267.

Romanoff, A. L. (1960). "The Avian Embryo." Macmillan, New York.

Romanoff, A. L. and Romanoff, Anastasia J. (1949). "The Avian Egg." John Wiley and Sons, New York.

Rosenfeld, M. G. and O'Malley, B. W. (1970). Steroid hormones: effects on adenyl cyclase activity and adenosine 3′, 5′-monophosphate in target tissues, *Science, N.Y.* **168**, 253–255.

Roth, T. F., Cutting, J. A. and Atlas, Susan B. (1976). Protein transport: a selective membrane mechanism, *J. Supramolecular Struct.* **4**, 527–548.

Rothchild, I. and Fraps, R. M. (1944a). Relation between light-dark rhythms and hour of lay of eggs experimentally retained in the hen, *Endocrinology* **35**, 355–362.

Rothchild, I. and Fraps, R. M. (1944b). On the function of the ruptured ovarian follicle of the domestic fowl, *Proc. Soc. exp. Biol. Med.* **56**, 79–82.

Rothchild, I. and Fraps, R. M. (1945). The relation between ovulation frequency and the incidence of follicular atresia following surgical operations in the domestic hen, *Endocrinology* **37**, 415–430.

Rothwell, B. and Solomon, Sally E. (1977). The ultrastructure of the follicle wall of the domestic fowl during the rapid growth phase, *Br. Poult. Sci.* **18**, 605–610.

Rowan, W. (1930). A unique type of follicular atresia in the avian ovary, *Trans. R. Soc. Can.* **24**, 157–164.

Salzgeber, B. (1950). Stérilisation et intersexualité obtenues chez l'embryon de poulet par irradiation aux rayons-X, *Bull. biol. Fr. Belg.* **84**, 225–233.

Sandoz, D., Ulrich, E. and Brard, E. (1971). Étude des ultrastructures du magnum des oiseaux. I. Évolution au cours du cycle de ponte chez la poule *Gallus domesticus*, *J. Microscopie* **11**, 371–400.

Sandoz, D., Boisvieux-Ulrich, Emmanuelle, Laugier, C. and Brard, E. (1975). Interactions du benzoate d'estradiol et de la progesterone sur le développement de l'oviducte de caille (*Coturnix coturnix japonica*). II. Étude ultrastructurale des deux types de response obtenues selon la doèse de benzoate d' estradiol injectée, *Gen. comp. Endocr.* **26**, 451–467.

Sathyanesan, A. G. (1960). On the occurrence of burst atretic follicles in the teleost, *Mystus seenghale*, *Naturwissenschaften* **47**, 310–311.

Sayler, A., Dowd, A. J. and Wolfson, A. (1970). Influence of photo-period on the localization of

Δ^5-3β-hydroxysteroid dehydrogenase in the ovaries of maturing Japanese Quail, *Gen. comp. Endocr.* **15**, 20–30.

Scheib, Denise (1970). Origine et ultrastructure des celles sécrétrices de l'ovaire embryonaire de Caille (*Coturnix coturnix japonica*), *C.r. hebd. Séanc. Acad. Sci., Paris* **271**, 1700–1703.

Scheib, D. and Haffen, K. (1968). Sur la localization histoenzymologique de la 3β-hydroxysteroide déhydrogénase dans les gonades de l'embryon de poulet: apparition et spécificité de l'activité enzymatique, *Ann. Embryol. Morphol.* **1**, 61–72.

Schimke, R. T., McKnight, G. S., Shapiro, D. J., Drew, S. and Palacois, R. (1975). Hormonal regulation of ovalbumin synthesis in the chick oviduct, *Recent Prog. Horm. Res.* **31**, 175–211.

Schimke, R. T., Pennequin, P., Robins, Diane and McKnight, G. S. (1977). Hormonal regulation of egg white protein synthesis in chick oviduct. *In* "Hormones and Cell Regulation" (J. Dumont and J. Nunez, Eds), Vol. 1, 209–221. North Holland Publishing Co., Amsterdam.

Schindler, H., Ben-David, E., Hurwitz, S. and Kempernich, O. (1967). The relation of spermatozoa to the glandular tissue in the storage sites of the hen oviduct, *Poult. Sci.* **46**, 1462–1471.

Schjeide, O. A. and McCandless, R. G. (1962). On the formation of mitochondria, *Growth* **26**, 309–321.

Schjeide, O. A. and de Vellis, J. (1970). "Cell Differentiation." Van Nostrand Reinhold, New York.

Schjeide, O. A., McCandless, R. G. and Munn, R. J. (1963). Possible participation of RNA in formation of mitochondria-like organelles, *Growth* **27**, 125–128.

Schjeide, O. A., McCandless, R. G. and Munn, R. J. (1964). Mitochondrial morphogenesis, *Nature, Lond.* **203**, 158–160.

Schjeide, O. A., Munn, R. J., McCandless, R. G. and Edwards, R. (1966). Unique organelles of avian oocytes, *Growth* **30**, 471–489.

Schjeide, O. A., Galey, F., Grellert, E. A., I-San Lin, R. I., de Vellis, J. and Mead, J. F. (1970). Macromolecules in oocyte maturation, *Biology Reprod.*, Suppl. **2**, 14–43.

Schmidt, W. J. (1958a). Über den Aufbau der Schale des Vogeleies nebst Bemerkungen über Kalkige Eischalen anderer Tiere, *Ber. Oberhess Ges. Nat.-u. Heilk.* **28**, 82–108.

Schmidt, W. J. (1958b). Schleim-, Kalk- und Farbstoff-Ablagerungen auf der Kalkschale von Vogeleiern, *Z. Zellforsch. mikrosk. Anat.* **47**, 251–268.

Schmidt, W. J. (1962). Liegt der Eischalenkalk der Vögel als submikroskopische Kristallite vor? *Z. Zellforsch. mikrosk. Anat.* **57**, 848–880.

Schönwetter, M. (1960). "Handbuch der Oologie" (W. Meisse, Ed.), Academie Verlag, Berlin.

Schwartz, R. (1969). Eileiter und Ei vom Huhn. Die Wechelbeziechungen von Morphologie und Funktion bei Gegenüberstellung von Sekretionsorgan und Sekretionsprodukt, *Zentbl. VetMed. A.* **16**, 64–136.

Scott, Helen A. (1972). Some reproductive abnormalities in the hen, *Br. Poult. Sci.* **13**, 327–328.

Scott, H. M. and Burmester, B. R. (1939). The effect of resection of the albumen tube on secretion of egg white. 7th Wld's Poult. Congr., Cleveland. 102–106.

Scott, H. M. and Huang, W. (1941). Histological observations on the formation of the chalazae in the hen's egg, *Poult. Sci.* **20**, 402–405.

Sell, J. (1959). Incidence of persistant right oviducts in the chicken, *Poult. Sci.* **38**, 33–35.

Senior, B. E. and Furr, B. J. A. (1975). A preliminary assessment of the source of oestrogen within the ovary of the domestic fowl, *Gallus domesticus*, *J. Reprod. Fert.* **43**, 241–247.

Shenstone, F. S. (1968). The gross composition, chemistry and physico-chemical basis of organisation of the yolk and white. *In* "Egg Quality: A Study of the Hen's Egg" (T. C. Carter, Ed.). Oliver and Boyd, Edinburgh.

Sibley, C. G. (1960). The electrophoretic patterns of avian egg-white proteins as taxonomic characters, *Ibis* **102**, 215–284.

Simkiss, K. (1958). The structure of the eggshell with particular reference to the hen. Ph.D. Thesis, University of Reading.

Simkiss, K. (1968). The structure and formation of the shell and shell membranes. *In* "Egg Quality: A Study of the Hen's Egg" (T. C. Carter, Ed.). Oliver and Boyd, Edinburgh.

Simkiss, K. (1975). Calcium and avian reproduction, *Symp. zool. Soc. Lond.*, **35**, 307–337.

Simkiss, K. and Taylor, T. G. (1971). Shell formation. *In* "Physiology and Biochemistry of the Domestic Fowl" (D. J. Bell and B. M. Freeman, Eds), Vol. 3, 1331–1343. Academic Press, London and New York.

De Simone-Santoro, I. (1967). Aspetti ultrastrutturali delle cellule interstiziali ovariche in embrione di pollo, *Boll. Soc. ital. Biol. sper.* **43**, 908–910.

Simons, P. C. M. (1971). Ultrastructure of the hen egg shell and its physiological interpretation. Agric. Res. Repts. (Versl. landbouwk. Onderz.) 758.

Simons, P. C. M. and Wiertz, G. (1963). Notes on the structure of membranes and shell in the hen's egg. An electron microscopical study, *Z. Zellforsch. mikrosk. Anat.* **59**, 555–567.

Simons, P. C. M. and Wiertz, G. (1965). Differences in measurement of membrane thickness in hens' egg shells, *Br. Poult. Sci.* **6**, 283–286.

Simons, P. C. M. and Wiertz, G. (1966). The ultrastructure of the surface of the cuticle of the hen's egg in relation to egg-cleaning, *Poult. Sci.* **45**, 1153–1162.

Simons, P. C. M. and Wiertz, G. (1970). Notes on the structure of shell and membranes of the hen's egg; a study with the scanning electron microscope, *Annls Biol. anim. Biochim. Biophys.* **10**, 31–49.

Von Sinowatz, F., Wrobel, K. and Friess, A. (1976). Zur histotopochemie der Uterovaginalregion bei der Wachtel (*Coturnix coturnix japonica*), *Acta histochem.* **57**, 55–67.

Skalko, R. G., Kerrigan, J. M., Ruby, J. R., and Dyer, R. F. (1972). Intercellular bridges between oocytes in the chicken ovary, *Z. Zellforsch. mikrosk. Anat.* **128**, 31–41.

Snapir, N. and Perek, M. (1970). Distribution of calcium, carbonic anhydrase and alkaline phosphatase activities in the uterus and isthmus of young and old White Leghorn hens, *Poult. Sci.* **49**, 1526–1531.

Soliman, K. F. A. and Walker, C. A. (1975). Avian post-ovulatory follicle homogenate effect on the genital tract motility, *Hormone Res.* **6**, 357–365.

Solomon, Sarah E. (1975a). Variations in phosphatases in plasma and uterine fluid and in the uterine epithelia of the domestic fowl, *Poult. Sci.* **49**, 1243–1248.

Solomon, Sarah E. (1975b). Studies on the isthmus region of the domestic fowl, *Br. Poult. Sci.* **16**, 255–258.

Sonnenbrodt, A. (1908). Die Wachstumsperiode der Oocyte des Huhnes, *Arch. mikrosk. Anat. EntwMech.* **72**, 415–480.

Stampfli, H. R. (1950). Histologische Studien am Wolff'schen Körper (Mesonephros) der Vögel und über seinen Umbau zu Nebenhoden und Nebenovar, *Revue suisse Zool.* **57**, 237–316.

Stanley, A. J. (1937). Sexual dimorphism in North American hawks. I. Sex organs, *J. Morph.* **61**, 321–339.

Stanley, A. J. and Witschi, E. (1940). Germ cell migration in relation to asymmetry in sex glands of hawks, *Anat. Rec.* **76**, 329–342.

Stein, G. S. and Bacon, W. L. (1976). Effect of photoperiod upon age and maintenance of sexual development in female *Coturnix coturnix japonica*, *Poult. Sci.* **55**, 1214–1218.

Stemberger, Bridget H., Mueller, W. J. and Leach, R. M. Jnr (1977). Microscopic study of the initial stages of egg shell calcification, *Poult. Sci.* **56**, 537–543.

Stewart, G. F. (1935). Structure of the hen's egg shell, *Poult. Sci.* **14**, 24–32.

Stieve, H. (1918). Über experimentell, durch veränderte äussere Bedingungen, hervorgerufene Rückbildungsvorgänge am Eierstock des Haushuhnes (*Gallus domesticus*), *Arch. EntwMech. Org.* **44**, 530–588.

Stieve, H. (1919). Die Entwicklung des Eierstockes der Dohle (*Colaeus monedulus*), *Arch. mikrosk. Anat. EntwMech.* **92**, 137–288.

Stieve, H. (1924). Die Eierstöcke und der Legedarm beim Hühnerhabicht (*Falco palumbarius*) and Sperber (*Accipiter nissus*), *Verh. dt. zool. Ges.* **29**, 67–78.

Stoll, R. (1944). Evolution normale des canaux de Müller de l'embryon de poulet dans la deuxième partie de l'incubation, *C.r. Séanc. Soc. Biol.* **138**, 7–8.

Sturkie, P. D. (1955). Absorption of egg yolk in the body cavity of the hen, *Poult. Sci.* **34**, 736–737.

Sturkie, P. D. (1965). "Avian Physiology", 2nd edn. Comstock Associates, New York.

Sturkie, P. D. and Weiss, H. S. (1950). The effects of sympathomimetic and parasympathomimetic drugs upon egg formation, *Poult. Sci.* **29**, 781.

Sturkie, P. D., Weiss, H. S. and Ringer, R. K. (1954). The effects of injections of acetylcholine and ephedrine upon components of the hen's egg, *Poult. Sci.* **33**, 18–24.

Surface, F. M. (1912). The histology of the oviduct of the domestic hen, *Bull. Maine agric. Expl Stn* **206**, 397–430.

Sutherland, R., Mešter, J. and Baulieu, E. (1977). Hormonal regulation of sex steroid hormone receptor concentration and subcellular distribution in chick oviduct. *In* "Hormones and Cell Regulation" (J. Dumont and J. Nunez, Eds), Vol. 1, 31–48. North Holland Publishing Co., Amsterdam.

Suzuki, S. and Strominger, J. L. (1959). Enzymic synthesis of sulphated mucopolysaccharides in the hen oviduct, *Biochim. biophys. Acta* **31**, 283–285.

Suzuki, S. and Strominger, J. L. (1960a). Enzymic sulfation of mucopolysaccharides in the hen oviduct. I. Transfer of sulphate from 3′-phosphoadenosine 5′-phosphosulphate to mucopolysaccharide, *J. biol. Chem.* **235**, 257–266.

Suzuki, S. and Strominger, J. L. (1960b). Enzymic sulfation of mucopolysaccharides in the hen oviduct. 2. Mechanism of the reaction studies with oligosaccharides and mucopolysaccharides as accepters, *J. biol. Chem.* **235**, 267–273.

Suzuki, S. and Strominger, J. L. (1960c). Enzymic sulfation of mucopolysaccharides in hen oviduct. 3. Mechanism of sulfation of chondroitin and chondroitin sulphate, *J. biol. Chem.* **235**, 274–276.

Swift, C. H. (1914). Origin and early history of the primordial germ cells in the chick, *Am. J. Anat.* **15**, 483–516.

Swift, C. H. (1915). Origin of the definitive sex-cells in the chick and their relation to the primordial germ-cell, *Am. J. Anat.* **18**, 441–470.

Swift, C. H. (1916). Origin of the sex-cords and definitive spermatogonia in the male chick, *Am. J. Anat.* **20**, 375–410.

Sykes, A. H. (1953a). Premature oviposition in the hen, *Nature, Lond.* **172**, 1098.

Sykes, A. H. (1953b). Some observations on oviposition in the fowl, *Q. Jl exp. Physiol.* **38**, 61–68.

Sykes, A. H. (1954). Reflex 'bearing-down' in the hen. 10th Wld's Poult. Congr., Edinburgh, pp. 184–187.

Sykes, A. H. (1955a). The effect of adrenaline on oviduct motility and egg production in the fowl, *Poult. Sci.* **34**, 622–627.

Sykes, A. H. (1955b). Further observations on reflex "bearing-down" in the hen, *J. Physiol., Lond.* **128**, 249–257.

Talbot, C. J. and Tyler, C. (1974). A study of the progressive deposition of shell in the shell gland of the domestic hen, *Br. Poult. Sci.* **15**, 217–224.

Talo, A. and Kekäläinen, R. (1976). Ovum promotes its own transport in Japanese Quail, *Biology Reprod.* **14**, 186–189.

Tamura, T. and Fujii, S. (1966a). Histological observations on the quail oviduct: histochemical observations on the secretions of the glands and the mucous cells, *J. Fac. Fish. Anim. Husb. Hiroshima Univ.* **6**, 373–393.

Tamura, T. and Fujii, S. (1966b). Histological observations on the quail oviduct: on the secretions in the mucous epithelium of the uterus, *J. Fac. Fish. Anim. Husb. Hiroshima Univ.* **6**, 357–371.

Tamura, T., Fujii, S., Kunisaki, H. and Yamane, M. (1965). Histological observations on the quail oviduct, with reference to pigment (*porphyrin*) in the uterus, *J. Fac. Fish. Anim. Husb. Hiroshima Univ.* **6**, 37–57.

Tanaka, K. (1976). Oviposition inducing activity in the ovarian follicle of different sizes in laying hens, *Poult. Sci.* **55**, 714–716.

Tanaka, K. and Goto, K. (1976). Partial purification of the ovarian oviposition inducing factor and estimation of its chemical nature, *Poult. Sci.* **55**, 1774–1778.

Tanaka, K., Mather, F. B., Wilson, W. O. and McFarland, L. Z. (1965). Effect of photoperiod on early growth of gonads and on potency of gonadotrophins of the anterior pituitary of *Coturnix*, *Poult. Sci.* **44**, 662–665.

Tarentino, A. L. and Maley, F. (1969). The purification and properties of a β-aspartyl *N*-acetylglucosamine amidohydrolase from hen oviduct, *Archs Biochem. Biophys.* **130**, 295–303.

Terepka, A. R. (1963a). Structure and calcification in avian egg shell, *Expl Cell Res.* **30**, 171–182.

Terepka, A. R. (1963b). Organic-inorganic relationships in an avian egg shell, *Expl Cell Res.* **30**, 183–192.

Thomson, A. (1859). "Cyclopaedia of Anatomy and Physiology" (R. B. Todd, Ed.), Vol. 5 (suppl.), 1–142. Gilbert and Piper, London.

Thomson, A. L. (1964). *In* "A New Dictionary of Birds". Thomas Nelson and Sons, London.

van Tienhoven, A. (1953). Further study of the neurogenic blockage of LH release in the hen, *Anat. Rec.* **115**, 374–375.

van Tienhoven, A. (1961). Endocrinology of reproduction in birds. *In* "Sex and Internal Secretion" (W. C. Young, Ed.), 3rd edn. Williams and Wilkins, Baltimore.

Tingari, M. D. and Lake, P. E. (1973). Ultrastructural studies on the uterovaginal sperm-host glands of the domestic hen. *Gallus domesticus*, *J. Reprod. Fert.* **34**, 423–431.

Tschantz, B. (1959). Zur Brutbiologie der Trottellume (*Uria aalge aalge Pont*), *Behaviour* **14**, 1–100.

Tullett, S. G., Lutz, P. L. and Board, R. G. (1975). The fine structure of the pores in the shell of the hen's egg, *Br. Poult. Sci.* **16**, 93–95.

Tuohimaa, P. (1975). Immunofluorescence demonstration of avidin in the immature chick oviduct epithelium after progesterone, *Histochemie* **44**, 95–101.

Tyler, C. (1964). A study of the egg shells of the Anatidae, *Proc. zool. Soc. Lond.* **142**, 547–583.

Tyler, C. (1965). A study of the egg shells of Sphenisciformes, *J. Zool., Lond.* **147**, 1–19.

Tyler, C. (1966). A study of the egg shells of the Falconiformes, *J. Zool., Lond.* **150**, 413–425.

Tyler, C. (1969). A study of the egg shells of the Gaviiformes Procellariiformes, Podicipitiformes and Pelecaniformes, *J. Zool., Lond.* **158**, 395–412.

Tyler, C. and Simkiss, K. (1959a). Studies on egg shells. XII. Some changes in the shell during incubation, *J. Sci. Fd Agric.* **10**, 611–615.

Tyler, C. and Simkiss, K. (1959b). A study of the egg shells of rattite birds, *Proc. zool. Soc. Lond.* **133**, 201–243.

Van Tyne, J. and Berger, A. J. (1958). "Fundamentals of Ornithology." John Wiley, New York.

Ulrich, E. and Sandoz, D. (1972). Étude des ultrastructures du magnum des oiseaux. II. Présence de cellules intermédiaires entre les cellules muqueuses et les cellules séreuses, *J. Microscopie* **13**, 217–234.

Vannini, E. (1945). Ripartizione dei gonociti primari e origine dell' assimmetria delle gonadi nell'embrione di pollo, *Atti Ist. veneto Sci.* **104**, 1–20.

Venzke, W. G. (1954). The morphogenesis of the indifferent gonad of chicken embryos, *Am. J. vet. Res.* **15**, 300–308.

Verma, O. P. and Cherms, F. L. (1964). Observations on the oviducts of turkeys, *Avian Dis.* **8**, 19–26.

Verma, O. P. and Cherms, F. L. (1965). The appearance of sperm and their persistency in storage tubules of turkey hens after a single insemination, *Poult. Sci.* **44**, 609–613.

Verma, O. P. and Walker, C. A. (1974). Adrenergic activity of the avian oviduct *in vivo*, *Theriogenology* **2**, 47–61.

Verma, O. P., Prasad, B. K. and Slaughter, J. (1976). Avian oviduct motility induced by Prostaglandin E_1, *Prostaglandins* **12**, 217–227.

Waldeyer, W. (1870). "Eierstock and Ei. Ein Beitrag zur Anatomie und Entwicklungsgeschichte der Sexualorgane." Engelmann, Leipzig.

Warren, D. C. and Scott, H. M. (1935). The time factor in egg production, *Poult. Sci.* **14**, 195–207.

Webster, H. D. (1948). The right oviduct in chicks, *J. Am. vet. med. Ass.* **112**, 221–223.

Wechsung, E. and Houvenaghel, A. (1976). A possible role of prostaglandins in the regulation of ovum transport and oviposition in the domestic hen, *Prostaglandins* **12**, 599–608.

Welty, J. C. (1962). "The Life of Birds." W. B. Saunders, Philadelphia.

Weniger, J. P. (1968). Sur la précocité de la secretion d'oestrogènes par les gonades embryonnaires de poulet cultivés *in vitro*, *C.r. hebd. Séanc. Acad. Sci., Paris* **266**, 2277–2279.

Westphal, U. (1961). Das Arteriensystem des Haushuhnes (*Gallus domesticus*), *Wiss. Z. Humbolt-Univ. Berl.* **10**, 93–124.

White, C. M. (1969). The functional gonads in peregrines, *Wilson Bull.* **81**, 339–340.

Wight, P. A. L. (1970). The mast cells of *Gallus domesticus*, *Acta Anat.* **75**, 100–113.

Williams, J. (1967). *In* "The Biochemistry of Animal Development" (R. Weber, Ed.), Vol. 2, 341–382. Academic Press, New York and London.

Williamson, J. H. (1964). Studies of cystic right oviducts in two strains of White Leghorns, *Poult. Sci.* **43**, 1170–1177.

Williamson, J. H. (1965). Cystic remnants of the right Müllerian duct and egg production in two strains of White Leghorns, *Poult. Sci.* **44**, 321–324.

Willier, B. H. (1925). The behaviour of embryonic chick gonads when transplanted to embryonic chick hosts, *Proc. Soc. exp. Biol. Med.* **22**, 26–30.

Willier, B. H. (1933). Potencies of the gonad-forming area in the chick as tested by chorio-allantoic grafts, *Arch. EntwMech. Org.* **130**, 616–649.

Willier, B. H. (1937). Experimentally produced sterile gonads and the problem of the origin of the germ cells in the chick embryo, *Anat. Rec.* **70**, 89–112.

Willier, B. H. (1939). The embryonic development of sex. *In* "Sex and Internal Secretion" (E. Allen, Ed.), 2nd edn. Baillière, Tindal and Cox, London.

Willier, B. H. (1950). Sterile gonads and the problem of the origin of germ cells in the chick embryo, *Archs Anat. microsc. Morph. exp.* **39**, 269–273.

Wilson, Susan C. and Sharp, P. J. (1976). Induction of luteinizing hormone release by gonadal steroids in the ovariectomised domestic hen, *J. Endocr.* **71**, 87–98.

Winter, H. (1958). Persistant right oviducts in fowls including an account of the histology of the fowl's normal oviduct, *Aust. vet. J.* **34**, 140–147.

Wisse, E. (1970). An electron microscope study and the fenestrated endothelial lining of rat liver sinusoids, *J. Ultrastruct. Res.* **31**, 125–150.

Witschi, E. (1935a). Die Amphisexualität der embryonalen Keimdrüsen des Hausperlings, *Passer domesticus* (Linnaeus), *Biol. Zbl.* **55**, 168–174.

Witschi, E. (1935b). Origin of asymmetry in the reproductive system of birds, *Am. J. Anat.* **56**, 119–141.

Witschi, E. (1956). "Development of Vertebrates." W. B. Saunders, Philadelphia.

Witschi, E. (1961). Sex and secondary sexual characters. *In* "Biology and Comparative Physiology of Birds" (A. J. Marshall, Ed.), Vol. 2, 115–168. Academic Press, New York and London.

Witschi, E. and Fugo, N. W. (1940). Response of sex characters of the adult female starling to synthetic hormones, *Proc. Soc. exp. Biol. Med.* **45**, 10–14.

Wolff, E. (1950). La différenciation sexuelle normale et le conditionnement hormonal des caractères sexuels somatiques précoses tubercule génital et syrinx chez l'embryon de Canard, *Bull. biol. Fr. Belg.* **84**, 119–193.

Wolff, E., Haffen, K. and Scheib, D. (1966). Sur la détection et le rôle d'hormones sexuelles dans les jeunes gonades embryonnaires des oiseaux, *Annls Histochim.* **11**, 353–368.

Wood-Gush, D. G. M. and Gilbert, A. B. (1965a). Use of a behaviour pattern in the study of ovarian activity in the domestic fowl, *J. Reprod. Fert.* **9**, 265–266.

Wood-Gush, D. G. M. and Gilbert, A. B. (1965b). Relationships between the ovary and oviduct in the domestic hen, *Nature, Lond.* **207**, 1210–1211.

Wood-Gush, D. G. M. and Gilbert, A. B. (1969). Observations on the laying behaviour of hens in battery cages, *Br. Poult. Sci.* **10**, 29–36.

Wood-Gush, D. G. M. and Gilbert, A. B. (1970). The rate of egg loss through internal laying, *Br. Poult. Sci.* **11**, 161–163.

Wood-Gush, D. G. M. and Gilbert, A. B. (1975). The physiological basis of a behaviour pattern in the domestic hen, *Symp. zool. Soc. Lond.* **35**, 261–276.

Woods, J. E. and Domm, L. V. (1966). A histochemical identification of the androgen-producing cells of the domestic fowl and albino rat, *Gen. comp. Endocr.* **7**, 559–570.

Woods, J. E. and Weeks, R. L. (1969). Ontogenesis of the pituitary-gonadal axis in the chick embryo, *Gen. comp. Endocr.* **13**, 242–254.

Woods, J. W., Woodward, M. P. and Roth, T. F. (1978). Common features of coated vesicles from dissimilar tissues: composition and structure, *J. Cell Sci.* **30**, 87–97.

Wyburn, G. M. (1974). Histological changes in the oviduct of hens laying watery whites, *Proc. Univ. Otago med. Sch.* **51**, 13–14.

Wyburn, G. M. and Baillie, A. H. (1956). Some observations on the fine structure and histochemistry of the ovarian follicle of the fowl. *In* "Physiology of the Domestic Fowl" (C. Horton-Smith and E. C. Amoroso, Eds). Oliver and Boyd, Edinburgh.

Wyburn, G. M., Aitken, R. N. C. and Johnston, H.S. (1965a). The ultrastructure of the zona radiata of the ovarian follicle of the domestic hen, *J. Anat.* **99**, 469–484.

Wyburn, G. M., Johnston, H. S. and Aitken, R. N. C. (1965b). Specialized plasma membranes in the preovulatory follicle of the fowl, *Z. Zellforsch. mikrosk. Anat.* **68**, 70–79.

Wyburn, G. M., Johnston, H. S. and Aitken, R. N. C. (1966). Fate of the granulosa cells in the hen's follicle, *Z. Zellforsch. mikrosk. Anat.* **72**, 53–65.

Wyburn, G. M., Johnston, H. S., Draper, M. H. and Davidson, Maida F. (1970). The fine structure of the infundibulum and magnum of the oviduct of *Gallus domesticus*, *Q. Jl exp. Physiol.* **55**, 213–232.

Wyburn, G. M., Johnston, H. S., Draper, M. H. and Davidson, Maida F. (1973). The ultrastructure of the shell-forming region of the oviduct and the development of the shell of *Gallus domesticus*, *Q. Jl exp. Physiol.* **58**, 143–151.

Wynne-Edwards, V. C. (1939). Intermittent breeding of the Fulmar (*Fulmarus glacialis*) with some general observations on non-breeding sea birds, *Proc. zool. Soc. Lond.* **109**, 127–132.

Yocom, H. B. (1924). Luteal cells in the gonad of the phalarope, *Biol. Bull., Woods Hole* **46**, 101–105.

Yu, J. Y.-L. and Marquardt, R. R. (1973a). Development, cellular growth and function of the avian oviduct. I. Studies on the magnum during a reproductive cycle of the domestic fowl (*Gallus domesticus*), *Biology Reprod.* **8**, 283–298.

Yu, J. Y.-L. and Marquardt, R. R. (1973b). Synergism of testosterone and estradiol in the development and function of the magnum from the immature chicken (*Gallus domesticus*) oviduct, *Endocrinology* **92**, 563–572.

Yu, J. Y.-L. and Marquardt, R. R. (1974). Hyperplasia and hypertrophy of the chicken *Gallus domesticus* oviduct during a reproductive cycle, *Poult. Sci.* **53**, 1096–1105.

Yu, J. Y.-L., Marquardt, R. R. and Hodgson, G. C. (1972). Development, cellular growth and function of the avian oviduct. II. A comparative study on the changes of cellular components in each region of the domestic fowl (*Gallus domesticus*) oviduct during a reproductive cycle, *Can. J. Physiol. Pharmac.* **50**, 689–696.

Yusko, S. C. and Roth, T. F. (1976). Binding to specific receptors on oocyte plasma membranes by serum phosvitin-lipovitellin, *J. Supramolecular Struct.* **4**, 89–97.

Zawadowsky, M. M. and Zubina, E. M. (1928). Cockfeathered pheasants and embryogenesis of sex glands in hens, *Trudy Lab. eksp. Biol. musk. Zoopk.* **4**, 175–200.

6
The blood cells

R. D. HODGES

Wye College, University of London, Near Ashford, Kent TN25 5AH, England

---

CONTENTS

I. Introduction . . . 361

II. Erythrocytes . . . 362

III. Thrombocytes . . . 365

IV. Leucocytes . . . 367
- A. Non-granular leucocytes . . . 367
- B. Granular leucocytes . . . 370

References . . . 376

I. Introduction

In general, avian blood cells are morphologically very similar to their mammalian counterparts with two major exceptions, the erythrocyte and the thrombocyte. Avian erythrocytes are characteristically nucleated cells but are functionally very similar to the non-nucleated mammalian red cell. Thrombocytes, on the other hand, differ considerably from the mammalian platelet, being entire, nucleated cells whose role in the function of blood coagulation is by no means clear. Otherwise, avian blood cells possess fairly typical structures and staining characteristics, except for the counterpart of the neutrophil which is termed a heterophil because of its acid-staining cytoplasmic granules.

Since most of the research into the structure and function of avian blood cells has been performed on domestic species, this chapter will be largely concerned with these birds.

II. Erythrocytes

Avian erythrocytes are flattened, ellipsoidal cells which, because they retain their nuclei throughout their functional life, are normally larger than their mammalian counterparts. Mean red cell size varies from $9{\cdot}75 \times 5{\cdot}04$ μm in the Golden-crowned Kinglet (*Regulus satrapa*) (Bartsch *et al.*, 1937) up to $15{\cdot}9 \times 9{\cdot}7$ μm in the Ostrich (*Struthio camelus*) (De Villiers, 1938), although larger values have been given for a cassowary by Gulliver (1846) and for the Greater Rhea (*Rhea americana*) by Venzlaff (1911). In general, the larger sized cells are found in the more primitive species of birds (Cleland and Johnston, 1912; De Villiers, 1938; Hartman and Lessler, 1963); also birds with a higher metabolic rate tend to have smaller erythrocytes, the increased surface area to size ratio resulting in greater efficiency of gaseous exchange (Hartman and Lessler, 1963). A number of surveys of red cell size have been published in the literature: e.g. Venzlaff (1911, 43 species), Cleland and Johnston, (1912, 98 species), Bartsch *et al.* (1937, 50 species), Hartman and Lessler (1963, 126 species), Bennett and Chisholm (1964, 158 species) and Balasch *et al.* (1973, six galliform species). Numerous measurements for individual species have also been published, and Sturkie (1976) and Hodges (1977) have collated many of the figures for domestic birds.

The erythrocyte's overall appearance may vary according to the type of preparation examined. In histological preparations they appear as irregularly ellipsoidal cells, with an uneven lenticular profile in longitudinal section. However, in a stained smear they are more flattened and more regularly ellipsoidal (Fig. 6.1a). Mature erythrocytes possess a clear, homogeneous cytoplasm which stains with a faint blue colour when treated with typical blood stains (e.g. Wright's stain). Little sign of internal structure is apparent normally, but staining with iron haematoxylin (Dehler, 1895) or with a supravital stain such as gentian violet (Meves, 1911) does demonstrate a narrow homogeneous ring about half a micrometer in thickness around the periphery of the cell (Fawcett, 1959). This marginal band consists of a skein of microtubules circumferentially aligned just beneath the cell membrane and having a supporting function during the development of the cell's ellipsoidal shape (Barclay, 1966). Precursors of avian erythrocytes are spheroidal cells which flatten out into discoidal and finally ellipsoidal shapes after the final mitotic division. This flattening process is at least helped by the microtubular marginal band but in the mature cell it plays little or no role in the maintenance of cell shape (Barrett and Dawson, 1974). The centrally-placed nucleus is a slightly more elongated ellipsoid when compared with the overall cell shape and consists of a uniform network of deeply-staining chromatin clumps. The appearance of the nuclear chromatin varies with the age of the cell, becoming more condensed and deeply staining as the erythrocyte matures and finally ages. The reticulocyte, the penultimate developmental stage of the red cell, is commonly found in normal blood and is so named because the cytoplasm contains a reticulum of fine granules specifically stainable by brilliant cresyl blue (Magath and Higgins, 1934; Coates and March, 1966). Detailed accounts of the histology of erythrocytes from *Gallus* have been given by Lucas and Jamroz (1961, pp. 17–24) and Hodges (1974, pp. 150–154).

When examined with the electron microscope (Fig. 6.2) the erythrocyte cytoplasm is seen to consist mainly of a homogeneous, granular, electron dense material, the haemoglobin content, which obscures other structural details including, frequently, the plasma and outer nuclear membranes. However, very

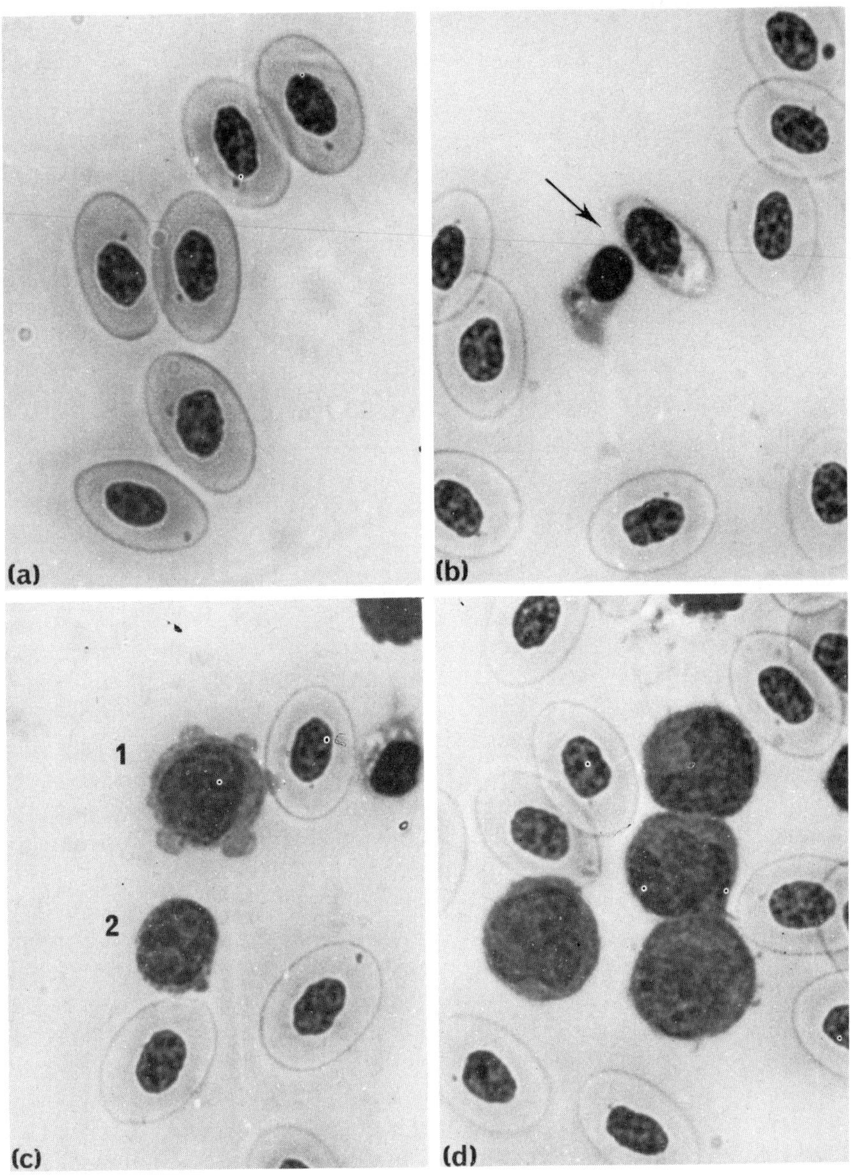

Fig. 6.1. Typical blood cells from an adult domestic fowl. Wright's stain. ×1600. (a) Erythrocytes. (b) Two thrombocytes (arrow). One is intact with reticulated cytoplasm and specific granules, the other is partially degenerate. (c) Medium-sized (1) and small (2) lymphocytes. (d) Four monocytes.

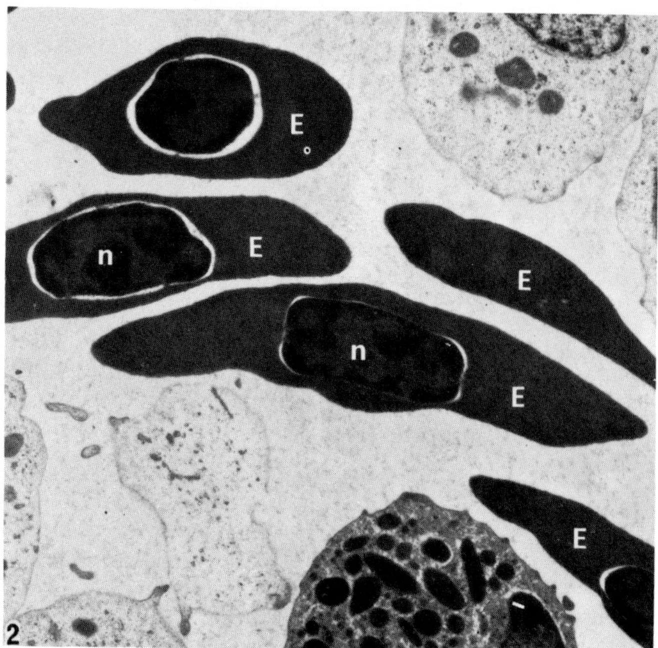

Fig. 6.2. Sections through several erythrocytes (E). n, Nucleus. *Gallus*. ×8 160.

few cell organelles remain in the mature erythrocyte, normally only a few mitochondria and ribosomes and, rarely, a Golgi complex being present (Harris, 1971) together with groups of peripheral microtubules (Behnke, 1970). These structures are more easily seen in red cell "ghosts" from which the haemoglobin has been extracted (Harris, 1971; Harris and Brown, 1971). The nucleus has a slightly irregular ellipsoidal shape and is separated from the cytoplasm by a clearly-defined gap which consists of an enlarged perinuclear cisterna, the gap being bridged by occasional nuclear pores. Many masses of heavily condensed heterochromatin lie both peripherally and throughout the nucleus, and the euchromatin bears some structural resemblance to the cytoplasm due to the presence of haemoglobin within the nucleus.

The avian erythrocyte performs an essentially identical function to that of the mammalian red cell in the transport of blood gases but it does differ from the latter in a number of ways. The haemoglobin content of adult erythrocytes in almost all avian species so far investigated consists of two main types (Allen, 1971; Bruns and Ingram, 1973; Sturkie, 1976, p. 64). Both haemoglobin types have been shown to have molecular weights of about 68 000. Type II is more acidic than Type I and they differ considerably in their amino acid compositions, particularly in the *N*-terminal amino acid sequences (Allen, 1971). Type I (Type A, Bruns and Ingram, 1973) comprises about 70% of the total and Type II (Type D, Bruns and Ingram, 1973) comprises about 30% of the total. However, Vandecasserie *et al.* (1973) have demonstrated in the penguin (*Aptenodytes forsteri*) and the parakeet (*Ara chloroptera*) that there is only a single haemoglobin type closely resembling the major type of other avian species. The oxygen dissociation curve for avian blood is generally considered to be located to the right of that of mammals, indicating the lower affinity of avian haemoglobins for oxygen (Sturkie, 1965, p. 182; Jones, 1972, p. 100; Wells, 1976), although Lutz *et al.* (1973, 1974) demonstrated no difference in oxygen affinity between avian and mammalian blood. Differences in oxygen affinity may be associated with higher concentrations of organic polyphosphate (inositol pentaphosphate) bound to the avian haemoglobin, rather than with actual differences between the haemoglobins (Wells, 1976). The ability of avian haemoglobins to give up oxygen readily to the tissues is probably of considerable functional significance in relation to the increased oxygen requirements of flight. However, embryonic haemoglobins compared with those of the adult, have a greater oxygen affinity and their dissociation curve is more to the left, which may reflect the differing respiratory efficiencies of the adult lung and the embryonic membranes.

The functional importance of the avian nucleated red cell as opposed to the mammalian non-nucleated erythrocyte, if any, is not understood (Schmidt-Nielsen, 1975, p. 85).

III. Thrombocytes

Thrombocytes bear a superficial resemblance to erythrocytes in both shape and general appearance (Fig. 6.1b). However, they tend to be smaller in size, with the length varying in *Gallus* from 6·1 to 11·5 μm and the width from 3·0 to 6·1 μm (Lucas and Jamroz, 1961, pp. 41 and 213), and to possess a larger but more rounded nucleus. There is considerable variation in the shape of thrombocytes, the typical cellular outline being oval but ranging from rounded to elongated and spindle-shaped. Unlike the erythrocyte the thrombocyte does not possess a homogeneous cytoplasm; instead it has a reticulated appearance normally staining a pale, dull blue with traces of purplish shades (Lucas and Jamroz, 1961, p. 42). The reticulation is at least partly due to the presence of cytoplasmic vacuoles (Carlson *et al.*, 1968); other cytoplasmic inclusions are the specific granules which possess considerable variability of number, size and position. Although the typical granular appearance is that of a single, reddish-purple, compact structure lying at one pole of the nucleus, there may frequently be two or more granules.

Ultrastructurally the thrombocyte can be described as a round, oval or somewhat elongated cell with an irregular outline extending into occasional fine pseudopodial processes (Fig. 6.3). The large, round to oval nucleus, which is frequently irregular and indented in outline, is sharply divided into blocks of heavily condensed, granular heterochromatin and intervening areas of pale euchromatin. The cytoplasm contains a few round mitochondria, a Golgi complex, much smooth endoplasmic reticulum and very small amounts of rough endoplasmic reticulum, and many groups of microtubules. The latter are partly arranged in a peripheral band, as in the erythrocyte, and partly as apparently random bundles throughout the cytoplasm. Large osmiophilic granules, varying in diameter from 0·2 to 3 μm (Sweeny and Carlson, 1968;

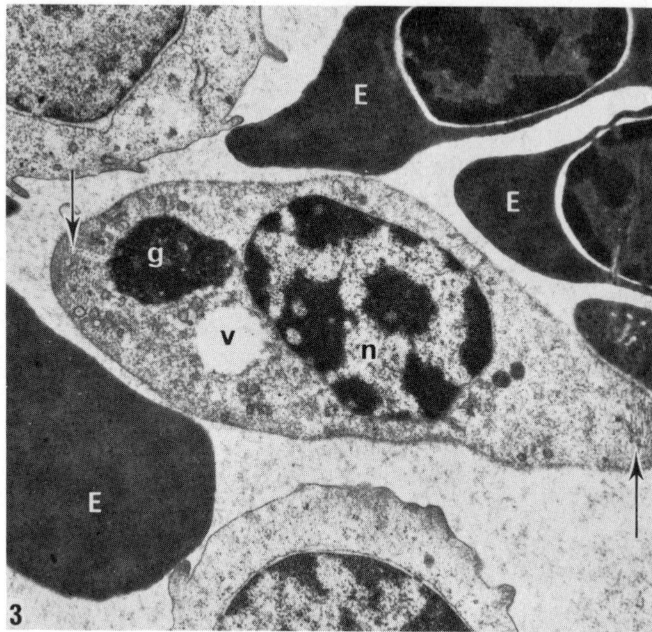

Fig. 6.3. Spindle-shaped thrombocyte containing a single dense granule (g), a large vacuole (v), and numerous peripheral microtubules (arrows). n, Nucleus; E, erythrocytes. *Gallus*. ×10 880.

Maxwell, 1974), are the most striking of the cytoplasmic organelles and are apparently identical to the specific granules seen with the light microscope, both granule types breaking down rapidly when blood is shed. Within the membrane-bound granules are normally found masses of dense osmiophilic material or closely-packed arrays of osmiophilic membranes (Simpson, 1968; Enbergs and Kriesten, 1968a). However, after blood has been shed the granules break down, giving first vacuolated structures containing a partially degenerate content and, finally, empty vacuoles, this process being accelerated by the action of reserpine (Kuruma *et al*., 1970). The ultrastructure of thrombocytes has also been studied by Enbergs and Kriesten (1968b), Maxwell and Trejo (1970), Enbergs (1973b) and Sterz and Weiss (1973).

Although the function of the avian thrombocyte is generally considered to be similar to that of the mammalian platelet (Olson, 1959; Lucas and Jamroz, 1961, p. 41), the part it plays in blood coagulation is by no means clear. Once blood is shed the thrombocytes physically aid haemostasis by clumping together to form plugs of cells (Stalsberg and Prydz, 1963), although the rate at which clumping takes place is not as fast as that of platelets. During the haemostatic process degenerative changes occur in the cells; the specific granules break down and the nucleus becomes pyknotic (Lucas and Jamroz, 1961, pp. 42 and 46). However, thrombocytes seem to play little part in the initiation of the clotting process (Archer, 1971), since they contain very little intrinsic thromboplastin. The osmiophilic granules seen with the electron microscope consist largely of 5-hydroxytryptamine (Kuruma *et al*., 1970) and are unlikely to be a source of thromboplastin. Most thromboplastin in avian shed blood is extrinsic in origin, coming from damaged tissues, and efficient clotting depends upon the mixing of shed blood with tissue juices derived from damaged cells. Blood coagulation in birds has been reviewed by Archer (1971) and Sturkie (1976, pp. 70–71).

A second function of the thrombocyte is that of phagocytosis. Living cells examined by phase contrast microscopy are incessantly moving and extending pseudopodia (Kuruma *et al.*, 1970) and are capable of phagocytosing bacteria (Glick *et al.*, 1964; Carlson *et al.*, 1968), dye particles (Carlson *et al.*, 1968) and viruses (Sterz and Weiss, 1973). The presence of large, acid phosphatase-positive granules has been demonstrated histochemically in thrombocyte cytoplasm by Sweeny and Carlson (1968) and these may be lysosomal structures associated with the phagocytic activity. However, although the osmiophilic cytoplasmic granules frequently have the appearance of secondary lysosomes (Sweeny and Carlson, 1968), no ultrastructural correlation between the granules and the enzyme activity has yet been made.

IV. Leucocytes

A. Non-granular leucocytes

(*1*) *Lymphocytes*

Lymphocytes are rounded cells which possess great variability in both size and number, although they are normally the most frequently-occurring leucocyte in the blood. Up to 75% of the total leucocyte count has been shown to consist of lymphocytes in the normal domestic fowl (Blount, 1939; Lucas and Jamroz, 1961, p. 216). The range of lymphocyte diameter varies from about 12 μm in large lymphocytes, through medium lymphocytes, down to about 6 μm in small lymphocytes. The large cells only occur rarely and are probably immature forms; it is the medium and small cells which are the mature, functional blood cells. Discrimination between small and medium lymphocytes is usually made on the basis of the amount of cell cytoplasm; in the small lymphocyte it may only consist of the thinnest margin peripheral to the nucleus but in the medium lymphocyte it forms a moderately wide band. The regular, rounded outline of lymphocytes may frequently be broken by pseudopodial projections. Lymphocyte cytoplasm is normally slightly basophilic and homogeneous in appearance. The nucleus is usually large, round to oval in shape, and centrally-placed within the cell (Fig. 6.1c). Nuclear chromatin is heavily condensed in small cells but less so in medium lymphocytes.

When examined with the electron microscope both small and medium lymphocytes are seen to be rounded cells (Fig. 6.4) with peripheral pseudopodia and a large nucleo-cytoplasmic ratio (Maxwell and Trejo, 1970; Maxwell, 1974). Relatively few cytoplasmic organelles are present; only some mitochondria, short strands of rough endoplasmic reticulum, many ribosomes and dense, membrane-bound granules up to 1·0 μm in diameter (Enbergs and Kriesten,

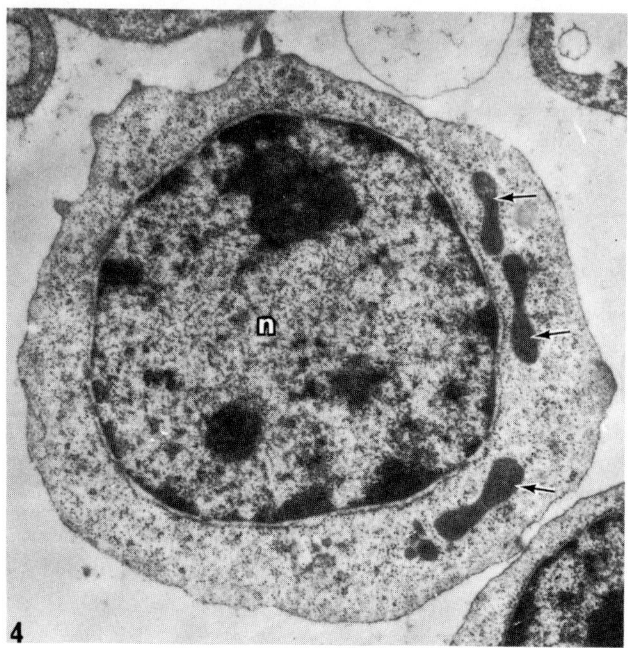

Fig. 6.4. Medium sized lymphocyte showing the cytoplasm relatively free from organelles other than mitochondria (arrows) and ribosomes. n, Nucleus. *Gallus*. ×13 600.

1968b; Maxwell, 1974). A Golgi complex is present in both cell types, being better developed in medium lymphocytes, and is frequently located opposite a nuclear indentation. The nuclear heterochromatin is more strongly condensed and more peripherally arranged in the small than in the medium lymphocytes. Other ultrastructural studies of lymphocytes have been carried out by Schumacher (1965) and Simpson (1968).

The thymus and cloacal bursa (*bursa cloacalis*) are the so-called central lymphoid organs in birds which are responsible for the development of adaptive immunity. In *Gallus* it has been demonstrated that lymphocytes derived from these organs can be functionally divided into thymus-dependent and bursa-dependent lines (Cooper *et al.*, 1966, 1967). The thymus-dependent line is represented morphologically by the small lymphocytes of the circulation and the white pulp type of tissue. These cells, which bear no obvious immunoglobulins on their surfaces, are concerned with the development of aspects of cellular immunity such as graft-versus host reactions, homograft reactions and delayed hypersensitivity reactions (Cooper *et al.*, 1966). The bursa-dependent line is represented morphologically by the larger lymphocytes of the germinal centres and by the plasma cells. These cells bear surface immunoglobulins or immunoglobulin receptors which appear to function as antibody receptors for various antigens (Cooper *et al.*, 1967; Kincade *et al.*, 1971). Experimental extirpation of the bursa-dependent line results in agammaglobulinaemia and an inability to produce detectable antibody when stimulated in various ways.

Nair (1973) has demonstrated that the circulating lymphocytes respond to an inflammatory stimulus by migrating out of the blood vessels about 6 h after the initial stimulus and eventually spreading into the tissue spaces after about 72 h.

(2) *Monocytes*

Monocytes are large, rounded cells with occasional peripheral cytoplasmic projections (Fig. 6.1d). In size they are the largest of the circulating leucocytes (Lucas and Jamroz, 1961, p. 65; Christoph and Borowski, 1961), although there is some similarity in both size and appearance between monocytes and large lymphocytes. In *Gallus* they average 12 μm in diameter (Lucas and Jamroz, 1961, pp. 66 and 214). The cytoplasm usually possesses a reticular appearance and may be divided into an outer, palely-staining mantle and an inner, more densely-staining core. Throughout the cytoplasm there may also be scattered delicate, pink-staining particles. Monocyte nuclei are round to elongated in shape with one side tending to be flattened or indented, resulting in a characteristic kidney shape. The degree of indentation is occasionally great enough for the nucleus to be considered bilobed. The nuclear chromatin is arranged as a reticulum of finer structure than that of the lymphocyte. Lying adjacent to the nuclear indentation is often seen the so-called Hof, an area of cytoplasm where the reticulation is more obvious, giving a vacuolated appearance. Within the Hof may be found orange-staining spheres.

The ultrastructure of monocytes has been investigated in *Gallus* by Maxwell and Trejo (1970) and Enbergs and Kriesten (1968b, 1969), and in *Anas*, *Anser*, *Meleagris*, *Columba*, *Coturnix* and *Numida* by Maxwell (1974). They are irregularly rounded to elongated cells with peripheral pseudopodia or lobopodia (Fig. 6.5). The indented nucleus contains more euchromatin than heterochromatin and the latter is not usually found in coarse aggregations. One or two nucleoli are found centrally. The relatively large cytoplasmic volume of the cell contains a well-developed population of organelles. Rounded mitochondria, moderate amounts of rough endoplasmic reticulum, ribosomes, pinocytotic vesicles and dense, membrane-bound granules up to 0·5 μm in diameter occur throughout the cytoplasm, whilst bundles of fine filaments and well-developed Golgi complexes are found close to the nucleus. The Golgi complexes normally occur in the cytoplasm adjacent to the nuclear indentation and thus would appear to be related to the Hof seen with the light microscope.

A number of tissue-culture studies in *Gallus* (Carrel and Ebeling, 1926; Weiss and Fawcett, 1953; Sutton and Weiss, 1966) have demonstrated that the blood monocyte and the tissue macrophage are functional variations of a single type of cell. Carrel and Ebeling (1926) concluded that macrophages were only monocytes in a more active metabolic state. As in the mammal (Nichols *et al.*, 1971; Nichols and Bainton, 1975), the avian monocyte appears to be a relatively transient cell type; after a short life in the blood stream the monocyte migrates into the tissues to become a macrophage. Detailed studies by Nair (1973) of the inflammatory reaction in *Gallus* have clearly demonstrated these processes of cell migration and change. The ultrastructure of macrophages has been described by Sutton and Weiss (1966) and Nair (1973). Although the monocyte appears

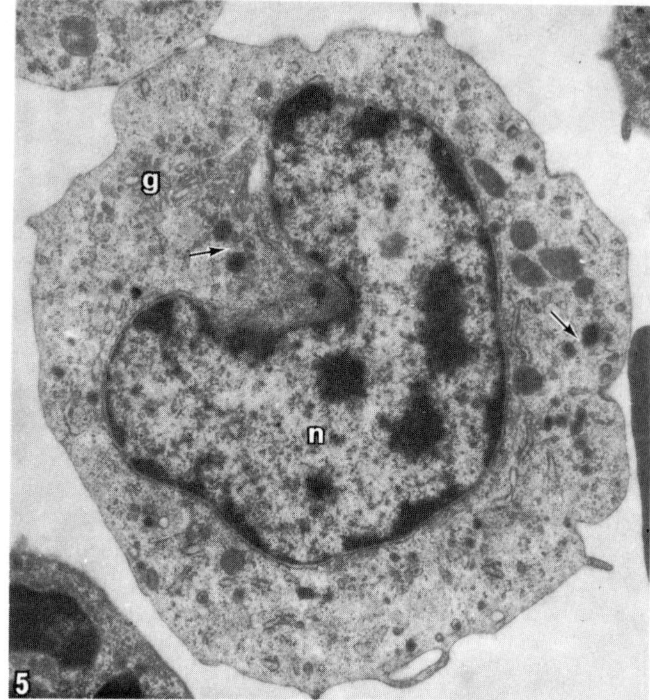

Fig. 6.5. Monocyte with a large, indented nucleus and well-developed cytoplasmic organelles, including a Golgi complex (g) and dense granules (arrows). n, Nucleus. *Gallus*. ×10 880.

as a rounded cell in fixed preparations, *in vivo* it is more elongated and ameboid (Carrel and Ebeling, 1926) and is actively motile, although not as motile as the heterophil (Nair, 1973). Monocytes are actively phagocytic cells which can engulf particulate material, bacteria or damaged and senescent cells, and possess lysosome-like bodies (Maxwell and Trejo, 1970; Maxwell, 1974) which are associated with the destruction of phagocytosed material.

B. Granular leucocytes

(*1*) *Heterophils*

Heterophils are active, amoeboid cells *in vivo*, particularly when outside the blood stream (Kelly and Dearstyne, 1935; Hirsch, 1962) but appear as rounded cells after fixation and staining. They are the functional equivalent of the mammalian neutrophil and closely resemble the latter, apart from the morphology and staining reactions of the specific granules. In *Gallus* heterophils are about 8·0–10·0 μm in diameter (Lucas and Jamroz, 1961, pp. 73 and 215) with palely-staining, almost colourless cytoplasm and a polymorphic nucleus consisting of between one and five lobes joined together by narrow chromatin strands. The nuclear chromatin is usually well condensed and deeply-staining.

The cytoplasm typically contains many, closely-packed, spindle-shaped specific granules which have an acidophilic, or eosinophilic, staining reaction and possess a central, spherical core (Fig. 6.6a). Cells containing vacuolated or rounded granules may be seen (Natt and Herrick, 1954; Lucas and Jamroz, 1961, p. 74); such appearances are frequently the result of partial granular dissolution during staining, often leaving only the central core.

Ultrastructural studies of heterophils in several domestic species have been carried out by Enbergs and Kriesten (1968b), Dhingra *et al.* (1969), Maxwell and Trejo (1970), Osculati (1970), Enbergs (1973b), Ericsson and Nair (1973), Maxwell (1973) and MacRae and Spitznagel (1975). Circulating heterophils appear as irregularly rounded cells with small peripheral pseudopodia (Fig. 6.7).

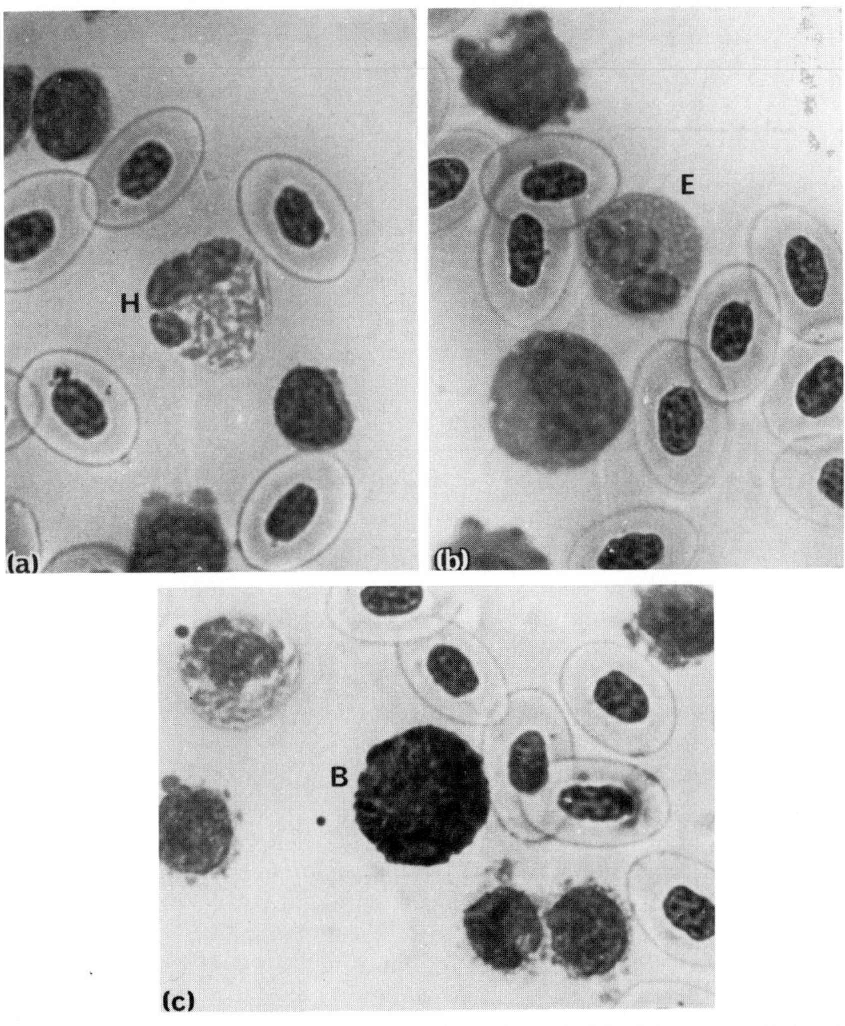

Fig. 6.6. Typical blood cells from an adult domestic fowl. Wright's stain. ×1600. (a) Heterophil granulocyte (H) with characteristic spindle-shaped granules and lobed nucleus. (b) Eosinophil granulocyte (E). (c) Basophil granulocyte (B).

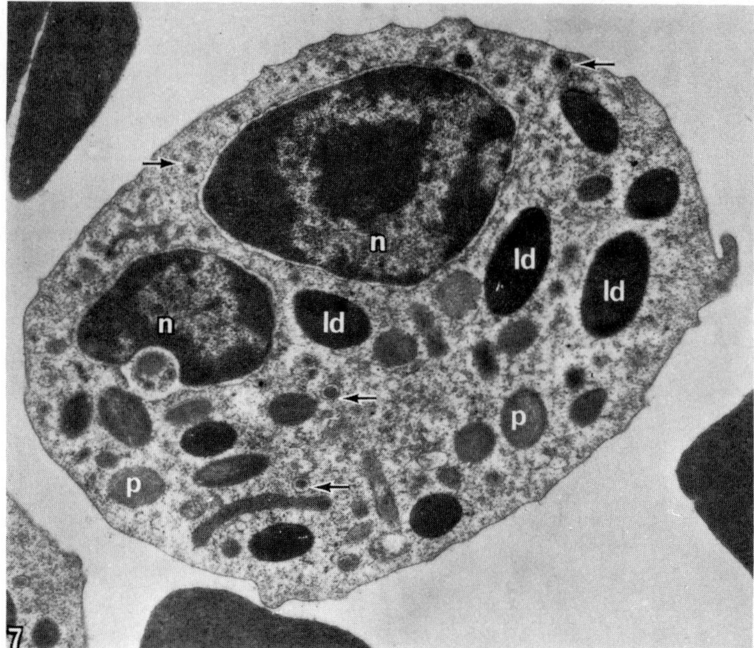

Fig. 6.7. Heterophil granulocyte showing two nuclear lobes (n) and the three characteristic types of cytoplasmic granule, large dense (ld), pale (p) and small, dense-cored (arrows). *Gallus*. × 13 600.

Apart from the specific granules, the cytoplasm contains relatively few, small organelles; in particular these include rounded mitochondria, endoplasmic reticulum, glycogen particles and a Golgi complex. Several types of cytoplasmic granule have been described in heterophils, the granular morphology varying slightly according to species. In *Gallus* there are three types of membrane-bound granule (Dhingra *et al.*, 1969; Ericsson and Nair, 1973): small dense granules up to 0·2 μm in diameter; large, spindle-shaped, dense granules up to 3·5 μm long and frequently possessing a central spherical core; and round to oval, pale granules possessing a fibrillar matrix and somewhat smaller than the large dense granules. Maxwell (1973) has described similar granule types in the other domestic species.

Heterophils are the most frequently-occurring of the granulocytes. They are highly motile, amoeboid cells (Hirsch, 1962) which actively phagocytose particulate matter and organisms such as *Escherichia coli* (Gross, 1961), *Bacillus megaterium* (Hirsch, 1962) and *Staphylococcus aureus* (Glick *et al.*, 1964; Topp and Carlson, 1972b) both in the blood stream and within the tissues (Carlson and Allen, 1969; Nair, 1973). Studies of the inflammatory reaction in *Gallus* (Nair, 1973) have demonstrated the great motility of the heterophils in that they are the first cells to enter the tissues from blood vessels. However, in *Coturnix*, Atwal and McFarland (1966) described the motility of heterophils as being feeble compared to monocytes. A cinemicrophotographic study of heterophils by Hirsch (1962) has demonstrated how the cell phagocytoses bacteria and how the spindle-shaped granules adjacent to the ingested organisms immediately undergo lysis. The presence of acid hydrolase enzymes has been demonstrated both histochemically (Topp and Carlson, 1972a) and cytochemically (Osculati, 1970; Ericsson and Nair, 1973) within some of the heterophil granules, suggesting that they are lysosomal structures, as are the granules of the mammalian neutrophil (Cline, 1975). The substance of the heterophil granules has been shown to take part in the cell's attack on ingested bacteria (Macrae and Spitznagel, 1975).

(2) *Eosinophils*

Eosinophils are basically rounded cells with clear ground cytoplasm and a lobed nucleus composed of dense chromatin clumps (Fig. 6.6b). The nucleus is usually bilobed, the number of lobes normally being less than that of the heterophil. In *Gallus* Lucas and Jamroz (1961, pp. 89 and 215) have reported a mean cellular diameter of 7·3 μm, with a wide range of between 4 and 11 μm. The characteristic eosinophil granule is a large, spherical, homogeneous structure which occurs in large enough numbers to obscure both the cytoplasm and nucleus in a stained preparation. Smaller or less regular and less homogeneous granules are sometimes seen but these are not the result of degeneration during staining since eosinophil granules, unlike those of heterophils, are stable structures. Because the eosinophilic staining affinity of the cytoplasmic granules does not differ greatly from that of the heterophil granules, these two cell types are more easily differentiated morphologically than by staining affinity.

The ultrastructure of eosinophil granulocytes has been investigated in *Gallus* by Enbergs and Kriesten (1968b, 1970), Kelényi and Németh (1969), Dhingra *et al.* (1969) and Maxwell and Trejo (1970); in *Anas* by Maxwell and Siller (1972) and Enbergs (1973a,b); in *Anser* by Enbergs and Beardi (1971) and Maxwell and Siller (1972); and in *Coturnix*, *Numida* and *Mealeagris* by Maxwell and Siller (1972). Eosinophils are generally described as irregularly rounded cells with small peripheral pseudopodia and occasional cytoplasmic lobules (Fig. 6.8). Apart from the specific granules the cytoplasmic content consists of small, round mitochondria, a moderate to well-developed Golgi complex, free ribosomes and frequently abundant rough endoplasmic reticulum. The granules themselves are membrane-bound structures of rounded, oval or elongated shape with maximum dimensions up to 1·6 μm, and with a densely-granular, homogeneous content. In *Anas* and *Anser* the granules possess a crystalline core (Maxwell and Siller, 1972). The nuclear lobes show heavy condensations of heterochromatin.

The functions of the avian eosinophil are not clearly understood. In mammals these cells occur in the tissues for much of their life span and interact with immune complexes, frequently by active phagocytosis. They are normally abundant in areas of chronic inflammation (Cline, 1975, p. 111). Attempts by Nair (1973) to produce a local eosinophilic response in *Gallus*, using agents which normally evoke such a response in mammals, were negative, suggesting that such agents do not operate in the bird in the same way as they do in the mammal. Avian eosinophil function may therefore be somewhat different to that of its mammalian counterpart.

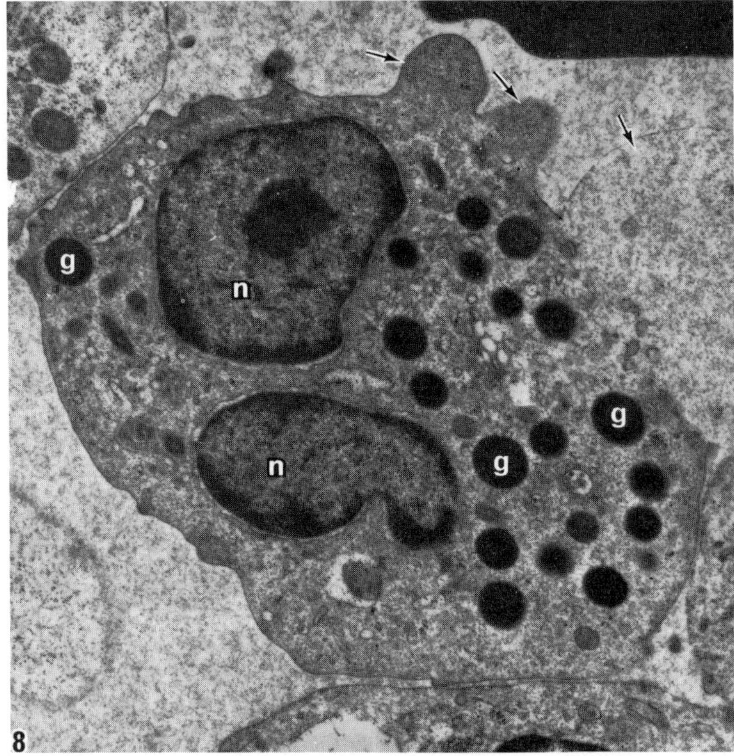

Fig. 6.8. Eosinophil granulocyte showing the typical round, dense granules (g), the lobed nucleus (n) and several cytoplasmic lobules (arrows). *Gallus*. × 13 600.

(3) *Basophils*

Basophils are rounded cells with a clear colourless ground cytoplasm and a diameter in the mature domestic fowl of about 8 μm (Lucas and Jamroz, 1961, p. 215). The nucleus is frequently found in a lateral rather than a central position within the cell and normally consists of a single lobe which may be somewhat indented, but rarely deep enough to be termed bilobed. The nuclear chromatin is not heavily condensed. In the intact cell both nucleus and cytoplasm are normally obscured by the presence of numerous spherical, deeply basophilic granules up to 0·8 μm in diameter (Fig. 6.6c). However, the granules are water soluble and frequently undergo at least partial dissolution when treated with aqueous blood stains and other processing techniques.

When examined with the electron microscope the basophil is normally seen to be a rounded cell possessing a few, small, peripheral pseudopodia. Within the cytoplasm is a rather sparse complement of organelles together with numbers of characteristic basophil granules (Fig. 6.9). The organelles consist of a few rounded mitochondria, a poorly developed Golgi complex, some rough endoplasmic reticulum, occasional small lipid droplets and glycogen granules (Maxwell and Trejo, 1970). The membrane-bound basophil granules may be

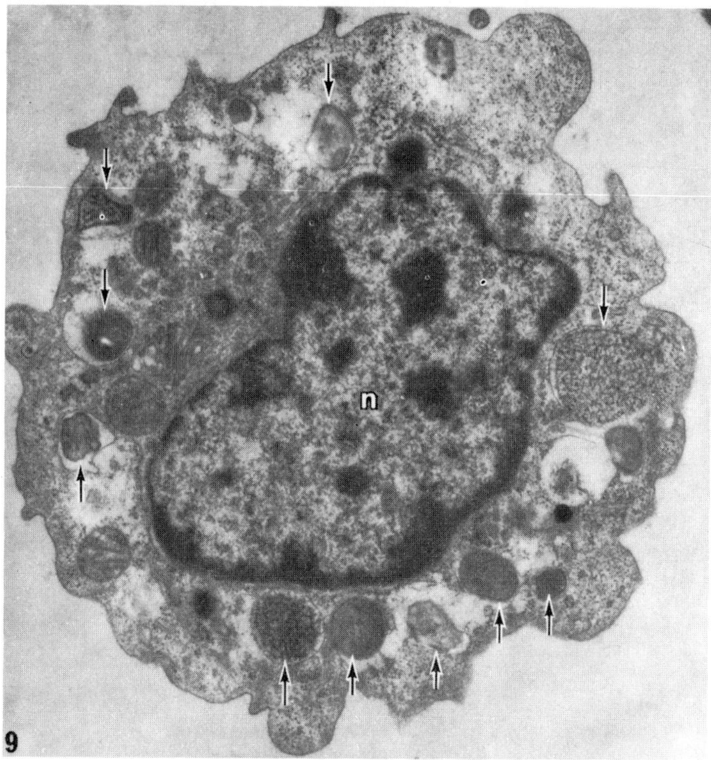

Fig. 6.9. Basophil granulocyte with a single-lobed nucleus (n) and numerous basophil granules (arrows) some of which are showing signs of dissolution. *Gallus*. × 17 680.

rounded, crescent-shaped or somewhat irregular in cross-section and their morphology varies slightly according to species and fixation (Maxwell, 1973). In *Gallus* there are four types of granule, the internal structure of which was described by Maxwell and Trejo (1970) as being dense, granular, webbed and with myelin whorls. In other species the granules generally fall into three groups (Maxwell, 1973). Dense granules are most numerous in *Coturnix*, *Numida* and *Columba*, whilst stippled granules are most dominant in *Anas*, *Anser* and *Meleagris*. The third type of granule has a honeycomb appearance and is more frequently found in gluteraldehyde fixed cells, particularly in *Coturnix*. No crystalloid structures have been seen in avian basophil granules (Maxwell, 1973). The nucleus is rounded with a variable degree of indentation and contains one or two irregular nucleoli.

Other electron microscope studies of basophils have been performed by Enbergs and Kriesten (1968b), Dhingra *et al.* (1969), Kriesten and Enbergs (1970) and Enbergs (1973b).

Nair (1973) has demonstrated that avian basophils, unlike those of mammals, regularly take part in inflammatory reactions, entering the tissues from the blood vessels at an early stage of the reaction. Basophils are thus motile cells (Burton and Higginbotham, 1966). Basophilic infiltration takes place when both immunogenic and non-immunogenic agents are used to promote inflammation (Nair, 1973), suggesting that they play a wider role than that proposed for mammalian basophils, which is connected with delayed hypersensitivity reactions (Wolf-Jürgensen, 1968) or certain types of tissue injuries induced by immune complexes (Cline, 1975, p. 129). It has also been suggested that avian basophils react in response to virus infections (Burton and Higginbotham, 1966; Cheville and Beard, 1972). Both Burton and Higginbotham (1966) and Nair (1973) believed that the basophil granules contain heparin, the release of which may act to inhibit viruses or to contain lymphocytes within the perivascular areas of inflammatory foci.

Basophils have been considered to be the blood component and mast cells to be the tissue component of a single cell line in both mammals (Selye, 1965, p. 360) and birds (Michels, 1938; Olson, 1959), although mammalian mast cells and basophils do differ in several important characteristics (Cline, 1975, p. 130). With avian cells, although there is much ultrastructural similarity between mast cells (Wight, 1970) and basophils (Maxwell and Trejo, 1970; Maxwell, 1973), particularly in the structure of the cells' specific granules, there are histological and histochemical differences between the two types (Hunt and Hunt, 1959; Wight and Mackenzie, 1970; Carlson and Hacking, 1972).

References

Allen, R. L. (1971). The properties and biosynthesis of the haemoglobins. *In* "Physiology and Biochemistry of the Domestic Fowl" (D. J. Bell and B. M. Freeman, Eds), pp. 873–881. Academic Press, London and New York.

Archer, R. K. (1971). Blood coagulation. *In* "Physiology and Biochemistry of the Domestic Fowl" (D. J. Bell and B. M. Freeman, Eds), pp. 897–911. Academic Press, London and New York.

Atwal, O. S. and McFarland, L. Z. (1966). A morphologic and cytochemical study of erythrocytes and leukocytes of *Coturnix coturnix japonica*, *Am. J. vet. Res.* **27**, 1059–1065.

Balasch, J., Palacios, L., Musquera, S., Palomeque, J., Jimenez, M. and Alemany, M. (1973). Comparative hematological values of several Galliformes, *Poult. Sci.* **52**, 1531–1534.

Barclay, N. E. (1966). Marginal bands in duck and camel erythrocytes, *Anat. Rec.* **154**, 313.

Barrett, L. A. and Dawson, R. B. (1974). Avian erythrocyte development: microtubules and the formation of the disk shape, *Devl Biol.* **36**, 72–81.

Bartsch, P., Ball, W. H., Rosenzweig, W. and Salman, S. (1937). Size of red blood corpuscles and their nucleus in fifty North American birds, *Auk.* **54**, 516–519.

Behnke, O. (1970). Microtubules in disk-shaped blood cells, *Int. Rev. exp. Path.* **9**, 1–92.

Bennett, G. F. and Chisholm, A. E. (1964). Measurements on the blood cells of some wild birds of North America, *Wildl. Dis.* **38**, 1–22.

Blount, W. P. (1939). The blood picture at birth in the chick, *Vet. J.* **95**, 193–195.

Bruns, G. A. P. and Ingram, V. M. (1973). The erythroid cells and haemoglobins of the chick embryo, *Phil. Trans. R. Soc. B* **266**, 225–305.

Burton, A. L. and Higginbotham, R. D. (1966). Response of blood basophils to *Rous sarcoma virus* infections in chicks and its significance, *J. Reticuloendothel. Soc.* **3**, 314–316.

Carlson, H. C. and Allen, J. R. (1969). The acute inflammatory reaction in chicken skin: blood cellular response, *Avian Diseases* **13**, 817–833.

Carlson, H. C. and Hacking, M. A. (1972). Distribution of mast cells in chicken, turkey, pheasant and quail and their differentiation from basophils, *Avian Diseases* **16**, 574–577.

Carlson, H. C., Sweeny, P. R. and Tokaryk, J. M. (1968). Demonstration of phagocytic and trephocytic activities of chicken thrombocytes by microscopy and vital staining techniques, *Avian Diseases* **12**, 700–715.

Carrel, A. and Ebeling, A. H. (1926). The fundamental properties of the fibroblast and macrophage. II. The macrophage, *J. exp. Med.* **44**, 285–305.

Cheville, N. F. and Beard, C. W. (1972). Cytopathology of Newcastle disease. The influence of bursal and thymic lymphoid systems in the chicken, *Lab. Invest.* **27**, 129–143.

Christoph, H.-J. and Borowski, G. (1961). Beiträge zur Hämatologie der Zootiere. IV. Das Blutbild von Greifvogeln (Accipitres) unter besonderer Berücksichtigung einiger in Deutschland noch heimischer kleinerer Arten, *Kleintier-Praxis* **6**, 71–76.

Cleland, J. B. and Johnston, T. H. (1912). Relative dimensions of the red blood cells of vertebrates, especially of birds, *Emu* **11**, 188–197.

Cline, M. J. (1975). "The White Cell." Harvard University Press, Cambridge, Massachusetts.

Coates, V. and March, B. E. (1966). Reticulocyte counts in the chicken, *Poult. Sci.* **45**, 1302–1305.

Cooper, M. D., Peterson, R. D. A., South, M. A. and Good, R. A. (1966). The functions of the thymus system and the bursa system in the chicken, *J. exp. Med.* **123**, 75–102.

Cooper, M. D., Gabrielson, A. E. and Good, R. A. (1967). Role of the thymus and other central lymphoid tissues in immunological disease, *Ann. Rev. Med.* **18**, 113–138.

De Villiers, O. T. (1938). The blood of the Ostrich, *Onderstepoort J. Vet. Sci. Anim. Husb.* **11**, 419–504.

Dehler, A. (1895). Beiträge zur Kenntnis des feineren Baues der roten Blutkörperchen beim Hühnerembryo, *Arch. mikrosk. Anat.* **46**, 414–430. Quoted by Fawcett, (1959).

Dhingra, L. D., Parrish, W. B. and Venzke, W. G. (1969). Electron microscopy of granular leukocytes of chicken (*Gallus domesticus*), *Am. J. Vet. Res.* **30**, 637–642.

Enbergs, H. (1973a). Die Feinstruktur der eosinophilen Granulocyten der Hausente (*Anas platyrhynchos dom.*), *Zbl. Vet. Med. A* **20**, 47–55.

Enbergs, H. (1973b). Die Feinstruktur der Leukozyten der Ente (*Anas platyrhynchos dom.*), *Berl. Münch. Tierärztl. Wschr.* **86**, 285–289.

Enbergs, H. and Beardi, B. (1971). Zur Feinstruktur der eosinophilen Granulocyten der Hausgans (*Anser anser dom.*), *Z. Zellforsch.* **122**, 520–527.

Enbergs, H. and Kriesten, K. (1968a). Zytoplasmatische Feinstrukturen der Thrombozyten des Haushuhns, *Experientia* **24**, 579–598.

Enbergs, H. and Kriesten, K. (1968b). Die weissen Blutzellen des Haushuhns im elektronenmikroskopischen Bild. *Dt. Tierärztl. Wschr.* **75**, 271–275.

Enbergs, H. and Kriesten, K. (1969). Zur Feinstruktur der Blutmonozyten des Haushuhns, *Z. Zellforsch.* **97**, 377–382.

Enbergs, H. and Kriesten, K. (1970). Zur zytoplasmatischen Feinstruktur der eosinophilen Granulozyten des Hühner bluts, *Zbl. Vet. Med. A*, **17**, 430–439.

Ericsson, J. L. E. and Nair, M. K. (1973). Electron microscopic demonstration of acid phosphatase activity in the developing and mature heterophils of the chicken, *Histochemie* **37**, 97–105.

Fawcett, D. W. (1959). Electron microscopic observations on the marginal band of nucleated erythrocytes, *Anat. Rec.* **133**, 379.

Gilbert, A. B. (1965). Sex differences in the erythrocyte of the adult domestic fowl, *Res. vet. Sci.* **6**, 114–116.

Glick, B., Sato, K. and Cohenour, F. (1964). Comparison of the phagocytic ability of normal and bursectomized birds, *J. Reticuloendothel. Soc.* **1**, 442–449.

Gross, W. B. (1961). Blood cultures, blood counts and temperature records in an experimentally produced air sac disease and uncomplicated *Escherichia coli* infection of chickens, *Poult. Sci.* **41**, 691–700.

Gulliver, G. (1846). Quoted by Hartman and Lessler (1963).

Harris, J. R. (1971). The ultrastructure of the erythrocyte. *In* "Physiology and Biochemistry of the Domestic Fowl" (D. J. Bell and B. M. Freeman, Eds), pp. 853–862. Academic Press, London and New York.

Harris, J. R. and Brown, J. N. (1971). Fractionation of the avian erythrocyte: an ultrastructural study, *J. Ultrastruct. Res.* **36**, 8–23.

Hartman, F. A. and Lessler, M. A. (1963). Erythrocyte measurements in birds, *Auk.* **80**, 467–473.

Hirsch, J. G. (1962). Cinemicrophotographic observations on granule lysis in polymorphonuclear leucocytes during phagocytosis, *J. exp. Med.* **116**, 827–833.

Hodges, R. D. (1974). "The Histology of the Fowl." Academic Press, London and New York.

Hodges, R. D. (1977). Normal avian (poultry) haematology. *In* "Comparative Clinical Haematology" (R. K. Archer and L. B. Jeffcott, Eds). Blackwell, Oxford.

Hunt, T. E. and Hunt, E. A. (1959). Blood basophils of cockerels before and after intravenous injection of compound 48/80, *Anat. Rec.* **133**, 19–33.

Jones, J. D. (1972). "Comparative Physiology of Respiration." Edward Arnold, London.

Kelényi, G. and Németh, Á. (1969). Comparative histochemistry and electron microscopy of the eosinophil leucocytes of vertebrates, *Acta biol. Acad. Sci. hung.* **20**, 405–422.

Kelly, J. W. and Dearstyne, R. S. (1935). Haematology of the fowl. A. Studies on normal chick and normal adult blood, *N. Carol. Agric. Exp. Sta. Tech. Bull.*, No. 50.

Kincade, P. W., Lawton, A. R. and Cooper, M. D. (1971). Restriction of surface immunoglobulin determinants to lymphocytes of the plasma cell line, *J. Immunol.* **106**, 1421–1423.

Kriesten, K. and Enbergs, H. (1970). Elektronenmikroskopische Untersuchungen zur Feinstruktur der basophilen Granulozyten des Haushuhns, *Blut* **20**, 282–287.

Kuruma, I., Okada, T., Kataoka, K. and Sorimachi, M. (1970). Ultrastructural observation of 5-hydroxytryptamine-storing granules in the domestic fowl thrombocytes, *Z. Zellforsch.* **108**, 268–281.

Lucas, A. M. and Jamroz, C. (1961). Atlas of avian hematology. Agriculture Monograph No. 25. U.S. Department of Agriculture, Washington.

Lutz, P. L., Longmuir, I. S., Tuttle, J. V. and Schmidt-Nielsen, K. (1973). Dissociation curve of bird blood and effect of red cell oxygen consumption, *Resp. Physiol.* **17**, 269–275.

Lutz, P. L., Longmuir, I. S. and Schmidt-Nielsen, K. (1974). Oxygen affinity of bird blood, *Resp. Physiol.* **20**, 325–330.

Macrae, E. K. and Spitznagel, J. K. (1975). Ultrastructural localization of cationic proteins in cytoplasmic granules of chicken and rabbit polymorphonuclear leukocytes, *J. Cell Sci.* **17**, 79–94.

Magath, T. B. and Higgins, G. M. (1934). The blood of the normal duck, *Folia Haematol.* **51**, 230–241.

Maxwell, M. H. (1973). Comparison of heterophil and basophil ultrastructure in six species of domestic birds, *J. Anat.* **115**, 187–202.

Maxwell, M. H. (1974). An ultrastructural comparison of the mononuclear leucocytes and thrombocytes in six species of domestic bird, *J. Anat.* **117**, 69–80.

Maxwell, M. H. and Siller, W. G. (1972). The ultrastructural characteristics of the eosinophil granules in six species of domestic bird, *J. Anat.* **112**, 289–303.

Maxwell, M. H. and Trejo, F. (1970). The ultrastructure of white blood cells and thrombocytes of the domestic fowl, *Br. vet. J.* **126**, 583–592.

Meves, F. (1911). Quoted by Behnke (1970).

Michels, N. A. (1938). The mast cells. *In* "Handbook of Hematology" (H. Downey, Ed.), Vol. 1, 231–372. Haffner Publishing Co., New York. (Reprinted in facsimile, 1965.)

Nair, M. K. (1973). The early inflammatory reaction in the fowl. A light microscopical, ultrastructural and autoradiographic study, *Acta vet. Scand.* Suppl. **42**, 1–103.

Natt, M. P. and Herrick, C. A. (1954). Variations in the shape of the rod-like granule of the chicken heterophil leucocyte and its possible significance, *Poult. Sci.* **33**, 828–830.

Nichols, B. A. and Bainton, D. F. (1975). Ultrastructure of mononuclear phagocytes. *In* "Mononuclear Phagocytes in Immunity, Infection and Pathology" (R. van Furth, Ed.), pp. 17–55. Blackwell, Oxford.

Nichols, B. A., Bainton, D. F. and Farquhar, M. G. (1971). Differentiation of monocytes. Origin, nature and fate of their azurophil granules, *J. Cell. Biol.* **50**, 498–515.

Olson, C. (1959). Avian hematology. *In* "Diseases of Poultry" (H. E. Biester and L. H. Schwarte, Eds), 4th edn, pp. 53–69. The Iowa State University Press, Ames, Iowa.

Osculati, F. (1970). Fine structural localization of acid phosphatase and arylsulfatase in the chick heterophil leucocytes, *Z. Zellforsch.* **109**, 398–406.

Schmidt-Nielsen, K. (1975). "Animal Physiology. Adaptation and Environment." Cambridge University Press, London.

Schumacher, A. (1965). Zur submikroskopischen Struktur der Thrombozyten, Lymphozyten und Monozyten des Haushuhnes (*Gallus domesticus*), *Z. Zellforsch.* **66**, 219–232.

Selye, H. (1965). "The Mast Cells." Butterworths, Washington.

Simpson, C. F. (1968). Ultrastructural features of the turkey thrombocyte and lymphocyte, *Poult. Sci.* **47**, 848–850.

Stalsberg, H. and Prydz, H. (1963). Studies on chick embryo thrombocytes. II. Function in primary hemostasis, *Thromb. Diath. Haemorrhag.* **9**, 291.

Sterz, I. and Weiss, E. (1973). Elektronenmikroskopische Untersuchungen zur Phagozytose und Vermehrung des Virus der Klassischen Geflügelpest (KP) in Thrombozyten infizierter Hühner, *Zbl. Vet. Med. B*, **20**, 613–621.

Sturkie, P. D. (1965). "Avian Physiology." 2nd edn. Balliere, Tindall and Cassell, London.

Sturkie, P. D. (1976). "Avian Physiology." 3rd edn. Springer-Verlag, New York.

Sutton, J. S. and Weiss, L. (1966). Transformation of monocytes in tissue culture into macrophages, epithelioid cells and multinucleated giant cells: an electron microscopic study, *J. Cell Biol.* **28**, 303–332.

Sweeny, P. R. and Carlson, H. C. (1968). Electron microscopy and histochemical demonstration of lysosomal structures in chicken thrombocytes, *Avian Diseases*, **12**, 636–644.

Topp, R. C. and Carlson, H. C. (1972a). Studies on avian heterophils. II. Histochemistry, *Avian Diseases* **16**, 369–373.

Topp, R. C. and Carlson, H. C. (1972b). Studies on avian heterophils. III. Phagocytic properties, *Avian Diseases* **16**, 374–380.

Vandecasserie, C., Paul, C., Schek, A. G. and Leonis, J. (1973). Oxygen affinity of avian haemoglobins, *Comp. Biochem. Physiol.* **44A**, 711–718.

Venzlaff, W. (1911). Über Genesis und Morphologie der roten Blutkörperchen der Vögel, *Arch. mikrosk. Anat.* **77**, 377–432.

Weiss, L. P. and Fawcett, D. W. (1953). Cytochemical observations on chicken monocytes, macrophages and giant cells in tissue culture, *J. Histochem. Cytochem.* **1**, 47–65.

Wells, R. M. G. (1976). The oxygen affinity of chicken haemoglobin in whole blood and erythrocyte suspensions, *Resp. Physiol.* **27**, 21–31.

Wight, P. A. L. (1970). The mast cells of *Gallus domesticus*. I. Distribution and ultrastructure, *Acta Anat.* **75**, 100–113.

Wight, P. A. L. and Mackenzie, G. M. (1970). The mast cells of *Gallus domesticus*. II. Histochemistry, *Acta Anat.* **75**, 263–275.

Wolf-Jürgensen, P. (1968). The basophilic leukocyte. *Ser. Haemet.*, **1**, 45–68.

7

The autonomic nervous system

A. R. AKESTER

Sub-Department of Veterinary Anatomy, University of Cambridge, England

---

CONTENTS

I. Introduction 382

II. Centres of autonomic activity in the brain 384
A. Feeding centres 385
B. Respiratory centres 386
C. Cardiovascular centres 386

III. Sympathetic outflow from the central nervous system 387

IV. The paravertebral chain of sympathetic ganglia 387

V. Parasympathetic outflow from the central nervous system (craniosacral) . 388

VI. Sympathetic supply to the head (cranial cervical ganglion) 388

VII. Sympathetic supply to the neck 391

VIII. Parasympathetic supply to the head and neck 391
A. Ciliary, sphenopalatine and ethmoidal ganglia 391
B. Glossopharyngeal, vagus and accessory nerves 398

IX. Autonomic nerves to the viscera of the body cavity 399
A. The cardiac nerves 400
B. Rotation—compensating reflexes 404
C. Aortic, coeliac and other visceral ganglia and plexuses 404
D. The intestinal ganglionated nerve trunk (Remak's nerve) 405
E. The pelvic plexus 407

X. Nature of a sympathetic ganglion 408
A. The cell population 408
B. Communication between cells 413
C. Cell movements 416
D. Myelination 417

XI. The postganglionic sympathetic neurone. Release and uptake of noradrenaline 421

XII. Synthesis of noradrenaline 423

XIII. Axonal flow. 423

XIV. Recognition of noradrenaline in postganglionic sympathetic neurones . . 425
A. The paraformaldehyde-induced fluorescence of catecholamines and indolamines 425
B. Characteristics of autonomic transmitter vesicles 426
C. Autoradiography of noradrenergic terminals 430
D. Uptake of 5- and 6-hydroxydopamine (5- and 6-OHDA) by noradrenergic terminals 432
E. Dopamine-β-hydroxylase immunofluorescence 433

XV. Modulation of noradrenergic neurotransmission 434

References 436

I. Introduction

The autonomic nervous system is that part of the overall nervous system of the body which controls such things as: the patency of blood vessels, the opening and closing of vascular shunts, the rate and force of the heart beat, the activity of smooth muscle throughout the entire length of the alimentary canal, the secretion from many glands (salivary, pancreatic, intestinal, lacrimal etc.), the activity of their secretory ducts as well as the vas deferens, the ureter and the oviduct; the smooth muscle of the trachea and bronchi and certain striated muscles inside the eye which control accommodation and the size of the pupil. Every individual feather can be moved by tiny smooth muscles and these are under autonomic nervous control (Langley, 1904).

Reproduction, moulting, migration, thermoregulation, nesting, diving, flight and even the first breath an unhatched chick takes inside its egg are all concerned in one way or another with the autonomic nervous system. Indeed it would be difficult to think of any aspect of a bird's normal life with which the autonomic nervous system is not associated.

The whole concept of an autonomic nervous system can be misleading. Its activity in controlling gut motility and respiration is no more autonomous than is the somatic nervous coordination of skeletal muscles which enables a stork to sleep whilst balancing on one leg. Perhaps we should emphasize the interdependence of "autonomic" and "somatic" activity, one of the most dramatic illustrations of which must be that of courtship and display in birds during the breeding season. An interaction between somatic sensory nerves and autonomic sympathetic motor neurones has been demonstrated experimentally by Leonard and Cohen (1975), who caused tachycardia in pigeons by stimulating the sciatic nerve near the hip. This somatosympathetic reflex could be spinally mediated, as it still occurred after high spinal transection but was abolished by cutting the cardioaccelerator nerve, thus proving that it was not due to a release of adrenal catecholamines. However, the precise route by which sciatic sensory fibres are linked to preganglionic sympathetic neurones at the level of the first thoracic vertebra is not known.

The term "autonomic" probably originated in the mistaken belief by Bichat (1801–3) and others that those nerves which controlled the viscera of the body were autonomous, that is independent of the nervous system which was concerned with the soma, that is, the musculo-skeletal system and skin. It is, of course, now well known that all parts of the nervous system can be influenced by the brain and several specific regions are well established as centres for the control of autonomic activity.

In a rather loose way the term "autonomic" has become synonymous with visceral, vegetative and involuntary, whilst the term "somatic" implies skeletal muscular, or voluntary. This is inappropriate even in humans where a great many skeletal muscular adjustments take place in an involuntary or automatic reflex way. In birds the concept of voluntary and involuntary nervous systems is even less appropriate.

The autonomic nervous system is classically divided into a sympathetic division (the fibres of which leave the spinal cord over the restricted thoraco-lumbar region), and a parasympathetic division (the fibres of which leave the spinal cord over the restricted craniosacral region). Preganglionic sympathetic fibres (which speed the heart, slow the gut and are vasomotor to blood vessels) synapse with postganglionic neurones in ganglia outside the central nervous system, at a considerable distance from the target tissue. Most preganglionic fibres are therefore relatively short and postganglionic fibres relatively long. The former are cholinergic, the latter are noradrenergic and the ratio of pre- to postganglionic fibres is something like 1:15 or 20.

On the other hand parasympathetic fibres (which slow the heart, speed the gut and are secretomotor to many glands) have long pre- and short post-ganglionic fibres, both of which are cholinergic. The ratio of one to the other is small, 1:1 or 2. Both types of preganglionic fibre are considered to be myelinated whilst the postganglionic fibres are said to be unmyelinated. However, preganglionic axons lose their myelin before they synapse and at least in some cases the postganglionic axons are known to be myelinated.

The first recorded recognition of the sympathetic trunks (in mammals) is by Galen (A.D. 130–200) who is also credited with the idea that there was a "sympathy" between all parts of the body and that this was organized by nerves. Another long-established view is that the autonomic nervous system (sympathetic and parasympathetic) is a purely motor (efferent) system and that, by definition, there are no autonomic afferent or sensory nerves. This is obviously a very restricted and functionally untenable concept and it is now generally accepted that sensory fibres associated with the gut, the heart, lungs etc. should be regarded as autonomic afferents. However, they are not normally divided into sympathetic and parasympathetic afferents.

A recent development in the complicated story of certain autonomic sensory nerves is the discovery of a reciprocal innervation between the terminals of these nerves and the sensory cells with which they are associated. This type of mutual stimulation has been demonstrated in the carotid body of the duck (Butler and Osborne, 1975) and of the domestic fowl (King *et al.*, 1975). In this organ the glomus (Type 1) cells are the chemoreceptors and according to the classical view of De Castro (1951) they respond to a reduction in the oxygen tension of blood flowing through the organ.

Dopaminergic synapses (transmission from glomus cell to nerve terminal) as well as cholinergic synapses (transmission from sensory nerve terminal to glomus cell) have been found side by side on the cell membranes of a single glomus cell and its associated axonal ending. A scheme by which this reciprocal stimulation may operate has been proposed by Butler and Osborne (1975). They suggested that, under conditions of normal oxygen tension, there is a steady release of dopamine by the glomus cells, which inhibits the discharge of impulses by the sensory nerves. When the level of oxygen in the blood falls, this release of dopamine is reduced and the discharge of impulses in the sensory nerves increases. In addition to this, the removal of dopamine inhibition of the sensory nerve terminals cause their (cholinergic) stimulation of the glomus cells. This results in a further reduction of dopamine release and an even greater discharge frequency in the sensory nerves.

The relatively simple scheme of a thoracolumbar (sympathetic) outflow with ganglia remote from the innervated organ and a craniosacral (parasympathetic) outflow with ganglia close to or even in the innervated organ is becoming progressively less acceptable; so also is the generalization that most organs receive a double innervation (sympathetic and parasympathetic) and that one tends to have the opposite effect of the other. This has led to the suggestion that the terms sympathetic and parasympathetic should be discarded altogether (Bennett, 1974). There may be some merit in this approach, but it is the opinion of this author that there is still sufficient anatomical and physiological accuracy in the traditional view to justify its retention.

It must be emphasized that although a significant amount of experimental work has been done in connection with the avian autonomic nervous system, much remains obscure. Communicating rami are described between ganglia and nerve trunks, and between one nerve and another, but the types of fibres contained within these rami, their destination, and the information they carry, are all too often unknown. A correlation of ultrastructural observations on the fibre content of these communications with unit recordings from individual axons inside them is long overdue.

All recent accounts of the avian autonomic nervous system have drawn heavily upon the remarkable series of dissections by Hsieh (1951) and this one is no exception. A useful historical review of the subject, although related to mammals, is given by Mitchell (1953) and there are probably more broad similarities between the autonomic nervous systems of the two groups of homeotherms than there are differences.

II. Centres of autonomic activity in the brain

Many regions of the brain are concerned with influencing autonomic activity, and it is here that the interdependence of "somatic" and "visceral" activity is most closely coordinated. It is also here that the artificiality of splitting off one large part of the nervous system from the rest is most apparent.

Nevertheless there are regions of the brain that appear to have a specific influence over such things as waking and sleeping, eating and drinking, heart rate and respiration, movement of the alimentary canal and thermoregulation, and these are generally considered to be "autonomic" activities.

A. Feeding centres

There appears to be a well established feeding centre in the hypothalamus in mammals and particularly in the rat (Anand, 1961). However, in spite of many experiments, the existence of a comparable centre in birds is at the moment controversial. Experiments may be either by electrical stimulation, which ideally would initiate feeding behaviour, or by placing electrical lesions in selected parts of the brain, which would abolish this behaviour. Claims that a "feeding centre" exists in birds have been made by Akerman *et al.* (1960) in the pigeon, Phillips (1964) in the Mallard and by Harwood and Vowles (1966) in the Ringed Turtle Dove (*Streptopelia risoria*).

On the other hand, a comprehensive survey of the effects of stimulating many different sites in the brain of Ringed Turtle Doves by Wright (1969) failed to demonstrate any clear-cut response. Negative results have also been reported by Phillips and Youngren (1971) in chickens and ducks, Putkonen (1967) in chickens and by Delius (1967, 1971) in Lesser Black-backed and Herring Gulls.

In this type of experimental work it is important to remember that "feeding behaviour" or "feeding movements" generally refer to pecking by chickens and to searching or shovelling by ducks. The ability to manipulate food with the beak and to swallow it is quite distinct, and although one would be rather pointless without the other in the normal animal, under experimental conditions these activities can be separated. In this way pigeons, with lesions in the main sensory nucleus of the fifth cranial nerve, were able to peck at food but were unable to manipulate it with the beak or transfer it to the pharynx for swallowing. However, if food were placed in the pharynx then it would be swallowed normally (Zeigler *et al.*, 1969). A useful survey of the neural control of food and water intake in birds is given by Wright (1973).

It seems likely that there are two regions in the hypothalamus involved in the control of normal feeding in mammals. Destruction of the ventro-medial hypothalamus (immediately below the floor of the III ventricle) causes overeating and increased body weight, whilst damage to the lateral hypothalamus prevents feeding and leads to starvation and death of the animal if not force-fed (Anand, 1961; Brobeck, 1946; Sclafani, 1971).

Evidence that a similar system exists in the White-throated Sparrow (*Zonotrichia albicollis*) has been presented by Kuenzel (1972). He chose a migratory bird because it normally eats more in the spring and autumn, when it lays down fat in preparation for its migratory flight. Kuenzel suggested that the hypothalamic-controlled balance between over and under-eating is tipped to favour overeating in preparation for migration.

B. Respiratory centres

Centres in the brain which influence respiration can be found in the medulla oblongata and have been reported by Cohen and Schnall (1970) in the pigeon, and by Richards (1971) in pigeons and chickens; whilst centres in the forebrain and thalamus have been described by Vediaev (1964) in doves and by Kotilainen and Putkonen (1974) in chickens.

Kotilainen and Putkonen claimed that complete and partial respiratory inhibition could be caused by stimulating a rostrolateral region in the thalamus, in front of and around the nucleus rotundus, although this prominent thalamic nucleus had no effect on respiration when stimulated itself. Stimulation in all parts of the quinto-frontal tract, as well as medial and lateral areas in the mesencephalon also caused apnoea.

Apnoea caused by electrical stimulation of the lateral thalamus was associated with bradycardia and hypotension, whilst that caused by stimulation of the quinto-frontal tract was associated with bradycardia and hypertension. Hyperpnoea resulted from stimulation in several places, including septal and pre-optic areas, medial archistriatum, the entire length of the occipito-mesencephalic tract and much of the mesencephalic lateral reticular formation.

The importance of CO_2 sensitive chemoreceptors (avian pulmonary chemoreceptors) in the bird's lung has been demonstrated by Fedde *et al.* (1974) in the duck and by Burger *et al.* (1974) in the chicken. Autonomic afferents in the vagus greatly increased their discharge rate in response to the removal of CO_2 from the perfused gas (Osborne and Burger, 1974; Leitner and Roumy, 1974). This makes an interesting comparison with CO_2 sensitive chemoreceptors in both the avian and the mammalian carotid body which respond to an increased level of circulating CO_2.

The interdependence of respiratory and cardiovascular control by the autonomic nervous system is important at all times, but it is amongst diving birds (and mammals) that it becomes most impressive. The "diving reflex" (Jones and Johansen, 1972) seems to be initiated by apnoea which may result from impulses reaching a respiratory inhibition centre in the lateral thalamus. These impulses probably travel in sensory fibres of the glossopharyngeal nerve and are associated with water receptors around the glottis. Following a reflex apnoea cardiovascular adjustments result in bradycardia and an intense vasoconstriction to practically all parts of the body except the brain, heart and certain endocrine glands.

C. Cardiovascular centres

Compared with mammals relatively little work on this topic has been done in birds (Pearson, 1972). However, an extensive and systematic study in the pigeon has recently been carried out by MacDonald and Cohen (1973) using stereotaxic stimulation whilst monitoring heart rate, arterial blood pressure and respiration. They found that many regions of the brain could influence heart rate and arterial blood pressure. Short latency tachycardia, hypertension and hyperpnoea were caused by electrical stimulation of the archistriatum, its principal outflow tract (the occipito-mesencephalic tract) and the hypothalamus.

Many midbrain sites also caused tachycardia, hypertension and hyperpnoea. These included the lateral reticular formation, ventrolateral tegmentum and several others. In the pons, tachycardia and hypertension resulted from stimulation of a ventrolateral tegmental region, the internal cerebellar nucleus (MacDonald and Cohen, 1973) and its principal outflow tract (the uncinate fasciculus).

In the medulla, tachycardia was caused by stimulation of the dorsal aspect of the medial longitudinal fasciculus, the medial aspect of the dorsal motor nucleus of the vagus and the central medullary region lateral to the VI nerve nucleus. This last site may correspond to the vasomotor centre in mammals. Regions causing bradycardia included the lateral aspect of the dorsal motor nucleus of the vagus just rostral to the obex, vagal rootlets, the solitary complex and the commisural nucleus of Cajal.

There are, of course, other autonomic centres in the brain. The hypothalamus is possibly the most obvious with its well known control over the pituitary gland. The pineal gland is receiving a considerable amount of attention at the moment largely because of its influence over reproduction.

III. Sympathetic outflow from the central nervous system

Preganglionic sympathetic fibres leave the spinal cord in the ventral roots of spinal nerves from the last cervical to the first or second synsacral segment (MacDonald and Cohen, 1970). There is thus a thoracolumbar outflow similar to its counterpart in mammals. The cell bodies of these axons are localized in two columns (of Terni) in the central part of the grey matter, one each side of the central canal. There is thus no mediolateral column as is found in mammals. There is also said to be a direct communication from the medulla oblongata to the interganglionic trunk close to the cranial cervical ganglion (Hsieh, 1951).

IV. The paravertebral chain of sympathetic ganglia

With the exception of the first and second cervical segments, birds have a chain of sympathetic ganglia, one for each of the spinal nerves, running on both sides of the vertebral column from the back of the skull to the base of the tail. Here the two chains unite in the midline, but unlike the mammal there is no midline or impar ganglion. The most rostral is the cranial cervical ganglion which probably represents the fusion of the first two cervical ganglia. The seventh and eighth synsacral ganglia may also be fused.

There are about 37 ganglia in each paravertebral chain in the domestic fowl. However, there is considerable variation between different species, some variation between individuals of the same species and sometimes small differences between left and right trunks of the same animal (Hsieh, 1951). The size of these ganglia is roughly proportional to the extent of the region to which their postganglionic fibres are distributed. The cranial cervical ganglion is the largest (about 2·5 mm) as it contains the cell bodies of the great majority of the postganglionic sympathetic fibres to the head. In the domestic fowl those ganglia which are associated with the spinal nerves of the brachial (C12–C16) and synsacral (S/S1-S/S8) limb plexuses, together with those thoracic ganglia which connect directly to the splanchnic nerves, are next in size (2 mm). The rest of the cervical ganglia are smaller still, and the smallest of all are those in the lower synsacral and coccygeal regions.

The proximity of the sympathetic ganglion to the dorsal root ganglion of the same segment varies. In the thoracic and upper lumbar regions of the domestic fowl and possibly other species, the two types of ganglia are fused together, whereas in the cervical and lower lumbar regions the sympathetic ganglia are attached either directly or by a communicating ramus to the medial side of the ventral division of the spinal nerve.

There has long been speculation as to whether any preganglionic sympathetic fibres leave the spinal cord by way of the dorsal roots in mammals (von Lenhossék, 1911; Mitchell, 1953), and it may be pure coincidence that the fusion of sympathetic and dorsal root ganglia in birds occurs mainly over that region of the cord from which there is a sympathetic outflow. On the other hand there is at least the possibility that preganglionic sympathetic axons may leave the cord by the dorsal root and pass through the dorsal root ganglion to reach the sympathetic part of the fused ganglion.

V. Parasympathetic outflow from the central nervous system (craniosacral)

Cell bodies of preganglionic parasympathetic neurones are associated with the nuclei of origin of cranial nerves III, VII, IX and X. Their axons leave the brain within these cranial nerves. At the sacral end of the spinal cord similar neurones with their cell bodies in the central part of the grey matter have axons which leave the spinal cord in the ventral roots of the more caudal spinal synsacral nerves.

VI. Sympathetic supply to the head (cranial cervical ganglion)

Sympathetic nerves are probably distributed to all blood vessels in the head and neck (arteries, veins, but not capillaries) and are responsible for determining the level of vasomotor tone. They probably also innervate the smooth muscles in the walls of the larger lymph ducts, and cause relaxation of the tracheal, extrapulmonary bronchial, oesophageal and crop smooth muscle. This is confirmed by the inhibitory effect of adrenaline on the upper oesophagus and crop. On the other hand adrenaline is said to have the opposite effect on the lower oesophagus and gizzard (Hassan, 1966; Everett, 1967) which may be explained by a large concentration of β-(inhibitory) adrenergic receptors in the upper oesophagus and of α-(excitatory) adrenergic receptors in the lower oesophagus and gizzard. (See Jenkinson, 1973 and Haber and Wrenn, 1976, for a discussion of adrenergic receptors.)

The cervical part of the paravertebral chain of sympathetic ganglia (*truncus paravertebralis cervicalis*) extends forwards from the thoracic region, passing through the transverse foramina of the cervical vertebrae accompanied by the vertebral artery and vein. Spindle-shaped sympathetic ganglia are attached to the ventral divisions of all the cervical spinal nerves except the first and second. They are situated just outside the intervertebral foramen and are connected together by a single interganglionic trunk (*connexus interganglionaris*). Right and left chains are generally considered to stop at the cranial cervical ganglion on each side.

The cranial cervical ganglion is one of the largest sympathetic ganglia in the body. It is about 2·5 mm long in the domestic fowl, roughly cone-shaped, and lies behind the external auditory meatus, close to the internal carotid artery just before this artery divides into cerebral carotid and external ophthalmic arteries. This ganglion is the last main site for preganglionic axons, from the thoracic part of the spinal cord, to synapse with the cell bodies of postganglionic neurones. From here postganglionic sympathetic axons are distributed to all parts of the head and to many parts of the neck.

The main direct connections of the cranial cervical ganglion are summarized in Fig. 7.1. They include the following: interganglionic cord from the paravertebral chain, hyomandibular branch of the facial nerve, main trunk of the vagus nerve, main trunk of the hypoglossal nerve, vascular branches to the internal carotid, external ophthalmic, cerebral carotid and occipital arteries (Hsieh, 1951).

In addition to these connections the cranial cervical ganglion is attached to the petrosal ganglion of the glossopharyngeal nerve at which point there may well be an exchange of axons. Although these connections are reasonably well established it is not necessarily clear in every case what types of axons are contained within them, nor in which direction the impulses pass. For example, whilst it is obvious that the vascular nerves are conducting sympathetic vasomotor activity away from the ganglion, the links with cranial nerves VII, IX, X and XII could contain postganglionic nerves leaving the ganglion for distribution with the cranial nerves; parasympathetic preganglionic axons entering the ganglion for distribution along the vascular plexuses; sensory fibres passing in either direction, or a combination of all three of these possibilities.

Whilst sympathetic nerves are vasomotor to most glands in the head it is possible that they may also have a direct influence on the enzymes responsible for synthesizing the pineal hormone, melatonin (Axelrod *et al.*, 1968; Klein *et al.*, 1970). Although the precise role of the pineal in avian reproduction is far from clear it seems that, in the fowl at least, the pineal produces a substance (melatonin) which delays gonad development (Wight, 1971). The pineal receives a rich innervation of (mostly) unmyelinated postganglionic sympathetic nerves whose cell bodies are in the cranial cervical ganglion. Removal of this ganglion is performed for various experimental purposes and an account of the technique is given by Lauber *et al.* (1972).

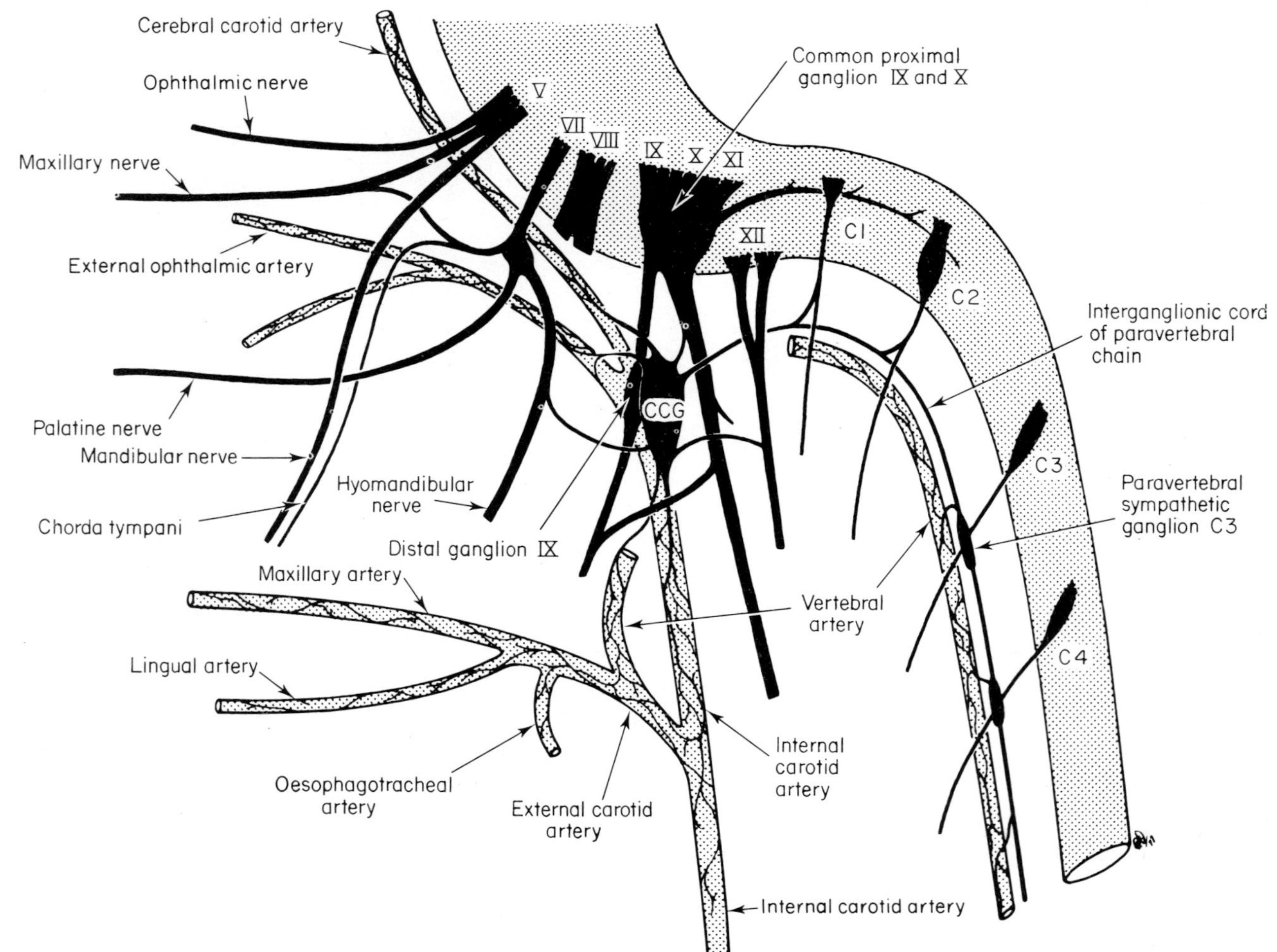

Fig. 7.1. Cranial cervical ganglion (CCG) showing its principal connections. *Gallus*.

VII. Sympathetic supply to the neck

The blood vessels of the neck, like all other parts of the body, are surrounded by a vasomotor plexus of sympathetic fibres and it is likely that the sympathetic (noradrenergic) supply to the smooth muscle of the oesophagus, crop and trachea is also derived from these vascular plexuses. There is said to be only a small amount of smooth muscle associated with the trachea, which receives a sparse sympathetic supply (Bennett and Malmfors, 1970).

VIII. Parasympathetic supply to the head and neck

Parasympathetic nerves are secretomotor, and possibly vasodilator, to the salivary, lacrimal, nasal (salt) glands; to the gland of the nictitating membrane (Harderian), and to many small scattered mucous glands in the lining of the nasal, buccal and pharyngeal cavities. They are stimulatory to oesophageal and crop peristaltic contractions and also to the intrinsic muscles of the eye.

We have seen that preganglionic parasympathetic fibres leave the brain with cranial nerves III, VII, IX and X. They either synapse with postganglionic neurones in one or other of the following ganglia: ciliary, sphenopalatine and ethmoidal; or they synapse in small unnamed ganglia which are scattered along the nerve trunks. Figure 7.3 gives the approximate position of the principal ganglia with their connections.

A. Ciliary, sphenopalatine and ethmoidal ganglia

The ciliary ganglion is a spindle-shaped swelling about 2 mm long in the fowl, lying on, or attached by a short branch to, the oculomotor nerve. It lies in the caudal part of the orbital cavity, between the ventral and lateral rectus muscles on the lateral side of the optic nerve. The cell bodies of preganglionic neurones which pass out with the oculomotor nerve and whose axons synapse in the ciliary ganglion, are located in the accessory part of the oculomotor nucleus. This was demonstrated by Narayanan and Narayanan (1976) who studied the retrograde degeneration that followed unilateral ciliary ganglionectomy in the day-old chick. After nine days they found almost complete degeneration of neurones in the accessory oculomotor nucleus, but none in other parts of the oculomotor nucleus. The accessory part of the oculomotor nucleus in the bird is thus homologous with the Edinger-Westphal nucleus in mammals.

An important role of the ciliary ganglion and of the autonomic nerves associated with it, is to maintain a variable degree of pupil constriction. Removal of the ganglion caused immediate and permanent ipsilateral pupil dilatation, after which the pupil was unable to constrict in response to a very bright light (Isomura, 1974).

The ciliary ganglion contains two quite distinct populations of neurones. One group (larger cell bodies) innervates the striated radial muscles of accommodation (the anterior and posterior corneal muscles) and the striated pupil constrictor muscles of the iris, and have been called "ciliary" neurones (their axons presumably travelling in the iridociliary nerve). The other group (smaller cell bodies) innervates the smooth muscle of the choroid and have been called "choroid" neurones (presumably passing via the choroidal nerves). There are about 2000 of each type and they are confined to different parts of the ganglion, the former to the proximal part and the latter to the distal part.

Sympathetic postganglionic axons, originating in the cranial cervical ganglion, pass by way of the external ophthalmic plexus, the dorsal ramus of the palatine (Vidian) nerve, and thence to the ophthalmic nerve, finally reaching the ciliary ganglion by the connection of the ophthalmic nerve to the ciliary ganglion. Alternatively they may pass directly to the ganglion from the palatine nerve. They may pass straight through the ganglion, become associated with post-ganglionic parasympathetic axons originating in it and together constitute the iridociliary and choroidal nerves (previously known as the long and short ciliary nerves respectively). The iridociliary and choroidal nerves penetrate the sclera of the eyeball and are distributed between the sclera and the choroid to reach the muscles of the ciliary body, iris and choroid (see Chapter 7, Vol. 2, by Bubien-Waluszewska for further details). Sensory fibres (pain and proprioception) which originate in the eye also travel in the iridociliary and choroidal nerves. They may bypass the ciliary ganglion or pass through it to join the ophthalmic branch of the fifth cranial nerve. Two types of myelinated axons have been demonstrated in the iris of the fowl (Saari *et al.*, 1974). Thick myelinated fibres were said to be motor to the striated sphincter muscle and thinner myelinated fibres in the anterior part of the iris were believed to be sensory. The dilator muscle of the pupil was innervated by unmyelinated nerves.

The dominant role and the remarkable efficiency of sight in many birds, particularly birds of prey, has made it a popular subject for investigation. The avian pupillary constrictor response to a bright light is very fast indeed and, in the pigeon, is said to be completed (in about 100 ms) before the pupil of a human eye would even begin to respond to the same light stimulus (Gundlach, 1934).

With this in mind it is interesting to note the distinctive features of the neurones in the ciliary ganglia which control these dramatic responses. In the late embryo and young chick the preganglionic axon terminates in the form of an expanded calyx around the cell body and axon hillock of the postganglionic neurone (de Lorenzo, 1960). Within the first two weeks after hatching, this calyx changes into a large number of terminal swellings which contact the axon hillock and neighbouring part of the cell body, as well as about the first 30 μm of the axon itself. There is said to be a 1 : 1 relationship between the preganglionic axons and the "ciliary" neurones in the ganglion (i.e. those forming the iridociliary nerve) (Lenhossék, 1911; Marwitt *et al.*, 1971).

Transmission across this synapse is considered to be both chemical and electrotonic (Landmesser and Pilar, 1972). The former is by acetylcholine whilst the latter is facilitated by the extremely large surface area of synaptic contact, estimated at 1600 μm^2 by Cantino and Mugnaini (1975). Electrotonic transmission is also helped by a number of low resistance sites at synaptic contacts around the axon hillock, and by a layer of myelin which covers the nuclear pole of the cell body. The myelin is on the side of the cell body opposite to that where the synaptic contacts are established. Presumably the myelin increases the electrical resistance in this non-synaptic part of the cell and helps to concentrate synaptic current in the region of the axon hillock. The fact that *in vivo* perfusion by tyrode solution, which disrupts myelin, has the effect of decreasing the electrical coupling, supports this view (Cantino and Mugnaini, 1975). On the other hand, transmission across the "choroid" neurones (i.e. those forming the choroidal nerve) is only chemical, although it may be influenced by both acetylcholine and noradrenaline.

There is evidence that some of the sympathetic (noradrenergic) fibres which enter the ciliary ganglion terminate there, in close association with the "choroid", but not with the "ciliary" neurones. These noradrenergic varicose terminals form basket-like meshes around the "choroid" neurones. Although they do not actually form a synaptic contact, the noradrenergic varicose terminals are close enough to the cell bodies for Cantino and Mugnaini to suggest that they may have a modulating influence on the preganglionic cholinergic terminals.

Pilar *et al.* (1973) have calculated that each ganglion cell body in the pigeon ciliary ganglion contains about 1·2 pg of acetylcholine. This is based on the results of gas chromatograph/mass spectrometer analysis of the ganglion and on the assumption that each of the estimated 6000 cell bodies (Marwitt *et al.*, 1971) contain the same amount of acetylcholine, although the number of cells was given as 4000 by Cantino and Mugnaini (1974). After denervation of the ganglion, by cutting the oculomotor nerve, the amount of acetylcholine present in those parts of the postganglionic neurones that lie within the ganglion, dropped considerably. This suggested that synthesis and storage of acetylcholine by the postganglionic parasympathetic neurones of the ciliary ganglion is influenced by, and may even be dependent on, the integrity of the preganglionic neurones. The mechanism by which this trans-synaptic influence takes place is not known.

Some interesting experiments have been performed on the ciliary ganglion of adult pigeons by Landmesser and Pilar (1970). Taking advantage of the two recognizably distinct cell populations in the ganglion, they cut the preganglionic axons and then studied the restoration of synaptic transmission by the regeneration of these axons. The preganglionic fibres were cut by severing the oculomotor nerve about 2 mm proximal to the ganglion. Regeneration of the axons and reestablishment of synaptic transmission was measured by recording from the iridociliary (their ciliary) and the choroidal nerves, following electrical stimulation of the oculomotor nerve.

The normal electrical response from the ciliary nerve has a double peak (bimodal). The first peak, with the shorter latency, was due to direct electrical coupling between pre- and postsynaptic structures (Martin and Pilar, 1963). The second and higher peak was the chemical response due to acetylcholine transmitter release from the preganglionic terminal. The latency of the chemical response in the iridociliary nerve was always less than that in the choroidal nerve, which of course only had a chemical response and was thus unimodal.

After cutting the oculomotor nerve, the preganglionic neurones in the proximal stump required two or three days to undergo the necessary metabolic changes for regeneration to start. They then took about ten days to grow across the 2 mm gap between the proximal and distal stumps. This was long enough for the complete degeneration of the distal ends of the cut axons and for the breakdown of their synaptic organization. Thirteen days after the operation the first responses were recorded in the iridociliary nerve, and after another two days responses could be recorded in both iridociliary and choroidal nerves.

Because the electrical responses in the two nerves were recognizably different in the normal animal, and because they regained their normal characteristics after preganglionic nerve section and regeneration, it was deduced that the regenerating preganglionic axons had each reestablished a synaptic relationship, almost exclusively, with their own type of ganglion cell. Of particular interest was the fact that the larger iridociliary preganglionic axons reached the ganglion first and when confronted with both types of ganglion cell established synaptic contact only with the appropriate ones. If reinnervation had occurred, to any significant extent, in an inappropriate way then long latency bimodal responses and short latency unimodal responses would have been expected, but this did not happen. How regenerating preganglionic parasympathetic axons exercise this selection of postganglionic cell bodies with which to establish their characteristic type of synapse, is not known.

Comparative studies have been made of the activity of choline acetyltransferase (which synthesizes acetylcholine) and acetylcholinesterase (which destroys it), at different stages of development, in the ciliary and cranial cervical ganglia of the embryo chick. In the ciliary ganglion, in which both pre- and postganglionic neurones are cholinergic, there is a remarkable increase in the activity of choline acetyltransferase starting at the 14th day of incubation and reaching a maximum at the 29th day after hatching. At this stage the total activity of the enzyme is 450 times greater than it was on the 7th day of incubation (the specific activity showed a 47 times increase over the same period). On the other hand the total activity of acetylcholinesterase increased only 7 times throughout the same period of development and showed no increase in specific activity at all (Sorimachi and Kataoka, 1974). Morphological and electrophysiological studies have shown that the formation of transmitter vesicles and the electrical activity of synapses also reached a peak at the 29th day after hatching (Landmesser and Pilar, 1972). No such differences in the development of enzyme activity were observed in the cranial cervical ganglion, in which both enzymes increased to the same extent.

Although, in these experiments, the development of choline acetyltransferase activity appeared to correlate well with the development of synapses and transmitter vesicles, it has been suggested that this may not always be the case (Sorimachi and Kataoka, 1974). Choline acetyltransferase activity has been found in many parts of the chick brain (Marchisio and Giacobini, 1969). At hatching and two months later the optic lobes were found to have a much greater amount of this enzyme than any other part of the brain. However, they were not able to correlate the development of regional differences in choline acetyltransferase activity with the onset of spontaneous and reflex electrical activity in those regions. On the other hand Marchisio and Consolo (1968) did find such a correlation in chick dorsal root ganglia.

Burt and Narayanan (1976) have demonstrated differences in the enzyme activity of the two types of postganglionic neurones which are found in the ciliary ganglion. The specific activity of choline acetyltransferase in the "choroid" cells was significantly higher than that in the "ciliary" cells after 16 days of incubation. From their observations it became clear that the major increase in choline acetyltransferase activity correlated with the maturation of synapses in the ganglion and with the establishment of functional connections at the periphery. They also showed that relatively low levels of acetylcholine synthetic capacity (29·5 fmol h^{-1} per cell) are sufficient for a functional cholinergic synapse. There is thus at least the possibility that choline acetyltransferase may have some function in the ciliary ganglion other than that of synaptic transmission. It is conceivable that it could contribute in some way to the development of synapses. This might explain why it can be found in ganglia before synapses have developed. However, the whole question of synaptic development and what controls it is under investigation at the moment and no clear answer is available.

The remaining parasympathetic ganglia are associated with various branches of the VII cranial nerve and are all concerned with glandular secretion (Fig. 7.3). The sphenopalatine ganglion is a swelling (about 1 mm long in the fowl) on the ventral division of the palatine (Vidian) nerve, situated above the caudal part of the palatine bone, in the floor of the orbital cavity. The ethmoidal (orbitonasal) ganglion is of similar size on the dorsal division of the palatine nerve, situated in the rostrodorsal part of the orbital cavity ventral to the ophthalmic nerve and rostral to the gland of the nictitating membrane.

Both these ganglia receive preganglionic parasympathetic fibres from the VII cranial nerve by way of the palatine nerve. They also receive sympathetic postganglionic fibres from the external ophthalmic plexus, fibres from which join with the greater superficial petrosal nerve as it leaves the geniculate ganglion to form the palatine nerve. The nerves leaving these parasympathetic ganglia are thus mixed, as they contain sympathetic as well as parasympathetic fibres and most probably sensory fibres as well. They are distributed to mucous salivary glands in the palate and to mucous secreting glands in the nasal cavity, including the specialized nasal (salt) gland.

The gland of the nictitating membrane (Harderian) is supplied by branches from the ethmoidal ganglion and possibly also by a branch from the trunk which connects the sphenopalatine and ethmoidal ganglia. This gland has been shown to contain small autonomic ganglia in its capsule and to have acetylcholinesterase-positive fibres associated with the acini and with the blood vessels, in the turkey and pigeon (Stammer, 1964) and in the duck (Fourman and Ballantyne, 1967). The lacrimal gland probably receives its autonomic innervation from the palatine nerve via the sphenopalatine ganglion, and thence the lacrimal branches of the maxillary nerve.

Small parasympathetic ganglia are also found along the chorda tympani. This is a nerve which carries parasympathetic fibres from the facial nerve to the rostral group of mandibular salivary glands in the floor of the mouth. The chorda tympani leaves the facial nerve either as a common trunk with the lesser superficial petrosal nerve, from which it soon separates, or independently from the region of the geniculate ganglion (Fig. 7.2). It passes down the rostral wall of the external auditory meatus through the lateral part of the pterygoid muscle, after which it unites with the sublingual nerve of mandibular V for distribution to the mandibular salivary glands (Fig. 7.3). The pre- to postganglionic relay is in the "mandibular ganglion". The latter consists either of scattered ganglionic

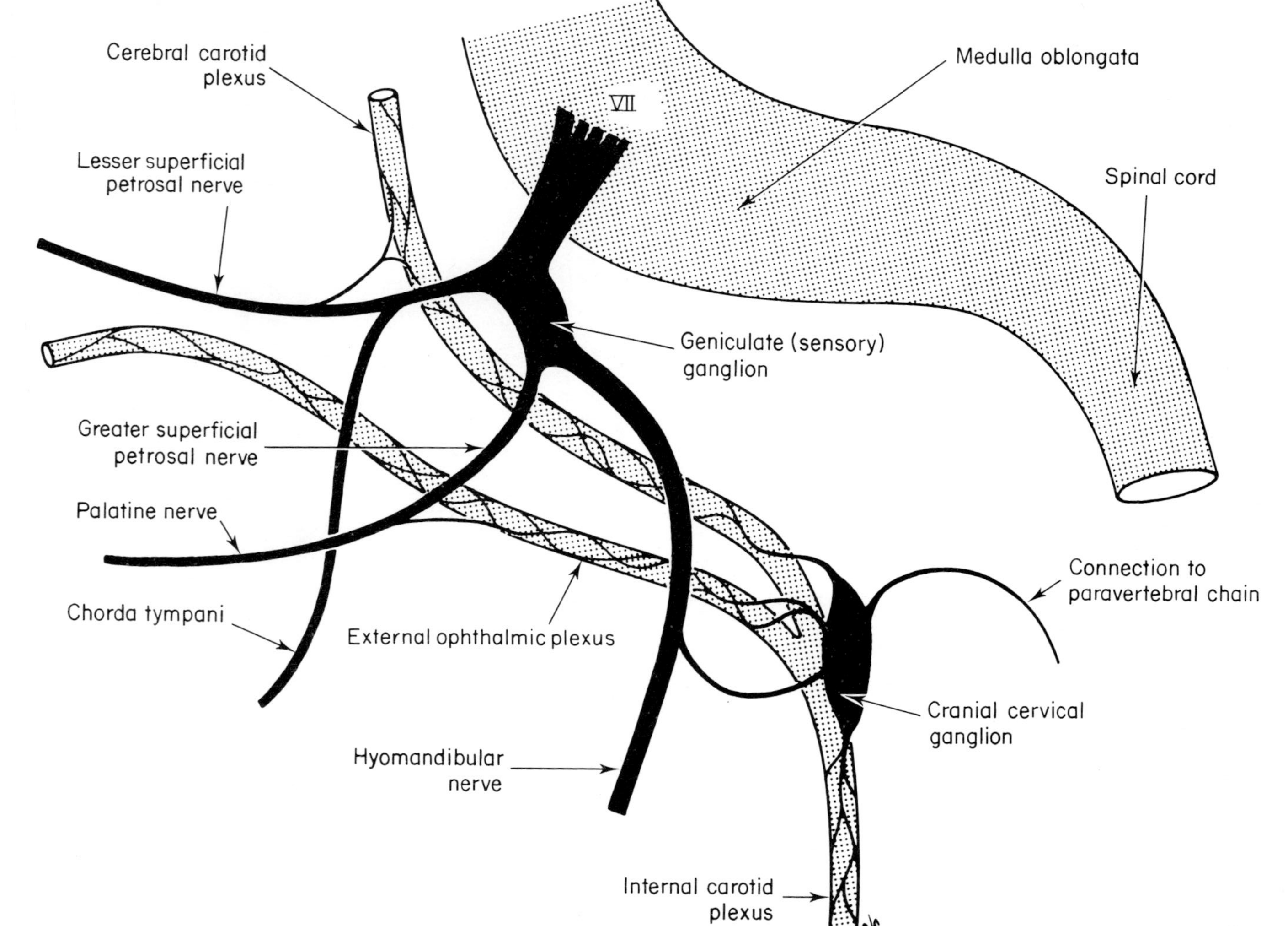

Fig. 7.2. Branches of the facial (VII) nerve carrying parasympathetic fibres and related sympathetic fibres from the cranial cervical ganglion. *Gallus*.

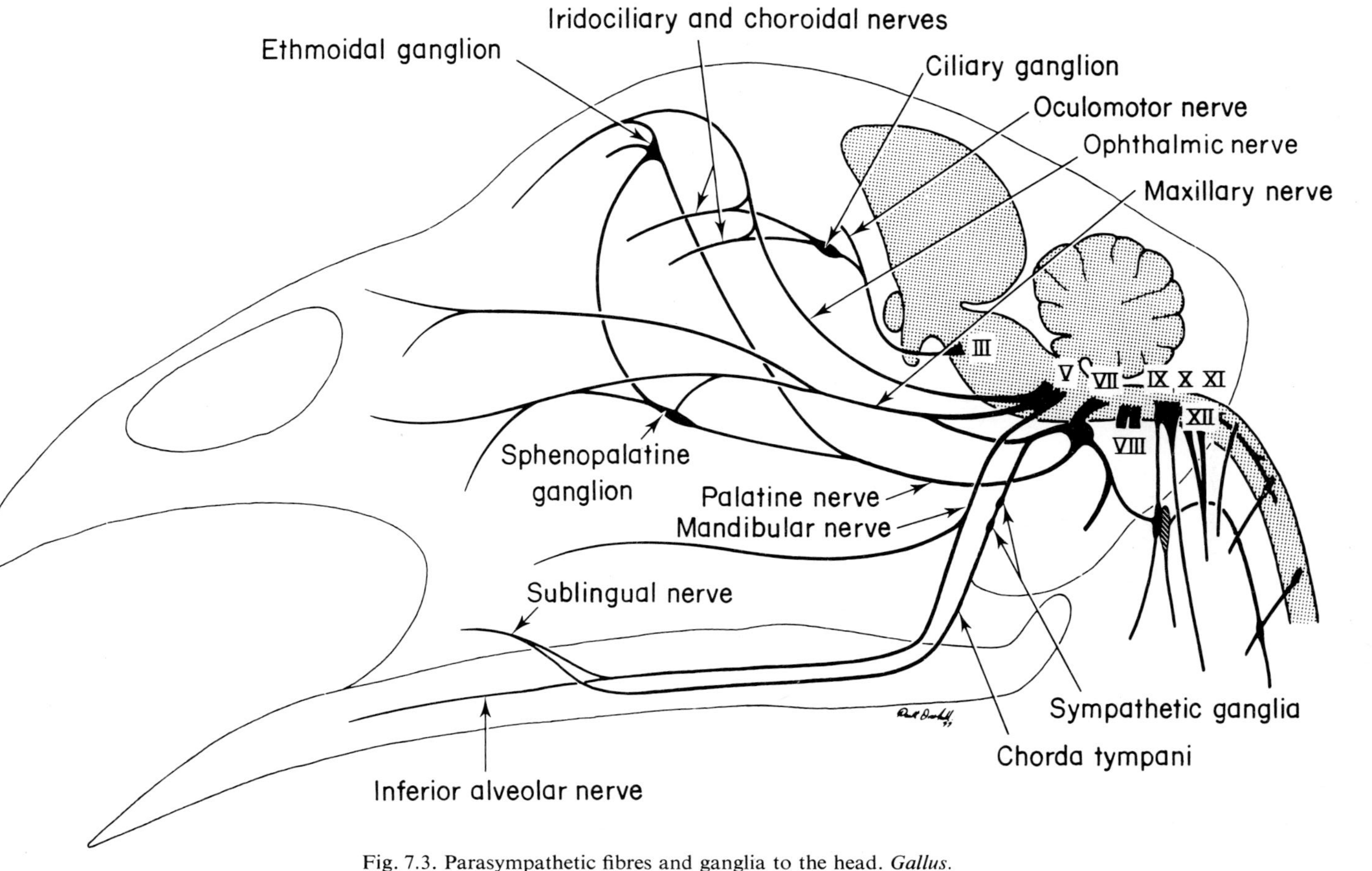

Fig. 7.3. Parasympathetic fibres and ganglia to the head. *Gallus*.

cell bodies within the chorda tympani, or of two to four small ganglia on the dorsal aspect of this nerve before it joins the sublingual nerve (see Waluszewska-Bubien, Chapter 7, Vol. 2).

B. Glossopharyngeal, vagus and accessory nerves

Cranial nerves IX and X are very closely related both topographically and functionally. Their motor nuclei lie in a continuous zone in the medulla oblongata and give origin to preganglionic parasympathetic fibres (general visceral efferent) which innervate smooth muscle and glandular tissue over a large area including most of the alimentary canal, heart, lungs, blood vessels and pancreas. An adjacent region gives rise to motor neurones which are distributed by IX, X and XI to the striated muscle of the pharynx, larynx and palate (special visceral efferent).

These fibres leave the side of the medulla by a series of rootlets which lead to the proximal ganglion shared by IX and X. Here they are joined by a trunk coming from the spinal cord down to the level of C_2. Of the medullary rootlets, the most cranial are associated with the glossopharyngeal, the middle ones lead to the vagus, whilst the last few, together with the spinal root, belong to the accessory nerve. The large common proximal ganglion which is shared by nerves IX and X contains cell bodies of neurones concerned with taste, general visceral sensation, and somatic sensation (see also Waluszewska-Bubien, Chapter 7, Vol. 2). Immediately distal to their proximal ganglion the glossopharyngeal and vagus nerves separate and enter the glossopharyngeal and vagal canals which pass through the exoccipital bone and open into the parabasal fossa. The accessory nerve also passes through the vagal canal.

The glossopharyngeal nerve is distributed to the tongue, pharynx, larynx, oesophagus and crop. It carries sensory fibres from taste buds, palate and floor of the pharynx in birds (Bubien-Waluszewska, 1968), and general visceral sensation, i.e. pain, pressure, mechanical irritation, chemosensitivity (other than smell) and temperature from the same area. We may regard these sensory fibres as autonomic afferents.

Motor fibres in the glossopharyngeal nerve innervate the intrinsic laryngeal muscles whilst its parasympathetic motor fibres stimulate motility in the oesophagus and crop and secretion of mucus by the caudal mandibular, lingual and cricoarytenoid salivary glands, as well as by many small mucous glands throughout the region. Small parasympathetic ganglia are scattered along some of the nerve branches. It appears that the fowl is particularly sensitive to temperature in the oral cavity. It will reject warmed water, in response to which the rate of impulse conduction in the lingual branch of the glossopharyngeal nerve increases (Halpern, 1962).

There is a large anastomosis between the glossopharyngeal and vagus nerves and many fibres which left the medulla in the vagus are believed to be distributed by the glossopharyngeal (see Bubien-Waluszewska, Chapter 7, Vol. 2). These include the laryngopharyngeal nerve which innervates the oesophagus, intrinsic muscles of the larynx, mucous membrane of the pharynx, larynx and trachea, as well as the lingual and cricoarytenoid salivary glands. One of the terminal branches of the glossopharyngeal nerve is the descending oesophageal branch which anastomoses with the recurrent branch of the vagus above the crop. From these branches parasympathetic fibres run to the muscle and mucous glands of the oesophagus and crop and a few pass to the trachea and syrinx.

The vagus nerve is the most widely distributed nerve in the body. It is a mixed parasympathetic motor and autonomic sensory nerve to most of the alimentary canal and to the heart, lungs, trachea, bronchi, blood vessels and many glands. Sensory fibres also pass to the pleura, peritoneum and pericardium and as we have seen, some of its fibres are distributed by branches of the glossopharyngeal nerve. After leaving the skull by the vagal canal, the main vagal trunks pass down the neck in close association with the jugular vein on each side. Just before entering the body cavity, at the level of the caudal pole of the thyroid gland, there is a small swelling on each vagus nerve which represents the distal vagal (thoracic or nodose) ganglion. Here are found the cell bodies of sensory neurones which are distributed to the heart, lungs, syrinx, bronchi and the alimentary canal below the crop.

The left and right vagus nerves enter the body cavity on the medial side of the jugular vein and cranial vena cava between the cervical and clavicular air sacs. They pass by the hilus of the lung and the heart to run down the ventral side of the proventriculus onto the gizzard. Within the body cavity branches from the vagus join with sympathetic nerves to form extensive mixed autonomic plexuses. These will be discussed later.

At the top of the neck the vagus is connected to the cranial cervical ganglion. There is a cutaneous branch (*r. externus*) to the skin above and behind the orbital cavity and to the cucullaris muscle. This branch is regarded as part of the accessory nerve which left the skull fused to the vagus.

A large anastomosing branch connects the vagus to the glossopharyngeal and another connects it to the hypoglossal nerve. Several small branches contribute to the autonomic plexus which surrounds the two common carotid arteries, and fibres from a point below the distal ganglion pass to the thyroid, parathyroid and ultimobranchial glands (Hsieh, 1951). These fibres may only be distributed to blood vessels.

Within the body cavity branches pass to the heart and pericardium in close association with sympathetic fibres. A recurrent branch from each vagus leaves the parent trunk in front of the heart. These branches do not carry motor fibres to the intrinsic laryngeal muscles, as they do in the mammal. Instead the fibres to the larynx pass from the vagus nerve to the glossopharyngeal nerve for their final distribution. The right recurrent branch passes round the aortic arch whilst the left one passes round the ligamentum arteriosum. They are distributed to the oesophagus, crop, trachea, syrinx and lung. It should be remembered that, unlike the mammal, it is the 4th pharyngeal arch artery on the right side which persists to form the aortic arch in birds. As well as the recurrent branch of each vagus nerve, there is also a descending branch which innervates the lower oesophagus and proventriculus.

IX. Autonomic nerves to the viscera of the body cavity

Preganglionic sympathetic motor fibres reach the viscera in all parts of the body cavity through a number of splanchnic nerves which leave the paravertebral chain and synapse in one or other of various visceral ganglia. Postganglionic motor fibres which leave these ganglia form extensive plexuses around the blood vessels and in association with the heart, lungs and alimentary canal, as well as with muscular ducts like the oviduct, duetus deferens, and ureters (Figs 7.4, 5). Parasympathetic motor fibres join these plexuses either from the terminal branches of the vagus nerve, or from the sacral nerves, so that probably all the viscera receive a double innervation.

Autonomic sensory fibres are also very widely distributed throughout the viscera and although the great majority of these fibres are likely to pass back to the brain in the vagi, their cell bodies being either in its distal or proximal ganglion, there is still the possibility that some may accompany the sympathetic nerves and enter the central nervous system via the dorsal roots of the spinal nerves.

A. The cardiac nerves

The first of the sympathetic splanchnic nerves is the cardiac sympathetic nerve (Fig. 7.4) which leaves the paravertebral chain at the level of the first thoracic ganglion in the chicken and travels along the caudal part of the jugular vein and cranial vena cava to the heart and pericardium. Here it is joined by vagal parasympathetic branches (cranial and caudal cardiac vagal nerves) and together with sensory fibres forms a cardiac plexus which is distributed to all parts of the pericardium, heart, aortic arch and its main arterial branches, the pulmonary vessels, the lungs and pleurae and possibly to the cranial and caudal thoracic air sacs as well.

By a combination of microdissection, electrical stimulation, selective transection of the sympathetic chain and retrograde degeneration following section of the cardiac sympathetic nerve, MacDonald and Cohen (1970) demonstrated that preganglionic cardioaccelerator fibres frequently originate from the last cervical segment, always from the first and second thoracic segments and occasionally from the third thoracic segment of the spinal cord in the pigeon. All these fibres join to form the cardiac sympathetic nerve on each side, but the right has a greater accelerator effect than the left. This also occurs in mammals and was probably first pointed out by Hunt (1899).

However, the bird differs from the mammal in that its sympathetic outflow to the heart originates over a more restricted region of the spinal cord. This is given as the first thoracic segment in the chicken (Hsieh, 1951; Tummons and Sturkie, 1968), the last cervical and the first three thoracic segments in the pigeon (MacDonald and Cohen, 1970) and the first six thoracic segments in the mammal (Saccomanno, 1943).

The supply to the lungs is provided by branches from the cardiac sympathetic nerve joining with three to seven pulmonary branches which arise from the vagus to form a plexus on the pulmonary blood vessels (Hsieh, 1951; McLelland and Abdalla, 1972). The smooth muscle of the various bronchi in the avian lung also receives an autonomic innervation but it seems as if the smooth muscle of the parabronchi only receives an acetylcholinesterase-positive innervation (which is likely to be cholinergic, parasympathetic) but no fluorescent noradrenergic supply (Akester and Mann, 1969b).

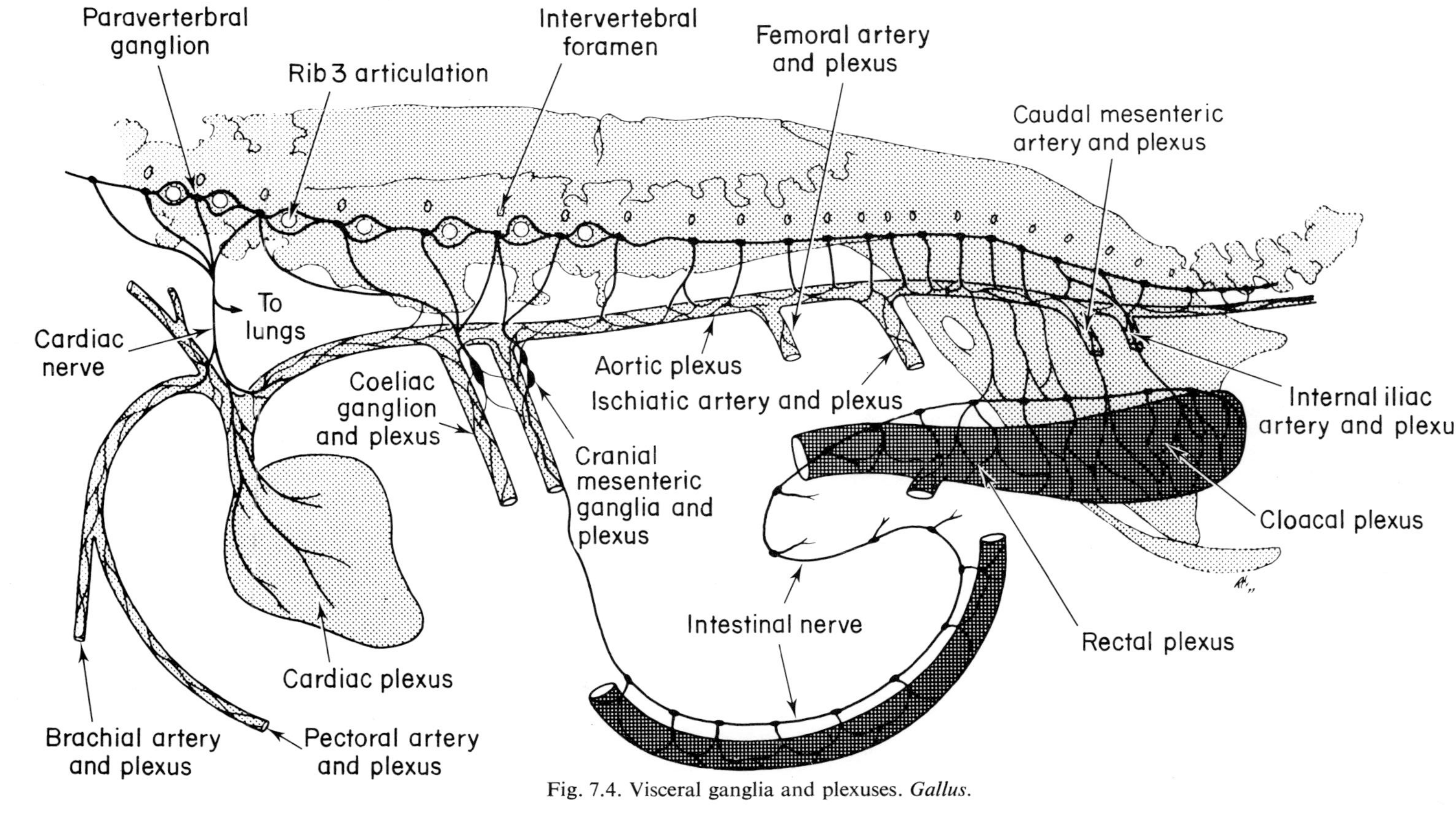

Fig. 7.4. Visceral ganglia and plexuses. *Gallus*.

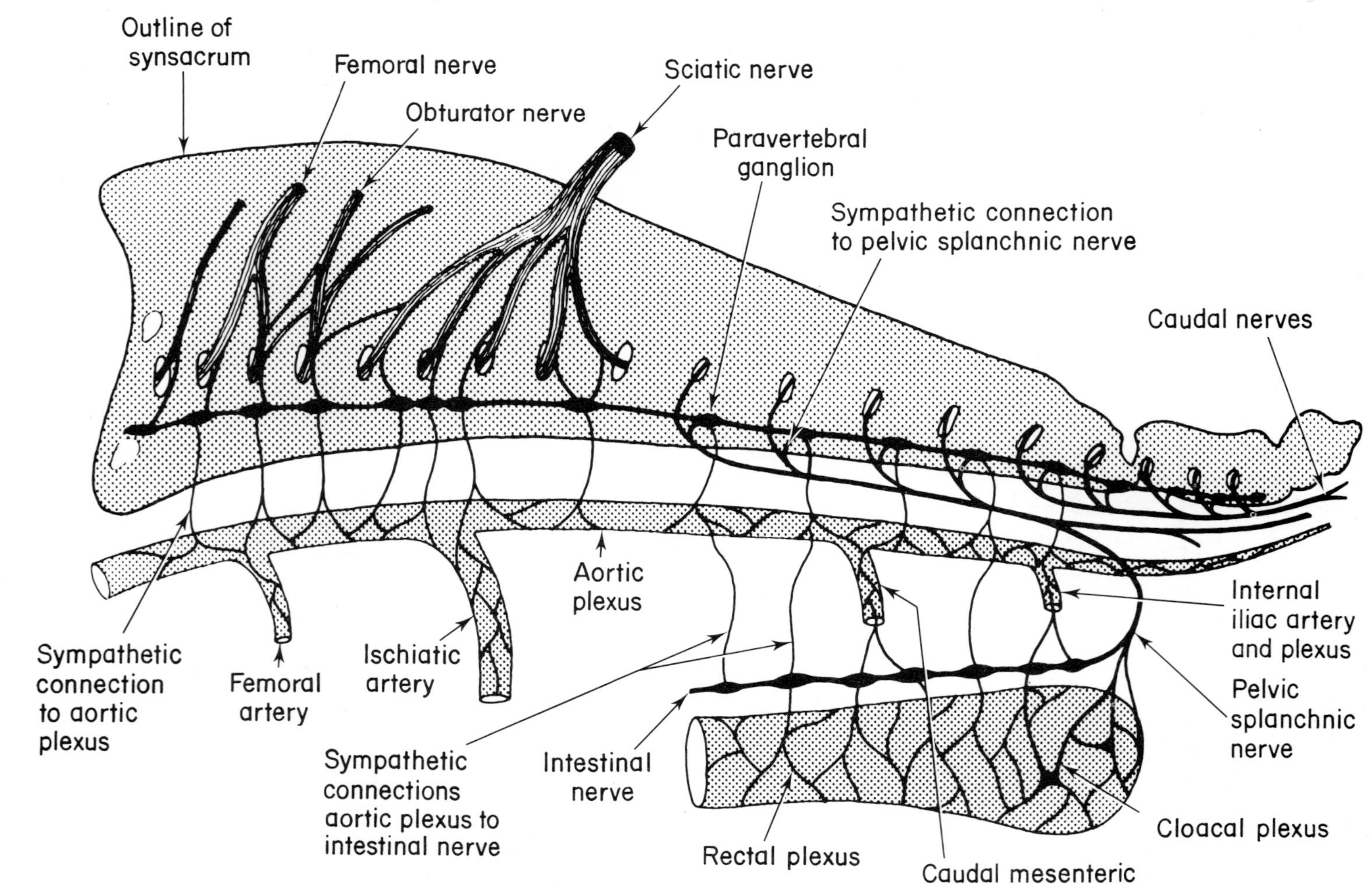

Fig. 7.5. Detail of pelvic plexus and intestinal nerve. *Gallus*.

The role of all these fibres is complicated, particularly those in the heart, where they are distributed to the sinuatrial and atrioventricular nodes, the myocardium in both atria and ventricles and possibly to the conducting cells beyond the nodes. They also provide a vasomotor innervation to the coronary blood vessels.

In addition to the sympathetic and parasympathetic motor supply to the heart there are many autonomic sensory nerves concerned with cardiac reflexes, many of which are basically similar to those in the mammal. However, one that appears to be very different is concerned with the control of salt secretion by the nasal gland which occurs in response to an increased osmotic pressure in the blood. This has been studied in the goose by Hanwell *et al.* (1972), who found that an injection of hypertonic saline into the pericardial cavity stimulated salt secretion by the nasal gland. This was prevented by cutting, or anaesthetizing, the vagus nerves above the heart but it was not prevented if the vagus nerves were cut below the heart. It thus appeared that osmoreceptors were situated in or near the heart and that their associated sensory nerves passed to the brain in the vagus.

The possibility of the osmoreceptors being in the region of the head or viscera, other than the heart, was ruled out by the fact that intra-arterial injection of hypertonic saline to the head, gizzard and proventriculus did not stimulate secretion from the gland any quicker than did intravenous injection of the saline. Further confirmation that the osmoreceptors and their associated sensory nerve endings did not lie in the head was provided by cross-circulation studies in which, effectively, an isolated (decerebrate) head of one (recipient) goose was perfused by blood from a second (donor) goose.

In this arrangement blood from the donor's brachial arteries was fed into the common carotids of the recipient, whilst venous blood from the recipient's jugular veins was returned to the brachial veins of the donor. The vertebral arteries of the recipient were ligated and all the tissues were severed at the base of the neck, except the spine and the vagus nerves. When hypertonic saline was injected intravenously into the donor its salt glands were stimulated to secrete but those of the recipient were not, although both heads were perfused by the same blood. When hypertonic saline was injected intravenously into the body of the recipient its isolated head, which was connected to its own body only by the spine and the vagus nerves, secreted salt normally.

The possibility that secretion by the salt gland might be stimulated by an increased blood volume was ruled out by the fact that intravenous injection of a large volume of homologus blood (enough to raise the blood volume by 16%) had no effect on the activity of the gland. So for the time being, it looks as if the osmoreceptors which control nasal gland salt secretion in the goose, must join the chemoreceptors which control respiratory movements in the chicken, as being physiologically localized but anatomically undetected. However, it should be mentioned that there does appear to be some vascular link between the body and the head of a bird, other than by way of the common* carotid and vertebral arteries. This was demonstrated by Akester and Brackenbury (1976) who showed that when both common* carotid and both vertebral arteries were ligated there was still a vascular link to the head which permitted a partial

* Strictly, the internal carotid (see Baumel, J., Nomina Anatomica Avium, Academic Press, 1980).

restoration of the resultant drop in intraocular pressure. This link, they presumed to be by way of the spinal arteries which would also have been intact in the cross-circulation experiments on the goose.

Sensory fibres are particularly difficult to recognize by histochemical methods either with the light or electron microscope. They are usually said to be myelinated, but so are many motor axons. Attempts to demonstrate sensory terminals in the lung have been made by injecting radioactive leucine into the distal ganglion on the vagus trunk. It has been claimed that radioactive tracers (C^{14}-labelled noradrenaline and leucine) can be injected into autonomic ganglia and identified by autoradiography at peripheral sites in the terminals of axons whose cell bodies were in the injected ganglion (Livett *et al.*, 1968). Uptake of the tracer is said to be only by the cell bodies and not by the axons passing through the ganglion. This has been tried recently (Bower *et al.*, 1977) by injecting H^3-leucine into the distal ganglion of the fowl in order to demonstrate the distribution of sensory fibres in the avian lung. If it proves to be a reliable procedure it could be a valuable method for recognizing sensory nerves.

B. Rotation-compensating reflexes

An interesting group of (possibly stretch) receptors in the mesenteries, which are associated with autonomic sensory nerves and which are responsible for initiating certain rotation-compensating reflexes in pigeons, have recently been reinvestigated by Biederman-Thorson and Thorson (1973). It has long been known that if a pigeon (and presumably other birds) is rotated to one side, it extends its wing and leg of that side and bends its tail towards the same side. These responses are neither dependent on the balance mechanism of the semi-circular canals (Singer, 1884; Trendelenburg, 1906), nor on sight as they still occur after labyrinthectomy and when the bird, inside a closed box, is rotated in the same way.

Biederman-Thorson and Thorson (1973) measured the electrical activity of the main muscles concerned with extending the wing (*m. triceps*) and bending the tail (*m. levator caudae*) as a convenient way of monitoring the reflexes. They confirmed that these rotation-compensating reflexes in normal birds are virtually unchanged by labyrinthectomy or by preventing the bird from seeing its surroundings. They also agreed with earlier suggestions that the sensory apparatus, which initiates these reflexes, probably responds to distortion or stretching of the mesenteries due to the weight of the viscera varying with the effective direction of gravity. However, the exact nature and position of these receptors and of the autonomic afferents associated with them is not known.

C. Aortic, coeliac and other visceral ganglia and plexuses

Splanchnic nerves containing mostly preganglionic sympathetic fibres pass from thoracic ganglia 2–5 in the domestic fowl and converge on the origin of the coeliac artery where the coeliac ganglion is situated (Fig. 7.4). Here post-ganglionic sympathetic fibres, which originate in the ganglion and which outnumber the preganglionic fibres by something like 20:1, join with branches of the vagus to form a vascular plexus around all branches of the coeliac artery supplying the spleen, liver, gizzard, proventriculus, part of oesophagus, part of duodenum and pancreas as well as the mesenteries associated with these structures and the smooth muscle of the alimentary canal from lower oesophagus to duodenum. Splanchnic nerves originating in the next group of sympathetic ganglia (thoracic 6–7 and synsacral 1–3) in the chicken, together with vagal fibres form an extensive vascular plexus along the aorta. This is extended along the major branches so that several plexuses have been named according to the vessels and organs they supply, including coeliac, cranial and caudal mesenteric, renal, adrenal, gonadal, oviducal, vas deferens and internal iliac (hypogastric) plexuses. In the kidney the smooth muscle of the unique avian renal portal valve (Fig. 7.6a,b) receives a particularly dense innervation of noradrenergic (sympathetic) and acetylcholinesterase-positive (probably parasympathetic) fibres (Akester and Mann, 1969a).

Sympathetic ganglia are numerous, variable in size and widely scattered throughout the region, lying in the adventitia of blood vessels, within mesenteries and quite likely embedded within some of the organs. The coeliac and cranial mesenteric ganglia are amongst the most prominent (Fig. 7.4).

D. The intestinal ganglionated nerve trunk (Remak's nerve)

So far the general distribution of autonomic nerves in the bird has been much the same as in mammals. However, the route by which these nerves reach the small and large intestine is different in birds, which have a prominent intestinal nerve (of Remak) running along the whole length of the alimentary canal from duodenum to cloaca (Figs 7.4, 5).

The intestinal nerve is a mixed sympathetic and parasympathetic nerve trunk running in the mesentery immediately above the duodenum, jejunum and ileum and at a rather greater distance above the rectum. It receives parasympathetic and sympathetic fibres from both ends, and at least sympathetic fibres from many points along its length. The latter pass from the aortic plexus, down the mesentery of the rectum to join the intestinal nerve which lies in the same mesentery. Cell bodies of postganglionic cells are scattered along this nerve in longitudinal clusters. Where these groups are large, there is a recognizable swelling on the nerve, which is then described as a ganglion. It is likely that both sympathetic and parasympathetic neurones have their cell bodies in these ganglia (Fig. 7.9). There are from 30 to 50 ganglia more or less uniformly distributed along the length of the intestinal nerve. However, many cell bodies can be found in this nerve outside the macroscopic ganglia and it is likely that many of the other sympathetic nerves and plexuses which do not carry macroscopic ganglia also have small groups of postganglionic cell bodies distributed within them. For most of its length the intestinal nerve is single, but above the rectum it divides into a double trunk. Its ganglia tend to be larger towards the caudal end. At its cranial end the intestinal nerve originates in the autonomic plexus (sympathetic and parasympathetic) associated with the cranial mesenteric artery. At its caudal end sympathetic fibres join the trunk from the caudal mesenteric and internal iliac (hypogastric) plexuses whilst parasympathetic fibres enter from the pelvic splanchnic nerve (*n. pudendus*) (Figs 7.4, 5).

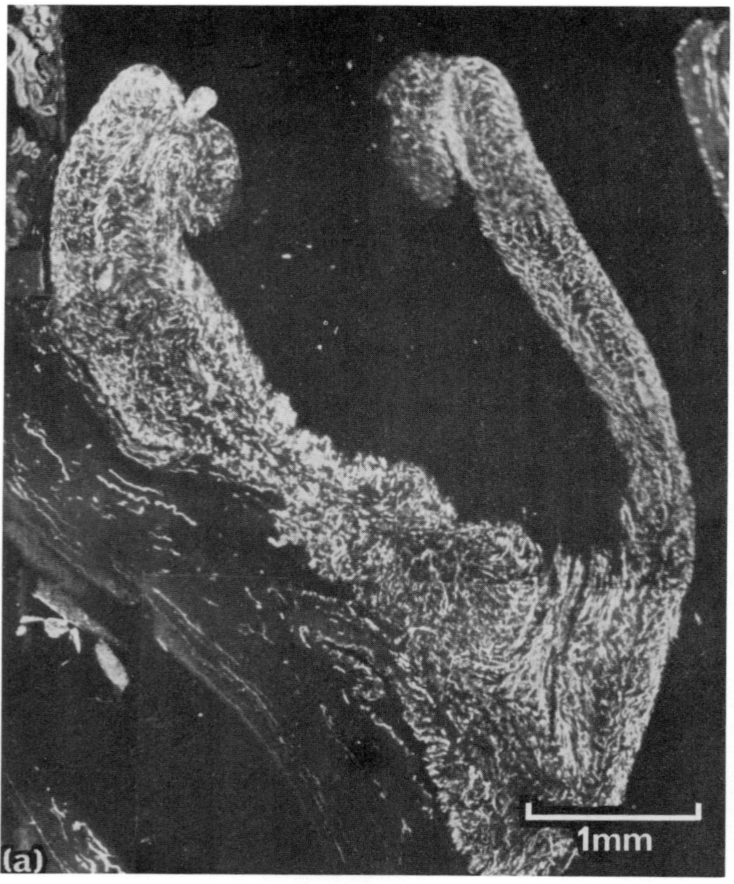

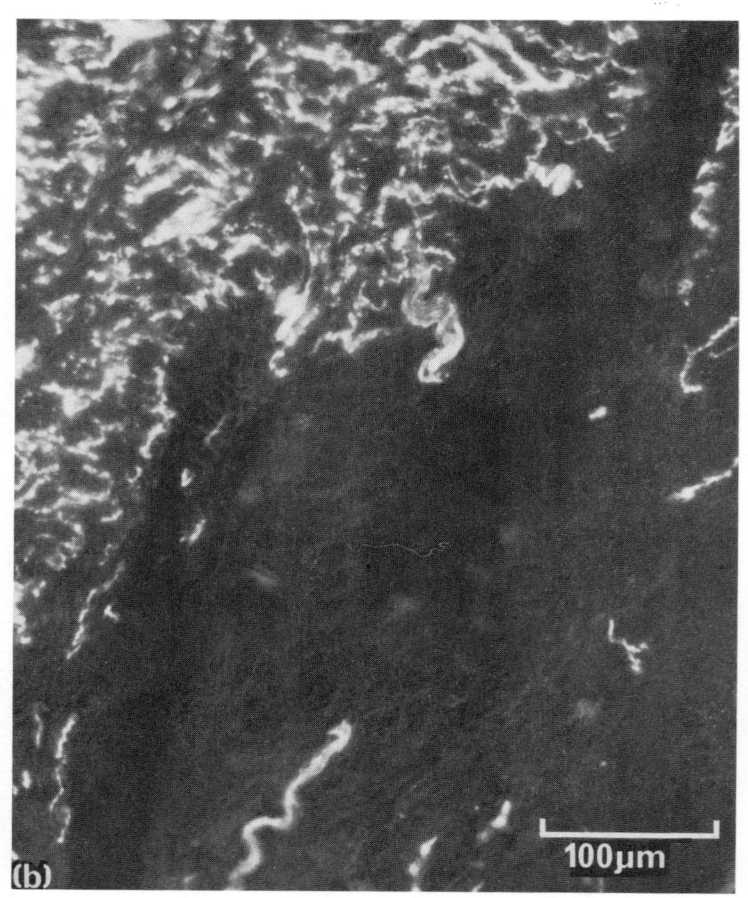

Fig. 7.6. (a) Noradrenergic fibres to the renal portal valve. *Gallus*. (b) Detail of (a).

E. The pelvic plexus

This is the name given to an extensive mesh of autonomic nerves, both sympathetic and parasympathetic (sensory as well as motor) which is situated largely within the mesentery supporting the rectum and cloaca (Fig. 7.5). The pelvic plexus is continuous with the rectal plexus (over the rectum) and the cloacal plexus (over the cloaca). Together with the intestinal nerve these nerve plexuses are concerned with control of movements in the distal parts of the alimentary canal, ureters, ductus deferens and oviduct; the erectile vascular bodies of the male, as well as visceral sensory responses throughout the region and somatic sensation from the skin around the vent.

These plexuses are not only extremely complicated but the precise distribution, origin and function of their various components is by no means properly understood. They contain pre- and postganglionic sympathetic and parasympathetic nerves, visceral and some somatic sensory nerves as well as a few macroscopic ganglia and numerous microscopic ganglia, some of which (possibly most) will be mixed sympathetic and parasympathetic.

Sympathetic fibres feed into these plexuses from the intestinal nerve and from vascular plexuses which are associated with the arteries that supply blood to the region. These arterial plexuses are extensions of the aortic plexus which follow all its branches. The principal arteries to the caudal (pelvic) viscera are the caudal mesenteric and internal iliac (hypogastric) arteries. We have already seen that veins have a greater innervation than arteries (Akester, 1971) so that venous autonomic plexuses may join those of arteries in their final distribution to the walls of the viscera.

Another source of sympathetic fibres to this region is the pelvic splanchnic (pudendal) nerve. This nerve contains the visceral fibres from synsacral spinal nerves 9–12. It is largely parasympathetic (sacral outflow) but receives sympathetic fibres from nearby chain ganglia and also carries many visceral sensory fibres.

It should be emphasized that the internal iliac (hypogastric) arterial plexuses in birds refer to the autonomic nerves which are associated with, and distributed by the internal iliac arteries and their branches. They are not the same as the hypogastric nerves in mammals which carry sympathetic fibres from the caudal mesenteric ganglion to the pelvic viscera and which run as discrete nerve trunks.

A considerable number of complex reflex movements take place along the alimentary canal and a great deal still remains to be learned about them. For example, what controls the filling and emptying of the caeca? What determines whether peristalsis in the rectum will be cranially or caudally directed? (Akester *et al.*, 1967). The cloaca is a very mobile and complicated region incorporating sphincters on either side of the coprodeum (A. R. Akester, unpublished) and has the ability to evert the urodeal region, which carries the terminal parts of the oviduct or ductus deferens, during copulation and egg laying.

X. Nature of a sympathetic ganglion

A. *The cell population*

It is easy to oversimplify the nature of a typical sympathetic ganglion. It tends to be regarded as a collection of postganglionic neurone cell bodies on which the preganglionic axons synapse, and not much more (Figs. 7.7, 8). In fact there is a good deal more and this soon becomes obvious when one looks at such a ganglion under the electron microscope.

One of the serious limitations of all electron microscope studies is the dead and completely static nature of the material. Chamley and her colleagues (1972) went some way towards overcoming these limitations by studying the cells which grew from sympathetic ganglia in tissue culture. These cells were observed by time-lapse cinematography as well as by various histochemical and microscopic techniques.

In cultures grown from the cranial cervical ganglion of the chick, nerve cells were recognized which varied considerably in size, morphology and in migratory activity. They also varied in their content of noradrenaline, as shown by the brightness of their paraformaldehyde-induced fluorescence (Fig. 7.9). Some neurones were associated with satellite cells whilst others were not, although in the normal intact ganglion they probably all are.

Many other non-nervous cells were recognized which included small intensely fluorescent cells (S.I.F. cells), Schwann cells, oligodendrocytes, astrocytes, sheath cells, macrophages, perineural epithelial cells, fibroblasts and endothelial cells. S.I.F. cells are so called because of the extreme brightness of their paraformaldehyde-induced fluorescence. They are dopaminergic (Eränkö and Härkönen, 1965; Björklund *et al.*, 1970; Libet and Owman, 1974) and are believed to have a modulating (inhibitory) influence on the large postganglionic neurones. It has long been known that catecholamines have an inhibitory effect on transmission through the mammalian cranial cervical ganglion (Marrazzi, 1939). Stimulation of the small preganglionic "C" fibres to this ganglion in the rat also has an inhibitory effect on the postganglionic response (Dunant and Dolivo, 1967).

Ultrastructural studies suggest that the chromaffin cells which have been found in sympathetic ganglia of mammals (Coupland, 1965; Siegrist *et al.*, 1968), and the more recently recognized S.I.F. cells are most probably one and the same thing (Eränkö and Eränkö, 1971). These cells are sometimes described as interneurones lying between the preganglionic "C" fibres and the large postganglionic neurones. Cholinergic synapses (from "C" fibre to chromaffin cell) and dopaminergic synapses (from chromaffin cell to postganglionic neurone) have both been recognized. In the terminal swelling of the "C" fibre (against the chromaffin cell) acetylcholine is contained in clear vesicles, whereas in the chromaffin cell process (against the postganglionic neurone) dopamine is contained in dense-core vesicles. The possibility of there being interneurones in sympathetic ganglia is not new and was first implied by Dogiel (1896).

Chromaffin (S.I.F.) cells in mammals, and they are probably similar in birds

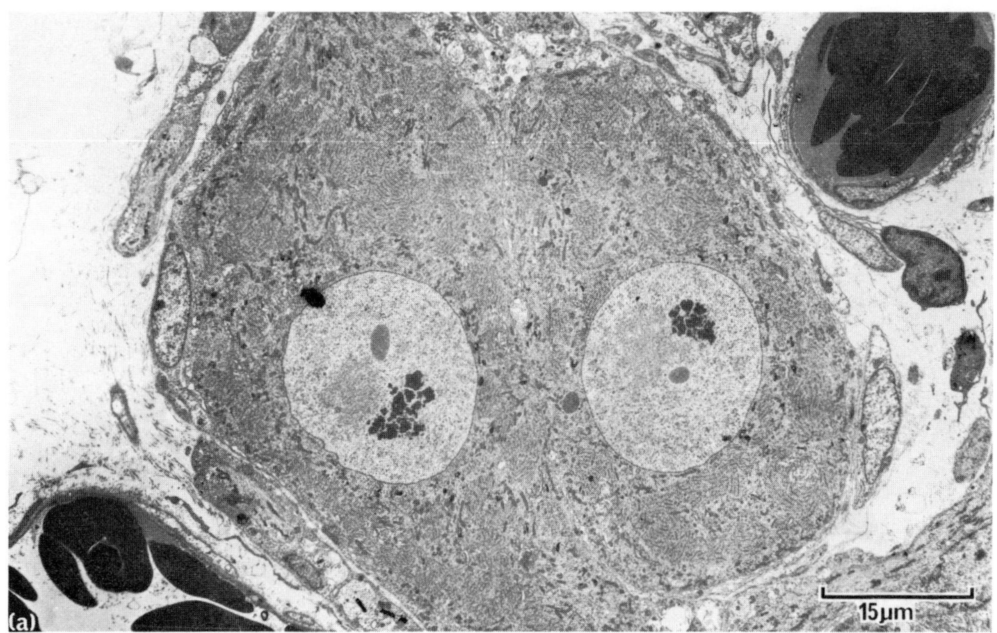

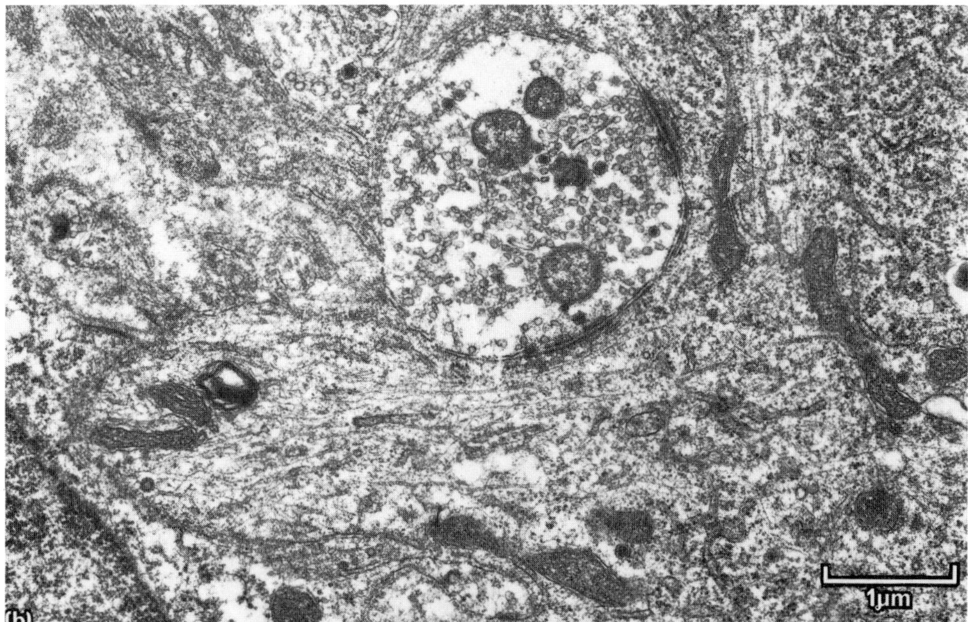

Fig. 7.7. (a) Two ganglion cell bodies close to blood capillaries (intestinal nerve). *Gallus*. (b) Cholinergic synapse in ganglion cell (intestinal nerve). *Gallus*.

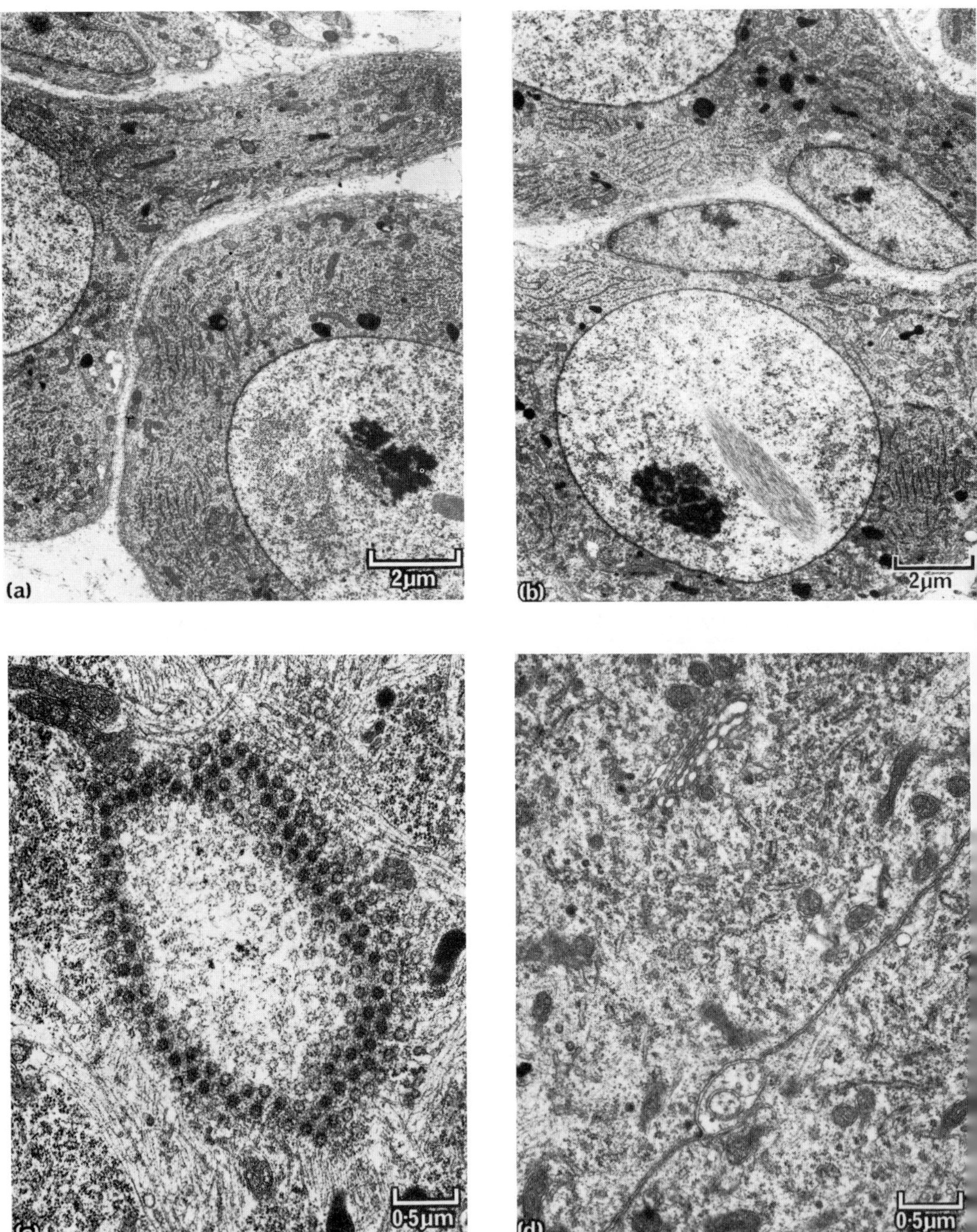

Fig. 7.8. (a) Two ganglion cells in cranial cervical ganglion. *Gallus*. (b) Two ganglion cells in cranial cervical ganglion. *Gallus*. Satellite cells and intranuclear rodlets are shown. (c) Nuclear pores in ganglion cells (cranial cervical ganglion). *Gallus*. (d) Contact zone between two adjacent cell bodies. No satellite cell cytoplasm between them. Golgi body possibly forming dense-core vesicle (intestinal nerve). *Gallus*.

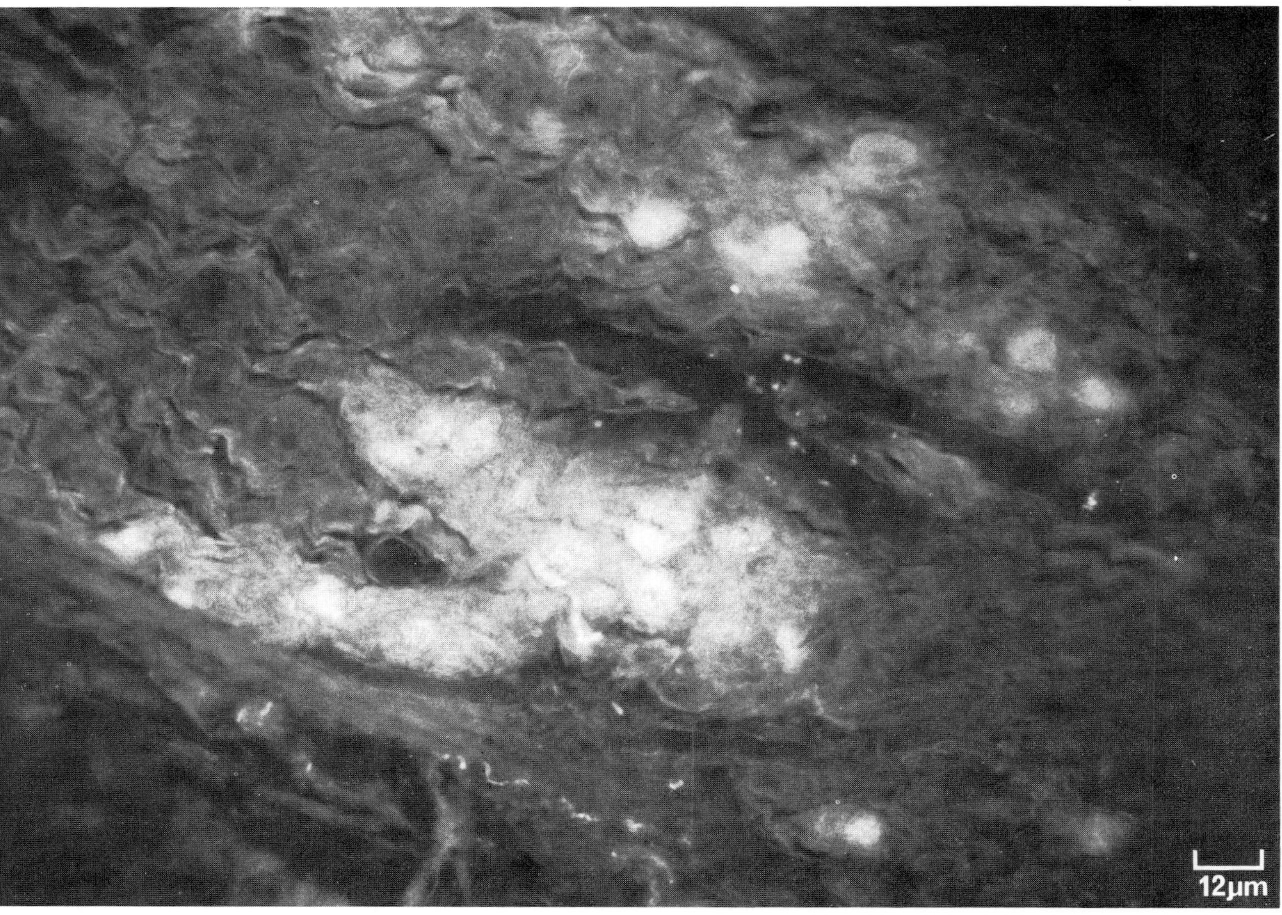

Fig. 7.9. Intestinal nerve, showing fluorescent and non-fluorescent ganglion cell bodies. *Gallus*.

(Akester and Akester, 1975), may be a kind of hybrid cell with features resembling both adrenal medullary endocrine cells and postganglionic sympathetic neurones. All three cell-types secrete catecholamines which are stored in dense-core vesicles of different sizes. The largest are in adrenal medullary cells (200 nm), the next in chromaffin cells (100 nm) and the smallest are in postganglionic sympathetic neurones (35 nm).

Chromaffin cells are about 10 μm in diameter and have relatively short (about 40 μm) but extensively branching varicose fibres. These form a mesh around the cell bodies of the large postganglionic neurones with which they may make somato–somatic, axo–somatic, axo–dendritic or axo–axonal synapses. They normally occur in groups of between 2 and 12 cells and there may be from 20 to 30 groups in a single ganglion. They seem to be localized in that part of the ganglion which is towards the outflow tract.

Although they are present in relatively small numbers their strategic position and the extent of their processes makes it feasible for them to influence all the postganglionic neurones in the ganglion. They have been grown in relatively pure culture from chick embryo sympathetic ganglia by Jacobowitz and Greene (1974). A thin layer of satellite cell cytoplasm surrounds groups of chromaffin cells but is said to be discontinuous at certain points, and particularly where a chromaffin cell lies close to the fenestrated endothelium of a blood capillary into which it could (at least theoretically) release dopamine (Mathews and Raisman, 1969). Further evidence that these cells are dopaminergic is provided by the fact that they do not contain dopamine-β-hydroxylase, which converts dopamine to noradrenaline in noradrenergic neurones (Fuxe *et al.*, 1971).

Satellite cells are a type of neuroglia which surround the large postganglionic cell bodies in sympathetic ganglia with a continuous thin film of cytoplasm (Figs 7.7a, 8b). As we have seen they also surround chromaffin cells. There may be four or five such cells associated with the cell body of a single neurone. Although their precise role is still not clear, information is accumulating from experiments on various mammals, as well as frogs, mud puppies and leeches, but as yet no birds. (For a recent summary see Kuffler and Nicholls, 1976.) These cells are obvious enough in the sympathetic ganglia of birds (Fig. 7.8b) and it is likely that their role in avian species will be shown to have much in common with similar cells in other vertebrates. They have long been assumed to support in some way the postganglionic neurones with which they are so closely associated, although there is no real evidence to prove it. They have recently been shown to respond to potassium ions which are released by the adjacent neurone, possibly by taking up, storing and subsequently releasing transmitter substance (proved for GABA, gamma aminobutyric acid, Schon and Kelly, 1975).

The increased potassium concentration which depolarizes the satellite cell membrane may stimulate it to produce something (as yet unidentified) that may be required by the neurone for its continued activity or recovery. Perhaps one of the most obvious roles these cells play is to isolate the cell body of one neurone from its neighbours. They may have some influence over the pre-postganglionic synapse, although they themselves have no synaptic relationship with other cells. On occasions they may be phagocytic in response to local cellular damage.

As one would expect, quite large parts of the ganglion are occupied by bundles of axons and these split up the cell bodies of the postganglionic neurones into clusters. There appears to be a much greater size range amongst the unmyelinated axons than amongst those that have a myelin sheath (Fig. 7.10).

B. Communication between cells

Nerve cells communicate with one another at synapses and a great deal of work has been carried out to establish the nature of these structures (Figs 7.7b, 11b,c,d). Pre- and postsynaptic membrane thickenings, axonal swellings and the localized concentration of different types of transmitter vesicles form the main parts of what are now well recognized as synaptic characteristics. The freeze-fracturing technique has confirmed the general impressions gained from transmission electron-microscopy. When fracturing occurs parallel to the plane of the synaptic cleft it does not separate one cell from the other as one might expect. Instead it splits the inner and outer layers (leaflets) of a single membrane (usually presynaptic) in such a way that the synaptic vesicles show as spherical profiles on the exposed surface (Branton, 1966). This type of fracturing suggests that there is a strong adhesion between the outer leaflets of the two apposed membranes across the synaptic cleft.

A recent study of freeze-fractured synaptic membranes in the central nervous system of several mammalian species, and also the pigeon, has been conducted by Pfenninger *et al.* (1972) and it is reasonable to assume that a similar arrangement will be found in sympathetic ganglia. They demonstrated a regular hexagonal pattern of small protuberances on the outer face of the inner leaflet of the presynaptic membrane. These projections had a centre to centre separation of 40–50 nm and have only been found on the presynaptic membrane face of a nerve terminal. They were interpreted as being the attachment sites of transmitter vesicles and have been referred to as the "presynaptic vesicular grid".

Evidence for communication between the axon and its surrounding Schwann cell has been provided by experiments on rats conducted by Weinberg and Spencer (1976). They performed two types of cross-anastomosis between myelinated and non-myelinated nerves. In the first experiment the proximal stump of a myelinated (sternohyoid) nerve was anastomosed to the distal stump of an unmyelinated (cervical sympathetic) nerve. The Schwann cells in the myelinated nerve were labelled with tritiated thymidine. In the second experiment the same anastomosis was made but on this occasion the Schwann cells in the unmyelinated nerve were labelled.

These experiments demonstrated that Schwann cells do not migrate from one trunk into the other, although their axons do. They also demonstrated that axons which are regenerating from the proximal stump of a predominantly myelinated nerve, into the distal stump of a predominantly unmyelinated nerve are able to stimulate Schwann cells in the latter to produce myelin, which they normally would not have done. This confirmed earlier work by Hillarp and Olivecrona (1946) who anastomosed the proximal stump of the phrenic nerve to the distal stump of the cervical sympathetic trunk.

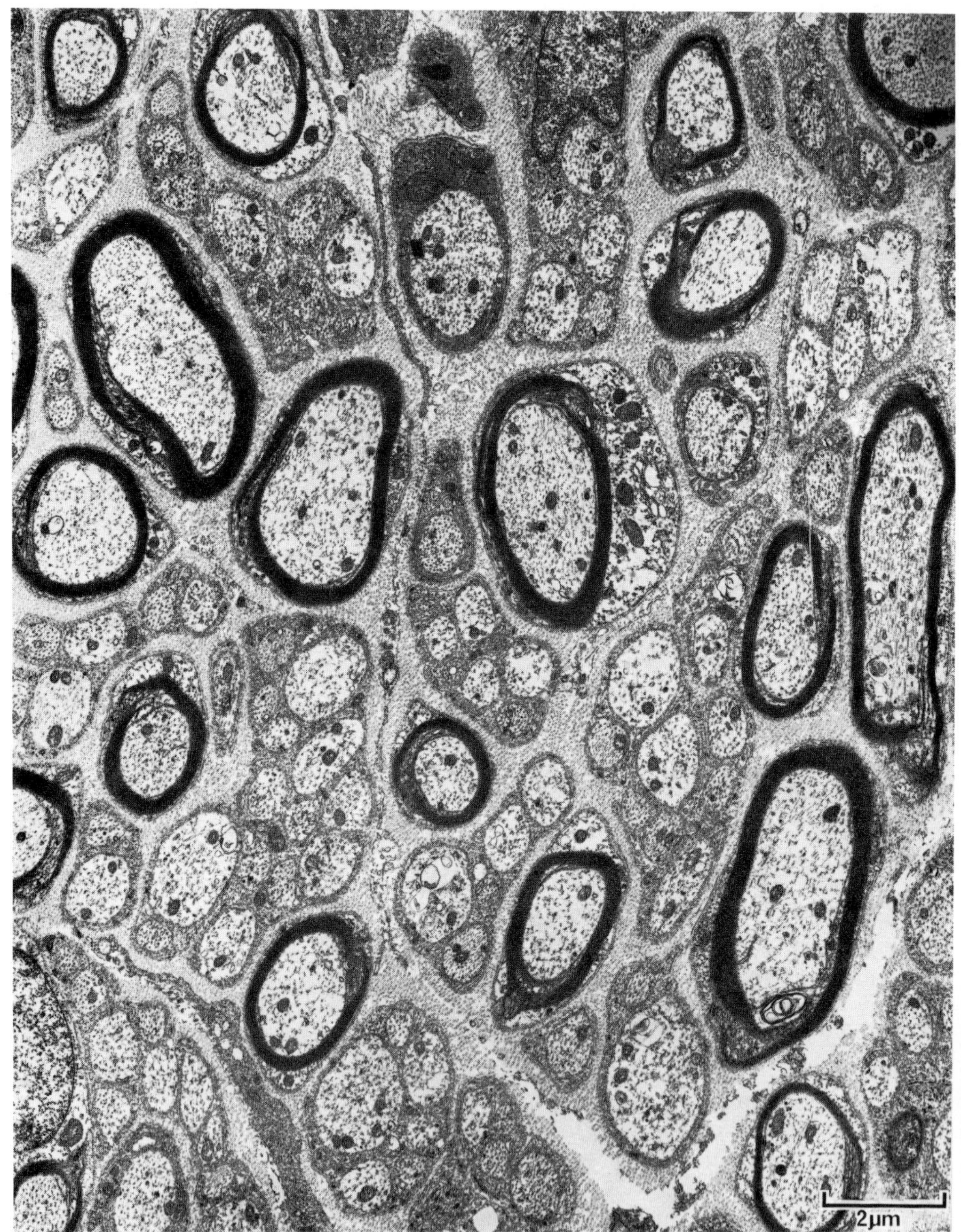

Fig. 7.10. Myelinated and non-myelinated axons in the cranial cervical ganglion. *Gallus.*

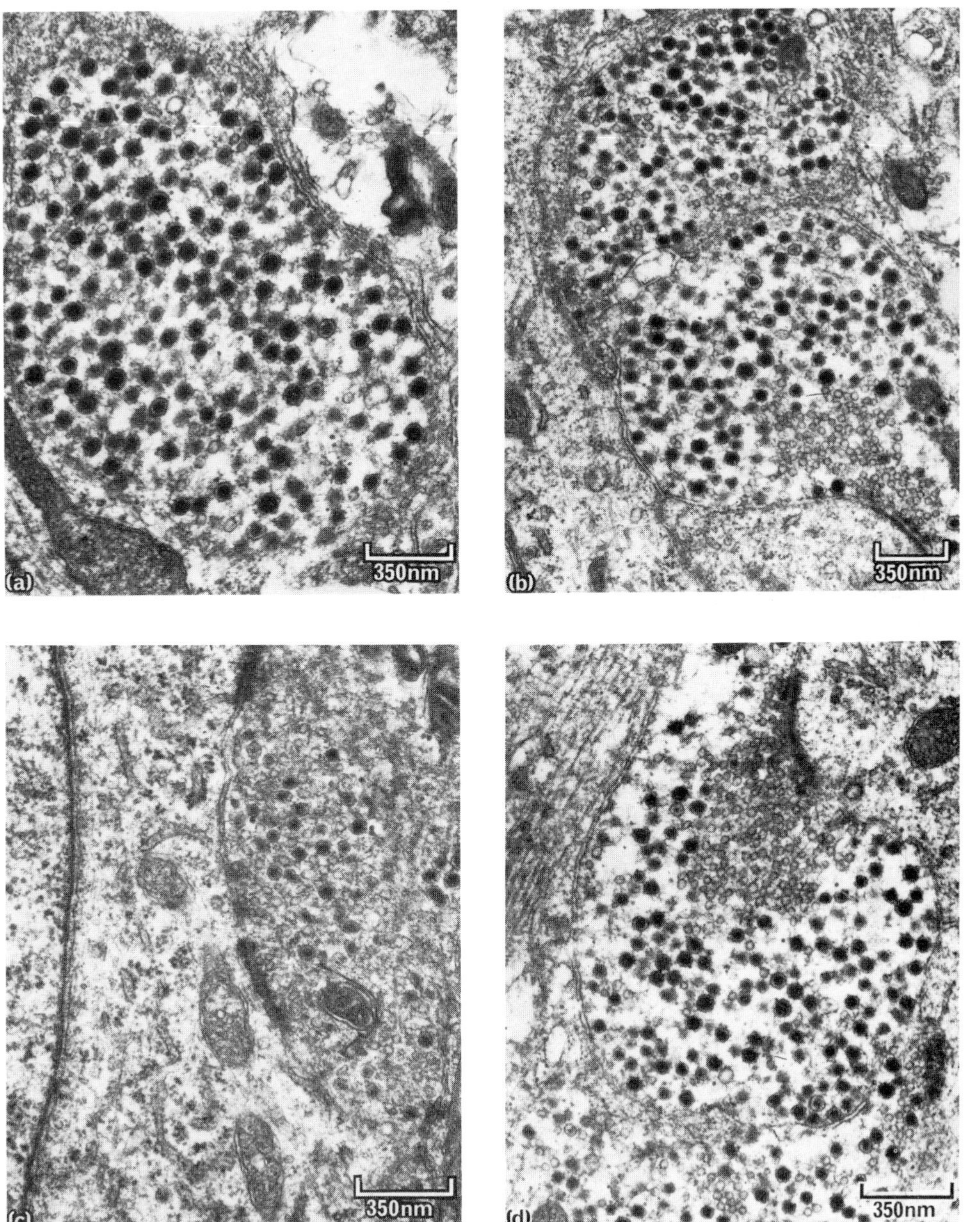

Fig. 7.11. Intestinal nerve, showing synaptic swellings loaded with dense-core vesicles, which may be catecholaminergic. *Gallus*.

C. Cell movements

As well as oversimplifying the cellular composition of a sympathetic ganglion it is easy to underestimate the cellular movements that are taking place within it. Apart from the obvious pulsatile nature of arterioles and the rush of blood through the capillaries, there will be macrophages pushing their way through the tissue spaces, as well as the rhythmic contractions of Schwann cells (Pomerat, 1959) which may even contribute to axonal flow. Intraneural movements are considerable with all the organelles performing a complex and interrelated dance. Direct recordings of these movements have mostly been by time-lapse cinematography of neurones in tissue culture. Under these conditions mitochondria appear to be "buzzing" around inside nerve terminals like bees round a hive. The movements of even such improbable structures as intranuclear rodlets have been filmed in tissue cultures of chick embryo and adult fowl sympathetic neurones (Masurovsky *et al.*, 1970). Very little is known about these rodlets, although they are obvious enough under the electron microscope (Fig. 7.8b) and were first described in birds towards the end of the last century by Holmgren (1899). According to Masurovsky and his colleagues intranuclear rodlets, which consist of bundles of protein filaments (about 7 nm diameter), appear to be linked to the nucleolus and undergo slow wave-like movements. Sometimes they appear to have penetrated the nuclear membrane and may contribute something to the cytoplasm of the cell.

In addition to these intracellular movements, which also include axonal flow (to be discussed later), another type of movement has been observed to occur on the surface of cell membranes of various tissues including fibroblasts (Abercrombie *et al.*, 1970), lymphocytes (Taylor *et al.*, 1971) and muscle cells (Edidin and Fambrough, 1973). Koda and Partlow (1976) have applied a similar technique to chick sympathetic neurones growing in tissue culture and have reported the movement of various markers which are known to attach, either specifically (red blood cells coated with concanavalin A) or nonspecifically (polystyrene beads), to the cell membrane (axolemma). Both types of marker were seen to move over the surface of the axon at approximately the same rate (50 μm h^{-1}) towards the cell body (somatopetally). It was suggested that the markers were attached to receptors, which might be either glycoprotein or glycolipids and were considered to be a part of the membrane structure.

These observations supported the fluid mosaic model of the cell membrane proposed by Singer and Nicolson (1972). They suggested that a typical membrane might consist of a viscous fluid bilayer of phospholipid molecules so arranged that the ionic (hydrophilic) heads of each layer faced outwards on either side of the membrane, and their fatty acid (hydrophobic) chains faced inwards towards the centre. Globular proteins would be scattered throughout the membrane in such a way that their ionic groups would protrude into the aqueous phase on either side of the membrane and their non-polar groups would be dissolved in the hydrophobic central part of the viscous phospholipid bilayer [see also Winkler, 1971, for a possible structure of the chromaffin (dense-core) vesicle membrane]. It would thus be possible for membrane proteins to diffuse, in the plane of the membrane, through the phospholipid layer in which they are dissolved. Some of these membrane proteins may be the reactor sites of antigens or enzymes. They may also be the basis of the experimentally observed retrograde movement of markers on the surface of chick sympathetic axons growing in tissue culture, the reason for which is not known although it may have something to do with growth and regeneration of nerve fibres.

D. Myelination

Anyone who takes an interest in the autonomic nervous system of birds, or any other higher vertebrate, must pay at least some attention to the process of myelination (Fig. 7.12), to the distribution of myelinated and unmyelinated axons and to their relative numbers in a nerve trunk (Figs 7.10, 13). Although it is a simple matter to prepare transverse sections of a nerve trunk and by light or electron microscopy to count the number of myelinated and unmyelinated axons contained within it, very few such analyses have been made of avian autonomic nerves. One of the few that has been made recently is by Brown *et al.* (1972) who counted about 10 000 myelinated and 5000 unmyelinated axons in the cervical vagus of the fowl. The great majority (87 %) of the myelinated axons were less than 3 μm in external diameter. This may be compared with the cat in which the cervical vagus is said to contain 5000 myelinated and 25 000 unmyelinated axons (Agostoni *et al.*, 1957).

These axons can be further subdivided by size, degree of myelination and to some extent by the nature of their transmitter chemical. They may also vary by the electrical characteristics of their action potentials, their enzyme content and, of course, their function. However, there are many complications and we are very far from being able to glance at an electron micrograph of a nerve trunk and to recognize the origin, destination and function of its axons (Fig. 7.10).

The first description of the general process of myelination in peripheral nerves to be based on electron microscopy, was made by Geren (1954), using the sciatic nerve of the chick embryo, and it is virtually certain that the same mechanism operates in the autonomic nervous system of the chick and of all other birds.

Groups of unmyelinated axons, which are growing into a developing tissue, become surrounded by a layer of Schwann cells which thus form a tube around them. Mitotic division of the Schwann cells enables them to keep pace with the extending axons and also to grow in amongst them, and so divide them into smaller bundles. Eventually certain axons come to be associated with individual Schwann cells (Fig. 7.12) and it is only under these conditions (one Schwann cell with one axon) that myelination can occur (Webster *et al.*, 1973).

The axon sinks into the Schwann cell cytoplasm, the cell membrane of which surrounds the axon and forms the mesaxon. Growth of the mesaxon, associated with rotation of the Schwann cell around the axon, is responsible for the development of the myelin lamellae. These vary in number but in the myelinated fibres of autonomic nerves of the fowl are generally about 10–20.

A closer look at the myelin lamellae shows that they consist of single, broad electron-dense bands, alternating with narrow electron-dense bands which may be double. The broad (major period) bands (centre to centre distance about 12·5 nm) represent the fused (thicker) inner leaflets of the Schwann cell unit

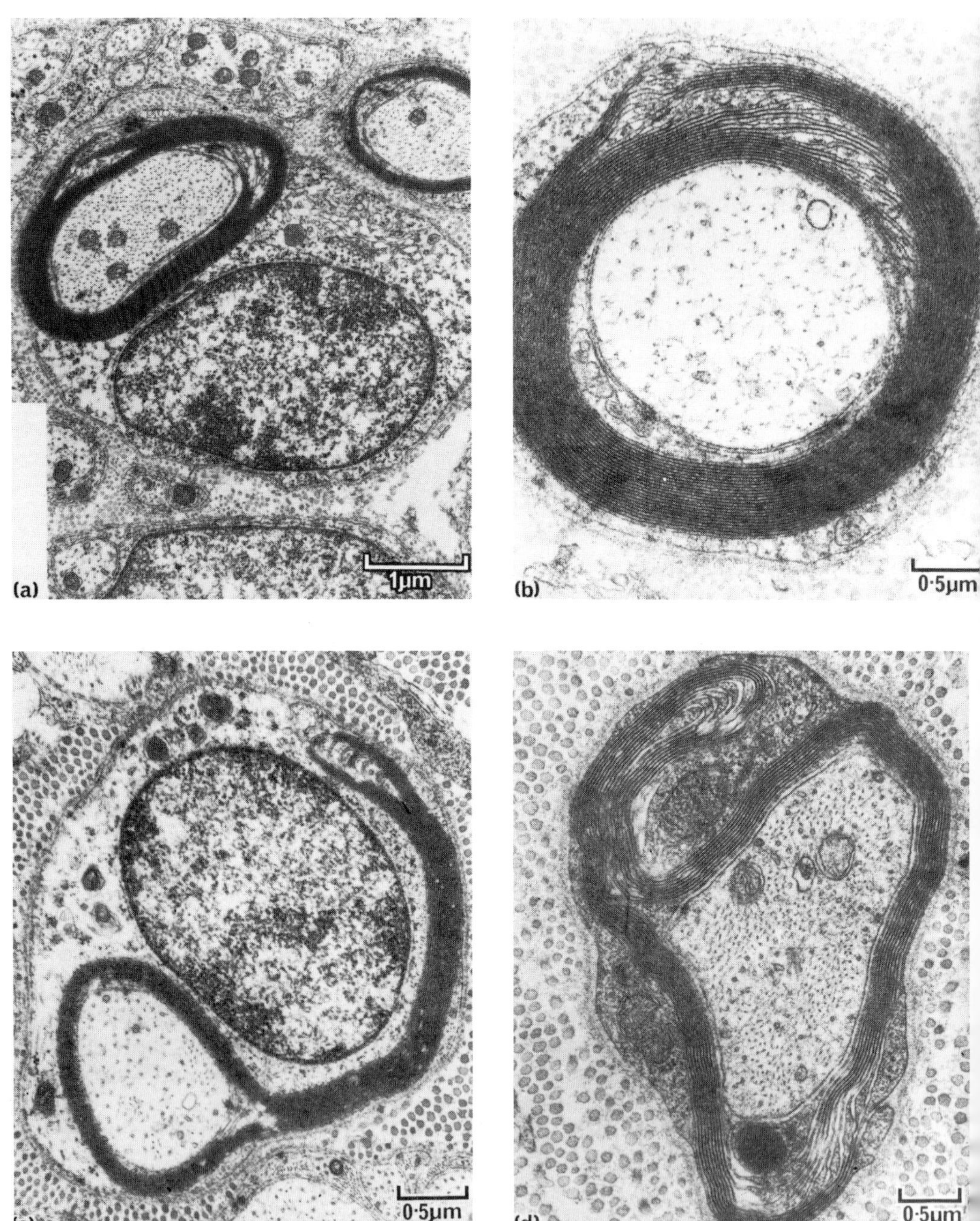

Fig. 7.12. Myelinated axons. *Gallus*.

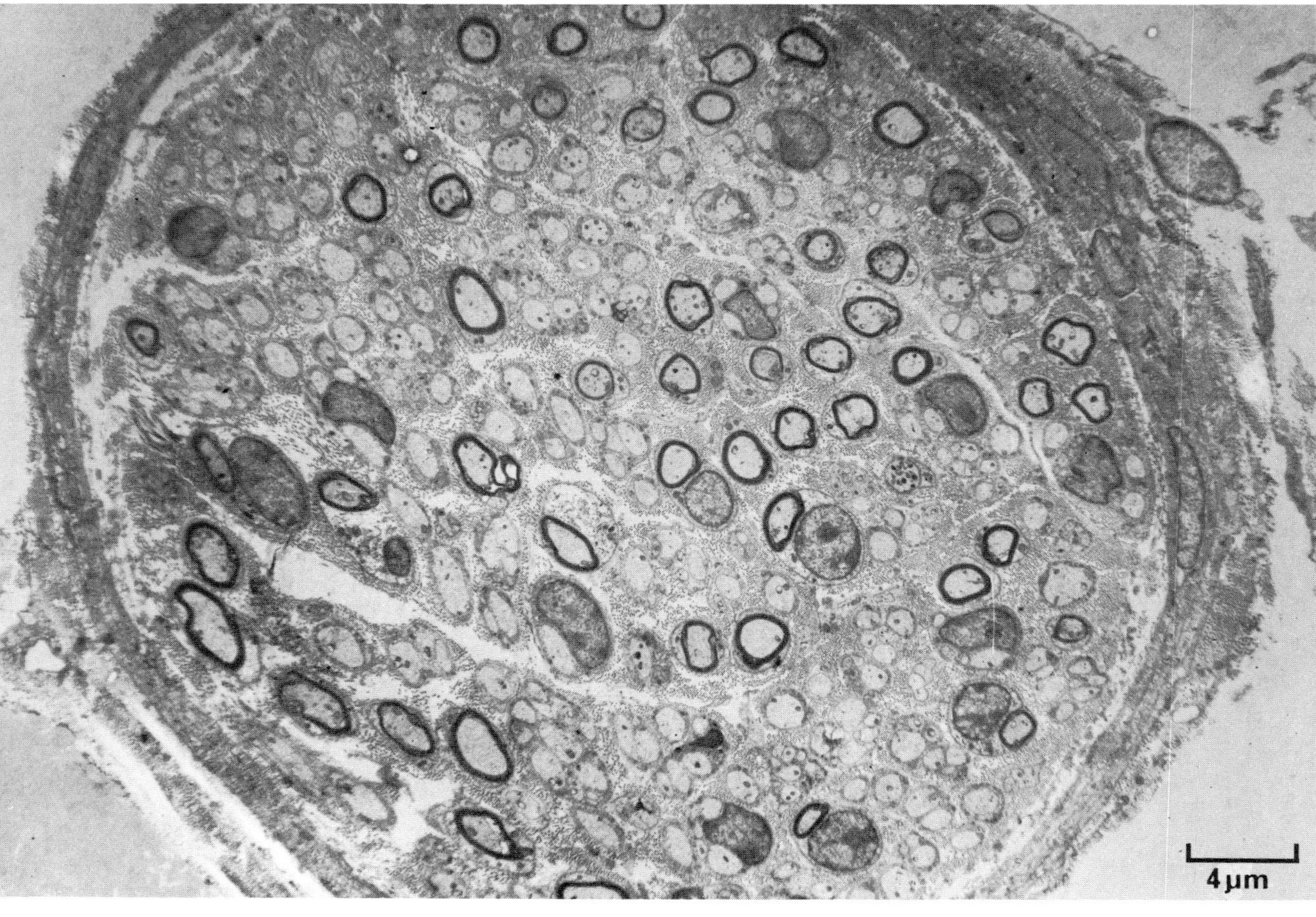

Fig. 7.13. Small branch of intestinal nerve. *Gallus*.

membrane which would have faced the cytoplasm. With the rotational folding of the cell membrane around the axon the cytoplasm has been squeezed out, thus making it possible for the inner leaflets of adjacent folds to fuse.

The narrow, less electron-dense (intraperiod) lines, which may appear single or double, represent the (thinner) outer leaflets of the Schwann cell unit membrane which face the extracellular space. Although in many preparations they appear to be fused into a single electron-dense line, when the tissue is treated with lanthanum nitrate (a substance which is electron dense and which diffuses into all parts of the extracellular space) it penetrates to the centre of the mesaxon. This substance then occupies the narrow extracellular space between the apposed thin outer leaflets of the concentrically rolled cell membrane which forms the myelin lamellae (Revel and Karnovsky, 1967). These lamellae are not always as symmetrical as they are often made out to be. They show varying degrees of separation of the two fused thick inner leaflets, with Schwann cell cytoplasm and even organelles (mitochondria) contained within the resulting helical tunnels. These separations are called Schmidt-Lanterman clefts and were once regarded as artefacts but are now considered to be quite normal. They are extremely common in chicken autonomic trunks and appear in every myelinated axon in Fig. 7.12.

Many observations (in mammals) have been made on the movements of Schwann cells and of their myelin, by using time-lapse cinematography on these cells in tissue culture, or on nerve fibres that have been teased out of freshly excised nerve trunks in a nutrient fluid (Gitlin and Singer, 1974). Schmidt-Lanterman clefts have been seen to open and close and there is speculation as to whether these movements may be associated with pulsatile activity of the Schwann cell and with the movement of Schwann cell cytoplasm. There is the further possibility that these movements may facilitate the exchange of substances between axon and Schwann cell cytoplasm and between axon and extracellular fluid. Certainly a great variety of folds, protrusions, indentations and flaps have been demonstrated in the myelin and they are unlikely to be artefacts (Fig. 7.12c,d). Many of our electron micrographs may represent ultrastructural "stills" of a more complicated activity than is sometimes expected from a myelin sheath.

By the cross-anastomosing experiments of Weinberg and Spencer (1976) we have seen that it appears to be the axon which "instructs" the Schwann cell to form myelin. We have also seen that the classical view of preganglionic autonomic axons always being myelinated and postganglionic axons being unmyelinated may be less true than it was once thought to be (Fig. 7.14f). A transverse section through a small branch of the intestinal nerve trunk is shown in Fig. 7.13. It contains approximately 400 unmyelinated axons and 50 myelinated axons.

Many visceral autonomic sensory fibres are myelinated and they are mixed in with the efferent autonomic fibres in all the main trunks. It should also be remembered that a given myelinated axon loses its myelin sheath, although it may retain its covering of Schwann cell cytoplasm a little longer, at some distance from the neuroeffector junction. Along the length of a myelinated axon there will be many Schwann cells which have formed its myelin. They may not all form the same number of lamellae, so that the extent to which an axon is myelinated may vary along its length. Useful information on the general subject of myelination is given by Davison and Peters (1970).

XI. The postganglionic sympathetic neurone. Release and uptake of noradrenaline

Although the majority of postganglionic sympathetic neurones have their cell bodies within a recognizable macroscopic ganglion, many of them do not. They may be found almost anywhere in small groups, scattered along the major autonomic trunks. As yet no comprehensive survey of these trunks has been

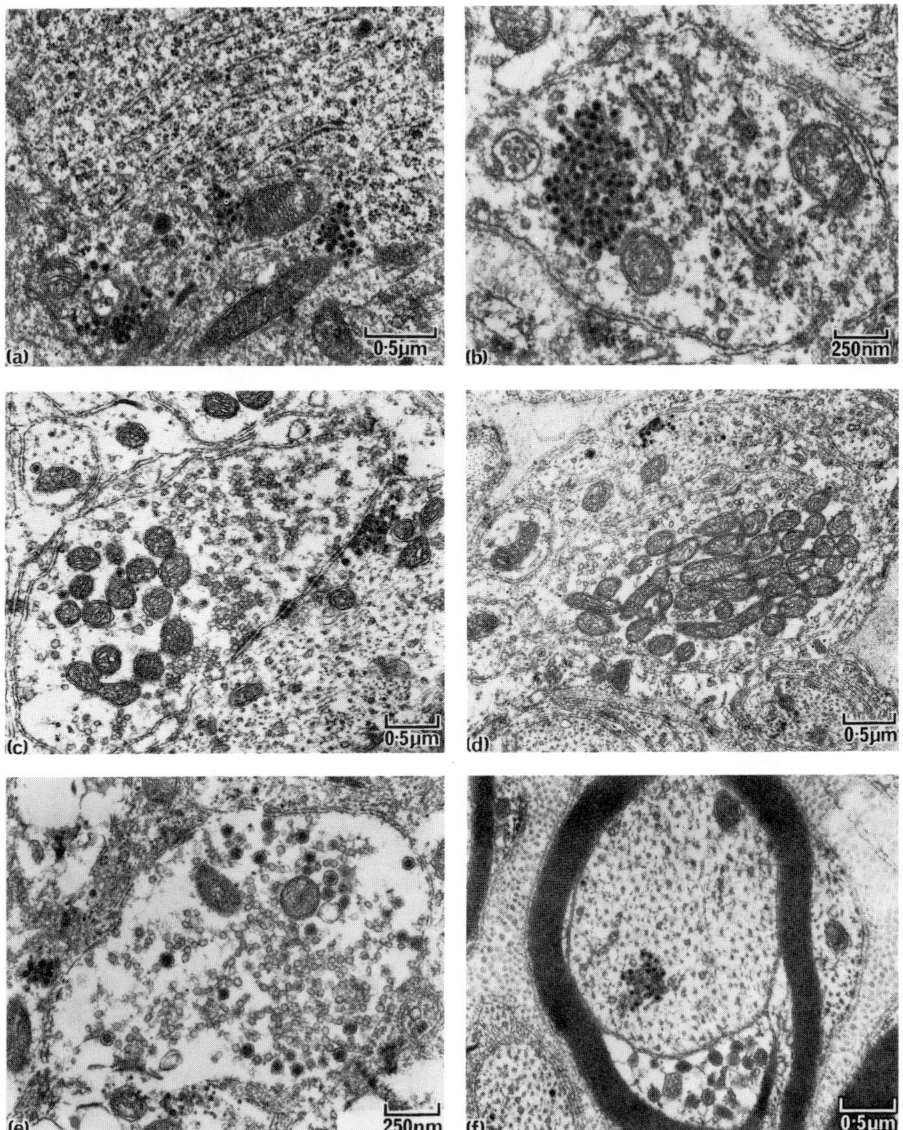

Fig. 7.14. Uptake of 5-OHDA by catecholaminergic neurones. *Gallus.*

carried out to find how many postganglionic sympathetic neurones have their cell bodies outside the major ganglia.

A typical ganglion cell body (of a postganglionic neurone) would be one of a group of such cell bodies separated within the ganglion by large numbers of nonmyelinated axons, scattered through which would be smaller numbers of myelinated axons (Fig. 7.10). The cell bodies are largely surrounded by a thin layer of satellite cell cytoplasm except where there are synapses with preganglionic nerve terminals (Fig. 7.8b).

Before making a synapse the preganglionic axon loses its myelin and its covering of Schwann cell cytoplasm. Synapses may be axo–somatic, axo–axonal or axo–dendritic. The direction of impulse transmission is indicated by the pre-synaptic terminal swelling of the preganglionic axon, which is loaded with small transmitter (acetylcholine) vesicles, and a more prominent synaptic thickening of the postganglionic than of the preganglionic cell membrane. The cell body of a postganglionic neurone is large (30 μm diameter) with a prominent nucleus and a great quantity of rough endoplasmic reticulum scattered throughout the cytoplasm (Figs 7.7a, 18a,b). Golgi bodies are also prominent. The postganglionic axon passes to the effector cells at the periphery which will frequently be smooth muscle. Before reaching its termination, such an axon divides into several pre-terminal branches so that the number of terminal processes produced by a single neurone is considerable. This is sometimes referred to as the sympathetic ground plexus. Along the entire length of all the terminal branches of the axon are regularly spaced swellings (varicosities). They are about 1 μm in diameter and about 3 μm apart. These swellings are the regions of the axon from which most noradrenaline is thought to be released. They contain transmitter vesicles, mitochondria, microtubules and microfilaments.

In some way which is not clearly understood the arrival of an impulse causes a few of the transmitter vesicles to fuse with the membrane of an axon swelling. A suggestion as to how this may happen has been given by Winkler (1971). Once fusion of the two membranes has taken place a rearrangement between the structural components of the axon and vesicle membranes results in the vesicle opening up and releasing the whole of its contents, which consists of various proteins (chromogranins) as well as noradrenaline. This procedure has been called exocytosis. The terminal swellings along an axon, which ramifies between the smooth muscle cells, have no covering of Schwann cell cytoplasm and are thus in very close contact with the sarcolemma of the muscle cell, from which they may be separated by a gap of about 50–70 nm containing a small amount of basement membrane. Noradrenaline passes across the gap to unite with active sites (α or β receptors) on the sarcolemma of the smooth muscle to cause contraction (α-receptors) or relaxation (β-receptors) according to their relative concentrations.

Once the neuro-muscular response is completed, surplus noradrenaline must be removed from the tissue spaces. One method by which this is done is by "uptake" into the surrounding tissue where some of it will be inactivated (by monoamine oxydase and catechol *O*-methyltransferase). However, the bulk of it returns to the axon from which it came, to be stored once again in dense-core vesicles ready for the next impulse-triggered release.

XII. Synthesis of noradrenaline

Although a growing list of substances is under consideration as possible neurotransmitters, noradrenaline (for postganglionic sympathetic fibres) and acetylcholine (for postganglionic parasympathetic, as well as both types of preganglionic neurones) are still the best understood.

Noradrenaline is the postganglionic sympathetic transmitter in birds, as it is in mammals, and the synthetic pathway is the same in the two groups. The sequence is as follows: L-tyrosine to L-DOPA (tyrosine hydroxylase); L-DOPA to dopamine (L-DOPA decarboxylase); dopamine to noradrenaline (dopamine β-hydroxylase) (Blaschko, 1973); dopamine β-hydroxylase, for the most part (80%) appears to be attached to the storage vesicles, possibly as a major part of the protein component of the membrane itself. For this reason it is sometimes called "chromomembrin" (Hörtnagl *et al.*, 1972). The other enzymes are not associated with storage vesicles or with any other major organelles, as they are found in the supernatant of centrifuged homogenates of nervous tissue.

Synthesis of noradrenaline may take place throughout the entire neurone, but it is likely to occur most at the periphery, within the terminal swellings (varicosities) where great concentrations of storage vesicles are to be found. Control of this synthesis is on both a short and a long term basis (Axelrod *et al.*, 1970). Short-term control is exercised by noradrenaline itself which is known to inhibit the activity of tyrosine hydroxylase. The conversion of L-tyrosine to L-DOPA is thus the rate-limiting step and is an example of "end product inhibition". Long-term control of noradrenaline production is exercised by the cell body and its ability to synthesize enzyme proteins. This in its turn is influenced by the frequency of nerve impulse transmission. In this way maximal sympathetic nervous activity would result in maximal release of noradrenaline from the terminals and hence minimal (if any) inhibition of the first step in the chain (L-tyrosine to L-DOPA) so that synthesis of the transmitter would continue within the storage vesicles.

The high frequency of nerve impulses would stimulate the ribosomes on the rough endoplasmic reticulum in the cell body to synthesize more enzyme protein, whilst the Golgi apparatus increased its production of vesicles. New vesicles and new enzymes would then flow down the axon to replace those lost at the periphery. In addition to these influences on the production of noradrenaline there is evidence that the level of circulating hormones also has an important part to play. Both hydrocortisone (Eränkö and Eränkö, 1972) and oestrogen (Sjoberg, 1968) have been shown to increase the concentration of catecholamines in adrenergic tissue in rats.

XIII. Axonal flow

The axons of neurones are something like pipes, down which many constituents move from the centrally placed cell body outwards to the terminals at the periphery. There is also movement in the reverse direction. Mitochondria, transmitter storage vesicles, macromolecules including enzyme proteins as well as all the smaller metabolic chemicals move at different rates and in some cases by different mechanisms along the axons of both myelinated and unmyelinated neurones.

Many experiments have been based on the finding that if a ligature is tied tightly round a nerve trunk, certain constituents in the axon pile up in abnormally high concentrations immediately above the ligature. There may also be a tendency for a much smaller accumulation to occur immediately below the ligature (Dahlström and Fuxe, 1964; Dahlström, 1971; Akester and Akester, 1975). The large accumulation above the ligature is regarded as indicating a peripherally directed movement of organelles which has been blocked by the ligature. The much smaller accumulation below the ligature has been interpreted either as an injury reaction or possibly as an indication of reverse movement. Noradrenaline is easily demonstrated in the axons, thanks to the paraformaldehyde-induced fluorescence technique, and a region of very high fluorescence can be shown immediately above the ligature. Electronmicrographs of the swollen axons in this region show a massive pile-up of mitochondria, storage vesicles and other membrane-bound structures some of which are difficult to interpret.

Naturally a great deal of interest has been focused on the movement of transmitter vesicles and in particular, on the mechanism by which their movement down the axon takes place. A useful mini-review on this subject in the mammal has recently been published by Kerkut (1974) whose own proposal for vesicle transport is based on the assumption that the vesicle membrane is coded in some way (possibly by a protein) according to what is inside it and also to the type of myosin-like filament the vesicle should attach itself to.

Running throughout the whole length of the axon are several microtubules (25 nm diameter). These may be associated with a mesh of actin-like filaments. Kerkut suggested that the coded vesicles may attach to myosin-like filaments of various lengths. Long myosin-like filaments with many heads for attachment to actin-like filaments would move faster than short filaments with fewer heads. He suggested that the attachment of vesicle membrane to myosin-like filament may resemble the antibody–antigen reaction. By a method similar to the contraction of striated muscle, myosin-like filaments with their heads attached to the mesh of actin-like filaments would slide down the microtubules dragging the storage vesicles with them. There is the further possibility that storage vesicles may be able to take up and release calcium ions, in a way similar to the sarcoplasmic reticulum of striated muscle, and so control the availability of active sites on the actin-like filaments. Energy is required for vesicle transport as it can be prevented by metabolic poisons (cyanide, dinitrophenol, azides) anoxia and cold temperature.

A pile-up of noradrenaline has been shown by paraformaldehyde-induced fluorescence to occur above a ligature in isolated lengths of chicken sciatic nerve, demonstrating that neither the cell body nor the passage of impulses is essential for the movement of transmitter substance down the axon. When a length of nerve trunk was suspended in nutrient fluid "the wrong way up" movement of transmitter substance took place against gravity as quickly as it did when the proximal end was uppermost (S. N. Silver, unpublished). Strong evidence for the involvement of the microtubules in vesicle transport is provided by the fact that colchicine, which destroys microtubules, also prevents vesicle transport and that tubulin, the principal protein of the microtubules, has a high affinity for colchicine.

Further evidence of axonal flow is provided by the rate and distribution of fluorescence recovery in sympathetic fibres after noradrenaline depletion by reserpine (in rats). Fluorescence (noradrenaline) returns to the depleted neurones in a precise pattern. It is first seen 12 h after reserpine injection as a narrow zone around the nucleus, which spreads during the next 12 h to include the cytoplasm of the whole cell body. By 15 h after the injection fluorescence could be detected above a high ligature, and by 18 h at a more distal point in the nerve (sciatic).

These concentrations increased with time until at about 36 h recovery in the axons appeared to be complete. However, recovery in the terminals required about a week. From these observations it has been suggested (Dahlström and Häggendal, 1970) that the basis of recovery after reserpine depletion is the production of new storage vesicles by the cell body (12 h) and their progressive peripheral transportation through the cytoplasm of the cell body (24 h) down the axons (36 h) and into the terminal swellings (1 week). The average life span of a storage vesicle has been estimated by the same authors to be about 3 weeks.

XIV. Recognition of noradrenaline in postganglionic sympathetic neurones

A. Paraformaldehyde-induced fluorescence of catecholamines and indolamines

One of the most important advances in neurohistochemistry in recent years has been the development of an extremely sensitive and specific technique by which monoamine transmitter substances can be demonstrated inside the neurones.

The technique was originally worked out by Falck and Hillarp (Falck *et al.*, 1962; Falck, 1962) and is based on the fact that when monoamines are exposed to formaldehyde gas (at about 70% r. h., for 1 h at 80°C), in the presence of dried protein (produced by freeze-drying the tissue) which catalyzes the second part of the reaction, intensely fluorescent substances (isoquinolines) are formed. These can easily be recognized and photographed under a fluorescence microscope. The colour of the fluorescence is a yellowish green and can readily be distinguished from the autofluorescence of collagen, elastic fibres and various other naturally occurring biological substances found in the tissues.

This procedure can be used to demonstrate the primary catecholamines (dopamine and noradrenaline) as well as the secondary catecholamine (adrenaline), the indolamine (5-hydroxytryptamine) and the immediate precursors DOPA (3,4-dihydroxyphenylalanine) and 5-hydroxytryptophan, so long as the exposure to formaldehyde gas is at least twice as long for the secondary catecholamine as it is for the primary catecholamines or the indolamine. Catecholamines have their peak emission at 480 nm (looks green with yellow filter) whilst that for 5-hydroxytryptamine is at 520 nm (yellow) so that microspectrofluorometric measurements can be made if necessary. General accounts of the method have been given by Falck and Owman (1965) and by Corrodi and Jonsson (1967).

In this way noradrenergic fibres have been demonstrated in practically all parts of the avian body (Akester and Mann, 1969a; Akester *et al.*, 1969; Bennett and Malmfors, 1970; Akester and Akester, 1971; Akester, 1971; Bennett, 1971a,b; Bennett *et al.*, 1973). Of equal importance is the ability to recognize the few tissues which do not have this type of innervation. In this context it is interesting to compare the noradrenergic innervation of arteries and arterioles with that of veins (Fig. 7.15a,b). Both are influenced by sympathetic nerves but only on very rare occasions have fluorescent terminals been found within the muscular tunica media of arteries, whereas they penetrate the entire muscular part of the vein wall right down to the subendothelial region (Akester and Akester, 1971), as they do in the ductus deferens and ureter.

An extension of this procedure is to embed the tissue in low-melting-point wax and cut it in a cryostat. This makes it possible to carry out paraformaldehyde-induced fluorescence on one section and acetylcholinesterase histochemistry on the adjacent one, so that the distribution of tissue structures in the two sections is almost identical. Figure 7.18 demonstrates the advantages and limitations of this adjacent section technique. Figure 7.18a tells us that there are noradrenergic axons in the adventitia of a small arteriole. Three small nerve trunks containing noradrenergic axons lie further out in the adventitia (although they could be parts of the same trunk spiralling round the arteriole and cut in three different places). Figure 7.19b tells us that there are acetylcholinesterase-positive axons which appear to have exactly the same distribution as the fluorescent ones. From this information we cannot say whether we have demonstrated two different types of axon with the same distribution or a single type of axon which is both noradrenergic and acetylcholinesterase-positive (Akester and Mann, 1969a).

At one time it was generally assumed that an acetylcholinesterase-positive neurone was cholinergic but this may not always be the case. However, there are occasions when one may feel justified in presuming that certain acetylcholinesterase-positive fibres are likely to be cholinergic, especially if the tissue is assayed for choline acetyltransferase and found to contain a considerable amount of this enzyme. This was done by Akester and Mann (1969a) who suggested that the renal portal valve in the fowl had a very considerable double innervation of fluorescent (noradrenergic) sympathetic fibres (Fig. 7.6a,b) and also acetylcholinesterase-positive (presumed cholinergic) parasympathetic fibres. Figure 7.17 shows the precipitation of acetylcholinesterase end product on the axolemma of myelinated and unmyelinated axons in the adrenal gland, which are probably cholinergic. It also shows end product precipitation in the endoplasmic reticulum of a ganglion cell. A comprehensive account of cholinesterases has recently been published (Silver, 1974).

B. Characteristics of autonomic transmitter vesicles

One of the limitations of the paraformaldehyde-induced fluorescence techniques is the fact that it can only be used at the level of the light microscope. No such simple and reliable procedure is available for use with the electron

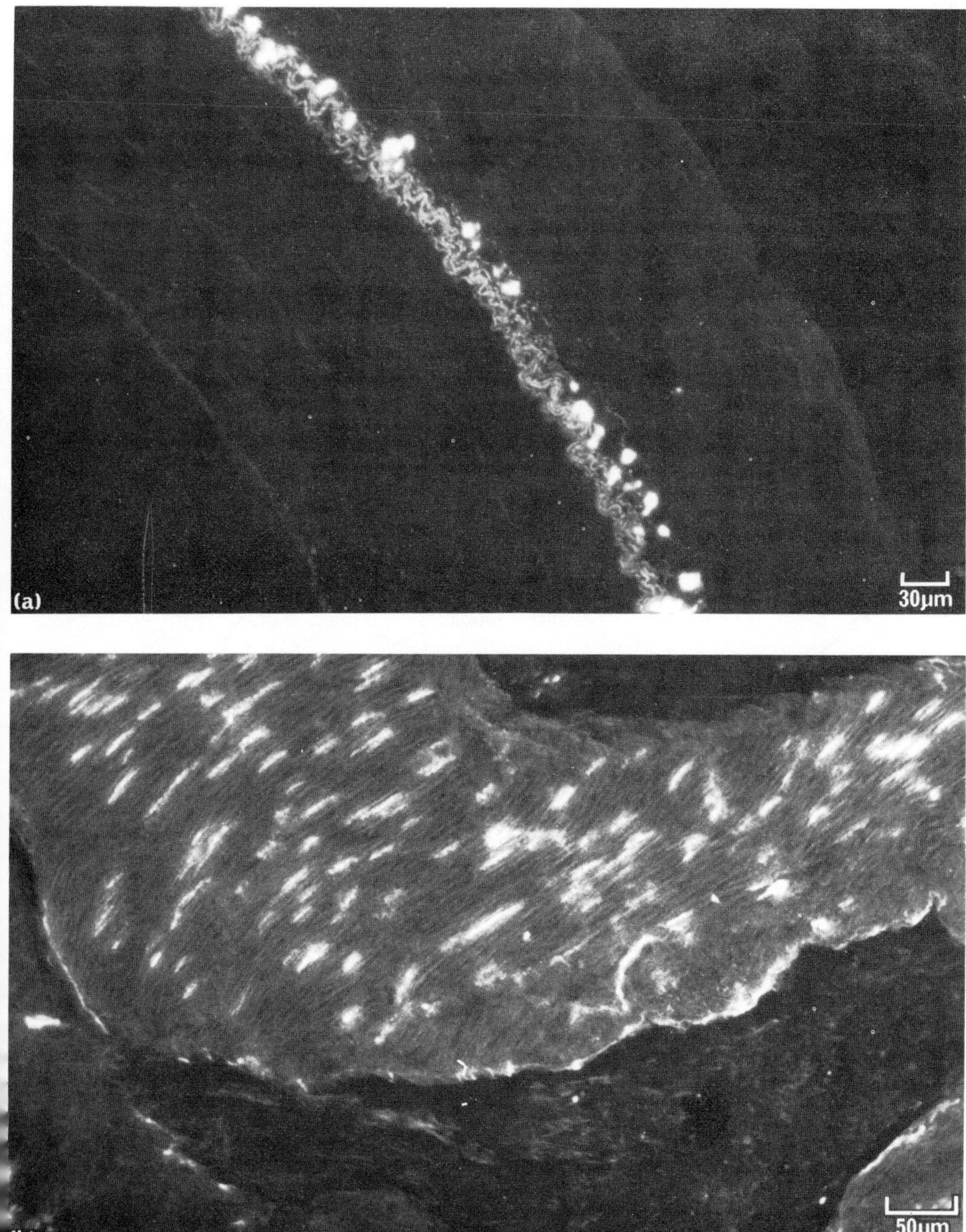

Fig. 7.15. Noradrenergic nerve distribution in a large artery (common carotid) and vein (caudal vena cava). *Gallus*.

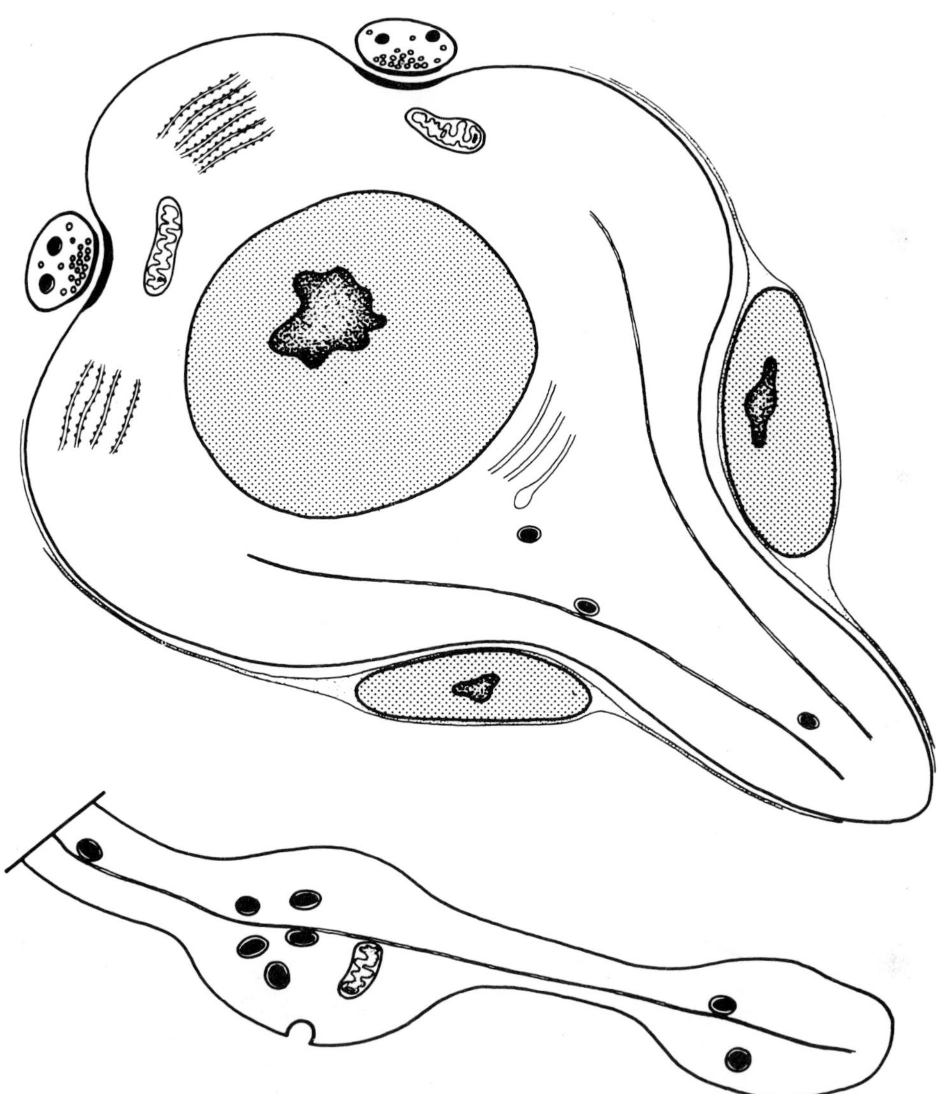

Fig. 7.16. Diagram to show formation and movement of dense-core vesicles. *Gallus*.

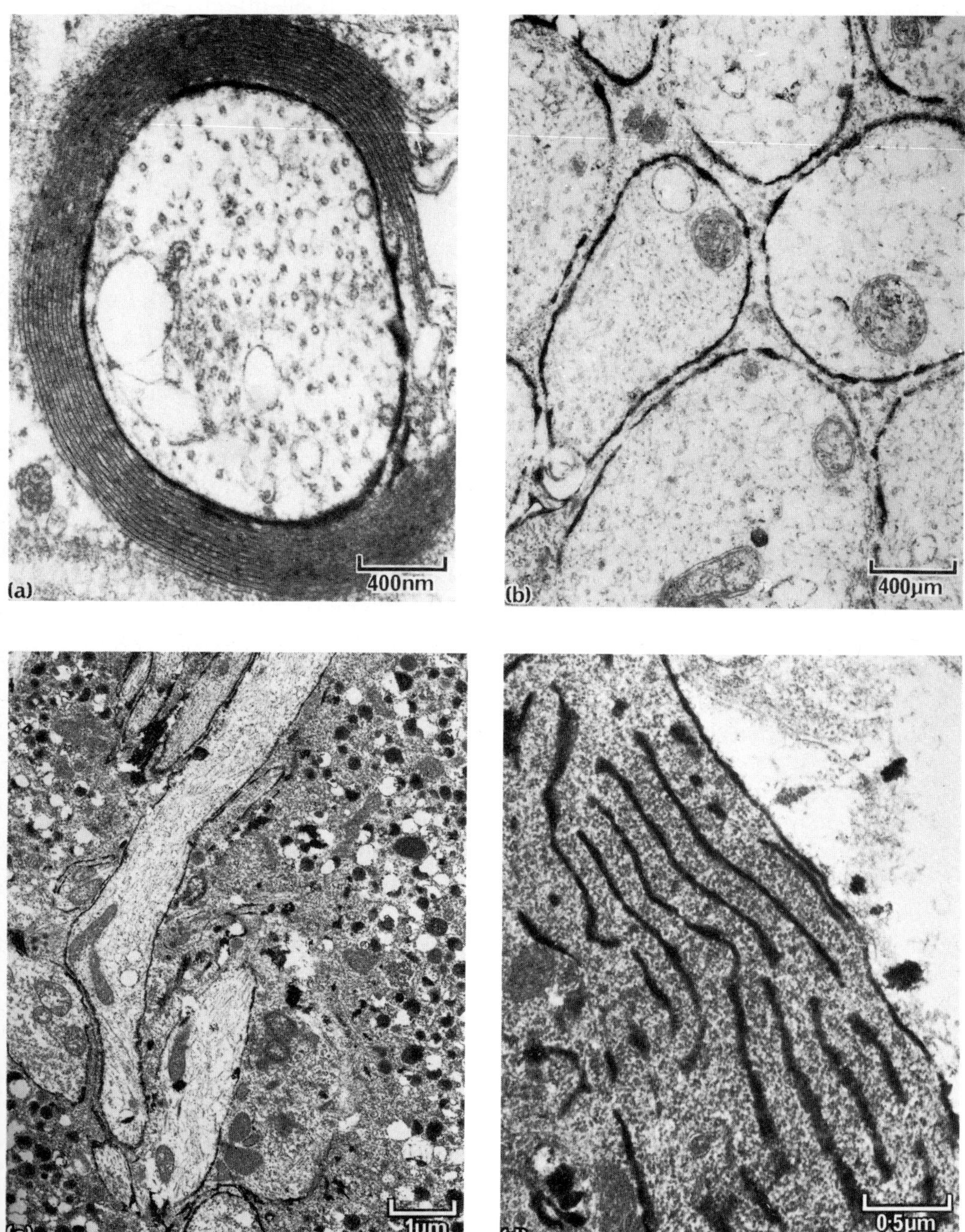

Fig. 7.17. Demonstration of acetylcholinesterase-positive neurones. *Gallus.*

microscope. A great deal of attention has been paid to the transmitter vesicles in mammalian postganglionic sympathetic terminals (Geffen and Livett, 1971; Blaschko and Smith, 1971) and a limited amount to their counterparts in birds (Bennett and Cobb, 1969; Akester, 1970; King *et al.*, 1975). The two groups are similar but not identical.

A typical noradrenergic terminal in the fowl contains a mixed population of large and small vesicles in approximately equal numbers, together with mitochondria and microtubules. The larger vesicles (70 nm) probably contain noradrenaline linked with protein in a stable and relatively nondiffusible form. When fixed with glutaraldehyde and stained with osmium this complex becomes electron-dense and produces the dark cores which tend to dominate the terminal as a whole (Fig. 7.11a,b,d).

The smaller vesicles (35 nm) also contain noradrenaline but in higher concentration than the larger ones. In the smaller vesicles the noradrenaline may be in a more diffusible form so that it, as well as various proteins, are lost during fixation and an electron-dense core does not develop when stained with osmium. It is the smaller vesicles that accumulate in the immediate vicinity of the synaptic thickening on the cell membrane (Fig. 7.11b,d) giving the impression that it is they, and not the large ones, which actually release noradrenaline from the axon during the passage of a nerve impulse. It is possible that the smaller vesicles are formed by division of the larger ones.

A cholinergic terminal, on the other hand, is characterized by large numbers of closely packed small clear vesicles (35 nm) and a small number of large dense-core vesicles. These do not carry any noradrenaline, but morphologically they are indistinguishable from those which are found in much greater numbers in the noradrenergic terminals.

In axon profiles remote from the terminal swellings, occasionally large dense-core vesicles are seen but not the smaller ones. This suggests that the larger vesicles may be formed in the cell body for transportation down the axon to accumulate in the varicosities, where they divide into the smaller ones. Although in most electron-micrographs the dense-core vesicles are circular in cross-section, they are frequently found to be elongated, and may even form short tube-like structures which are indented at intervals giving the impression that a tube is breaking up into spherical vesicles (Akester, 1970, 1971; Akester and Akester, 1975). The transmitter vesicle situation in avian autonomic nerves is by no means fully explained, and it is probably wise to retain an open mind on exactly what they contain and where they come from.

C. *Autoradiography of noradrenergic terminals*

Ever since it was shown that the main method for removing surplus noradrenaline from the tissue spaces, after nerve excitation and transmitter-release, was to take it back into the terminals from which it originally came ("uptake"), attempts have been made to present the terminals with substances chemically similar to the natural transmitter, in the hope that they would be taken up by the terminals as well. In this way radioactive noradrenaline has been used and when followed by autoradiography the distribution of silver grains supported

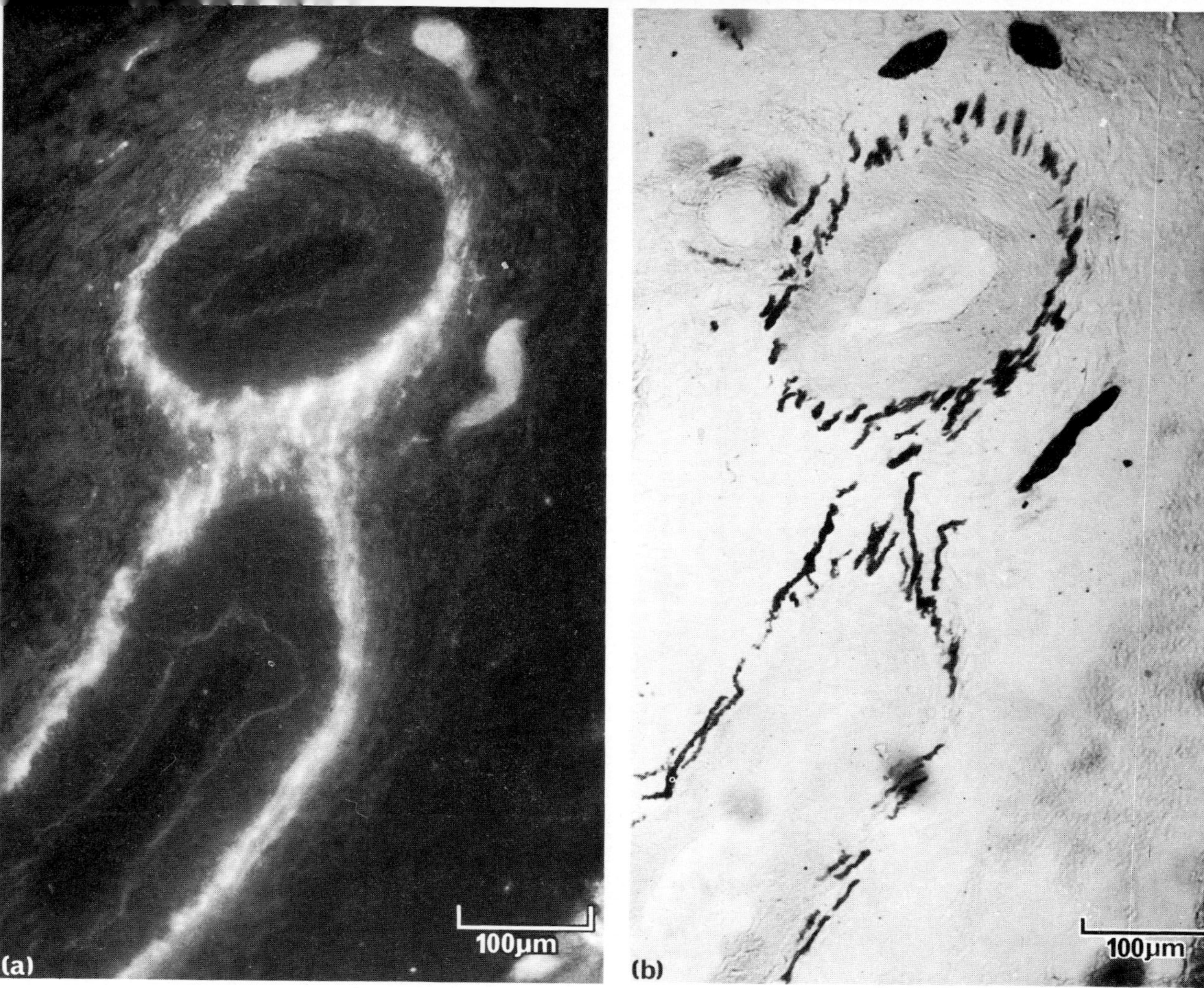

Fig. 7.18. Demonstration of noradrenergic and acetylcholinesterase-positive fibres on adjacent sections. *Gallus*.

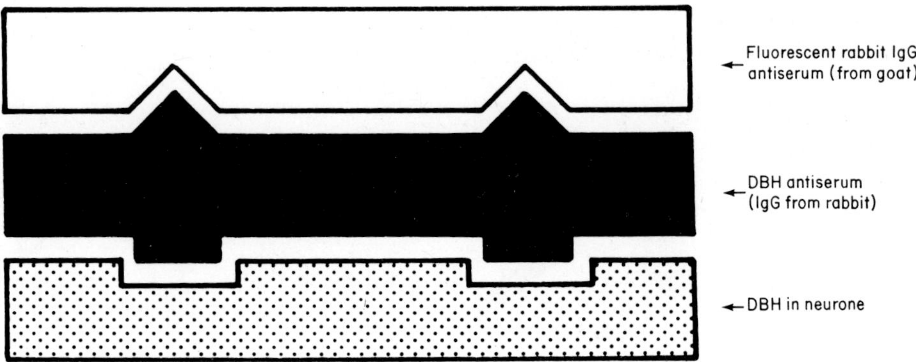

Fig. 7.19. Diagram to illustrate principle of the "sandwich" technique of immunofluorescence. *Gallus*.

the idea that terminals loaded with dense-core vesicles were noradrenergic. As yet this has not been done on avian tissue (but see Bower *et al.*, 1977). Catecholamines, such as noradrenaline under normal circumstances and 5-hydroxydopamine experimentally, are both taken up and released more readily by the smaller vesicles than by the larger ones. This supports the idea that the former are the more metabolically active and that the noradrenaline they contain is in a more diffusible form.

D. Uptake of 5- and 6-hydroxydopamine (5- and 6-OHDA) by noradrenergic terminals

Both 5- and 6-OHDA are taken up by postganglionic sympathetic neurones. The former is retained in all parts of the neurone and can be recognized in the form of small intensely dark granules which occur individually and in clusters and which were not present in the untreated controls. Neurones in which one can recognize these granules are thus noradrenergic (Akester and Akester, 1974).

A glance at Fig. 7.15 shows what 5-OHDA looks like when it has been taken up into noradrenergic neurones. It is presumed that none is taken up by cholinergic neurones. However, we cannot assume that the absence of small electron-dense granules (about 35 nm diameter) in an electronmicrograph of tissue that has been treated with 5-OHDA indicates a non-noradrenergic neurone. Obviously the uptake will be patchy in distribution and the size of the clusters will vary considerably. Figure 7.14a shows three small clusters of these granules in the cell body of a neurone in the cranial cervical ganglion of a chicken. Figure 7.14b shows a small process from a ganglion cell with a larger cluster of 5-OHDA granules. Figure 7.14c shows a nerve terminal making a synaptic contact with the cell body of a ganglion cell. In this case the ganglion cell has a small group of 5-OHDA granules close to its cell wall and is thus noradrenergic. The axon terminal beside it has no 5-OHDA and so we may presume that it is non-adrenergic, but we have no proof. However, if we assume that it is a cholinergic terminal (which it is likely to be) then we see that it has many small clear vesicles (35 nm) and a small number of larger dense-core vesicles (70 nm). Figure 7.14e shows a synapse similar to that in Fig. 7.14c. The cell body has taken up 5-OHDA but the terminal appears not to have done so. Here again we have many small clear vesicles and several larger dense-core vesicles (what are they doing?). Figure 7.14f shows a group of 5-OHDA granules in an axon, so we would expect it to be part of a noradrenergic postganglionic neurone; but the axon is myelinated, and postganglionic neurones are usually said to be nonmyelinated. Once again we realize that a good deal more clarification is required.

6-OHDA is also taken up by these neurones, but in this case the reaction is so severe that the neurones are destroyed and it has been claimed that this can be used as an experimental means of chemical sympathectomy (Thoenen and Tranzer, 1968; Bennett *et al.*, 1970; Bennett, 1971c). By combining this type of sympathectomy with the visual demonstration of the noradrenergic nerves Bennett *et al.* (1973), were able to follow the degeneration and subsequent regeneration of noradrenergic nerves in chicks of different ages. They found a greater destruction of nerves in older than younger birds (possibly due to greater uptake efficiency in the former) and a more rapid regeneration in the younger animals.

E. Dopamine β-hydroxylase immunofluorescence

The enzyme dopamine β-hydroxylase (DBH) is responsible for converting dopamine into noradrenaline, and is contained in the transmitter vesicles of noradrenergic neurones. If it could be visualized then we would have another method for demonstrating the distribution of these nerves. Such a technique (DBH-immunofluorescence) has been developed and it is said to give results comparable with those produced by paraformaldehyde-induced fluorescence (Hartman, 1973). The procedure involves the attachment of a fluorescent dye to the DBH enzyme in the neurones. To make the dye stick to the enzyme specifically, the intervention of two types of antibodies is necessary. The technique is therefore called an indirect method of immunofluorescence or the "sandwich" technique (Fig. 7.19). The following are required: (1) DBH-antiserum. This is prepared by injecting purified bovine DBH (from bovine adrenal medulla) into rabbits. Rabbit DBH-antiserum is then collected and its immunological purity is checked. This will be used to react with the DBH in the neurones. (2) Antiserum against rabbit antiserum. This is prepared by injecting rabbit immunoglobin (I_gG) into goats. Anti-rabbit I_gG is then collected. (3) The goat antiserum against rabbit I_gG is conjugated with a fluorescent dye (fluorescein isothyocynate), so that we now have a fluorescent antibody which can be seen.

Tissue slices are incubated with the DBH-antiserum (from rabbits) and with the fluorescent rabbit I_gG-antiserum (from goats), so that the former reacts with DBH enzyme in the neurones; the latter reacts with the former and we finish up with a sandwich of DBH-antiserum lying between, and holding together, the DBH enzyme and the fluorescent antibody. This means that wherever there was DBH (i.e. in the noradrenergic neurones) it will have fluorescent dye attached to it and can thus be seen and photographed using a fluorescence microscope.

It has been claimed that this procedure is just as precise and specific as that of paraformaldehyde-induced fluorescence. Cell bodies, axons and terminal varicosities have been demonstrated, as has the accumulation of DBH above a ligature. For noradrenergic nerves it is possible that immunofluorescence is more specific than paraformaldehyde-induced fluorescence. Certainly it is useful to be able to demonstrate the same type of neurone by different methods, one based on the visualization of the transmitter itself and the other making use of the enzyme which synthesizes the transmitter. As yet this method has not been applied to birds.

XV. Modulation of noradrenergic neurotransmission

Neurotransmission by postganglionic sympathetic neurones is controlled by the interaction of several factors. These include impulse firing rate along the axon, various chemical substances which are released in the immediate vicinity of the neuroeffector junction and receptors that may be pre- or postsynaptic and which may have a stimulatory or an inhibitory influence on the release of noradrenaline.

Some of the complexities of this control system have recently been summarized by Westfall (1977) and although this work is based on mammals it is probable that a broadly similar situation will be found in birds. It has been suggested that both stimulatory and inhibitory receptors are situated on the cell membrane of the axon varicosities. These respond in such a way that a balance of stimulatory and inhibitory influences will determine whether more or less noradrenaline is released.

Presynaptic axon receptors may be influenced by a feedback mechanism involving noradrenaline itself (autoregulation), or by chemical substances (prostaglandins) which may be released from the (postsynaptic) neuroeffector cells. Other chemicals may be released from nearby axons (acetylcholine-muscarinic) brought from a distant organ by the blood (angiotensin) or formed locally by cell metabolism. Any, or all, of these substances may influence the release of noradrenaline from the vesicles within the terminal varicosities of postganglionic sympathetic neurones, and so modulate neurotransmission.

The exact nature of the axon receptors is unknown but experiments with stimulating and blocking drugs suggest that one type may be some form of α-adrenergic receptor similar to, but not identical with, those on smooth muscle cells. The mechanism of their action is not known either but it is possible that, when stimulated, they reduce the availability of Ca^{2+} upon which exocytosis is thought to be dependent. In this way excess noradrenaline in the synaptic space would stimulate α-receptors on the axon varicosities, which would inhibit further release of noradrenaline (negative-feedback) by reducing the availability of Ca^{2+}.

It is for this reason that blockade of α-adrenergic receptors, by such drugs as phenoxybenzamine or dibenamine, increases the amount of surplus noradrenaline (overflow) which results from nerve stimulation. Such a blockade also increases the outflow of dopamine β-hydroxylase, which is released from the storage vesicles at the same time as noradrenaline, and for which there is no known uptake mechanism. This strongly supports the idea that noradrenaline stimulates axon α-receptors to reduce further release of noradrenaline by inhibiting exocytosis, rather than to increase noradrenaline removal by stimulating uptake (Westfall, 1977).

Another negative-feedback system is based on the release of prostaglandins (of the E series, mostly E_2), by the neuroeffector cells. Prostaglandins are a widespread group of biologically active acid lipids, all of which are derivatives of prostanoic acid. They have been isolated from many mammalian species and some birds, and have been found in several different organs. In mammals, prostaglandins of the E series have been shown to depress the normal response of the heart, blood vessels, spleen, ductus deferens and oviduct to noradrenaline and also to sympathetic nerve stimulation.

Support for the idea of a prostaglandin-dependent negative-feedback system has come from the use of inhibitors of prostaglandin synthesis (indomethacin). If this substance is perfused through an organ, like the spleen or uterus, there is a vasoconstriction caused by a reduction in the level of prostaglandin synthesis. There is also an enhanced response to sympathetic nerve stimulation by the heart, ductus deferens and spleen after infusion by inhibitors of prostaglandin synthesis. This is due to the greater release of noradrenaline by sympathetic nerve terminals in response to a reduction of prostaglandin inhibition.

It has further been suggested that a third negative-feedback system, by which release of noradrenaline from sympathetic terminals may be inhibited, is based on the release of acetylcholine from nearby cholinergic (parasympathetic fibres). By this mechanism postganglionic parasympathetic activity would not only stimulate its own type of response by the release of acetylcholine, but would simultaneously inhibit that of neighbouring sympathetic axons by a muscarine-inhibitory system. In this way acetylcholine would stimulate muscarine receptors on the sympathetic axon varicosities which, in turn, would inhibit the release of noradrenaline.

This raises the interesting possibility that an important role of parasympathetic nerves in some tissues (mammalian ductus deferens) may be to modulate the activity of sympathetic nerves which run beside them. It is also possible that the three types of inhibitory receptors thought to be present on sympathetic axon varicosities (α-adrenergic, prostaglandin and acetylcholine-muscarinic) may all work in the same way, that is by reducing the availability of Ca^{2+} and hence inhibiting the release of noradrenaline by exocytosis.

Still more inhibitory systems have been proposed, mostly on the grounds of pharmacological studies. We have already seen that dopamine is released by small intensely fluorescent cells (SIF cells or interneurones) in sympathetic ganglia with an inhibitory effect on the postganglionic neurones and it is now thought that inhibitory dopaminergic receptors may also occur on the axon terminals of these neurones. Histamine, ATP, ADP, adenosine and enkephalin have all been shown, in one tissue or another, to inhibit the release of noradrenaline from sympathetic nerves. For a more detailed consideration of this work the reader is again referred to Westfall (1977).

In addition to the many negative-feedback mechanisms already considered there may well be positive-feedback mechanisms which increase the release of noradrenaline from sympathetic terminals. Much less information is available about these stimulatory systems, but what there is suggests that β-adrenergic, angiotensin, acetylcholine-nicotinic and 5-hydroxytryptamine may be involved.

From this very brief summary of the many factors which are thought to modulate sympathetic noradrenergic neurotransmission it will be clear that a mechanism has evolved (in mammals and most probably in birds), the complexity of which appears well suited to provide the subtleties of nervous control. However, in spite of all this, the dominant control system appears to be the most obvious, that of negative-feedback by noradrenaline itself.

References

Abercrombie, M., Heaysman, J. E. M. and Pergrum, S. M. (1970). The locomotion of fibroblasts in culture. III. Movements of particles on the dorsal surface of the leading lamella, *Expl. Cell Res.* **62**, 389–398.

Agostoni, E., Chinnock, J. E., Daly, M. de B. and Murrary, J. G. (1957). Functional and histological studies of the vagus nerve and its branches to the heart, lungs and abdominal viscera in the cat, *J. Physiol.* **135**, 182–205.

Akerman, B., Anderson, B., Fabricius, E. and Svenssen, L. (1960). Observations on central regulation of body temperature and of food and water intake in the pigeon (*Columba livia*), *Acta. Physiol. Scand.* **50**, 328–336.

Akester, A. R. (1970). Dense-core vesiculated axons in the renal portal valve of the domestic fowl, *J. Anat.* **106**, 185–186.

Akester, A. R. (1971). The blood vascular system. *In* "Physiology and Biochemistry of the Domestic Fowl" (D. J. Bell and B. M. Freeman, Eds), Vol. 2. Academic Press, London and New York.

Akester, A. R. and Akester, B. V. (1971). Double innervation of the avian cardiovascular system, *J. Anat.* **108**, 618–619.

Akester, A. R. and Akester, B. V. (1974). Appearance of autonomic nerves in adrenal gland and superior cervical ganglion of the domestic fowl, after treatment with 5-hydroxydopamine (5-OHDA), *J. Anat.* **117**, 660–661.

Akester, A. R. and Akester, B. V. (1975). Ultrastructure and experimental studies of Remak's autonomic nerve trunk in the fowl, *J. Anat.* **119**, 416–417.

Akester, A. R., Akester, B. V. and Mann, S. P. (1969). Catecholamines in the avian heart, *J. Anat.* **104**, 591.

Akester, A. R., Anderson, R. S., Hill, K. J. and Osbaldiston, G. W. (1967). A radiographic study of urine flow in the domestic fowl, *Br. Poult. Sci.* **8**, 209–212.

Akester, A. R. and Brackenbury, J. H. (1976). The avian pecten and intra-ocular pressure responses to systemic drugs and arterial occlusion, *J. Anat.* **122**, 742–743.

Akester, A. R. and Mann, S. P. (1969a). Adrenergic and cholinergic innervation of the renal portal valve in the domestic fowl, *J. Anat.* **104**, 241–252.

Akester, A. R. and Mann, S. P. (1969b). Ultrastructure and innervation of the tertiary-bronchial unit in the lung of *Gallus domesticus*, *J. Anat.* **105**, 202–204.

Anand, B. K. (1961). Nervous regulation of food intake, *Physiol. Rev.* **41**, 677–708.

Axelrod, J., Mueller, R. A. and Thoenen, H. (1970). *In* "New Aspects of Storage and Release Mechanisms of Catecholamines" (H. J. Schümann and G. Kroneberg, Eds), pp. 212–219 (Bayer Symposium II). Springer-Verlag, Berlin.

Axelrod, J., Shein, H. and Wurtman, R. J. (1968). "The Pineal", p. 59. Academic Press, New York and London.

Bennett, T. (1971a). The neuronal and extra-neuronal localizations of biogenic amines in the cervical region of the domestic fowl (*Gallus gallus domesticus*), *Z. Zellforsch. mikrosk. Anat.* **112**, 443–464.

Bennett, T. (1971b). The adrenergic innervation of the pulmonary vasculature, the lung and the thoracic aorta, and on the presence of aortic bodies in the domestic fowl (*Gallus gallus domesticus*), *Z. Zellforsch. mikrosk. Anat.* **114**, 117–134.

Bennett, T. (1971c). Fluorescence histochemical and functional studies on adrenergic nerves following treatment with 6-hydroxydopamine. *In* "6-Hydroxydopamine and Catecholamine Neurons" (T. Malmfors and H. Thoenen, Eds), pp. 303–314. North-Holland Publishing Company, Amsterdam.

Bennett, T. (1974). Peripheral and autonomic nervous systems. *In* "Avian Biology" (D. S. Farner and J. R. King, Eds), Vol. 4. Academic Press, London and New York.

Bennett, T., Burnstock, G., Cobb, J. L. S. and Malmfors, T. (1970). An ultrastructural and histochemical study of the short-term effects of 6-hydroxydopamine on adrenergic nerves in the domestic fowl, *Br. J. Pharmac.* **38**, 802–809.

Bennett, T. and Cobb, J. L. S. (1969). Studies on the avian gizzard: Auerbach's plexus, *Z. Zellforsch. mikrosk. Anat.* **99**, 109–122.

Bennett, T. and Malmfors, T. (1970). The adrenergic nervous system of the domestic fowl (*Gallus gallus domesticus*), *Z. Zellforsch. mikrosk. Anat.* **106**, 22–50.

Bennett, T., Malmfors, T. and Cobb, J. L. S. (1973). Fluorescence histochemical observations on catecholamine-containing cell bodies in Auerbach's plexus, *Z. Zellforsch. mikrosk. Anat.* **139**, 69–81.

Bennett, T., Malmfors, T. and Cobb, J. L. S. (1973). A fluorescence histochemical study of the degeneration and regeneration of noradrenergic nerves in the chick following treatment with 6-hydroxydopamine, *Z. Zellforsch.* **142**, 103–130.

Bichat, X. (1801–3). "Traité d'anatomie descriptive". J. B. Baillière et Fils, Paris.

Biederman-Thorson, M. and Thorson, J. (1973). Rotation-compensating reflexes independent of the labyrinth and the eye. Neuromuscular correlates in the pigeon, *J. comp. Physiol.* **83**, 103–122

Björklund, A., Cegrell, L., Falck, B., Ritzen, M. and Rosengren, E. (1970). Dopamine-containing cells in sympathetic ganglia, *Acta Physiol. Scand.* **78**, 334–338.

Blaschko, H. K. F. and Smith, A. D. (1971). Organizers, a discussion on subcellular and macromolecular aspects of synaptic transmission, *Phil. Trans. Roy. Soc. Lond. B.* **261**, 273–437.

Blaschko, H. K. F. (1973). Catecholamine biosynthesis, *Br. med. Bull.* **29**, 105–109.

Bower, A. J., Parker, S. and Malony, V. (1977). Receptor sites in the extrapulmonary airways of the domestic hen, *J. Anat.* **124**, 254.

Branton, D. (1966). Fracture faces of frozen membranes, *Proc. Nat. Acad. Sci., U.S.A.* **55**, 1048–1056.

Brobeck, J. R. (1946). Obesity in animals with hypothalamic lesions, *Physiol. Rev.* **26**, 541–559.

Brown, C. M., Molony, V., King, A. S. and Cook, R. D. (1972). Fibre size and conduction velocity in the vagus of the domestic fowl (*Gallus gallus domesticus*), *Acta Anat.* **83**, 451–460.

Bubien-Waluszewska, A. (1968). Le groupe caudale des nerfs crâniens de la poule domestique (*Gallus domesticus*), *Acta Anat.* **69**, 445–457.

Burger, R. E., Osborne, J. L. and Banzett, R. B. (1974). Intrapulmonary chemoreceptors in *Gallus domesticus*: adequate stimulus and functional localization, *Resp. Physiol.* **22**, 87–97.

Burt, A. M. and Narayanan, C. H. (1976). Choline acetyltransferase, choline kinase, and acetylcholinesterase activities during the development of the chick ciliary ganglion, *Exp. Neurol.* **53**, 703–713.

Butler, P. J. and Osborne, M. P. (1975). The effect of cervical vagotomy (decentralization) on the ultrastructure of the carotid body of the duck, *Anas platyrhynchos*, *Cell. Tiss. Res.* **163**, 491–502.

Cantino, D. and Mugnaini, E. (1974). Adrenergic innervation of the parasympathetic ciliary ganglion in the chick, *Science* **185**, 279–280.

Cantino, D. and Mugnaini, E. (1975). The structural basis of electrotonic coupling in the avian ciliary ganglion . A study with thin sectioning and freeze-fracturing, *J. Neurocytol.* **4**, 505–536.

Chamley, J. H., Mark, G. E., Campbell, G. R. and Burnstock, G. (1972). Sympathetic ganglia in culture. I. Neurones, II accessory cells, *Z. Zellforsch.* **135**, 287–327.

Cohen, D. H. and Schnall, A. M. (1970). Medullary cells of origin of vagal cardioinhibitory fibres in the pigeon. II. Electrical stimulation of the dorsal motor nucleus, *J. comp. Neurol.* **140**, 321–342.

Corrodi, H. and Jonsson, G. (1967). The formaldehyde fluorescence method for the histochemical demonstration of biogenic monoamines, *J. Histochem. Cytochem.* **15**, 65–78.

Coupland, R. E. (1965). "The Natural History of the Chromaffin Cell." Longmans Green, New York.

Dahlström, A. (1971). Axoplasmic transport (with particular respect to adrenergic neurons), *Phil. Trans. Roy. Soc. Lond. B.* **261**, 325–358.

Dahlström, A. and Fuxe, K. (1964). A method for the demonstration of adrenergic nerve fibres in peripheral nerves, *Z. Zellforsch.* **62** 602–607.

Dahlström, A. and Häggendal, J. (1970). Biochemistry of simple neuronal models. *In* "Advances in Biochemical Psychopharmacology", (E. Costa and E. Giacobini, Eds), Vol. 2. Raven Press.

Davison, A. N. and Peters, A. (1970). "Myelination." C. C. Thomas, Springfield, Illinois.

De Castro, F. (1951). Sur la structure de la synapse dans les chemoreceptuers: leur mécanisme d'excitation et role dans la circulation sanguine locale, *Acta Physiol. Scand.* **22**, 14–43.

Delius, J. D. (1967). Displacement activities and arousal, *Nature* **214**, 1259.

Delius, J. D. (1971). Foraging behaviour patterns of herring gulls elicited by electrical forebrain stimulation, *Experientia* **27**, 1287.

de Lorenzo, A. J. D. (1960). The fine structure of synapses in the ciliary ganglion of the chick, *J. Biophys. Biochem. Cytol.* **7**, 31–36.

Dogiel, A. S. (1896). Zwei Arten sympathischer Nervenzellen, *Anat. Anz.* **11**, 679–687.

Dunant, Y. et Dolivo, M. (1967). Relations entre les potentiels sympatiques lents et l'excitabilité du ganglion sympathique chez le rat, *J. Physiol., Paris* **59**, 281–294.

Edidin, M. and Fambrough, D. (1973). Fluidity of the surface of cultured muscle fibres, *J. Cell Biol.* **57**, 27–37.

Eränkö, O. and Eränkö, L. (1971). Small intensely fluorescent, granule-containing cells in the sympathetic ganglion of the rat, *Progr. Brain Res.* **34**, 39–52.

Eränkö, L. and Eränkö, O. (1972). The effect of hydrocortisone on histochemically demonstrable catecholamines in the sympathetic ganglia and extra-adrenal chromaffin tissue in the rat, *Acta Physiol., Scand.* **84**, 125–133.

Eränkö, O. and Härkönen, M. (1965). Monoamine-containing small cells in the superior cervical ganglion of the rat and an organ composed of them, *Acta Physiol. Scand.* **63**, 511–512.

Everett, S. D. (1967). Pharmacological studies on the alimentary canal of the domestic fowl. Ph.D. Thesis, University of London.

Falck, B. (1962). Observations on the possibilities of the cellular localization of monoamines by a fluorescence method, *Acta Physiol. Scand.* **56**, Suppl. 197.

Falck, B., Hillarp, N. Å., Thieme, G. and Torp, A. (1962). Fluorescence of catecholamines and related compounds condensed with formaldehyde, *J. Histochem. Cytochem.* **10**, 348–354.

Falck, B. and Owman, C. (1965). A detailed methodological description of the fluorescence method for the cellular demonstration of biogenic amines, *Acta Univ. Lund.*, Sectio 11, No. 7.

Fedde, M. R., Gatz, R. N., Slama, H. and Scheid, P. (1974). Intrapulmonary CO_2 receptors in the duck: 1. Stimulus specificity, *Resp. Physiol.* **22**, 87–97.

Fourman, J. and Ballantyne, B. (1967). Cholinesterase activity in the Harderian gland of *Anas domesticus*, *Anat. Rec.* **159**, 17–27.

Fuxe, K., Goldstein, M., Hökfelt, T. and Joh, T. H. (1971). Cellular localization of dopamine-β-hydroxylase and phenylethanolamine-*N*-methyl transferase by immunohistochemistry, *Progr. Brain Res.* **34**, 127–138.

Geffen, L. B. and Livett, B. G. (1971). Synaptic vesicles in sympathetic neurons, *Phys. Rev.* **51**, 98–157.

Geren, B. B. (1954). The formation from the Schwann cell surface of myelin in the peripheral nerves of chick embryos, *Expl. Cell Res.* **7**, 558–562.

Gitlin, G. and Singer, M. (1974). Myelin movements in mature mammalian peripheral nerve fibres, *J. Morph.* **143**, 167–186.

Gundlach, R. H. (1934). The speed of pupillary contraction in response to light in pigeons, cats and humans, *J. Genet. Psychol.* **44**, 250–252.

Haber, E. and Wrenn, S. (1976). Problems of identification of the beta-adrenergic receptor, *Phys. Rev.* **56**, 317–338.

Halpern, B. P. (1962). Gustatory nerve responses in the chicken, *Am. J. Physiol.* **203**, 541–544.

Hanwell, A., Linzell, J. L. and Peaker, M. (1972). Nature and location of the receptors for salt-gland secretion in the goose, *J. Physiol.* **226**, 453–472.

Hartman, B. K. (1973). Immunofluorescence of dopamine-β-hydroxylase. Application of improved methodology to the localization of the peripheral and central noradrenergic nervous system, *J. Histochem. Cytochem.* **21**, 312–332.

Harwood, D. and Vowles, D. M. (1966). Forebrain stimulation and feeding behaviour in the ring dove (*Streptopelia risoria*), *J. Comp. physiol. Psychol.* **62**, 388–396.

Hassan, T. (1966). A pharmacological study of the alimentary tract of the fowl. M.Sc. Thesis, University of Edinburgh.

Hillårp, N. A. and Olivecrona, H. (1946). Role played by axons and Schwann cells in degree of myelination of peripheral nerve fibres, *Acta Anat.* **2**, 17.

Holmgren, E. (1899). Weitere Mitteilungen über den Bau der Nervenzellen, *Anat. Anz.* **16**, 388–397.

Hörtnagl, H., Winkler, H. and Lochs, M. (1972). *Biochem. J.* **129**, 187–195.

Hsieh, T. M. (1951). The sympathetic and parasympathetic nervous system of the fowl. Ph.D. Thesis, Edinburgh.

Hunt, R. (1899). Direct and reflex acceleration of the mammalian heart with some observations on the relations of the inhibitory and accelerator nerves, *Am. J. Physiol.* **2**, 395–470.

Isomura, G. (1974). Nerve centres for sphincter muscles of the iris in the fowl, *Anat. Anz. Bd.* **135**, S. 178–190.

Jacobowitz, D. M. and Greene, L. A. (1974). Histofluorescence study of chromaffin cells in dissociated cell cultures of chick embryo sympathetic ganglia, *J. Neurobiol.* **5**, 65–83.

Jenkinson, D. H. (1973). Classification and properties of peripheral adrenergic receptors, *Br. Med. Bull.* "Catecholamines" **29**, 142–147.

Jones, D. R. and Johansen, K. (1972). *In* "Avian Biology" (D. S. Farner and J. R. King, Eds), Vol. 2. Academic Press, London and New York.

Kerkut, G. A. (1974). Mini-Review. Axoplasmic Transport, *Comp. Biochem. Physiol.* **51A**, 701–704.

King, A. S., King, Zoe D., Hodges, R. D. and Henry, J. (1975). Synaptic morphology of the carotid body of the domestic fowl, *Cell Tiss. Res.* **162**, 459–473.

Klein, D. C., Berg, G. R. and Weller, J. (1970). Melatonin Synthesis: adenosine 3′, 5′-Monophosphate and norepinephrine stimulate *N*-acetyltransferase, *Science* **168**, 979–980.

Koda, L. Y. and Partlow, L. M. (1976). Membrane marker movement on sympathetic axons in tissue culture, *J. Neurobiol.* **7**, 157–172.

Kotilainen, P. V. and Putkonen, P. T. S. (1974). Respiratory and cardiovascular responses to electrical stimulation of the avian brain with emphasis on inhibitory mechanisms, *Acta Physiol. Scand.* **90**, 358–369.

Kuenzel, W. J. (1972). Dual hypothalamic feeding system in a migratory bird, *Zonotrichia albicollis*, *Am. J. Physiol.* **223**, 1138–1142.

Kuffler, S. W. and Nicholls, J. G. (1976). "From Neuron to Brain." Sinauer Associates, Massachusetts.

Landmesser, L. and Pilar, G. (1970). Selective reinnervation of two cell populations in the adult pigeon ciliary ganglion, *J. Physiol.* **211**, 203–216.

Landmesser, L. and Pilar, G. (1972). The onset and development of transmission in the chick ciliary ganglion, *J. Physiol., Lond.* **222**, 691–713.

Langley, J. N. (1904). On the sympathetic system of birds, and on the muscles which move the feathers, *J. Physiol.* **30**, 221–252.

Lauber, K. J., Boyd, J. E. and Boyd, T. A. S. (1972). Sympathetic denervation effects on avian eye development and aqueous fluid dynamics, *Proc. Soc. exp. Biol. Med.* **140**, 351–356.

Leitner, L. M. and Roumy, M. (1974). Vagal afferent activities related to the respiratory cycle in the duck: sensitivity to mechanical, chemical and electrical stimuli, *Resp. Physiol.* **22**, 41–56.

Lenhossék, M. von (1911). Das Ganglion Ciliare der Vögel, *Archiv. für Mikroskopische Anat.* **76**, 745–769.

Leonard, R. B. and Cohen, D. H. (1975). Responses of sympathetic postganglionic neurons to peripheral nerve stimulation in the pigeon (*Columba livia*), *Exp. Neurol.* **49**, 466–486.

Libet, B. and Owman, Ch. (1974). Concomitant changes in formaldehyde-induced fluorescence of dopamine interneurones and in slow inhibitory post-synaptic potentials of the rabbit superior cervical ganglion, induced by stimulation of the preganglionic nerve or by a muscarinic agent, *J. Physiol.* **237**, 635–662.

Livett, B. G., Geffen, L. B. and Austin, L. (1968). Prioxmodistal transport of ^{14}C-noradrenaline and protein in sympathetic nerves, *J. Neurochem.* **15**, 931.

MacDonald, R. L. and Cohen, D. H. (1970). Cells of origin of sympathetic pre- and post-ganglionic cardioacceleratory fibres in the pigeon, *J. Comp. Neur.* **140**, 343–358.

MacDonald, R. L. and Cohen, D. H. (1973). Heart rate and blood pressure responses to electrical stimulation of the control nervous system in the pigeon (*Columba livia*), *J. Comp. Neur.* **150**, 109–136.

Marchisio, P. C. and Consolo, S. (1968). Developmental changes in choline acetyltransferase activity in chick embryo spinal and sympathetic ganglia, *J. Neurochem.* **15**, 759–764.

Marchisio, P. C. and Giacobini, G. (1969). Choline acetyltransferase activity in the central nervous system of the developing chick, *Brain Res.* **15**, 301–304.

Marrazzi, A. S. (1939). Adrenergic inhibition at sympathetic synapses, *Am. J. Physiol.* **127**, 738–744.

Martin, A. R. and Pilar, G. (1963). Dual mode of synaptic transmission in the avian ciliary ganglion, *J. Physiol.* **168**, 443–463.

Marwitt, R., Pilar, G. and Weakly, J. N. (1971). Characterisation of two ganglion cell populations in avian ciliary ganglion, *Brain Res.* **25**, 317–334.

Masurovsky, E. B., Benitez, H. H., Kim, S. U. and Murray, M. R. (1970). Origin, development and nature of intranuclear rodlets and associated bodies in chicken sympathetic neurons, *J. Cell. Biol.* **44**, 172–191.

Mathews, M. R. and Raisman, G. (1969). The ultrastructure and somatic efferent synapses of small granule-containing cells in the superior cervical ganglion, *J. Anat.* **105**, 255–282.

McLelland, J. and Abdalla, A. B. (1972). The gross anatomy of the nerve supply to the lungs of *Gallus domesticus*, *Anat. Anz. Bd.* **131**, S. 448–453.

Mitchell, G. A. G. (1953). "Anatomy of the Autonomic Nervous System." E. and S. Livingstone, Edinburgh.

Narayanan, C. H. and Narayanan, Y. (1976). An experimental inquiry into the central source of preganglionic fibres to the chick ciliary ganglion, *J. Comp. Neur.* **166**, 101–110.

Osborne, J. L. and Burger, R. E. (1974). Intrapulmonary chemoreceptors in *Gallus domesticus*, *Resp. Physiol.* **22**, 77–85.

Pearson, R. (1972). "The Avian Brain." Academic Press, London and New York.

Pfenninger, K., Akert, K., Moor, H. and Sandri, C. (1972). The fine structure of freeze-fractured presynaptic membranes, *J. Neurocytology* **1**, 129–149.

Phillips, R. E. (1964). "Wildness" in the Mallard duck: effects of brain lesions and stimulation on "escape behaviour" and reproduction, *J. Comp. Neurol.* **122**, 139–195.

Phillips, R. E. and Youngren, O. M. (1971). Brain stimulation and species typical behaviour: activities evoked by electrical stimulation of the brains of chickens (*Gallus gallus*), *An. Behav.* **19**, 759–779.

Pilar, G., Jenden, D. J. and Campbell, B. (1973). Distribution of acetylcholine in the normal and denervated pigeon ciliary ganglion, *Brain Res.* **49**, 245–256.

Pomerat, C. M. (1959). Rhythmic contractions of Schwann cells, *Science* **130**, 1759–1760.

Putkonen, P. T. S. (1967). Electrical stimulation of the avian brain. *Ann. Acad. Scien. Fenn.*, Series A, Sec. V, Med.-anthrop., pp. 9–95.

Revel, J. P. and Karnovsky, M. J. (1967). Hexagonal array of subunits in intercellular junctions of the mouse heart and liver, *J. Cell. Biol.* **33**, C7.

Richards, S. A. (1971). Brain stem control of polypnoea in the chicken and pigeon, *Resp. Physiol.* **11**, 315–326.

Saari, M., Johansson, G. and Huhtala, A. (1974). Organization of the myelinated nerve fibres in the iris of the white leghorn hen, *Acta Anat.* **90**, 480–494.

Saccomanno, G. (1943). The components of the upper thoracic sympathetic nerves, *J. Comp. Neur.* **79**, 355–378.

Schon, F. and Kelly, J. S. (1975). Selective uptake of [$^{-3}$H]-alanine by glia: association with glial uptake system for GABA, *Brain Res.* **86**, 243–257.

Sclafani, A. (1971). Neural pathways involved in the ventromedial syndrome in the rat, *J. Comp. Physiol.* **77**, 70–96.

Siegrist, G., Dolivo, M., Dunant, Y., Foroglou-Kerameus, C., Ribaupierre, Fr. De and Roullier, Ch. (1968). Ultrastructure and function of the chromaffin cells in the superior cervical ganglion of the rat, *J. Ultrastr. Res.* **25**, 381–407.

Silver, A. (1974). "The Biology of Cholinesterases." North Holland, Amsterdam.

Singer, J. (1884). Zur kenntnis der motorischen Funktionen des Lendenmarks der Taube. *S.-B. Akad. Wiss, Wien, math.-nat.* Kl. **89** (III), 167–185.

Singer, S. J. and Nicolson, G. L. (1972). The fluid mosaic model of the structure of cell membranes, *Science* **175**, 720–731.

Sjöberg, N. O. (1968). Increase in transmitter content of noradrenergic nerves in the reproductive tract of female rabbits after oestrogen treatment, *Acta endocr. Copenh.* **57**, 405–413.

Sorimachi, M. and Kataoka, K. (1974). Developmental change of choline acetyltransferase and acetylcholinesterase in the ciliary and the superior cervical ganglion of the chick, *Brain Res.* **70**, 123–130.

Stammer, A. (1964). Ein Beidrag zur Struktur und mikroskospischen Innervation der Harderschen Drüse der Vögel, *Acta Universit. szegediensis* **10**, 99–107.

Taylor, R. B., Duffus, W. P. H., Raff, M. C. and De Petris, S. (1971). Redistribution and pinocytosis of lymphocyte surface immunoglobulin molecules induced by anti-immunoglobulin antibody, *Nature, New Biol.* **233**, 225–229.

Thoenen, H. and Tranzer, J. P. (1968). Chemical sympathectomy by selective destruction of adrenergic nerve endings with 6-hydroxydopamine, *Naunyn-Schmeidebergs Arch. Pharmac. exp. Path.* **261**, 271–288.

Trendelburg, W. (1906). Über die Bewegung der Vögel nach Durchschneidung hinterer Rückenmarkswurzeln, *Arch. Anat. Physiol.* (*Lpz.*), 1–126.

Tummons, J. and Sturkie, P. D. (1968). Cardioaccelerator nerve stimulation in chickens, *Life Sci.* **7**, 377–380.

Vediaev, F. P. (1964). Role of striatal and thalamic structures in CNS of birds in control of respiration and functional characteristics, *Fed. Proc.* **23** T, 1075–1079.

Webster, H. De F., Martin, J. R. and O'Connell, Maureen F. (1973). The relationships between interface Schwann cells and axons before myelination: a quantitative electron microscopic study, *Devl Biology* **32**, 401–416.

Weinberg, H. J. and Spencer, P. S. (1976). Studies on the control of myelinogenesis. II. Evidence for neuronal regulation of myelin production, *Brain Res.* **113**, 363–378.

Westfall, T. C. (1977). Local regulation of adrenergic neurotransmission, *Physiol. Rev.* **57**, 659–728.

Wight, P. A. L. (1971). The pineal gland. *In* "Physiology and Biochemistry of the Domestic Fowl" (D. J. Bell and B. M. Freeman, Eds), Vol. 1. Academic Press, London and New York.

Winkler, H. (1971). The membrane of the chromaffin granule, *Phil. Trans. Roy. Soc. Lond. B* **261**, 293–303.

Wright, P. (1969). Physiology of feeding and drinking in the Barbary dove (*Streptopelia risoria*). D.Phil. Thesis, Oxford University.

Wright, P. (1973). The neural basis of food and water intake in birds, *Ind. J. Physiol. Pharmac.* **17**, 1–16.

Zeigler, H. P., Green, H. L. and Karten, H. J. (1969). Neural control of feeding behaviour in the pigeon, *Psychon. Sci.* **15**, 156–157.

Index

Heavy type or ff. indicate pages where a subject is fully described. ff. means that more than one page is involved.

A

Abdominal air sac 47, 52, 57, 61, 66, 103, 270, 307
Absorption 149
Accessory reproductive organs **267** ff., **304** ff.
Acetylcholine 393, 408, 423, 435
Acetylcholinesterase 394, 426
Acid phosphatase 318
Acrocoracoid process 21
Adenosine 435
Adrenal gland 270, 273, 337
Adrenal medullary cells 412
Adrenaline 388
Adrenergic endings 98, 149, 216, 273, 289
Aerofoil 19, 21
Aetosaurus 6
Afferent autonomic fibres 383, 386, 398, 399, 404
Afferent glomerular arteriole 208, 213
Afferent renal veins 190, 195
Air capillaries 31
Air sac of egg 253
Air sac wall 66
Air sacs 17, 18, 22, 31, 46, 47, 52, 100
Albumen 241, 243, **250** ff., 320, 321
  densum 251
  formation 320, 321
  ligament 251
  polare 251
  proteins 317
  rarum 251
Aldosterone 209
Alkaline phosphatase 318
α-adrenergic receptors 389, 422, 434, 435
Altitude 31, 32, 33
Alula 9, 24, 29
Alveoli of mammal 47
Amphibians 2, 16, 17, 20
Ampulla 305
Amylase 83, 150, 155
Anapsida 2
Androgens 301, 302, 308
Angiotensin 209, 434
Ansa
  axialis 137
  duodenalis 135
  supracecalis 137
  supraduodenalis 137
Ansae
  ileales 137
  jejunales 137
  nephronorum 189
Anticoelous loops 141
Antipericoelous loops 142
Aortic pressure 17
Apex
  carinae 28
  ceci 140
  linguae 73
Apex of caecum 140
Apnoea 386
Apophysis furculae 28
Apparatus hyobranchialis 74, 79
Arboreal theory of flight 20, 34
Archaeopteryx **5**, **6** ff., **19**, 20, **21** ff., 26, 34
Archistriatum 387
Archosauria 2, 4, 11, 12
Arcuate arteries of kidneys 211
Area
  nuda 42, 54, 61
  opaca **247**, 263
  pellucida **247**, 263
Arginine vasotocin 229
Argonite 256
Arterial PCO_2 31
Arterial pressure 17, 386
Arterial supply of
  gizzard 127
  intestines 148
  liver 158
  oral cavity and pharynx 73
  ovarian follicle 287
  ovary 270
  proventriculus 127
  rectum 148
Arteries
  a. vaginalis 315
  aa. intralobulares 213
  aa. oviductales **315** ff.
  ascending oesophageal 97
  caudal mesenteric 148, 407
  cerebral carotid 389
  coeliac 97, 127, 148, 155, 158, 404
  common carotid 97
  cranial mesenteric 132, 148
  descending oesophageal 73, 97
  duodenal 155
  duodenojejunal 148, 155
  external iliac 186, 211, 315

Arteries—*contd.*
external ophthalmic 389
gastroduodenal 148
hepatic 158
hypogastric 407
ileal 148
ileocaecal 148
ingluvial 97
interlobar of kidney 211, 212
interlobular of mammalian kidney 189
internal carotid 389, 403
internal iliac 407
intralobar of kidney 212, 213
intralobular of kidney 213
ischiadic 186, 211, 315
jejunal 148, 155
laryngeal 73
lig. arteriosum 399
lingual 73
mandibular 73
marginal intestinal 149
maxillary 73
oesophageal 97
oesophagotracheobronchial 97
ovarian 270
ovario-oviductal 270, **315**
oviductal **315** ff.
palatine 73
pancreatic 155
pancreaticoduodenal 148, 155
pterygopharyngeal 73
pudendal 315, 316
pulmonary 32
renal 211, 270, 315
sublingual 73
vaginal 315, 316
vertebral 389, 403
Arteriola glomerularis afferens 213
Arteriola glomerularis efferens 213
Arteriolae rectae 189, **217**
Ascending part of thick limb 207, 210
Aspect ratio 24
Asymmetry of ovary **258**, 264
Asymmetry of oviduct **258**, 267, 335
Allanto-chorionic placenta 238
Atresia
in post-laying period **295** ff.
of follicles 266, 277, **292** ff., **295** ff., 302
Atretic follicles 266
Attraction sphere 278
Autonomic, meaning 383
Autonomic afferent fibres 383, 386, 398, 399, 404
Autonomic nervous system 382
Autoradiography 404, 430
Avidin 308, 317, 321
Axial intestinal loop 137
Axonal varicosities 422, 434, 435

B

Bacterial fermentation 150
Balance 34
Balbiani body 278, 290
Basal lamina of follicle 281, **286**
Base of caecum 140
Basis ceci 140
Basophils 374
Bats 22, 23, 24, 25, 28, 29, 30, 32, 33, 238
Beak 73, **84**
Begging display 73
β-adrenergic receptors 388, 422
Biceps tubercle 21
Bilateral ovaries 258
Bilateral oviducts 258
Bile 161, **164**
canaliculi 159, 160, 161
ducts 135, 154, 158, 159, **162**
pigments 160, **164**
Biliary system 161
Biliary system, microanatomy 163
Bill 73, **84**, 85
functions **85** ff., 385
mechanism of flamingos 86
microanatomy 86
structural adaptations **85** ff.
Bioenergetics 11, **13**
Biomechanics 11, **13**, 15
Birds nest soup 83
Blastoderm 247
Blastodisc 247
Blood coagulation **366** ff.
Blood islands 263
Blood supply of
intestines 148
kidney 211
liver 158
oesophagus 97
ovarian follicle 281, **287**
ovary **270** ff.
oviduct 307, **315**
pancreas 154
rectum 148
sperm-host glands 331
stomach 127
ureter 211
Blood-oxygen affinity 32, 365
Body form and flight 24
Body of caecum 140
Body size 23
Body weight and flight 23
Bradycardia 386
Brain 9, 15, 16, 22, **33**, 34
Brain and flight 33
Branches of ureter 189, 197, **217**
Breeding cycle 268

Breeding season 268, 269, 270
Bronchial tree 47
Brontosaurus 4
Broodiness 268, 295
Bursa cloacalis 368
Bursa-dependent lymphocytes 368
Bursting atresia 297

C

Caeca 8, 103, 132, **138**, 144, 145, 146, 149, 150, 222, 311
Caecal droppings 150
Caecal lymphoid nodules 147
Caecal sphincter 148
Caecal tonsils 147
Calcification of shell 318, 324, 329
Calcium 131
Canaliculi testae 256
Cancellous bone 12
Capsula glomerularis 203
Carbonic anhydrase 318
Cardiac accelerator fibres 400
Cardiac output 33
Cardiovascular centres 386
Cardiovascular system 16, **32**
Carina 21, 22, 28, 30
Carinates 8, 9, 22
Carnosauria 4
Carotenoid pigments 250
Carotid body 383, 386
Carpals 9, 21, 29
Carpometacarpus 22, 29
Casque of hornbills 85
Catechol *O*-methyltransferase 422
Catecholamines 425
Caudal thoracic air sac 55, 57, 400
Cavitas
  pericardialis 50
  peritonealis 52
  peritonealis intestinalis 52
  pharyngealis 71
Cavitates peritoneales hepaticae 52
  pleurales 50
Cell membrane 416
Cella aeria 253
Cellulae juxtaglomerulares 208
Cellular immunity 368
Central nervous system 15
Central vein of renal lobule 190, 195, 211, 213
Central yolk mass **241** ff.
Centrum latebrae 248
Cere 84
Cerebellum 16, 22, 34
Cerebral hemispheres 16
Cervical air sac 55, 57, 92, 399
Chalazae 250, 320
Chalaziferous layer 250
Chelonia 2, 5, 12
Chewing 18, 28
Choana 70, 71
Cholesterol 301
Choline acetyltransferase **394** ff.
Cholinergic endings 273, 289, 384, 392, 393, 394, 395, 408, 426, **430**, 433, 435
Cholinergic nerve fibres 98, 128, 149, 216, 383
Choroid 392
Chromaffin cells 408, 412
Chromogranins 422
Chromomembrin 423
Chylomicra 151
Ciliary body 391, 392
Ciliated cells of oviduct 312
Clavicle 5, 7, 8, 94
Clavicular air sac 54, 55, 57, 92, 399
Cleavage divisions 243, 247
Cleidoic egg 237, 238, **304**
Cloaca 137, 149, 221, 267, 307, 331, 407
Cloacal absorption 222
Cloacal bursa 368
Closed intestinal loop 135, 137, 141
Clutch size 269, 295, 335
CO_2
  receptors 386
  tension, arterial 31
Coagulation of blood **366** ff.
Coelom 39
Coelomic cavities 39
  of birds 49
  of reptiles 40
Coelomic septa 44
Coelophysis 4
Coelurosaurs **4**, **6** ff, 11, 15, 16, 21, 28
Collecting duct systems of mammalian kidney 189
Collecting ducts 189, 217
Collecting tubules 189, 195, 197, 199, 209
Collum latebrae 248
Colon 132
Columns (of Terni) 387
Common hepatoenteric duct 161
Compact bone 12
Complexus juxtaglomerularis 208
Compliance 43, 46, 47
Compsognathus 4, 6, 7
Conalbumin 308, 321
Concanavalin A 416
Connecting tubules 201
Constriction of pupil 391, **392**
Continuous layer 248
Coprodeum 137, 151, 222, 407

Copulation 407
Copulatory organs 8, 267
Coracoid 5, 8, 21, 22, 28, 94
Corona radiata 281
Corpora aurea 299
Corpus
  albicans 299
  ceci 140
  luteum 292, 295
Corpusculum renale 203
Cortex
  of embryonic gonad 265, 266
  of kidney 189, **190** ff., 199, 207
  of left ovary 273
  ovarii 273
  renalis 189
Cortical bone 12
Cortical nephrons 199
Cortical organization of kidney **190** ff.
Cortical portion of renal lobule 189
Cortical sex cords 265
Cortical units of kidney 190, **193**, 197
Cortico-medullary relationships 196
Cotylosauria 2, 11
Countercurrent multiplier 226
Courtship 268
Cover of cuticle 257
Cranial group of air sacs 47, 50, 52
Cranial nephric folds 42, 44, 48, 66
Cranial renal veins 213, 214
Cranial thoracic air sac 54, 55, 57, 92, 400
Craniosacral outflow 384
Crocodilia 4, 11, 12, 18
Crop 28, **92** ff., **99**, 388
  diverticula 94
  fundus 97
  glands 97
  milk 99
  microanatomy 97, 99
  movements 98
  opening 97
  structural adaptations **92** ff., 97
Cross-current pulmonary flow 32
Crura of diaphragm 17
Cryptic colouring of eggs 239
Crypts of Lieberkühn 146
Culmen 84, 85
Cursorial theory of flight 20
Cursorial animals 15
Cuticle 241, 251, 256, **257**, 327, 329
  formation 329
Cuticula 257
Cyclocoelous loops 141
Cynodonts 16, 17
Cysticoenteric duct 162
Cytolemma ovocyti 243

D

DBH 433
Decerebrate pigeon 34
Deglutition 79, **89**, 385
Dendritic crystals 254
Dense-core vesicles 422, **430**, 432
Density 27
Dentary 85
Diameter of lymphocytes 367
Diaphragm 17
Diarthrognathus 3
Dibenamine 434
Dicynodonts 28
Diet and intestinal form 133, 137
Dinosaurs **4**, 6, 11, 12, 13, 14, 15, 16, 18, 19, 22
Disc of the latebra 248, 292
Discus germinalis 243
Discus latebrae 248
Display 73, 81, 100
  chambers 71, 79
Distal convoluted tubule 189, 199, 201, **207**
Diverticulum ingluviale 94
Diverticulum of oral cavity 81
Diverticulum vitellinum 137
Diving reflex 386
Divisiones renales **185**
Divisions of kidney **185**, 211, 214, 305
Dopamine 423, 425
Dopamine β-hydroxylase 433, 434
Dopaminergic endings 384, 408, 412, 435
Dorsal mesentery 42, 48, 52, 59, 61, 66, 134, 135
Dorsal mesogastrium 41
Dorsal mesopneumonium 41
Dorsal motor nucleus of vagus 387
Dorsal root of spinal nerve 388
Dorsum linguae 71
Double innervation of organs 384
Drag 24
Drinking 90
Dromaeognathous palate 8
Dromiceiomimus 15, 16
Ductuli
  biliferi 161
  hepatici 161
Ductus
  colligens 189, 217
  deferens 219, 400, 407, 435
  hepatocysticus 162
  hepatoentericus communis 161
  hepatoentericus dexter 162
  vitellinus 137
Duodenal
  enzymes 150
  glands 148
  loop 132, 135, 141, 151, 154
Duodenum 66, 102, 106, 107, 113, 125, 126, 132, **135** ff., 154, 161, 162

E

Ecology 11
Ectothermy **11** ff.
Edinger-Westphal nucleus 391
Efferent glomerular arteriole 213, 217
Egestion of pellets 130
Egg 237, 238, **239**
  colour 239
  laying 268, 295, 334, 407
  number 269
  shapes 240, 324
  size 240
  structure 240
  tooth 85
Elbow joint 29
Electrical coupling 393
Electrotonic transmission 392
Embryology of reproductive organs **261** ff.
End expired air 31
Endocasts 16, 22
Endothermy 11 ff., **14** ff.
Endothermy and morphology **14** ff.
Enkephalin 435
Enzymes of intestine 150
Enzymes of oviduct 317
Eosinophils 373
Eosuchia 4
Epicardium 54
Epoophoron 268
Erythrocytes 361, **362** ff.
Euparkeria 4, 6
Euryapsida 2
Evolution of
  birds 2
  coelomic cavity 40
  endothermy 11
  flight 20
Exchange tissue 18, 45
Exoccipital bone 398
Exocytosis 422, 435
External nares 85
Extraglomerular mesangium 208, 209
Extravitelline layer 248
Eye 16

F

Feathers 5, 8, 9, 18, 25
Feeding centre of brain 385
Female gamete 238, 261
Female pronucleus 243
Fertilization 238, **243**, 304, 320
Fibre spectrum of vagus 417
Fimbriae infundibulares 318
First maturation division 278
First polar body 278
Fissura infundibularis 71
Flapping 21, 22, 23, 25
Flight 20, **23**, 29
  evolution of 20
  structural adaptations 23
Flightless birds 8, 20
Fluorescent neuronal structures 149, 408, 424, **425**, 426
Folds of crop 94
Follicle of ovary 273, **279**
Follicular epithelium 285
Folliculi atretici 297
Folliculus ovaricus 279
  postovulationem 292
Foot 5
Fossa vesica fellea 158
Fossae glandulares infundibulae 320
Fossulae spermaticae 331
Freeze fracture 413
Frenulum lingualis 71
Frontal shield 85
Functional adaptations of kidney 188
Fundus ingluvialis 97
Fundus of crop 97
Funiculus musculosus 311
Furcula 5, 28

G

GABA 412
Gall bladder 44, 52, 59, 66, 158, **162**, 163
Gallornis 7
Ganglion
  cervical sympathetic 387, 388, 389
  ciliary **391**
  coccygeal sympathetic 388
  coeliac 405
  cranial cervical 387, 388, **389**, 392, 394, 408, 432
  cranial mesenteric 405
  distal vagal 399
  dorsal root 388, 394
  ethmoidal 391, **395**
  geniculate 395
  impar 387
  mandibular 395
  nodose 399
  orbitonasal 395
  paravertebral 387
  petrosal 389
  proximal glossopharyngeal 398
  proximal vagal 398
  sphenopalatine 391, **395**
  sympathetic chain 273, 387, 389
  synsacral sympathetic 387, 388
Gas exchange 32, 43

Gaster 100
Gastralia 5
Gastric mill 18, 28
Gemmae gustatoriae 73, 84
Genital tubercle 267
Germinal crescent 263
Germinal disc 241, 243, 248
Germinal epithelium 264
Germinal vesicle 278
Gizzard
  angular fold 117
  angular notch 114
  annular surfaces 114
  blood supply 127
  body 114
  caudal groove 116
  caudal sac 114
  centrum tendineum 114
  conical processes 121
  corpus ventriculi 114
  cranial groove 116
  cranial sac 114
  curvatura major 114
  curvatura minor 114
  cuticle 117, 118, 120, 121, **124** ff.
  cuticula gastrica 117, **124**
  digestion 129
  facies annulares 114
  facies tendineae 114
  folds 118
  gastric mill 18, 28
  general 18, 28, 66, 102, 103, 106, **113** ff., 126
  glands 121, 122, 124
  greater curvature 114
  grinding plates 118, 121
  grit **125**, 126, 127
  incisura angularis 114
  interior **117** ff.
  lesser curvature 114
  microanatomy **121** ff.
  motility **130** ff.
  movements 98
  muscles 115, 123, 124
  ostium ventriculopyloricum 114
  plica angularis 117
  plicae ventriculares 118
  procc. conicales 121
  rugae and sulci 118
  saccus caud. 114
  saccus cran. 114
  stones 10, 18, 28, **125**
  structural adaptations **113** ff.
  sulcus caud. 116
  sulcus cran. 116
  surface features 113
  tendinous centre 114
  tendinous surfaces 114
  ventriculopyloric opening 114, 120
  ventriculus **113** ff.
Gland of nictitating membrane 391, 395
Glands of
  isthmus 324
  shell gland 327
  tubular part of infundibulum 320
Glandulae isthmi 324
Glandulae magni 321
Glandulae tubi infundibularis 320
Glandular grooves of infundibulum 307, **320**
Glenoid cavity 15, 21, 22, 29
Gliding 21, 29
Gll.
  duodenales 148
  esophageales 96
  ingluviales 97
  intestinales 146
  uterinae 327
Globular leucocytes 147
Glomerular capsule (of Bowman) 203, 204
Glomerulus **203** ff., 213
  corpusculi renalis 203
Glossopharyngeal canal 398
Glottis 73, 80, 100, 386
Gnathotheca 85
Goblet cells 147, 312
Gonad, right 258, 266, **302** ff.
Gonadotrophins 289, 308
Gonads 52, 61, 66
Gonadum dextrum 258, **302** ff.
Gondwanaland 12
Gonys 85
Grandry corpuscle 86
Granular endocrine cells 112, 126, 146
Granular leucocytes 370
Granulosa 247, 279, 281, 285, 299, 300, 302
Growth phases of oocyte **289** ff.
Growth rings 12

H

Haemoglobin content of erythrocytes 364
Haemotocrit 32
Hair 18, 19
Haploid zygote 277
Haversian bone 12
Heart 17, **33**, 42
Heart rate 33, 386
Hepar **156** ff.
Hepatic coelomic cavity 44
Hepatic ducts 161, 162
Hepatic ligaments 52, 59
Hepatic lobule 160
Hepatic peritoneal cavities 52, 54, **59**
Hepatocystic duct 162

Hepatocytes 159, **160**, 164
Hepatoduodenal ligament 61
Hepatoenteric duct, right 162
Hepatopericardial ligament 54
Herbst corpuscle 79, 86
Heterogametic sex 243
Heterophils 370
Hibernation 12
Hilus of
  lung 399
  ovary 270, 273
Hilus ovarii 270
Himalayas 31
Histamine 435
Hof 369
Homeothermy 12, 13
Homologies with mammalian kidney 197
Horizontal septum 46, 47, 50, **51**, 52, 66
Hovering flight 21, 26, 29
Humerus 22
Hydrocortisone 423
Hydroxydopamine, 5- and 6- 432, 433
Hyobranchial apparatus 74
Hypernoea 386, 387
Hypertension 386, 387
Hypobaric chamber 31
Hypotension 386
Hypothalamus 385, 387

I

Ichthyosauria 2, 5
Ileal loops 137
Ileo-rectal junction 137, 150
Ileorectal sphincter 148
Ileum 103, 132, **135** ff., 138, 146
Immunofluorescence 301, 433
Incubation 238, 268, 295
Indolamine 425
Indomethacin 435
Induced power 24
Infundibular folds 71
Infundibulum 243, 305, 307, 311, 312, 313, 315, **318** ff.
Infundibulum pharyngotympanicum 71
Ingluvies 92
Initial collecting ducts 201
Initial collecting tubule 201
Inner zone of renal cortex 207
Innervation
  alimentary canal 398
  aortic arch 400
  blood vessels 426
  bronchi 399
  choroid 392
  ciliary body 391, 392
  coronary vessels 403
  crop 388, 391, 398
  ductus deferens 407
  duodenum 405
  eyeball 392
  follicle of ovary 289
  gizzard 388, 389, 405
  gland of nictitating membrane 391
  heart 398, 399, 400
  intestines 149
  intrinsic muscles of eye 391
  iris 391, 392
  lacrimal gland 391, 395
  larynx 398
  liver 405
  lungs 398, 399, 400
  nasal (salt) gland 391, 395
  nasal mucosal glands 391, 395
  oesophagus 97, 98, 388, 389, 391, 398, 399, 405
  oral cavity 73
  ovary 273
  oviduct 316, 407
  palate 398
  pancreas 398, 405
  parabronchial muscle 400
  pericardium 399, 400
  peritoneum 399
  pharyngeal mucosal glands 391
  pharynx 73, 398
  pleura 399, 400
  pulmonary vessels 400
  rectum 149
  renal portal valve 405
  salivary glands 391, 395, 398
  shell gland 317, 334
  spleen 405
  stomachs 128, 399, 405
  syrinx 399
  taste buds 398
  thoracic air sacs 400
  tongue 79, 398
  trachea 388, 391, 398, 399
  ureter 220, 407
  vagina 317, 334
  vascular body 407
Insects 15, 19, 20
Insula juxtavascularis 208
Insulation 18
Intermediate palatine ridges 71
Intermediate segment 207
Intermediate zone of stomach 102, **112**
Internal cerebellar nucleus 387
Internal fertilization 238
Internal laying 258, 311
Interpleural septum 55
Interstitial cells **266**, 286, 289, 301, 302
Interstitial gland cells 266

Interstitiocyti ovarii 266, 286
Intestinal
  absorption 133, **150**
  blood supply 148
  diameter 133
  enzymes 150
  functions 149
  juice 150
  length 132
  loops 132, 135, **137**, 138, **141**
  nerve supply 149
  peritoneal cavity 47, 52, 59, **61**, 66
  surface area 133, 134, 150
Intestines **132** ff., **146** ff.
Intestinum
  crassum 132
  tenue 132
Intranuclear rodlets 416
Intraperiod line 420
Invasion atresia 299
Iris 391, 392
Isocoelous loops 141
Isopericoeclous loops 142
Isthmus 243, 247, 305, 307, 313, **321**, 324
  gastris 112
  of stomach 112

J

Jejunal loops 137, 151
Jejunum 103, 132, **135** ff., 146
Junctional tubules 201
Junctional zone glands of shell gland 327
Juvenile ovary **269** ff., 300
Juxtaglomerular cells 208
Juxtaglomerular complex **208** ff.
Juxtamedullary nephrons 199

K

Keel 5, 9, 30
Kidney 61, 66, 184, **185** ff., 207, 305
Kinesis 28

L

L-dopa 423
Labyrinthodonts 2
Lacrimal gland 391, 395
Lacteals 147
Lamellae of
  bill 86
  Lamellae uterinae 327
Lamina
  basalis folliculi 281
  continua 248
  extravitellina 248
  perivitellina 241, 281
Large intestine 132
Large lymphocytes 367, 368
Laryngeal mound 73
Latebra 248
Lateral palatine groove 71
Lateral palatine ridge 71
Lateral thalamus 386
Length of breeding oviduct 308
Lepidosauria 2, 4
Leucine 404
Leucocytes **367** ff.
Lice 9
Lig.
  dorsale oviductus 305
  hepatoduodenale 61
  hepatopericardiacum 54
  ventrale oviductus 305
Ligg. hepatica 52
Ligaments of oviduct 305
Lightness of avian bones 27
Lingua 71, **74** ff.
Lingual cushion 75
Lingual frenulum 71
Lingual groove 75
Lipases 155
Lipoidal atresia 300
Lips of infundibulum 318
Liquifaction atresia 299
Liver **156** ff.
  bile canaliculi 159, 160, 161
  bile pigments 160, **164**
  biliary system 161
  blood supply 158
  cardiac impressions 158
  central vein 160
  colour 156
  duodenal impression 158
  facies parietalis 158
  facies visceralis 158
  form 156
  functions 160
  gall bladder **162**
  gizzard impression 158
  hepatic arteries 158
  hepatic lobule 160
  hepatic portal vein 158
  hepatocytes 159, **160**, 164
  hilus 158
  impressio cardiacus 158
  impressio duodenalis 158
  impressio jejunalis 158
  impressio proventricularis 158
  impressio splenalis 158
  impressio testicularis 158
  impressio ventricularis 158

Liver—*contd.*
incisurae interlobares 157
incisures 157
interlobar part 157
intermediate processes 158
jejunal impression 158
lobes 157
lobi hepatici 157
lobule 160
lobulus hepaticus 160
microanatomy **159** ff.
muralium 159
papillary process 158
parenchymal sheets 159
parietal surface 158
pars interlobaris 157
perisinusoidal space (of Disse) 160
phagocytic cell (of Kupffer) 160
physical characteristics 156
porta hepatis 158
portal tracts 160, 163
proc. papillaris 158
procc. intermedii 158
proventricular impression 158
relations to coelomic cavities 41, 42, 50, 52, 54, 57, 59, 61, 66
sinusoidal network 159, 160
sinusoids 160
size and diet 156
splenic impression 158
testicular impression 158
v. centralis 160
visceral surface 158
weight 156
Lobar bile ducts 161
Lobus renalis 197
Locomotion 15, 34
Longisquama 19
Lung 18, 41, 42, 43, 46, 50, 54, 57, 61, 270
Luteal cells 266
Luteinizing hormone 289
Lymphocytes 367
Lymphoid nodules of
caeca 147
intestine 147
oesophagus 97
oviduct 317
Lymphonoduli
esophageales 97
pharyngeales 73
Lysozyme 251, 321

M

Macrophage, monocyte relationships 369
Macula densa 208
Magnum 312, 313, 316, **320** ff.
functions 321
Major period band 417
Male pronucleus 243
Mammalian-type nephrons 199, 203, 204, 208, 210, 211, 229
Mammals 14
Mammillae 254
Mammillary cores 254, 256, 324
Mammillary layer 254
Manus 29
Mating 268
Meal to pellet interval 130
Medial archistriatum 386
Medial longitudinal fasciculus 387
Median palatine groove 71
Median palatine ridge 71
Medium lymphocytes 367
Medulla oblongata 386, 387, 398
Medulla of
embryonic gonad 265, 305
kidney 189, 196, 197, **209** ff.
left ovary 266, 273
mammalian kidney 189
Medulla
ovarii 273
renalis 189
Medullary
collecting tubules 189, **209**
cones **189**, 197, 199, 203, **209**, 217, **225** ff.
interstitial cells 266
nephron 199
portion of lobule 199
pyramid of mammal 188, 197
ray of mammal 188, 195
Megalecithal egg 238
Meiotic divisions 278
Melatonin 389
Membranae testae 251
Meroblastic division 243
Mesangial cells 204
Mesangiocytes 204
Mesangium 204
Mesencephalic lateral reticular formation 386, 387
Mesencephalon 386
Mesenterium ventrale 52
Mesoduodenum 66
Mesogastrium 41
Mesogyrous loops 142
Mesogyrous spiral 142
Mesoileum 66
Mesojejunum 66
Mesonephric duct 261, 267, 304, **337**
Mesonephros 261, 267, 268
Mesopneumonium 41
Mesorchia 52, 66
Mesorectum 66

Mesosalpinx 42, 44, 52, 66
Mesovarium 52, 66, 270
Mesoviductus dorsalis 42, 52, 66
Metabolic rate 14, 17, 30, 31, 188
Metacarpal bones 21, 29
Metacarpus 5, 9
Metapatagium 25
Microfilaments 422
Microtubules 422, 424
Migration 31, 321, 268, 385
Minute volume 31
Monoamine oxydase 422
Monocytes 369
Monotremes 304
Mons laryngealis 73
Moths, flight 15
Moulting 302
Movements of neuronal elements 416
Mucosal intestinal projections 143
Müllerian duct 267, 304
Muscarine receptors 435
Muscle tunic of intestine 148
Muscles
  antigravity 14
  chewing 28
  coordination 16
  corneal 391
  costoseptal 47, 48, 51, 55
  diaphragmatic 46
  genioglossal 80, 81
  intercostal 46
  m. cucullaris 94, 399
  m. diaphragmaticus 45
  m. latissimus dorsi 25
  m. levator caudae 404
  m. linguales 79
  m. septi obliqui 59
  m. sphincter cecale 148
  m. sphincter ilealis 148
  m. sphincter vaginae 331
  m. sternotrachealis 59
  m. tensor patagialis 25
  m. triceps 404
  mm. costoseptales 51
  mylohyoid 80, 81
  of accommodation 391
  of crop 94
  of diaphragm 48
  of flight 5, 23, **29**, 30, 34
  of gizzard 115, 123, 124
  of insects 15
  of iris 391
  of propulsion 29
  of swallowing 90
  of tongue 79
  pectoral 21, 22, 27, 28, **29**, 30
  power output 15
  pterygoid 395
  rectus muscles of eyeball 391
  supracoracoid 22, 28, 29, 30
  tensor of oblique septum 48, 59
Myelin 417

N

Nail of beak 84
Nasal (salt) gland 391, 395, 400
Neb 84
Neck of latebra 248, 292
Neognathae 9, 10
Neognathous palate 8
Neornithes 9
Nephron **199** ff.
Nephronal loops (of Henle) 189, 199, **205**, 209
Nephronum corticale 199
Nephronum medullare 199
Nerve
  accessory 398
  cardiac sympathetic 400
  chorda tympani 395, 398
  choroidal 392, 393, 394
  connexus interganglionaris 389
  descending oesophageal of IX 398
  descending ramus of X 399
  facial 389, 391, 395
  follicular 273
  glossopharyngeal 97, 386, 389, 391, **395** ff.
  glossopharyngeal-vagal anastomosis 398, 399
  greater superficial petrosal 395
  hyomandibular branch of VII 389
  hypoglossal 97, 389
  intestinal 405, 407
  intramural, of follicle 289
  iridociliary 392, 393, 394
  lacrimal of VII 395
  laryngopharyngeal of IX 398
  lingual of IX 398
  long ciliary 392
  mandibular V 395
  maxillary V 395
  n. pudendus 405
  nn. folliculares 273
  nn. intramurales 289
  nn. pedunculares 289
  oculomotor 391, 393
  ophthalmic V 392, 395
  optic 391
  palatine of VII 392, 395
  pelvic splanchnic 405, 407
  pudendal 405, 407
  r. externus of X 399
  recurrent of X 97, 399
  Remak's 405

Nerve—*contd.*
short ciliary 392
splanchnic 149, 388, 399, 404
sublingual of V 395, 398
synsacral sympathetic 149
thoracic sympathetic 149
truncus paravertebralis 389
vagus 97, 128, 149, 273, 386, 389, 391, **395** ff., 400, 417
Vidian 392, 395
Nest cement 83
Nest site 239, 240
Nesting 268
Nesting behaviour 295
Neuraxis 15
Neutrophils 370
Non-granular leucocytes 367
Noradrenaline 393, 408, 412, 422, 423, 424, 425, 430, 434, 435
Noradrenergic endings 128, 383, 391, 393, 404, 405, 412, 426, **430**, 433, 434
Nostrils 85
Nuc. rotundus 386
Nucleus of Pander 248, 292
Number of lymphocytes 367

O

Obex 387
Oblique septum **46**, 47, 48, 50, 52, 54, 55, 57, 59, 61, 66
Occipitomesencephalic tract 386, 387
Occluding plate 307
Oculomotor nuclei 391
Oesophageal
blood supply 97
folds 92
glands 96
lymphatic nodules 97
nerve supply 97, 98
sac 92, **100**
Oesophagus 28, 52, 59, 71, 90, 101, 107, 388
microanatomy **96**, 99
functions **97** ff.
structural adaptations 92
Oestrogens 295, 301, 302, 307, 308
Olfaction 33
Olfactory lobe 33, 34
Oocyte 239, 277, **278**, 279, 281, 285, **289** ff., 295, 311
cytolemma 243, 247, 279, 281, 285, 299
diameter 292
growth phases **289** ff.
Oogenesis 277
Oogonium 261, 277
Open intestinal loop 135, 141
Optic lobes 33, 34, 394
Oral cavity 70, 73
Oral sacs 71, 80
Orbit 33
Orbital cavity 395
Ornithischia 4, 28
Ornitholestes 7
Ornithomimidae 15, 28
Orthocoelous loops 141
Osmoreceptors 403
Osmoregulation 224
Ostia of air sacs 55, 57, 61
Ostium
cloacale oviductus sinistri 331
infundibuli 311
ingluviale 97
ureteris 219
Outer zone of renal cortex 207
Ovalbumin 308, 321
Ovarian follicle 273, **279**
Ovarian lymphoid tissue 272
Ovarian pocket 311
Ovarium sinistrum 258, 270
Ovary
inter-breeding period 300
juvenile **269** ff., 300
left 66, 103, 238, 258, **263** ff., **268** ff.
Oviduct
left 103, 219, 238, 243, 248, 258, 267, 295, **304** ff.
length 308
breeding season **307** ff.
non-breeding condition 305
right 258, 267, **335** ff.
Oviductal lymphoid tissue 317
Oviductus
dexter 258, **335**
sinister 258, **304** ff.
Oviparity 238
Oviposition 241, 295, 317, 334
Oviraptor 7
Ovocytus
primarius 241
secundarius 241
Ovomucoid 321
Ovotestis 303
Ovotransferrin 251, 321
Ovoviviparity 238
Ovulation 243, 247, 277, 279, **289**, 293, 302, 304, 320, 334, 335
rate 335
Ovum 239, 243
Oxygen
capacity of blood 32
consumption 17, 30, 31, 33
dissociation curve 32, 365
saturation of arterial blood 33
transport 32, 33

P

Palaeognathiformes 10
Palaeognathous palate 8, 9
Palate 8, 9, 70
Palatine
  bones 71
  papillae 71
  prominence 71
Palatum 70
Paleognathae 9, 10
Palisade layer of shell 254
Pancreas
  blood supply 154
  colour 151
  ducts 135, 152, **153**, 162
  ductus pancreaticus 154
  exocrine component 152
  exocrine secretion 155
  external appearance 151
  general 103, 106, 135, **151** ff.
  islets 151, 155
  juice 155
  lobes 152
  lobus pancreatis 152
  nerve supply 155
  secretory granules 152
  size 151
  splenic lobe 152
Pancreatic ducts 135, 152, **153**, 162
Pancreozymin 155
Paneth cells 147
Papilla ductus vitellini 137
Papillae
  palatinae 71
  pharyngeales 71
  remigiales 21
Parabasal fossa 398
Parabronchi 31
Paramesonephric duct 267, 304, 335
Parasympathetic division 383
Parasympathetic supply to head and neck 391
Parathyroid gland 399
Parietal peritoneum **52**, 59, 61, 66
Parietal pleura 51, 55
Parietal serous pericardium 54
Paroophoron 268
Pars
  ascendens ansae 207
  conjungens 201
  cranialis uteri 327
  descendens ansae 205
  major uteri 327
  pylorica 102, 125
  translucens isthmi 321
Parenchymatous zones of ovary 273
Pectoral crest 22, 28
Pectoral girdle 28
Pedicle of follicle 281
Pedunculus folliculi 281
Pellets in raptors 129
Pelvis of ureter 189
Pelycosauria 3, 11
Penetration by sperm 279
Pepsinogen 111
Pericardial cavity 41, 50, **52**
Pericardium 41, 42, 45, 50, **54**, 57, 61
  serosum parietale 54
  serosum viscerale 54
Pericoelous loops 142
Perilobular collecting tubules 189, 197, 199
Peritoneal cavity 41, 54
Peritoneum
  parietale 52
  viscerale 52
Peritubular capillary plexus 190, 192, **211**, 213, 214, 216, 217
Perivitelline layer 241, **247**, 248, 281, **285**, 292, 320
Peyer's patches 148
Phalanges 29
Phallus 8, 267
Pharyngeal
  cavity 71
  cleft 71
  innervation 73, 398
  lymphatic nodules 73
  papillae 71
Pharyngotympanic infundibulum 71
Pharyngotympanic tube 71
Pharynx 70
Phenoxybenzamine 434
Phrenic nerve 48, 49
Pigment, formation 329
Pigmentation of shell 256, 327, 329
Pineal 387, **389**
Pituitary 387
Placenta 238
Plagiocoelous loops 141
Platelets 361, 366
Pleura
  parietalis 51
  visceralis 51
Pleural cavity 42, 46, 47, 50, 51, **55**, 57
Pleuroperitoneal cavity 41, 44
Pleuroperitoneal folds 48
Plexus
  adrenal 273, 405
  aortic 317, 404, 405, 407
  caudal mesenteric 405
  cloacal 407
  coeliac 97, 128, 155, **404**, 405
  cranial mesenteric 405
  external ophthalmic 392, 395
  gonadal 405
  hypogastric 405, 407

Plexus—*contd.*
internal iliac 405, 407
mesenteric 128
mucosal 97, 149
muscle 97, 149
myenteric 97, 128, 149
oesophageal 97
ovarian 273, 317
oviducal 405
pancreaticoduodenal 155
pelvic 317, 407
perivascular 149
pudendal 149
pulmonary 400
rectal 407
renal 317, 405
submucosal 97, 149
Plicae
esophageales 92
infundibulares 71
ingluviales 94
intestinales villosae 143
primariae 313
secundariae 313
Plumping 327, 329
Pneumatic bones 17, 22, 26, 27
Pneumatic foramina 18
Podocytes 203
Polocytus
primarius 278
secundarius 243
Polyspermy 243, 247
Pons 16
Pores of shell 256
Pori testae 256
Portal tracts of liver 160, 163
Postganglionic parasympathetic fibres 383
Postganglionic sympathetic neurones 383, **421**
Posthepatic septum 44, 45, 47, 49, 52, **59**, 61, 66
Postovulatory follicle **292** ff., 302
Postpulmonary septum **42**, 43, 44, 48, 49, 52, 57
Posture 14, 15
Pouch-like part of shell gland 327
Pre-ampulla 305
Precursors of erythrocytes 362
Preganglionic parasympathetic fibres **383**
Preganglionic sympathetic fibres **383**
Prepatagium 25
Preproavis 21
Presynaptic vesicular grid 413
Prey to predator ratio 13
Primary bronchi 41, 43, 47, 52
Primary folds of oviduct 313
Primary oocyte 241, 243, 278
Primary septum transversum 41, 44, 48
Primary sex cords 265, 266
Primordial germ cells **263** ff., 277, 303
Proavis 20, 21, 34
Procompsognathus 4
Progesterones 295, 301, 302, 307, 308
Prolactin assay 99
Pronephric duct 261, 267
Pronephros 261
Proprioceptive input 34
Prostaglandins 295, 334, 434, **435**
Proteinases 112, 155
Proventriculus
blood supply 112
diverticulum proventriculare 110
endocrine cells 112
gastric juice 112, 131
gastrin 112
general 52, 57, 61, 66, 92, 95, 101, **103** ff., 113
glands 107, **108** ff.
glandular patches 109, 110
interior 107
microanatomy **107** ff.
mucosal papillae 107
mucosal ridges 107
oxynticopeptic cell 111
papillae proventriculares 107
pepsinogen 111
plicae proventriculares 107
rugae and sulci 107
size and shape 103
stomach oil 111
structural adaptations 103, 108, 109, 111
Proximal convoluted tubule 204
Pseudosuchia 4, 6, 7
Pteranodon 23, 24, 27
Pterosaurs 4, 16, 17, 18, 22, 23, 25, 26, 27, 28, 29, 30, 33, 34
Pterygoid 8
Pubis 5
Pulmonary fold 42
Pulmonary nerves 400
Pygostyle 8, 26
Pygostylus 26
Pyloric part of
stomach 102, **125**
bulbus pyloricus 126
cuticle 126
endocrine cells 126
gll. pyloricae 125
torus pyloricus 126
ostium pyloricum 125
papillae 127
papillae filiformes pyloricae 127
plicae pyloricae 126
proteolysis 131
pyloric bulb 126
pyloric glands 126
pyloric orifice 125
structural adaptations **125** ff.

Q

Quill knobs 21, 29
Quintofrontal tract 386

R

R. uretericus 189
Radius 29
Rami communicantes 388
Ratites 8, 9
Rdx. renalis efferens 211
Recessus uterinus 327
Reciprocal synapses 383
Rectal absorption 151, 221
Rectal droppings 150
Rectum 61, 132, **137**, 144, 145, 149, 150, **221** ff., 307, 407
Red-region (isthmus or uterus) 324, 327, 329
  functions 329
Reduction divisions 278
Regression of
  follicles **292** ff.
  oviduct 307
Regurgitation 98, 99, 111, 129
Ren 184
Renal
  corpuscle 199, **203**, 207
  lobe 185, 188, **197**, 213
  lobule 188, 196, **197**, 213
  papilla 184
  portal circle 214
  portal shunts **214** ff.
  portal system 184, 189, 196, **213** ff.
  portal valve 214
  portal veins 190, **213** ff.
  veins 190, 211
Renculus kidney of cetaceans 197
Renin 209
Reptilian-type nephrons 199, 203, 204, 207, 210, 211, 213, 229
Reserpine 425
Respiratory centres **386**
Respiratory frequency 31
Respiratory system 17, **30**, 40
Rete capillare peritubulare corticale 190
Rete cords 265
Reticulocyte 362
Retroperistalsis 149, 150
Rhamphotheca 84, 85
Rhinotheca 84
Ribs 17, 28, 55
Right gonad 258, 266, **302** ff.
Right ovary, functional 303
Right oviduct 258, 267, **335** ff.
Rostral colliculus 33
Rostrum 73, 84
Rostrum mandibulare 84
Rostrum maxillare 84
Rotation of
  gut 135
  egg 329
  reflexes 404
Ruga palatina
  lateralis 71
  mediana 71
Rugae palatinae intermediales 71

S

Sacci orales 71, 80
Saccus esophagealis 92, 100
Saccus vitellinus 137
Salivary glands 73, 74, 81, 391, 395
  "lime stick" 83
  cheeks 82
  corner of mouth 82
  cricoarytenoid 82
  gl. anguli oris 82
  gll. cricoarytenoideae 82
  gll. linguales 82
  gll. mandibulares 82
  gll. maxillares 81
  gll. oris 81
  gll. palatinae 81
  gll. pharyngis 81
  gll. sphenopterygoideae 81
  lingual 82
  mandibular 82
  maxillary 81
  microanatomy 82
  palatine 81
  secretions 83
  sphenopterygoid 81
Salt glands 188
Satellite cells 408, 412, 422
Saurischia 4
Sauropodomorpha 4, 18
Sauropsida 5
Saurornithoides 16
Scales 19
Scapula 8, 28
Schmidt-Lanterman cleft 420
Schwann cells 413, 417, 420
Scleromochlus 19
Second maturation division 243, 279
Second polar body 243
Secondary folds of oviduct 313
Secondary sex cords 265, 266
Secondary oocyte 241, 243, 247, 279
Secretin 155
Secretory cells of oviduct 312

Seed extraction 89
Seed opening 88
Segisaurus 7
Segmenting intestinal movements 150
Segmentum intermedium 207
Semen 238
Septum
  horizontale 50
  obliquum 50
  posthepaticum 52
  transversum 41, 42, 44, 48, 61
Sex chromosomes 243
Sex cords 265
Sexual display 301
Shell 240, 241, **251** ff., 327
  canaliculi 256
  formation 329
  gland 247, 305, 312, 316, 313, **324** ff., 334
  gland epithelium 327
  gland functions 327, 329
  knobs 254
  membranes 247, 251, 257, 324
Shelled egg 238, **239**
Shoulder joint 15, 21, 26, 28, 29
SIF cells 408, 435
Sinocodon 3
Sinus transversus pericardialis 54
Sinusoids of liver 160
Size of
  erythrocytes 362
  monocytes 369
Skull 27
Small intestine 132, 144
Small lymphocytes 367, 368
Soaring 29, 30
Soft palate 70
Somatic motor pathways 34
Special sense organs 33
Speed 15
Sperm-host glands of
  infundibulum 312, 313, **320**
  vagina 331, 334
Spermatozoa 238, 243, 247, 303, 312, 313, 331
  transport 312, 313
Spherulitic crystals 254, 256
Sphincter of vitelline duct 148
Spleen 61, 103
Spongy layer 254
Sprawlers 14
Squamata 12
Stability in flight 20
Stalling 29
Stenonychosaurus 16
Sternum 5, 9, 21, 28, 94
Steroidogenesis 266, 279, 286, **301** ff., 308, 337
Stigma 281, 289
Stigma folliculare 281
Stomach motility **129** ff.
Stomachs 101
  structural adaptations **102** ff.
  topographical anatomy 103
Straining mechanism of bill 86
Strata of yolk 250
Stratum
  chalaziferum 250
  granulosum 281
  mamillarium 254
  spongiosum 254
Streamlining 24
Stretch receptors in mesenteries 404
Stroke volume 33
Structural adaptations for water economy **224** ff.
Struthiolipeurus 9
Sub-germinal cavity 247
Submucosal intestinal projections 145
Subpulmonary cavity 47, 51, 52, **57**, 59
Sulcus
  lingualis 75
  palatinus lateralis 71
  palatinus medianus 71
Supracaecal loop 137
Supraduodenal loop 137
Surface area of intestine 133, 143, 150
Surface tension 44
Sutures of skull 8, 9
Swimming 24, 43
Sympathetic division 383
Sympathetic outflow 387
Sympathetic supply to head 388
Sympathetic supply to neck 391
Synapsida 2, 11, 12
Synaptic morphology 413, 422
Synsacral fossae 185
Synsacrum 185
Systema portale renale 184

T

Tachycardia 386, 387
Tail 20, 26
Take-off 21, 23, 30
Taste buds 73, 84, 398
Teeth 18, 27, 28
Telogyrous loops 142
Telogyrous spiral 142
Telolecithal egg 238
Teratornis incredibilis 23
Testa 241, 251, **253** ff., 257, 327
Testis 66, 103, 303
Thalamus 386
Theca externa 279, 281, 282, **287**
Theca interna 266, 279, 281, 286, 299, 302
Thecal cells **286**, **287**, 292, 293, 300, 301, 302

Thecal glands 286
Thecal interstitial cells 302
Thecal luteal cells 293
Thecodontia 4, 6, 11, 13
Therapsida **3**, 4, 11, 12, 13, 14, 16, 28
Thermoregulation 12
Theropoda 4, 21
Theropsida 5
Thick albumen 251, 321
Thick limb of nephronal loop 206, 210
Thin albumen 251, 321
Thin limb of nephronal loop 205, 210
Thoracic air sacs 47, 57, 59
Thoracoabdominal cavity 42, 51
Thoracolumbar outflow 384, 387
Thrombocytes 361, 365
Thromboplastin 366, 367
Thymus 368
Thymus-dependent lymphocytes 368
Thyroid gland 399
Tidal volume 31
Tomia 84, 88
Tongue 71, **74** ff.
  functions **74** ff.
  innervation 79
  structural adaptations **74** ff.
Topography of oviduct 308
Torus
  lingualis 75
  palatinus 71
Trachea 52, 59, 100
Translucent zone of isthmus 313, 321
Transport of
  calcium 318, 329
  egg 312, 313, 317, 321
  spermatozoa 312, 313
Transverse pericardial sinus 54
Triconodonts 3
Triosseal canal 22, 29, 30
Tritiated thymidine 413
Trunk 24
Tubae pharyngotympanicae 71
Tubal ridge 267
Tubular glands of oviduct 307, 312, **313**
Tubular part of infundibulum 248, 320
Tubular secretion of uric acid 216
Tubuli
  colligentes medullares 189, **209**
  colligentes perilobulares 189
  convoluti distales 189
  epoophori 337
Tubulin 425
Tubulus convolutus proximalis 204
Tubus infundibularis 320
Tunica albuginea
  definitive 265, 273
  primary 265, 266
  secondary 265
Tunica muscularis intestini 148
Tunica superficialis 281
Tyrannosaurus 4, 13

U

Ulna 22, 29
Ultimobranchial gland 399
Uncinate fasciculus 387
Uncinate process 5, 28
Unicameral lungs 41
Unicellular glands of oviduct 312
Unidirectional air flow 18, 31
Urea 216
Ureter 184, 189, **217** ff., 400
  microanatomy **219** ff.
Ureteral
  branches 189, 197
  mucus 219
  orifice 219
  sphincter 219
Uric acid 216, 220
Urine 151, 184, 222
Urodeum 407
Urogenital ridge 263, 267
Utero-vaginal
  junction 312
  sphincter 331, 334
Uterus 305, 324

V

Vagal canal 398
Vagina 305, 311, 312, 313, **331** ff.
Vaginal opening into cloaca 331
Valva portalis renalis 214
Van't Hoff's generalization 15
Varicosities 422, 423, 434, 435
Vasa recta 189, **217**
Vascular zones 273
Vasomotor centre 387
Vasomotor fibres 389, 391
Vasomotor tone 388
Veins
  adrenal 272
  caudal mesenteric 149, 214
  caudal renal **213** ff.
  caudal vena cava 41, 42, 54, 61, 158, 211, 272, 273, 303, 316
  central vein of liver 160
  coccygeomesenteric 214
  common iliac 211, 213, 214, 272
  cranial mesenteric 137, 149
  cranial renal **213** ff.

cranial vena cava 54, 127, 399, 400
efferent renal radix 211
efferent renal veins 190, 192, 194
external iliac 186, 213
gastropancreaticoduodenal 149
hepatic portal 127, 149, 155, 158, 159, 316
hepatic veins 61, 158, **159**
interjugular anastomosis 73
interlobular (portal) 213
internal iliac 214
internal vertebral venous sinus 214
intralobular of kidney 190, 192, 199, 207, 211, 213
intramural, of follicle 287
ischiadic 186, 214
jugular 97, 399, 400
mandibular 97
ovarian 272
oviductal 272, 316
pancreaticoduodenal 155
peduncular, of follicle 287
postcardinal 268
pulmonary 54
renal portal 213, 214, 316
rostral cephalic 73
sinusoids of liver 160
v. intralobularis 190
v. portalis renalis 213
vv. interlobulares 213
vv. oviductales 316
vv. renales (*cran., caud.*) 190
Velociraptor 7
Ventilation rate 31
Ventral mesentery 41, **52**, 54, 59, 61, 66
Ventriculus 18, 102
Ventrolateral tegmentum 387
Ventrum linguae 73
Venulae rectae 189, **217**
Vesica fellea 158
Vestibular apparatus 34
Villi **143** ff., 222
embryonic growth 145
number 146
intestinales **143** ff.
Visceral pericardium 54
Visceral peritoneum 52, 61
Visceral pleura 51, 55
Viscosity of blood 32
Vision 15
Visual regions of brain 33
Vitelline diverticulum 132, **137**
Vitelline membrane 248, 281
Vitellogenesis 277, **289** ff.
Vitellus 248
albus 248
aureus 250
Viviparity 238, 304
Vocalization 43
Vomer 8

W

Water absorption in rectum 151, 221
Water economy 221, 224
Weight of kidney 186
Weight reduction 26
White of egg 240, 241, 250
White yolk 248, 250
Wing forms 24
Wing loading 23, 24
Wing slots 24
Wolffian duct 261, 267, 304, **337**

Y

Yellow bodies 299
Yellow yolk 250
Yolk 160, 240, 241, **248** ff., 263, 279, 281, 285, 286, **289** ff., 297, 299, 300
deposition 292
duct 137
membranes 247, 248
sac 137, 238
sac placenta 238
spheres 241, 250
transport 281, 286, 287, **289** ff.
spindles 250

Z

Zona
intermedia gastris 102, **112**
pellucida 285
radiata 281
Zonae
parenchymatosae 273
vasculosae 273